Advances in Collaborative Civil Aeronautical Multidisciplinary Design Optimization

Advances in Collaborative Civil Aeronautical Multidisciplinary Design Optimization

Edited by
Ernst Kesseler
Katern 09
Amsterdam, The Netherlands

Marin D. Guenov
Cranfield University
Cranfield, Bedfordshire, United Kingdom

Volume 233
PROGRESS IN
ASTRONAUTICS AND AERONAUTICS

Frank K. Lu, Editor-in-Chief
University of Texas at Arlington
Arlington, Texas

Published by the
American Institute of Aeronautics and Astronautics, Inc.
1801 Alexander Bell Drive, Reston, Virginia 20191-4344

American Institute of Aeronautics and Astronautics, Reston, Virginia

1 2 3 4 5

ISBN 978-1-60086-725-5

To my partner

E.K.

To my parents

M.D.G.

A note from our Editor-in-Chief...

This volume represents the culmination of a recent European effort in developing new techniques in aircraft design, proving once again that aircraft design is an evolutionary process. Enabling technologies, concepts, and the creativity of the human mind are all present in the chapters of this volume. The editors have assembled experts to write the chapters that highlight the life cycle of aircraft design, from a "predesign" stage through what is made possible nowadays—virtual enterprise collaboration. The final chapter on data management shows the necessity of this process throughout the design process, particularly in the context of multiple teams in a virtual enterprise. This volume is necessary reading for anyone interested in understanding the modern aircraft design process, in implementing such processes or in managing such teams. The volume can also serve as a useful volume for any researcher or graduate student who needs a ready reference.

Frank K. Lu
September 2009

Foreword

THE VIVACE (Value Improvement through a Virtual Aeronautical Collaborative Enterprise) research project gathered 63 partners from 12 countries around Europe to form a consortium including a wide range of manufacturers and suppliers from the European aeronautical industry, as well as universities, research organizations, and information technology vendors.

The main ambition of VIVACE was to create a new aeronautical collaborative design environment. This environment enables the aeronautical extended enterprise to reduce the overall development time and cost of new aircraft and engines (and their subsystems and components).

The four-year project, completed at the end of 2007, was recognized as a success. This was achieved through researching, testing, and validating the use of more efficient and effective collaborative product engineering processes, models, and methodologies.

Results were delivered in seven main areas: design simulation solutions, virtual testing solutions, business and supply chain modeling solutions, knowledge management solutions, decision support solutions, collaboration in the extended and virtual enterprise solutions, and design optimization solutions.

This book presents the contributions made by the VIVACE consortium to advance and establish the new state of the art in collaborative aeronautical multidisciplinary design optimization. These innovative results were created and validated through many and varied industrial (aircraft, engine, and equipment manufacturers) use cases.

It was my great pleasure and privilege to work with this book's editors and contributors during the VIVACE research project. I am happy that the time has come to share these results with the wider design optimization community, and I hope that you find the content to be interesting, informative, and useful.

Philippe Homsi
VIVACE Project Coordinator
AIRBUS

Table of Contents

Preface

THE objective of this book is to contribute to state-of-the-art multidisciplinary design optimization (MDO) with results from a major recent industry-led European research project on the virtual collaborative civil aeronautical enterprise.

This multiauthored publication is the end-product of an effort to systematize, put into context, and disseminate the outcome of several interrelated streams of work for the benefit of the wider scientific community and practitioners in the MDO field. A guiding principle during the compilation process was to follow the phases of a generic design cycle as closely as possible. As such, the book starts with chapters covering MDO issues at initial design stages and then moves to more detailed design optimization. Similarly, wherever possible, MDO applications are ordered by product complexity, from whole aircraft and engine to single component optimization.

This approach is complemented by the presentation of MDO infrastructure and business-related topics that contribute to the efficient functioning of the virtual collaborative enterprise. These include, among others, automated multi-partner workflows, product life cycle management, data security, and protection of intellectual property rights. Such factors are of paramount importance when considering the scale and complexity of an evolving multinational multipartner consortium. The introductory chapter contains a more extensive overview of the contents and can serve as a guide for readers with specific interests.

Real world design cases have been used to verify all presented methods and tools. As the majority of these are fairly generic, we believe that the text will be of interest to a wider readership, beyond the aeronautical domain.

We would like to acknowledge the support of the Sixth Framework Programme of the European Community, priority 4, "Aeronautics and Space" under which the Value Improvement through a Virtual Aeronautical Collaborative Enterprise (VIVACE) Integrated Project (AIP3 CT-2003-502917) was partly sponsored.

Also, we would like to extend our gratitude to all project partners who contributed to this monumental effort. Finally, our deepest thank you goes to our partners, family, and friends for putting up with us during this process.

Ernst Kesseler
Marin D. Guenov
July 2009

Chapter 1

Overview

Ernst Kesseler*
Katern 09, Amsterdam, The Netherlands

and

Marin D. Guenov†
Cranfield University, Bedfordshire, England, United Kingdom

Abbreviations

COTS = commercial-off-the-shelf
EU = European Union
IT = instructional technology
MDO = multidisciplinary design optimization
NSGA = nondominated sorting genetic algorithm
SCC = strongly connected components
TSCP = Transglobal Secure Collaboration Program
VEC hub = virtual-enterprise-collaboration hub
VIVACE = value improvement through a virtual aeronautical collaborative enterprise
WMD = workflow management device

I. Introduction

THE objective of this book is to contribute to the state-of-the-art in multidisciplinary design optimization (MDO), presenting results from a major European research project on the collaborative civil aeronautical enterprise. Because the majority of the presented methods and tools are fairly generic,

*Director.
†Professor.

we believe that the text will be of interest to a wider readership, beyond the aeronautical domain. Also anticipating that the book will find an audience outside Europe, we begin this introductory chapter with a concise overview of our European industry-led research collaboration. This is followed by a summary of the objectives of the collaboration and the work related to MDO in particular. The final section outlines the structure of the book, also highlighting the major contributions of each chapter.

II. European Industry-Led Research Collaboration

The European Union's (EU) long-term strategic research agenda in aeronautics was set recently in a report entitled "European Aeronautics—a Vision for 2020" [1]. The underlying thesis of this report was that air transport would play a key role in the global transport infrastructure of the 21st century. Vision 2020 put forward two high-level objectives: to meet society's needs for efficient air transport and to make Europe a world leader in aeronautics. Resulting from this vision were several dedicated research initiatives under the European Union Framework 6 Program. More information on the organization, requirements, and procedures of this framework program can be found on the Community R&D Information Service website (http://cordis.europa.eu/fp6/stepbystep/home.html). Here, we concentrate on a particular large-scale *integrated* project entitled "Value Improvement through a Virtual Aeronautical Collaborative Enterprise," or VIVACE for short. Table 1 provides an overview of some of the major characteristics of the collaboration. For organizational reasons most of the MDO work was performed in a dedicated work-package. However, a significant amount of MDO research and development was performed in several other work-packages, where it constituted part of those activities.

VIVACE was no exception to the European Union policy to stimulate transnational collaboration between industry, research institutes, and universities, with special attention to small- and medium-size enterprises. The distribution of the consortium members shown in Fig. 1 uses this partner classification

Table 1 Major characteristics of VIVACE EU collaboration

	Total collaboration	MDO related
Parameters of the project		
Value	€72.8 M	€30.3 M
Number of partners	63	39
Number of work-packages	17	7
Duration	4 years	4 years
Duration per 1st/2nd/3rd iteration	18/12/18 months	18/12/18 months
Output		
Number deliverables	383 + 101 public	203 + 51 public
Identified reusable results	141	55
Number of presentations, forum1/forum2/forum3	48/61/35	15/18/11
Number of publications	85	27

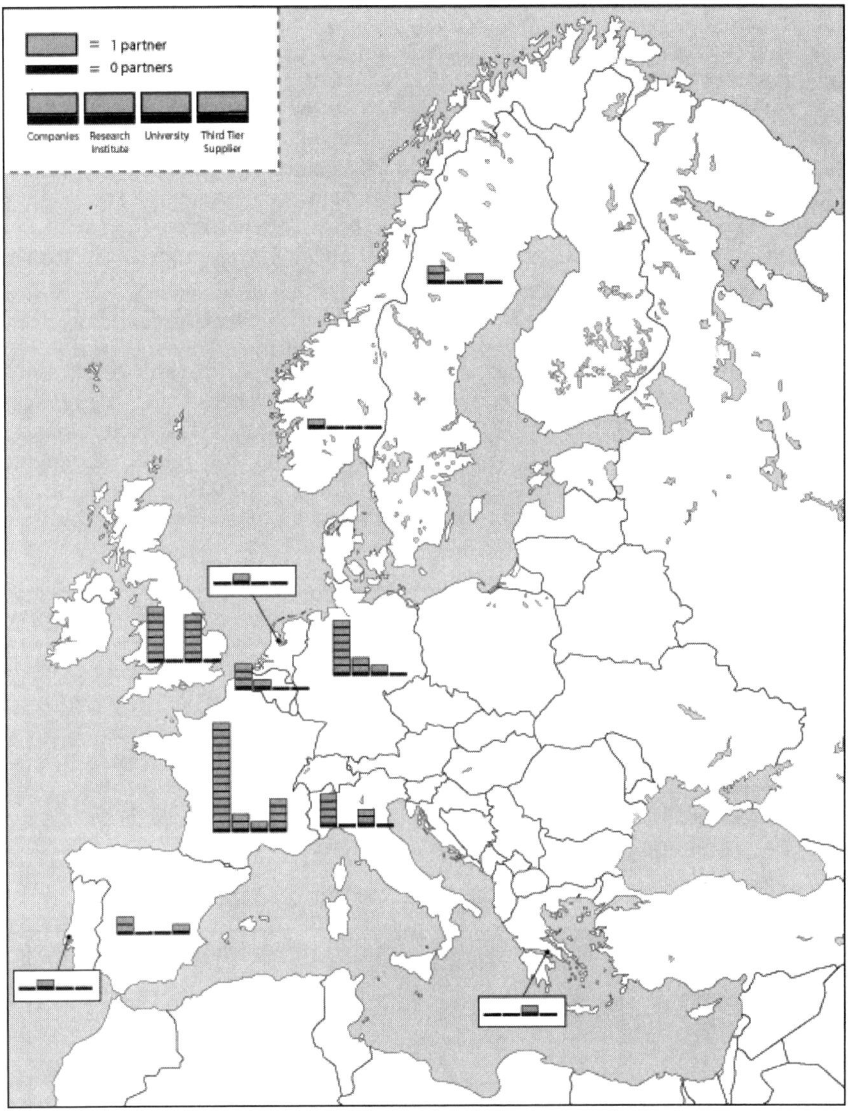

Fig. 1 Geographic overview collaboration partners per organization type.

with the clarification that separate business units belonging to multinational companies such as Airbus and Rolls Royce are shown in their respective country of origin. Appendix A provides a full list of all partners. It is perhaps worth mentioning that the geographical distribution of the partners in Fig. 1 reflects a program that started when the EU had only 15 member states, compared to 27 at present.

Under the EU regulations for industry-led projects such as VIVACE, each partner, with the exception of universities, contributes 50% of its share of the total funding. As a result, the consortium management has been based on consensus. That is, each partner had the right of veto on decisions impacting its business interests. The consortium was governed by a steering committee, composed of one representative from each partner. The committee met once a year to agree on the strategic agenda. Because of the atmosphere of trust created within the consortium, most other issues could be decided by electronic vote, for example, for releasing public information such as conference and journal papers. The executive management of the project was performed by a smaller core team of 10 partners, with more focus on technical progress. The core team met around five times per year. This organization proved efficient in reducing the number of consortium meetings while maintaining effective and responsive consortium management.

From a program-management point of view, the work was divided in three subprojects, as shown in Fig. 2. The mandate of the aircraft and engine subprojects was to define the business-driven issues that needed to be researched. These are referred to as the 81 use cases. In addition, the aircraft and engine subprojects could also verify the solutions provided by the advanced capabilities

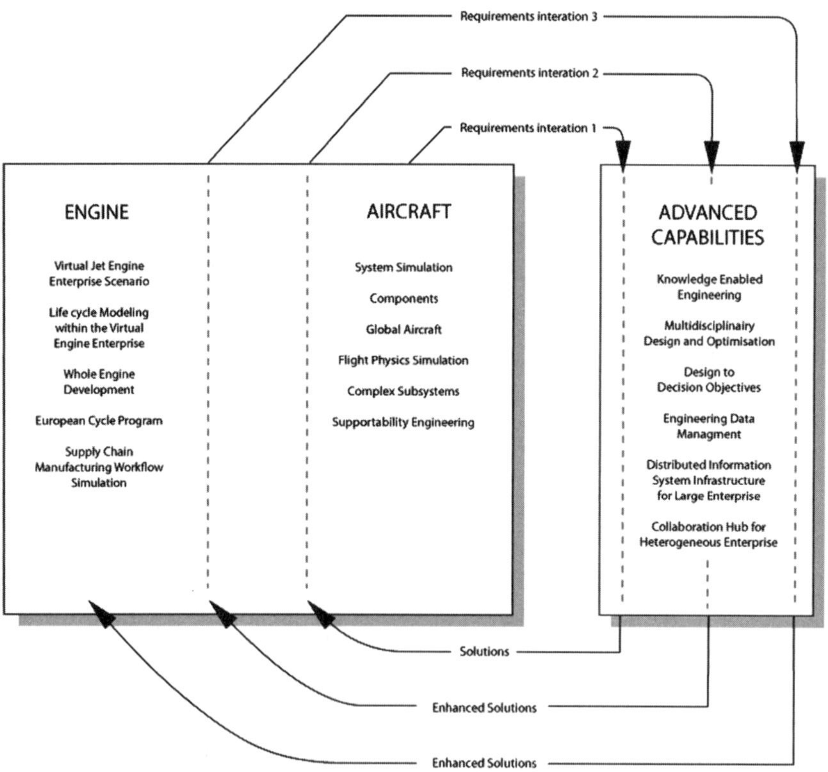

Fig. 2 Overview of subprojects, work-packages, and iterations in VIVACE.

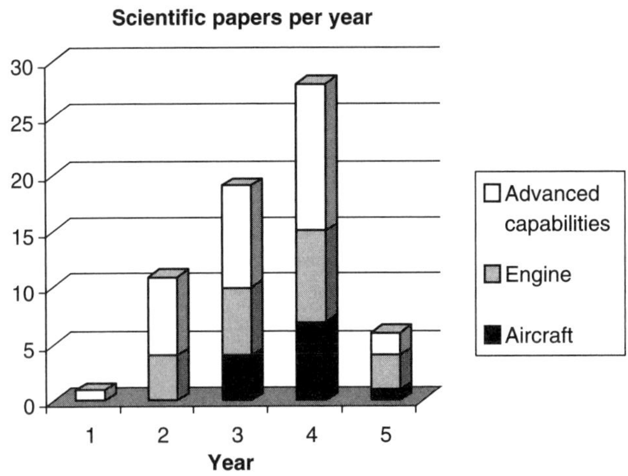

Fig. 3 Number of scientific papers per year per subproject.

subproject, which concentrated on solutions with the potential of satisfying more than one use case. For certain dedicated use case solutions, work might be performed also in the originating aircraft or engine subprojects. This project approach was aimed at focussing the research on results with relevance to the aeronautical practice. The MDO activities were concentrated in a dedicated work-package of the advanced capabilities subproject (Fig. 2). Inputs were obtained from four additional work-packages belonging to the aircraft and engine subprojects (refer to Table 1) with further support provided by the Engineering Data Management and Collaboration Hub for Heterogeneous Enterprise work-packages.

The use of public European research funding implied an active sharing of the results. Three dedicated public forums were held during the life of the project to disseminate knowledge to a wider audience. (Table 1 provides the number of dedicated presentations.) The final forum was accompanied by a public booklet with technical achievements information [2]. Additionally, 85 conference and journal papers were written and published as shown in Fig. 3. This public exposure, especially through the review process, allowed us to assess our work against the fast moving state-of-the-art in the MDO field.

In summary, following the evolutionary approach [3] and dividing the four-year duration into three iterations proved beneficial for managing the program. Within this framework, each work-package, including MDO, had to define independently verifiable objectives for every three-month period. The resulting earned-value assessment was considered instrumental in meeting the deadlines and the budgets. Strict adherence to the review procedure helped to achieve the intended quality of the deliverables. That is, all MDO deliverables were subjected to an optional company review, a review by all partners in the related work-package, and a final review by an independent expert, before being submitted to the client (EU) review.

Now, after the project has been completed on time and on budget, having demonstrably satisfied its high-level objectives with the envisaged quality, it can be concluded that the consortium management has been successful [4].

III. Project Objectives

A. High-Level Project Objectives

The overall aim of VIVACE was to enable the virtual product concept in a collaborative environment (the virtual enterprise) through design, simulation, and integration, starting from the early phases of aircraft conception. In this context the virtual product refers to all components that comprise an aircraft—the structure, the systems, and the engines.

The project was expected to contribute to the following three specific targets of the Strategic Research Agenda [1]:

1) *Halve the time to market for new products* with the help of advanced design, manufacturing, and maintenance tools, methods, and processes.

2) *Increase the integration of the supply chain* into a network.

3) *Maintain a steady and continuous reduction in travel charges* through substantial cuts in operating costs.

From a technical perspective, VIVACE had to deliver solutions that should significantly reduce the development cost of new aircraft and engines through the delivery of a *virtual product design and validation platform* based on a distributed *concurrent engineering methodology* supporting the virtual enterprise.

To achieve this global objective, the work in VIVACE was organized around *use cases*, that is, real industrial simulations of a part of the aircraft or the engine or part of a development process. The use cases were chosen to reflect both the virtual product and the virtual enterprise. Each of these included, on the one hand, requirements for early product simulation, and on the other hand, requirements for distributed working methods. The integration of these two aspects was the core of the VIVACE project.

Thus, the expectation was that the tools developed during the project would provide the European aeronautics industry with the best knowledge possible about the complete product prior to its development, thus reducing the final cost and accelerating the response time to market while increasing the integration of the supply chain.

B. Objectives of the Multidisciplinary Design Optimization Advanced Capability Work-Package

There is no universally accepted single definition of multidisciplinary design optimization (MDO). Throughout this book we adopt one of the AIAA MDO Technical Committee's definitions [5], namely, "*Optimal design of complex engineering systems which requires analysis that accounts for interactions amongst the disciplines (or parts of the system) and which seeks to synergistically exploit these interactions.*"

MDO was recognized during the conception of VIVACE as one of the promising approaches, the industrialization of which was expected to contribute to the achievement of the high-level project objectives. The activities in this work-package were therefore aimed at addressing some of the main obstacles to the effective application of MDO in industry, namely, the following:

1) Deployment on a case-by-case basis where a specific solution is developed for accessing MDO resources, such as storage, access, and exchange of information between the tools, and setup of the design and optimization processes, limits the synergy between disciplines as a result of the application of specific solutions to general problems.

2) The case-specific solutions often involve time-consuming, effort-consuming, and error-prone activities. The resulting process is often highly complex, requiring extensive skills within each of the disciplines involved and expert knowledge of the tools involved. Thus, instead of concentrating on the design process which is their core activity, the designers need to be part computer scientist and part IT specialists.

In this view (see also Fig. 4) the work had to explore the state-of-the-art methods and technologies in order to identify effective solutions for setup and automation of the MDO processes in a common way. These technologies had to be assessed and, if needed, developed further. Thus the MDO work-package focused on the development and construction of an advanced, generic, flexible,

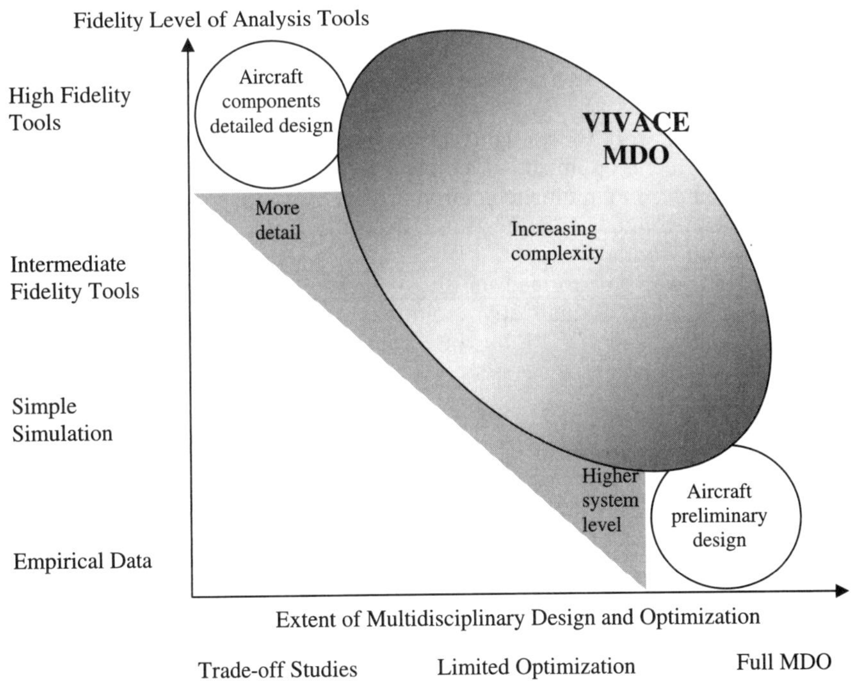

Fig. 4 Context of the VIVACE MDO activities.

multisite MDO framework for process support and integrated product design and optimization.

The main objective was to provide advanced MDO support to the work-packages in both the aircraft and engines domains (Fig. 2). The MDO work-package had to ensure the consistency of the MDO activities of both sub-projects. Moreover, innovative solutions and trends enabling improvements in aeronautical design processes had to be assessed by the aircraft and engine subprojects so that the development of the tools was always subject to a proper steering to ensure the usability of the tools in industry.

Underpinning the objective was the provision of common services and capabilities in the following key areas (see also Fig. 4):

1) Process definition, which is the specification of the product design and optimization processes for which the MDO framework has to operate;

2) Research, development, and validation of generic MDO schemes, including the specification of multidisciplinary, multilevel, multifidelity optimization processes and strategies;

3) Management and control of the design, optimization, and engineering analysis processes, for example, intuitive operation using standardized graphical user interfaces;

4) Generic, discipline independent tools for MDO involving, definition, construction, and integration of a coherent set of submodels and generic interfaces for effective communication and information exchange;

5) Discipline specific (high-fidelity) tools for MDO, involving assessment of technical disciplines needed within the MDO process depending on the design stage, thus enabling monodisciplinary tools to be linked to relevant tools from other disciplines; and

6) MDO framework functionality to link the cooperating tools and services, involving design and optimization template construction; uniform identification of MDO resources in a multisite environment; transparency of MDO resources; automated data communication and tool execution; and securing an automated multipartner collaboration.

The MDO work performed in the dedicated and other work-packages delivered a number of innovative solutions to the preceding requirements, which are summarized in the following chapters.

IV. Outline of the Book

A guiding principle during the compilation of this manuscript was to follow as best as we could the phases of a generic design cycle. That is, the book starts with chapters covering MDO issues at initial design stages and then gradually moves to more detailed design optimization (thereby covering the ellipse area in Fig. 4). Similarly, wherever possible, the MDO applications are ordered by product complexity, from whole aircraft and engine to single component optimization. Last, but not least, MDO infrastructure related topics contributing to the efficient functioning of the virtual collaborative enterprise are presented. These include engineering data management, product life-cycle management, security and automated workflows among others.

The chapter descriptions provided below are based on abstracts provided by the corresponding chapter authors.

A. Chapter 1: Overview

As just described, this chapter presents a concise overview of our European industry-led research collaboration, followed by a summary of the objectives and concluding with an outline of the book structure.

B. Chapter 2: MDO at Predesign Stage

Early design is a crucial stage of the product development process because the decisions taken during this period commit the majority of the life-cycle costs (Fig. 5). Although most of these decisions are taken under a great deal of uncertainty, it is also true that this design stage offers the greatest scope for innovation. In this context the objective of the work presented in this chapter was to research and to develop a prototype workflow management device (WMD). The main intended function of the workflow manager was the dynamic assembly of aircraft conceptual computational studies composed of hierarchical computational processes, which in turn are composed from atomic models (i.e., equations or compiled code referred to as *black boxes*). In addition, the workflow manager was intended to enable the dynamic application of relevant treatments to the computational processes, such as multi-objective optimization, sensitivity analysis, and uncertainty management. The realization of these treatments includes distinct research contributions, which are also described in this chapter.

The relation of this work to the classical notion of MDO, that is, *optimization of a system consisting of coupled disciplines*, becomes apparent when considering the objective of predesign, which is to define the characteristics on an aircraft given its properties, that is, to determine the design parameters, given performance and operational parameters derived from stakeholders' requirements, or to modify existing aircraft for the satisfaction of a different or a stretched requirement. In any case a workflow consisting of hundreds of models (black boxes)

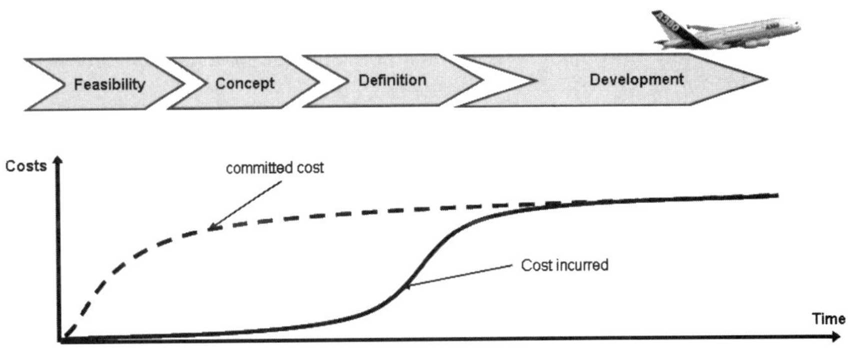

Fig. 5 Illustration of costs commitment during aircraft development life cycle (adapted from [6–8]).

and thousands of variables needs to be assembled and "(re)wired" every time a variable is added to or removed from the input set. During this process, subsets of models can become coupled through shared variables. These coupled or *strongly connected components* (SCCs) correspond to the "disciplines" in the classical MDO. The difference is that the predesign MDO process has to be configured *on the fly*, with hundreds, albeit low-fidelity models. Thus the novel computational framework presented in this chapter enables the designers to build, understand, manipulate, and share complex processes and studies. It also allows applying a number of advanced numerical treatments, including multi-objective optimization, sensitivity analysis, and uncertainty management with emphasis on design robustness. The work demonstrated for the first time that such capabilities can be used in a coordinated way to enhance the efficiency of the computational process and the effectiveness of the decision making at predesign stage.

C. Chapter 3: Helicopter Predesign Strategy: Design-to-Mass or Design-to-Cost

Considered in this chapter is the predesign of rotary-wing aircraft. Experience shows that helicopter configurations and operations tend to be more diverse compared to fixed-wing aircraft. Usually for helicopters the lowest mass is used as the predesign criterion, based on the rationale that a lower mass will result in lower operating costs. However, for the diverse spectrum of helicopter operations this is not necessarily the best design criterion. One of the innovative approaches behind the work reported here is the coupling of a total life-cycle cost model with a helicopter predesign tool. The total life-cycle cost covers all major cost categories of the operational life of the helicopter. This model has been validated by a major European industrial partner while the predesign tool has been validated by an established European research institute. By assessing the total life-cycle cost consequences of design choices, optimization for the specific mission profile(s) of an operator becomes possible. From a manufacturer's point of view, optimization for various mission profiles of various potential customers also becomes possible. A generic optimization approach allowing for any mix of continuous parameters (such as mission range) and discrete parameters (e.g., the type of main rotor hub) has been applied. As designs for minimum total life-cycle costs differ from designs with minimum weight, this innovative approach adds value to the early design process.

D. Chapter 4: Aeroelasticity and Sizing in a Multilevel Modeling Approach for Preliminary Design

Models with different fidelity levels can be used during the preliminary design phase. This necessitates the appropriate handling of associated model characteristics such as computational loads, amount and representation of additional input/output information, and so forth. This chapter presents a generic modular framework allowing the flexible coupling and exchange of models with different fidelity levels. Three levels of model fidelity are considered: the basic aircraft model, also addressed in Chapter 2, the physical aircraft model, and the disciplinary model. These model implementations are heterogeneous, ranging from commercial-off-the-shelf to in-house solutions. The proposed framework is referred to as the *simulation toolbox*. The feasibility of the toolbox is demonstrated for aeroelastic simulations.

A time simulation loop based on an innovative coupling of a multibody dynamics model with a beam analysis model has been used to demonstrate structural sizing of a wing. A single case of a 2.5-*g* pull-up maneuver is considered. However, this could easily be extended to cover any load case relevant to wing sizing. The obtained results show that for some designs lower local stresses can be expected in dynamic simulation with respect to standard static analysis. This provides valuable insight, even at the preliminary design stage. Being able to perform these simulations as early as possible during the design cycle demonstrates that the presented multilevel simulation toolbox approach is feasible.

E. Chapter 5: Multidisciplinary Wing Design Optimization

Aircraft wing design is a good example of an inherent multidisciplinary activity. Disciplines involved include aerodynamics, structural mechanics, propulsion, space allocation, and total life-cycle costs among others. Various constraints and often conflicting design objectives hinder the achievement of a balanced design. As a result, many design variants need to be analyzed. This chapter reports on how to tackle such design challenges. In the first place, the respective design analysis tools need to be integrated into a design workflow, which can be automatically executed while ensuring that the exchanged design information allows the analysis of a consistent wing design. The other MDO aspects addressed here are the wing geometry parameterization and the measures taken to ensure efficiency (speed) and effectiveness (accuracy) allowing the evaluation of a sufficient number of designs.

The proposed approach and its generic features are illustrated by providing results for designs as varied as blended wing bodies, supersonic civil transport aircraft, supersonic business jets, and low fuel-burn open-rotor driven novel aircraft concepts.

F. Chapter 6: Meta-Modeling and Multi-Objective Optimization in Aeronautic MDO

Although the preceding chapter provides a good example of the increasing accuracy of the models used during the earlier design stages, it also highlights that this accuracy comes at a price. That is, such higher-fidelity models consume significantly more computational resources. This, combined with the need to explore larger design spaces and to use more robust optimization algorithms, which require higher numbers of design evaluations, necessitates an approach based on meta models.

The objective of this chapter therefore is to describe a set of approximation and interpolation methods to create meta models, also referred to as surrogate models (and response surface models in particular). The key of the approach is the decoupling of the (computationally expensive) aircraft design analyses from the automated search and optimization process. In this approach a suite of different approximation functions is available, each of which can be selected easily. This approach has been demonstrated using the wing multidisciplinary analysis capability of the preceding chapter, the helicopter total life-cycle cost optimization of Chapter 3, and the engine optimization of Chapter 9. An implementation of the ε-NSGA (nondominated sorting genetic algorithm), based on a combination of existing algorithms, has been produced for the optimization. This algorithm has proven efficient in generating a Pareto surface for the aforementioned examples. Displaying the results in the design parameter space as well as in the objective space provides valuable information for design decisions by the design experts involved. The computational efficiency of the various approximation approaches allows one to choose among several options. This can be very useful in cases where the most appropriate (i.e., the most effective) fitting function cannot be determined a priori. Deploying this approach to several distinct case studies exemplifies its applicability to a wider range of optimization cases.

G. Chapter 7: Fuselage Structure Optimization

The subject of this chapter is a flexible optimization framework dedicated to stiffened panel optimization, which has been implemented and demonstrated on aircraft fuselage covers. The framework incorporates three methodologies that have been developed with consideration of the design stage. For the early design stages a rapid sizing approach has been demonstrated, which utilizes a neural-network-based surrogate model. For preliminary sizing approaches the optimization is based on exact stress responses obtained from semi-analytical stress tools. The selected approach is built upon independent local optimizations. Even if not fully optimal in terms of internal load redistribution and design continuity, the proposed approach has the advantage of being very easy to parallelize with the possibility to launch every structural element optimization on an independent processor.

The optimization with regard to detailed sizing approaches is based on advanced stress responses obtained either from semi-analytical stress tools or from nonlinear finite element analysis (buckling/postbuckling) for which a gradient-based optimization approach has been developed. Few publications so far demonstrate such a capability. Innovative measures have been introduced to guarantee the robustness of the optimization process, both with respect to linear buckling and with respect to nonlinear postbuckling.

Overall, the results of this work demonstrate a robust methodology that is fairly efficient, given the provided accuracy. The still significant computational requirements emphasize the need for high-performance computing. Although demonstrated on a structural optimization problem, the proposed framework can be an integral component of a wider MDO scheme.

H. Chapter 8: Pylon Multidisciplinary Optimization

From a design life-cycle perspective this chapter on pylon MDO is positioned further towards detailed design. In this context, an optimization framework has been researched, implemented, and demonstrated on various multidisciplinary scenarios. These have been restricted to two disciplines in order to follow a gradual step-by-step approach, which is currently more feasible and therefore preferable to an otherwise too ambitious integration of all disciplines.

Stress and loads optimization was developed and demonstrated for two types of loads: steady maneuver and fan blade-off loads. A complex integration was performed for maneuver loads chaining external load sensitivities with structure sensitivities in a direct gradient-based optimization scenario. An indirect approach was developed for fan blade-off loads using response surface models. In each case a load-looping scenario was compared with a scenario considering sensitivities. The outcome of the second scenario was quantified in terms of weight savings for an industrial test case. This quantified result yielded an important lesson learned on how to improve the industrial process for aerostructure optimization.

Stress and aerodynamic optimization was developed and demonstrated for simultaneous weight and drag optimization of a power plant. A multilevel optimization approach using direct optimization solvers and intermediate response surface models (polynomial and Kriging models) was developed. The use of response surfaces proved to be a simple and efficient way to start the implementation of a numerical MDO process in an industrial context. The advantage of the developed process is in keeping the autonomy of each discipline while exploiting their respective best-suited optimization processes.

A significant progress has been made in the numerical robustness of these processes, which is now close to automation and which, in turn, is necessary for the multilevel optimization. Such advancement is particularly relevant to tackling aerodynamic optimization with complex three-dimensional Navier–Stokes analyses. The optimization around fan blade-off simulation is also to be highlighted because of few publications on this subject. Furthermore the interaction between disciplines was well understood and modeled, thus allowing the most efficient integration scheme to be selected for each scenario.

The overall contribution of this research work is seen in giving practical directions for the implementation of MDO in an industrial context and tracing the path to a more integrated multidisciplinary optimization process, which incorporates the application of up-to-date numerical analysis methods.

I. Chapter 9: Engine MDO Deployed on a Two-Stage Turbine

Large design efforts, such as those put into the design of aeronautical engines, are typically shared between various partners. The resulting collaboration where no partner can impose the tool suite to be used by the collaboration can be organized as a virtual enterprise. Within such a virtual enterprise each partner uses its own design framework, which usually combines commercial-off-the-shelf (COTS) and proprietary tools. The ensuing design framework contains the expertise of the partners involved; hence, there is a balance between the need to share the capability during a collaborative design and protecting the intellectual property rights of each partner's core assets. Traditionally such multipartner collaboration is accomplished by exchanging limited design information between partners via paper or e-mail.

This chapter describes an automated integration of the various partners' design frameworks within the collaboration. Each partner contributes according to the integrated design needs and its specific expertise including also local multidisciplinary design optimization. By facilitating a fully automated exchange of design information, a design process was realized for coupled preliminary design and detailed design. Thus the innovation claimed is the capability to perform an automated robust MDO on the entire design, based on the coupled design frameworks of each contributing partner. Using a critical engine assembly, a parameterized two-stage high-pressure turbine design, a collaborative design optimization of two design variants, has been performed.

On a technical level, the design results indicate feasible alternative designs, optimized at global level while taking the locally optimized characteristics into account. An even more important contribution is the demonstration of a capability for collaborative multicompany multilevel multidisciplinary design based on the integration of each partner's individual MDO framework. According to the authors, this is the most complex integrated collaborative design they were aware of at the time of writing.

J. Chapter 10: Using the Virtual-Enterprise-Collaboration Hub for Distributed Engine Optimization

The multinational multicompany collaborative MDO topic is further extended in this chapter. The requirements for such a setting, as specified in Chapter 9, are illustrated here with an engine MDO example, typical for the European aeronautical context. A key requirement is that each partner provides its specific advanced design capabilities based on its company-specific MDO framework. These frameworks and the company specific IT policies, enterprise architectures, and security policies need to be respected, resulting in a heterogeneous IT environment where each environment evolves independently at each partner's discretion.

To address these needs, the virtual-enterprise-collaboration hub (VEC hub) has been developed. The VEC hub is based on a loose coupling of the local design systems, where the existing MDO processes are wrapped into Web services and exposed to the VEC hub. This chapter describes the VEC-hub concept and the way it supports the multinational multicompany MDO collaboration. The main components of the VEC hub are a secure collaboration area and workflow services. The secure collaboration area allows sharing design information between the partners. The top-level workflow allows the automatic activation of each company's collaborative MDO design capabilities. Typically this top-level workflow will be executed by the partner performing the integrator role; however, technically the workflow can be executed by any other partner possessing the appropriate permissions. The flexibility of the chosen approach is demonstrated by using the engine collaboration example of the preceding chapter. In this example each of the four partners has implemented a different complying mechanism. The resulting workflow is claimed to be the most complex the IT solution provider has encountered at the time of writing.

K. Chapter 11: Securing Virtual-Enterprise Collaboration

In the virtual enterprise as just described, each partner shares certain critical company assets such as proprietary data and design capabilities which differentiate that partner from its competitors. Therefore the security of these business critical assets is of paramount importance. In practice the group of collaborating partners will change during the collaboration as a result of business opportunities. In this view, a solution to these security concerns must accommodate such partner evolution. This chapter reports on a solution, based on the work of the collaborative engine MDO example of Chapter 9. [For topics related to trust and identity management, collaboration with the Transglobal Secure Collaboration Program (TSCP) at http://tscp.org/has been initiated.]

The proposed access control architecture manages the critical assets in the virtual enterprise, allowing each organization to stay independent, and provides each partner with technologies to manage its own employees and assets. The demonstrated access control solution is based on three concepts. Attributes-based access control models form the first concept. They offer flexibility during the specification phase of the access control policies. The privilege management infrastructures controlling the management of these attributes form the second concept. To harmonize the authentication and accreditation mechanisms, the third concept is a TSCP-influenced identity federation. The resulting prototype demonstrates that the collaborating partners' security requirements can be met. However, the prototype is partly based on proprietary solutions. Available standards-based alternatives are identified in order to arrive at the required scalability of the solution.

L. Chapter 12: EDMRF to Ensure Partners' Collaboration for Multidisciplinary Optimization

After describing the virtual enterprise collaboration hub in Chapter 10 and the security aspects of the collaborative design environment in Chapter 11, this chapter focuses on the interoperability aspects of the collaboration. The heterogeneity of the software used by the collaborating partners results in a diversity of shared data, both between partners and during the various phases of the product life cycle. The resulting lack of interoperability impedes collaboration by incurring significant costs as well as delays in the design and hence product development. This chapter provides an integrated reference framework to ensure multipartner multidomain collaboration. As discussed in Chapter 10, shared semantics needs to be agreed and implemented along with the mechanisms for data sharing. Traditional product life-cycle solutions address the production phase, but are less suited to include the engineering phases starting at the preliminary phases. This chapter introduces the engineering data-management framework. It is based on a multilayer architecture. The data-exchange concept is based on the digital mock-up. The engineering data-management framework has been implemented for a typical design simulation study, demonstrating the feasibility of the approach. The results are currently being implemented at major European manufacturers.

In conclusion, we believe that the innovative methods and infrastructure solutions presented in this manuscript represent a radical step forward towards the advancement and industrialization of the MDO concept. In this view we trust that the book will be useful to both researchers and practitioners in the field of complex systems design.

References

[1] Argüeles, P., et al., "Report of the Group of Personalities," *European Aeronautics: A Vision for 2020*, European Commission, Jan. 2001, http://ec.europa.eu/research/growth/aeronautics2020/en/personalities.html [retrieved July 2009].
[2] "VIVACE—Value Improvement Through a Virtual Aeronautical Collaborative Enterprise Final Technical Achievements (2004–2007)," VIVACE Consortium Members, 2007, http://www.vivaceproject.com/technical_leaflet_final.pdf [retrieved Jan. 2009].
[3] Gilb, T., *Competitive Engineering*, Elsevier, New York, Chap. 10, 2005, pp. 1–26.
[4] Martin Hernandez, J. M., "Opportunities in Framework 7," *VIVACE Forum-3*, Toulouse, France, Oct. 2007, http://www.vivaceproject.com/content/forum3/forum3.php [retrieved July 2009].
[5] AIAA Multidisciplinary Design Optimization Technical Committee, AIAA, Reston, VA, http://www.aiaa.org/portal/index.cfm?GetComm=80&tc=tc [retrieved Jan. 2009].
[6] Roskam, J., *Airplane Design, Part VIII: Airplane Cost Estimation: Design, Development, Manufacturing and Operating*, DAR Corp., Lawrence, KS, 1991, pp. 3–13.
[7] Badufle, C., and Homsi, P., "Value Improvement Through a Virtual Aeronautical Collaborative Enterprise," *Symposium on Applied Aerodynamics and Design of Aerospace Vehicles*, SAROD 2005, Bangalore, India, Dec. 2005.
[8] Sträter, O., *Cognition and Safety: An Integrated Approach to Systems Design and Assessment*, Ashgate, Farnham, U.K., 2005, pp. 12–14.

Chapter 2

MDO at Predesign Stage

Marin D. Guenov,* Paolo Fantini,† Libish Balachandran,† Jeremy Maginot,† and Mattia Padulo‡

Cranfield University, Cranfield, England, United Kingdom

Abbreviations

AD = automatic differentiation
AOF = aggregate objective function
CE = calculation engine
CFD = computational fluid dynamics
CWMD = Cranfield workflow management device
DHCBI = double hyper cone boundary intersection method
DM = dependence matrix
DSM = design structure matrix
EMO = evolutionary multi-objective optimization
FAST = Fourier amplitude sensitivity test
FPI = fixed point iteration
GA = genetic algorithm
GFCL = generate first choose later
GSA = global sensitivity analysis
GUI = graphical user interface
IM = incidence matrix
IMM = incidence matrix method
MDO = multidisciplinary design optimization
MM = Taylor-based method of moments

*Professor.

†Research Fellow.

‡Doctoral Research Student.

MO = model object
MOO = multi-objective optimization
NBI = normal boundary intersection
NC = normal constraint
NSGA = nondominated sorting genetic algorithm
OAT = one at a time
PP = physical programming
RHS = right-hand side
RO = robust optimization
SA = sensitivity analysis
SCC = strongly connected components
SLO = system-level optimizer
SP = sigma-point
USMAC = ultrasimplified model of aircraft
VBM = variance-based methods
WMD = workflow management device

Nomenclature

$\boldsymbol{a}_i$ = anchor point i
c = column number
F = robust function
f_i = objective function i
G = robust constraint function
g_k = kth inequality constraint
h_p = pth equality constraint
incm = numerical incidence matrix
incmf = numerical foundation incidence matrix
incmprod = product of nonzero elements of column c of *incmf* matrix
K_i = number of inequality constraints
$\boldsymbol{l}$ = normal to utopia plane
L = number of models
M = number of objectives
m = total number of columns in the *incm* matrix
N = number of design variables
n = total number of rows in the *incm* matrix
$nFdb$ = number of feedback loops
nMm = number of modified models
P_e = number of equality constraints
$\boldsymbol{p}$ = utopia plane point
Q = total number of variables
r = row number
$valc(c)$ = product of nonzero elements of column c of *incm* matrix
$valcf(c)$ = product of the nonzero elements of column c of matrix *incmf*, if *incmprod* is not equal to 2; otherwise, the value is 3
$valc2(r, c)$ = decision-making variable for element (r, c), for replacing its value with 2, based on the elements in column c
$valc3(r, c)$ = decision-making variable for the element (r, c), for replacing its value with 3, based on the elements in column c

$valr(r)$ = product of nonzero elements of row r of matrix *incm*
$valrf(r)$ = product of nonzero elements of row r of matrix *incmf*
$valr2(r, c)$ = decision-making variable for element (r, c), for replacing its value with 2, based on the elements in row r
$valr3(r, c)$ = decision-making variable for element (r, c), for replacing its value with 3, based on elements in row r
X = design space
x_i = ith design variables
Y = criterion or objective space
y_i = ith output variable
Γ_y = kurtosis of y
γ_y = skewness of y
μ_y = mean of the variable y
σ_y = variance of the variable y

I. Introduction

PREDESIGN is a crucial stage of the product development process because the decisions taken during this period commit the majority of the life-cycle costs. Although most of these decisions are taken under a great deal of uncertainty, it is also true that this design stage offers the greatest scope for innovation.

Our initial research indicated that although involving a lot of talent and producing great results, the predesign processes can be significantly improved. For example, there is a need for enhanced homogeneity and formalism. Currently the relevant process information is spread in tools, manuals, brains, and sites, and in many cases the procedural coding incorporates "hard-wired" assumptions from other disciplines (e.g., weight-estimation procedure incorporating assumptions on aerodynamic loading). Thus the result of a design study might depend on the way it was produced. On the other hand, innovation might be restrained by the confinement to known cases because the design starts from existing configurations, implicit in the computational code. Industrial need was therefore identified for 1) a new approach, allowing the merging of numeric and geometric design and the facilitating of a component-driven modularization of the disciplines; 2) a collaborative environment to support the exchange of models and simulation data in order to perform an overall design optimization process with robust, flexible, and dynamic workflows; and 3) a more discrete fragmentation of existing modules into elementary parts in order to enable the flexible assembly of new processes.

In this context, Cranfield University undertook to research and develop a prototype workflow management device (WMD) enabling a simple formalism and easily understandable description of the computational workflow, which would allow to store not only the results, but also the way the results were computed. This means that the computational workflow should be stored in an executable format and also be editable by the user in order to develop dynamic solutions. The WMD should be capable of dynamic assembly of hierarchical computational processes from other processes and/or from atomic models (i.e., equations or compiled code referred to as black boxes). In addition, the WMD should enable the dynamic application of relevant treatments to the computational processes, such as multi-objective optimization, sensitivity analysis, and uncertainty management. The realization of these treatments includes distinct research contribution, which is also described in this chapter.

The relation of this work to the classical notion of multidisciplinary design optimization (MDO), that is, *optimization of a system consisting of coupled disciplines*, becomes apparent when considering the objective of predesign, which is to define the characteristics on an aircraft given its properties, that is, to determine the design parameters, given performance and operational parameters derived from stakeholders' requirements, or to modify existing aircraft for the satisfaction of a different or a stretched requirement. In any case a workflow consisting of hundreds of models (black boxes) and thousands of variables needs to be assembled and "(re)wired" every time a variable is added to or removed from the input set. During this process, subsets of models can become coupled through shared variables. These coupled or *strongly connected components* (SCCs) correspond to the "disciplines" in the classical MDO. The difference is that the predesign MDO process has to be configured "on the fly," with hundreds, albeit low-fidelity models.

These problems, as part of the WMD specification, are tackled in the next section. The calculation engine (CE) is described in Section III. The CE is based on a novel method for multi-objective optimization, and in particular, for finding the local and global pareto surfaces. Section IV discusses sensitivity analysis. The importance of this subject to MDO is associated not least with the possibility to identify design variables whose range of variation does not have significant effect on the objectives and constraints. This would allow fixing the values of these variables, which in turn could significantly reduce the dimensionality of the design space and the computational effort, respectively. Uncertainty management is discussed in Section V with emphasis on robust multi-objective optimization. The results are presented in Section VI. These are based on the coordinated application of all aforementioned tools on a significantly simplified, but still a representative aircraft predesign test case, supplied by our industrial partners. Finally, conclusions are drawn and future work outlined.

II. Computational Workflow Management

As outlined in the introductory section, a complex workflow consisting of hundreds of models (black boxes) and thousands of variables needs to be assembled every time a variable is added to or removed from the input set. This has been one of the main challenges of our research, given the aim to extend the flexibility with which the designer can operate on the computational process while reducing the execution time. This section describes how we tackled that challenge. We first briefly summarize the state-of-the-art in design computational process modeling and then present our novel computational process modeling method.

A. State-of-the-Art in Computational Process Modeling

The methods for organizing and rearranging models into a computational system can be classified into variable flow modeling, decomposition, and scheduling methods.

Variable flow modeling is the process of identifying the data flow among the models depending on the system input variables selected by the designer. Constraint propagation approaches have been utilized by several researchers for variable flow modeling in conceptual design systems where only equations are present. In this approach the equations are represented as constraints between the variables. The changes in the variables' values are propagated across the constraint network. Serrano [1] developed a software tool named Concept Modeller in which a graph theoretical approach was chosen for constraint management. The constraint networks were modeled as directed graphs, where nodes represent parameters and arcs represent constraint relationships. Serrano's graph theoretical approach was developed primarily for systems consisting of algebraic equations. The approach did not address the computational complications that would have been encountered if the system contained models. Bouchard et al. [2] used directed constraints between design variables and numerical solution approaches to allow rapid production of trade-off studies. The limitation of this approach is that the designer has to decide in advance the input and output variables. Buckley et al. [3] developed a conceptual design tool, Design Sheet, in which a bipartite graph method was applied for variable flow modeling. The edges in the graph connect equation nodes to variable nodes, which indicate that a variable is present in the equation. The directing of the graph based on the known variables is accomplished using a variant of the Ford–Fulkerson algorithm [4] for finding maximal matching on bipartite graphs. Buckley et al. [3] focused on obtaining variable flow models for algebraic equations and not for models. Ramaswamy and Ulrich [5] have developed an adjacency-matrix-based heuristic algorithm for variable flow modeling. Their algorithm is restricted to functional forms that must be either algebraic or transcendental functions. Recursive functions and iterative computer programs (models) were explicitly excluded.

System decomposition is the process of decomposing a complex system into a number of subproblems, and scheduling is the process of sequencing the models for the purpose of executing them after eliminating or reducing the feedback loops among the models. Most of the decomposition and scheduling methods have evolved from graph theory [6]. One of the earliest available tools for scheduling was PERT [6]. PERT network is a directed, weighted, acyclic graph. The weights of the edges in a PERT network represent the time needed to complete that task. The PERT tool is applicable only to sequential activities and cannot handle nonsequential activities. Steward [7] developed a tool for displaying and sequencing processes, called the design structure matrix (DSM). DSM serves as a highly efficient tool for decomposition and scheduling. The advantage of DSM compared to PERT is the ability to group and display the iterative subcycles found in a design project. Tang et al. [8] introduced a DSM-based method for decomposition and a method for the scheduling of decoupled processes both of which have application in concurrent engineering.

Rogers [9] developed a software tool, named Design Manager's Aid for Intelligent Decomposition (DeMAID), for system decomposition and sequencing of models based largely on DSM. Initially, a knowledge-based tool was developed for sequencing the models [10]. The knowledge-based approach could only examine a limited number of orderings of the models, which are part of iterative subcycles (SCC). To overcome this limitation, Rogers [9] introduced a genetic-algorithm (GA)-based method for rearranging the models in the iterative subcycles.

Among the various tools and methods reviewed, Design Sheet and DeMAID have been identified as the most advanced tools for computational conceptual design studies. However, whereas Design Sheet focused on variable flow modeling only for equations, DeMAID had its focus only on decomposition and scheduling of models. Hence, there was a requirement for generating a novel method that combines the state-of-the-art methods for variable flow modeling, decomposition, and scheduling for both equations and models, which thereby should significantly increase the flexibility with which the designer is able to operate on a system of models. The next section describes our novel method that has been developed to address this issue.

B. Computational Process Modeler

Computational process modeling is the process of organizing a complex system of models in order to efficiently compute the output variables according to the specific independent (input) variables selected by the designer.

The computational process modeler is presented in Fig. 1. A brief summary of the flowchart is given next, and a detailed description of the associated techniques is presented in the following subsections.

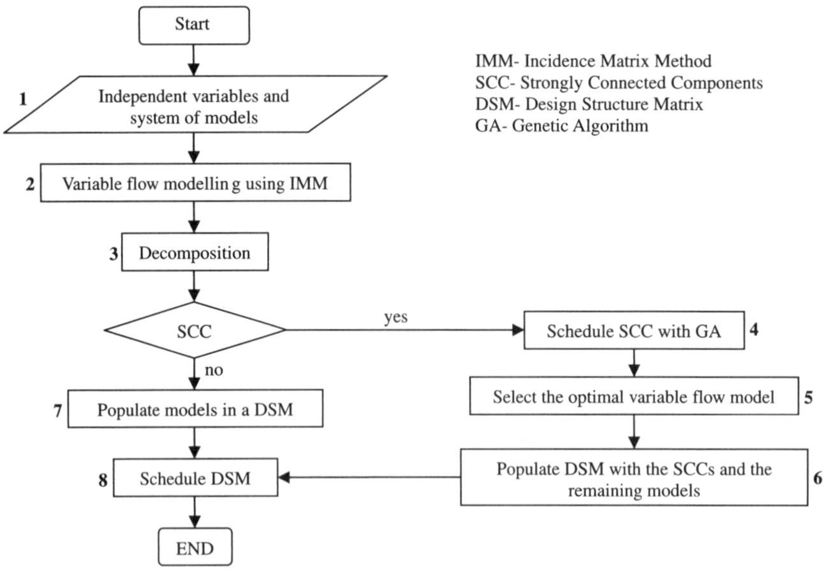

Fig. 1 Computational process modeler.

START

Step 1—Independent variables and system of models: Initially the designer needs to provide the system of models with a choice of independent (input) variables.

Step 2—Variable flow modeling using IMM: Variable flow modeling is performed using the incidence matrix method (IMM) in order to determine the information (data) flow among the models. All feasible variable flow models of the system are explored in this step.

Step 3—Decomposition: Each variable flow model generated is separated into hierarchically decomposable and nonhierarchically decomposable systems of models. Nonhierarchically decomposable systems are also known as SCC.

Step 4—Schedule SCCs with GA: Given an SCC, its constituent models are rearranged by means of GA.

Step 5—Select the optimal variable flow model for the SCC: The selection of the optimal variable flow model is based on the value of objective function that combines criteria such as number of modified models and number and length of the feedback loops.

Step 6—Populate DSM with the SCCs and the remaining models: Each of the rearranged SCCs is regarded as a single model and is reintroduced in the DSM along with the remaining models.

Step 7—Populate models in a DSM: If SCCs do not exist, then the models are populated directly in a DSM based on data flow obtained from the variable flow model.

Step 8—Schedule DSM: The DSM is rearranged into a lower triangular form using a graph theoretical algorithm. This rearrangement eliminates the feedback loops, and thus the final computational plan is obtained for the system.

END

The following subsections explain in detail the novel techniques used in the algorithm.

1. Variable Flow Modeling

As stated earlier, the existing methods for variable flow modeling are applicable to algebraic equations only. These methods need modification in order to be applicable to the current research context, where models are used instead of algebraic equations. This is because models produce multiple outputs compared to a single output generated by the equations. Furthermore, most of the methods reviewed focused on obtaining a single feasible variable flow model for solving a system while there can be multiple ones. This limits the chances for obtaining the feasible variable flow model, which could lead to a shorter execution time.

A novel IMM that dynamically obtains the information flow within the system is proposed here. The variable flow modeling using IMM corresponds to Step 2 in the computational process modeler flow diagram shown in Fig. 1.

An incidence matrix has models represented by the rows and variables by the columns. The association of a variable in the column with a model in the row is denoted by $*$ marked in the corresponding element (Fig. 2b). Populating the incidence matrix signifies substituting the $*$ in each element with either an i (input) or an o (output), depending on whether the variable in the column should be an input

a)

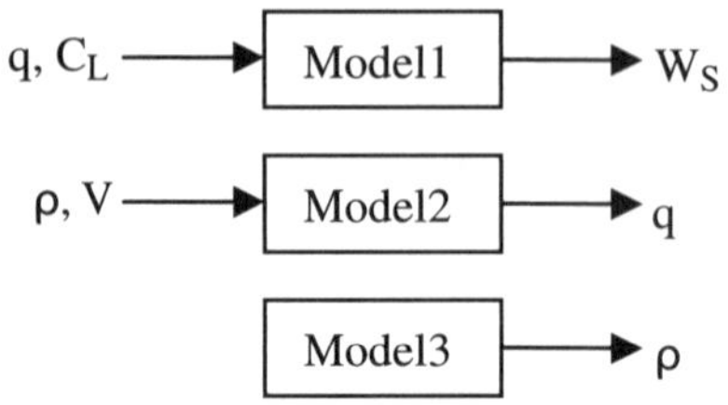

b)

	Ws	C_L	q	ρ	V
Model1	*	*	*		
Model2			*	*	*
Model3				*	

Fig. 2 a) Models balancing the weight of aircraft with its lift. b) Corresponding incidence matrix.

to or an output from the model in the row. This substitution is based on the five intuitive rules [11] as follows:

1) An independent variable should be always an input to a model. (This rule implies that all the * in the columns of the independent variables should be replaced with *i*.)

2) If a variable is associated with only one model and if it is not an independent variable, then it should be the output of that model. (This rule implies that if the elements of a column are all empty except a single * entry, and if the corresponding variable, which represents the column, is not an independent variable, then that * should be replaced with an *o*.)

3) If a model is associated with only one variable and if the latter is not an independent variable, then it should be the output of that model. (This rule implies that if the elements of a row are all empty except a single * entry, and if the corresponding data variable in the column is not an independent variable, then the * should be replaced with an *o*.)

4) Each variable should be an output of only one model in the system. (This rule implies that except for the columns of the independent variables, all other columns should have *o* marked in exactly one element.)

5) The number of outputs identified through variable flow modeling, for a particular model, should correspond to the number of outputs of the original model. (This rule implies that every row should have the same number of *o* as the number of outputs of the associated model.)

The population of the incidence matrix is demonstrated using a simple example presented in Fig. 2a. It shows a simple set of models for balancing the weight of an aircraft with its lift. Data variables entering the models are the inputs, and data variables leaving the models are the outputs. Variables *Ws* and *V* are assumed here to be the independent variables selected by the designer.

The set of models is represented in an incidence matrix as shown in Fig. 2b. The next step is to substitute the * in the incidence matrix with either *i* or *o* by applying the five rules just stated.

The final populated incidence matrix obtained by applying the rules is shown in Fig. 3. The numbers in the braces represent the sequence in which the matrix was populated. The rules applied in each step are as follows {1}-Rule1,

	Ws	C_L	q	ρ	V
Model1	i{1}	o{6}	i{5}		
Model2			o{4}	i{3}	i{1}
Model3				o{2}	

Fig. 3 Final arrangement of the populated incidence matrix.

{2}-Rule2, {3}-Rule3, {4}-Rule4, {5}-Rule3, and {6}-Rule4. Thus, the final data flow obtained by applying the incidence matrix method is as follows: model1 has *Ws* and q as input and C_L as output, model2 has ρ and V as input and q as output, and finally model3 produces ρ as output.

After performing the variable flow modeling, model1 has *Ws* and q as input and C_L as output. However, the actual model1, as shown in Fig. 2a, has q and C_L as input and *Ws* as output. In this case the input and output variables of the model1 are swapped, and therefore model1 is considered a modified model. That is, a modified model is a model that has some of its input and output variables swapped as a result of variable flow modeling. The modified models are solved using iterative methods (e.g., Newton method [12]) and hence add to the computational cost of the system.

Until now the incidence matrix method was illustrated in terms of five intuitive rules. An improved formal incidence matrix method that populates a numerical matrix instead of a character matrix is described next. The method operates in a similar fashion, except that the rules are transformed into a mathematical procedure.

Figure 4 shows the incidence matrix with its corresponding new representation, *incm*. Here the presence of a variable in the model is represented with 1 (instead of $*$ used earlier), an input by a 2 (i used earlier) and output with a 3 (o used earlier). Also here 0 denotes no relation between the variable and the model. The population of the incidence matrix in this case signifies replacing 1 with either 2 or 3. The algorithm for populating the numerical incidence matrix is given in the flowchart shown in Fig. 5.

At the start *incm* will have only 1 and 0. The elements marked with 1 in the *incm* matrix are further replaced with either 2 or 3, based on the values of the decision-making variables *valr*2, *valr*3, *valc*2, and *valc*3 calculated for these elements as shown in Fig. 5. The parameters used in the flowchart in are

a)

	W_S	C_L	q	ρ	V
Model1	i	o	*		
Model2			o	i	*
Model3				o	

b) incm

$$\begin{bmatrix} 2 & 3 & 1 & 0 & 0 \\ 0 & 0 & 3 & 2 & 1 \\ 0 & 0 & 0 & 3 & 0 \end{bmatrix}$$

Fig. 4 Incidence matrix: character to numeric conversion.

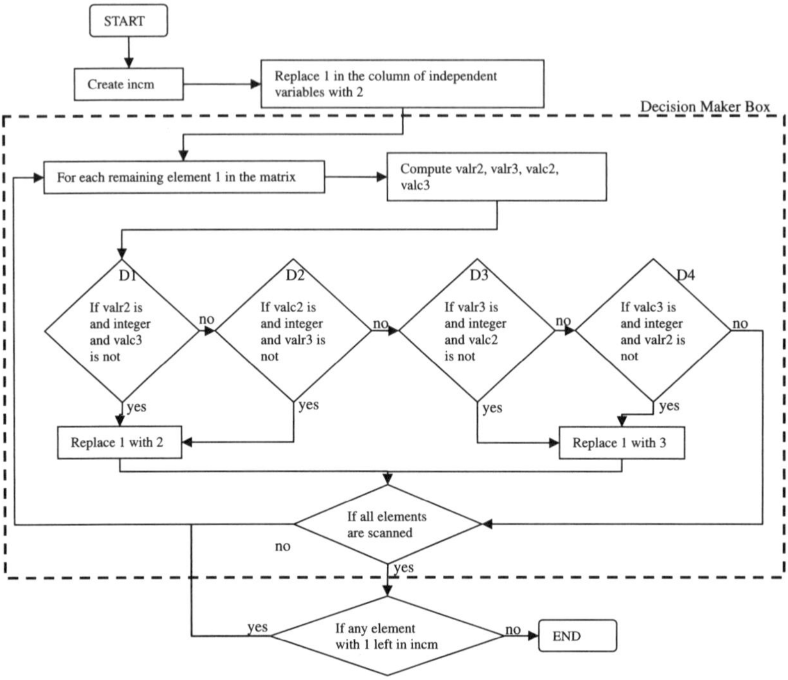

Fig. 5 Flowchart of the incidence matrix method.

given in Eqs. (1–4):

$$valr2(r,\, c) = \log\left(\frac{valrf(r)}{valr(r)}\right)\Big/\log(2) \tag{1}$$

$$valr3(r,\, c) = \log\left(\frac{valrf(r)}{valr(r)}\right)\Big/\log(3) \tag{2}$$

$$valc2(r,\, c) = \log\left(\frac{valcf(c)}{valc(c)}\right)\Big/\log(2) \tag{3}$$

$$valc3(r,\, c) = \log\left(\frac{valcf(c)}{valc(c)}\right)\Big/\log(3) \tag{4}$$

The unknown variables in the right-hand side (RHS) of Eqs. (1–4) are given in Eqs. (5–8):

$$valrf(r) = \prod_{c=1}^{m} incmf(r,\, c) \qquad incmf(r,\, c) \neq 0 \tag{5}$$

$$valcf(c) = \begin{cases} 3, & incmprod = 2 \\ incmprod, & incmprod \neq 2 \end{cases} \qquad \begin{array}{c} incmprod = \prod_{r=1}^{n} incmf(r,\, c) \\ incmf(r,\, c) \neq 0 \end{array} \tag{6}$$

$$valr(r) = \prod_{c=1}^{m} incm(r, c) \qquad incmf(r, c) \neq 0 \quad (7)$$

$$valc(c) = \prod_{r=1}^{n} incm(r, c) \qquad incm(r, c) \neq 0 \quad (8)$$

In the preceding equations *incmf* stands for the foundation incidence matrix, which corresponds to the true (canonical) inputs and outputs of the models in the system. The foundation matrix has the elements filled with 2 and 3 depending on inputs and outputs of the original model. For example, the *incmf* for the system in Fig. 2a will be

$$incmf = \begin{bmatrix} 3 & 2 & 2 & 0 & 0 \\ 0 & 3 & 0 & 2 & 2 \\ 0 & 0 & 0 & 0 & 3 \end{bmatrix}$$

Values of *valr*2, *valr*3, *valc*2, and *valc*3 determine whether the 1 in the *incm* matrix should be replaced with either 2 or 3. This can be easily seen if Eq. (1) is rewritten in the form:

$$2^{valr2(r,c)} = \frac{valrf(r)}{valr(r)} \quad (9)$$

The RHS of the preceding equation calculates the product of the values of the elements of row r of the *incm* matrix with current value 1, as if these were replaced with combinations of 2s and 3s. If $valr2(r, c)$ is an integer, this means that the RHS can be represented as multiples of 2. It signifies that the values which can replace the 1s in row r should all be 2s. This explanation also extends to Eqs. (2–4).

The queries in the decision box (D1 to D4) of Fig. 5 check whether the replacement of 1 in the *incm*, with either 2 or 3, is achievable. The first clause in the queries ensures that element 1, which is in consideration for substitution, can be replaced with either 2 or 3 based on the remaining elements in the column (row). This is indicated by the value calculated by the corresponding equation. The second clause of the queries makes sure that during the replacement the orthogonal row (column) of the element can accommodate the changes. The equations and the queries together implicitly satisfy the five rules stated earlier. The example used for demonstrating the IMM (Fig. 2a) is reused next to demonstrate the improved incidence matrix method. As before, *Ws* and *V* are chosen as independent variables. The initial incidence matrix *incm* and foundation incidence matrix *incmf* are given here:

$$incm = \begin{bmatrix} 1 & 1 & 1 & 0 & 0 \\ 0 & 0 & 1 & 1 & 1 \\ 0 & 0 & 0 & 1 & 0 \end{bmatrix}$$

$$incmf = \begin{bmatrix} 3 & 2 & 2 & 0 & 0 \\ 0 & 0 & 3 & 2 & 2 \\ 0 & 0 & 0 & 3 & 0 \end{bmatrix}$$

The preceding matrix representation is based on the layout of the models and variables in Fig. 4. In the next step, as outlined in the flowchart of Fig. 5, the nonzero elements of the corresponding columns of the independent variables (*Ws* and *V*) are replaced with 2s. The updated *incm* matrix is given here:

$$incm = \begin{bmatrix} 2 & 1 & 1 & 0 & 0 \\ 0 & 0 & 1 & 1 & 2 \\ 0 & 0 & 0 & 1 & 0 \end{bmatrix}$$

Each element 1 in the matrix is now scanned and analyzed to check whether it could be replaced with a 2 or a 3.

For element *incm*(1, 2):

$$valrf(1) = \prod_{c=1}^{5} incmf(1, c) \qquad \text{for } incmf \neq 0$$
$$= 3 \times 2 \times 2 = 12$$

$$valcf(2) = 3;\ \text{since } incmprod = \prod_{r=1}^{m} incmf(r, 2);\ \ incmf(r, 2) = 2$$

$$valr(1) = \prod_{c=1}^{5} incm(1, c) \qquad \text{for } incm \neq 0$$
$$= 2 \times 1 \times 1 = 2$$

$$valc(1) = \prod_{r=1}^{3} incm(r, 2) \qquad \text{for } incm \neq 0$$
$$= 1$$

$$valr2(1, 2) = \log\left(\frac{valrf(r)}{valr(r)}\right) \Big/ \log(2) = \log\left(\frac{12}{2}\right) \Big/ \log(2) = 2.5850$$

$$valr3(1, 2) = \log\left(\frac{valrf(1)}{valr(1)}\right) \Big/ \log(3) = \log\left(\frac{12}{2}\right) \Big/ \log(3) = 1.6309$$

$$valc2(1, 2) = \log\left(\frac{valcf(2)}{valc(2)}\right) \Big/ \log(2) = \log\left(\frac{3}{1}\right) \Big/ \log(2) = 1.5850$$

$$valc3(1, 2) = \log\left(\frac{valcf(2)}{valc(2)}\right) \Big/ \log(3) = \log\left(\frac{3}{1}\right) \Big/ \log(3) = 1$$

Now the queries in the decision boxes (see Fig. 5) D1 to D4 are checked:
Decision box D1:

$$valr2 = 2.508 \ \longrightarrow\ \text{noninteger}$$
$$valc3 = 0.6309 \ \longrightarrow\ \text{noninteger}$$

Therefore, D1 is unsatisfied.

Decision box D2:

$$valc2 = 1.5850 \longrightarrow \text{noninteger}$$
$$valr3 = 1.6309 \longrightarrow \text{noninteger}$$

Therefore, D2 is unsatisfied.

Decision box D3:

$$valr3 = 1.6309 \longrightarrow \text{noninteger}$$
$$valc2 = 1.5850 \longrightarrow \text{noninteger}$$

Therefore, D3 is unsatisfied.

Decision box D4:

$$valc3 = 1 \longrightarrow \text{integer}$$
$$valr2 = 2.5850 \longrightarrow \text{noninteger}$$

D4 is satisfied.

Decision box D4 is satisfied; therefore, the element (1, 2) of *incm* is replaced with 3. The updated incidence matrix is

$$incm = \begin{bmatrix} 2 & 3 & 1 & 0 & 0 \\ 0 & 0 & 1 & 1 & 2 \\ 0 & 0 & 0 & 1 & 0 \end{bmatrix}$$

Following the same procedure for other relevant elements of *incm*, the step-by-step population of *incm* is given in Table 1.

The final matrix indicates that model1 has Ws and q as input and C_L as output, model2 has ρ and V as input and q as output, and model3 has ρ as output. The populated matrix is obtained here in a single iteration through the decision-maker step in the flowchart of Fig. 5. However, for larger and more complicated cases iteration might be required to arrive at a solution. Furthermore, the IMM algorithm shown in Fig. 5 does not always lead to a converged solution. This means that some of the 1s in the *incm* could not be replaced even after populating the matrix by IMM. The conditions leading to nonconvergence and the means for handling such cases are explained next.

Generally, situations can arise where the independent variables specified by the designer are either too many or too few for the system to produce a fully populated *incm*. In the former case the system is overdetermined, and in the latter case the system is underdetermined. Both over- and underdetermined systems lead to partially populated (i.e., to nonconverged) *incm* after applying the IMM. Underdetermined systems can be resolved by defining additional variables as independent and overdetermined systems by removing variables from the set of independent variables.

There are situations where *incm* remains partially populated even though the system is determined, that is, the number of given independent variables is sufficient to obtain the data flow among the models. This situation arises because of the presence of SCCs in the system. Those models whose corresponding

Table 1 Population of *incm*

incm	*r*	*c*	*valr*2	*valr*3	*valc*2	*valc*3	Decision box number that is satisfied	Element 1 replaced with	Updated *incm*
incm(1, 3)	1	3	1	0.6309	2.5850	1.6309	D1	2	$\begin{bmatrix} 2 & 3 & 2 & 0 & 0 \\ 0 & 0 & 1 & 1 & 2 \\ 0 & 0 & 0 & 1 & 0 \end{bmatrix}$
incm(2, 3)	2	3	2.5850	1.6309	1.5850	1	D4	3	$\begin{bmatrix} 2 & 3 & 2 & 0 & 0 \\ 0 & 0 & 3 & 1 & 2 \\ 0 & 0 & 0 & 1 & 0 \end{bmatrix}$
incm(2, 4)	2	4	1	0.6309	2.5850	1.6309	D1	2	$\begin{bmatrix} 2 & 3 & 2 & 0 & 0 \\ 0 & 0 & 3 & 2 & 2 \\ 0 & 0 & 0 & 1 & 0 \end{bmatrix}$
incm(3, 4)	3	4	1.5850	1	1.5850	1	D3 and D4	3	$\begin{bmatrix} 2 & 3 & 2 & 0 & 0 \\ 0 & 0 & 3 & 2 & 2 \\ 0 & 0 & 0 & 3 & 0 \end{bmatrix}$

rows in the *incm* still have 1 remaining after applying IMM are considered part of an SCC.

The partially populated *incm* in the presence of an SCC is resolved by guessing the input and output variables of any one of the models belonging to the SCC. Guessing the input and output variables can be done in a number of alternative ways, and therefore each different guess for each model will lead to a different variable flow model. Because our objective is to reduce the number of modified models as much as possible (because, as mentioned earlier, modified models add to the computational cost of the system), the criteria for which a model is chosen for guessing its inputs and outputs should reflect this objective. A new rule is therefore introduced to account for modified models:

6) Among the models that are part of an SCC for which not all 1s have been replaced after applying the preceding five rules, the models for which the new inputs differ from the original ones are selected for guessing. If no such model exists, the incidence matrix is populated with the original inputs and outputs of the models.

Rule 6 limits the unnecessary generation of modified models and also the creation of alternative variable flow models to the ones that have minimum modified models. If there is more than one model that has its variables already modified, then any of those models can be selected for guessing, one at a time. The additional steps required for resolving SCC for the formal IMM are shown in Fig. 6.

Further populating the partially populated matrix using the just-described method will generally lead to multiple variable flow models for an SCC. The following example of a system with an SCC explains the procedure in detail. The corresponding partially populated incidence matrix (after applying IMM) is shown in Fig. 7. In this example X_3 is the independent variable.

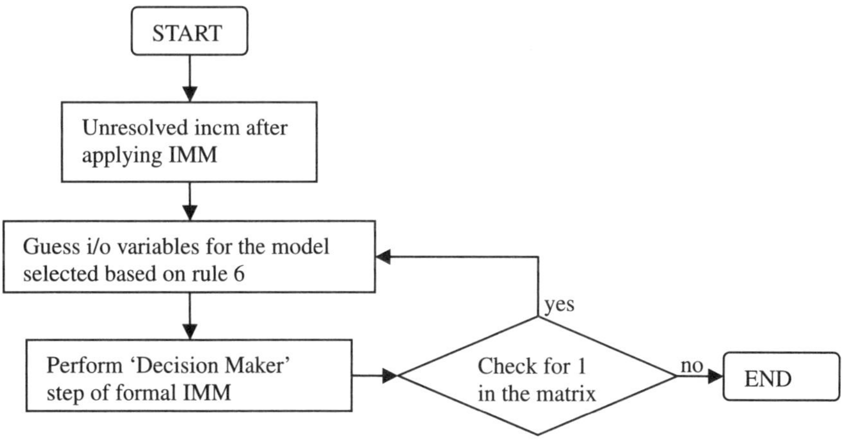

Fig. 6 Additional steps for formal IMM in the presence of SCC.

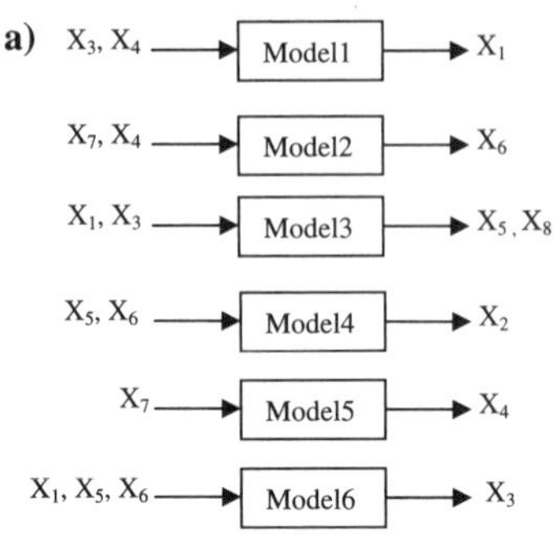

b)

	X_1	X_2	X_3	X_4	X_5	X_6	X_7	X_8
Model1	1		2	1				
Model2				1		1	1	
Model3	1		2		1			3
Model4		3			2	2		
Model5				1			1	
Model6	1		2		1	1		

(0s not shown for clarity)

Fig. 7 a) System of models. b) Corresponding populated incidence matrix with X_3 as the independent variable.

According to Fig. 7b, variable X_3 is input to model6, but the real model6 has X_3 as output (see Fig. 7a). Such a modification is not present in any other model. Hence, based on rule 6, model6 is chosen for guessing the input and output variables. The three different possible guesses for model6 are shown in Fig. 8.

Each arrangement for model6 replaces the sixth row of model6 of the populated matrix shown in Fig. 7b. The results obtained for each case, by further populating based on the flowchart in Fig. 5, are shown in Fig. 9. The guess for model6 shown in Fig. 8c generated two solutions after population. These are shown in Fig. 9c and Fig. 9d, respectively. In the figures, model4 is not displayed because its corresponding row in the incidence matrix was already fully populated by applying IMM (refer to Fig. 7b), and hence is not part of an SCC.

a)

	X_1	X_2	X_3	X_4	X_5	X_6	X_7	X_8
Model6	3		2		2	2		

b)

	X_1	X_2	X_3	X_4	X_5	X_6	X_7	X_8
Model6	2		2		3	2		

c)

	X_1	X_2	X_3	X_4	X_5	X_6	X_7	X_8
Model6	2		2		2	3		

Fig. 8 Alternative guesses for *i/o* variables of model6.

a)

	X_1	X_2	X_3	X_4	X_5	X_6	X_7	X_8
Model1	2		2	3				
Model2				2		3	2	
Model3	2		2		3			3
Model5				2			3	
Model6	3		2		2	2		

b)

	X_1	X_2	X_3	X_4	X_5	X_6	X_7	X_8
Model1	2		2	3				
Model2				2		3	2	
Model3	3		2		2			3
Model5				2			3	
Model6	2		2		3	2		

c)

	X_1	X_2	X_3	X_4	X_5	X_6	X_7	X_8
Model1	3		2	2				
Model2				3		2	2	
Model3	2		2		3			3
Model5				2			3	
Model6	2		2		2	3		

d)

	X_1	X_2	X_3	X_4	X_5	X_6	X_7	X_8
Model1	3		2	2				
Model2				2		2	3	
Model3	2		2		3			3
Model5				3			2	
Model6	2		2		2	3		

Fig. 9 Populated incidence matrix with the three guessed inputs and outputs for model6 (0s not shown in the figures for clarity).

Thus by applying IMM, four variable flow models are obtained for solving the system. The next step explains the procedure for selecting the optimal choice from the multiple variable flow models obtained.

2. *System Decomposition*

The next step, after the variable flow models have been generated, is to perform system decomposition. This is the process of decomposing a complex system into a number of subproblems. In the context of solving a computational system of models, the system decomposition corresponds to identifying the models that are strongly connected (i.e., the SCCs). The system decomposition is step 3 of the computational process modeler flow diagram shown in Fig. 1.

It was just shown that for determined systems during variable flow modeling the models that had unresolved $*$ (or 1 in case of formal IMM) after applying the IMM were considered as strongly connected. This dynamic identification of the SCCs is a significant advantage. However, because IMM cannot identify mutually exclusive SCCs, we have to employ an additional method for dealing with this problem. Here we adopt an algorithm [8] used for identifying set of coupled activities in a manufacturing environment. In this algorithm the design activities are represented in a binary design structure matrix (DSM). In the DSM rows and columns represent a design activity (models in our case). In the matrix an element 1 denotes that the model representing the column of the element has an input from the model representing the corresponding row. A 1 marked above the diagonal denotes a feed forward loop, and 1 below the diagonal denotes a feedback loop. The problem of recognizing a coupled activities set is translated into the problem of seeking SCCs in a directed graph as summarized next.

Let D denote the DSM. Then we define accessibility matrix $P = \sum_{n=1}^{j} D^n$, where j is the number of design activities. In the accessibility matrix the values of the elements that are greater than 1 are replaced with 1. Then the

Hadamard (entry wise) product of P and P^T is performed.

$$P \circ P^T = \begin{bmatrix} p_{11} & p_{12} & \cdots & p_{1n} \\ p_{21} & p_{22} & \cdots & p_{2n} \\ \cdots & \cdots & \cdots & \cdots \\ p_{n1} & p_{n2} & \cdots & p_{nn} \end{bmatrix} \circ \begin{bmatrix} p_{11} & p_{21} & \cdots & p_{n1} \\ p_{12} & p_{22} & \cdots & p_{n2} \\ \cdots & \cdots & \cdots & \cdots \\ p_{1n} & p_{2n} & \cdots & p_{nn} \end{bmatrix}$$

$$= \begin{bmatrix} p_{11}^2 & p_{12} \cdot p_{21} & \cdots & p_{1n} \cdot p_{n1} \\ p_{21} \cdot p_{12} & p_{22}^2 & \cdots & p_{2n} \cdot p_{n2} \\ \cdots & \cdots & \cdots & \cdots \\ p_{n1} \cdot p_{1n} & p_{n2} \cdot p_{2n} & \cdots & p_{nn}^2 \end{bmatrix}$$

In the matrix $P \circ P^T$, if the nonzero elements in the ith row are in the j_1th, j_2th, ..., j_kth column, then the design activities representing the rows 1, 2, ..., k are strongly coupled. (For details of the proof, refer to [8].)

An example, based on the system shown in Fig. 7a, is given next to explain the procedure in the current context. The variable flow model obtained for this system, given X_6 as the independent variable, is shown in Fig. 10.

The binary DSM, D, for the variable flow model in Fig. 10 and the computed accessibility matrix P are given here:

$$D = \begin{array}{r} \\ \text{model1} \\ \text{model2} \\ \text{model3} \\ \text{model4} \\ \text{model5} \\ \text{model6} \end{array} \begin{array}{c} \begin{array}{cccccc} 1 & 2 & 3 & 4 & 5 & 6 \end{array} \\ \begin{bmatrix} 1 & 0 & 1 & 0 & 0 & 1 \\ 1 & 1 & 0 & 0 & 1 & 0 \\ 1 & 0 & 1 & 0 & 0 & 1 \\ 0 & 0 & 0 & 1 & 0 & 0 \\ 0 & 1 & 0 & 1 & 1 & 0 \\ 0 & 0 & 1 & 1 & 0 & 1 \end{bmatrix} \end{array}$$

	X1	X2	X3	X4	X5	X6	X7	X8
Model1	o		i	i				
Model2				o		i	i	
Model3	i		i		o			o
Model4		o			i	i		
Model5				i			o	
Model6	i		o		i	i		

Fig. 10 Populated incidence matrix for the system shown in Fig. 7a with X_6 given as the independent variable.

$$P = D^1 + D^2 + D^3 + D^4 + D^5 + D^6 = \begin{bmatrix} 144 & 0 & 232 & 138 & 0 & 232 \\ 144 & 63 & 169 & 138 & 63 & 169 \\ 144 & 0 & 232 & 138 & 0 & 232 \\ 0 & 0 & 0 & 6 & 0 & 0 \\ 88 & 63 & 81 & 94 & 63 & 81 \\ 88 & 0 & 144 & 94 & 0 & 144 \end{bmatrix}$$

The nonzero elements of the matrix are now replaced with 1.

$$P = \begin{bmatrix} 1 & 0 & 1 & 1 & 0 & 1 \\ 1 & 1 & 1 & 1 & 1 & 1 \\ 1 & 0 & 1 & 1 & 0 & 1 \\ 0 & 0 & 0 & 1 & 0 & 0 \\ 1 & 1 & 1 & 1 & 1 & 1 \\ 1 & 0 & 1 & 1 & 0 & 1 \end{bmatrix}$$

Now the Hadamard product of P and P^T is

$$P \circ P^T = \begin{bmatrix} 1 & 0 & 1 & 1 & 0 & 1 \\ 1 & 1 & 1 & 1 & 1 & 1 \\ 1 & 0 & 1 & 1 & 0 & 1 \\ 0 & 0 & 0 & 1 & 0 & 0 \\ 1 & 1 & 1 & 1 & 1 & 1 \\ 1 & 0 & 1 & 1 & 0 & 1 \end{bmatrix} \circ \begin{bmatrix} 1 & 1 & 1 & 0 & 1 & 1 \\ 0 & 1 & 0 & 0 & 1 & 0 \\ 1 & 1 & 1 & 0 & 1 & 1 \\ 1 & 1 & 1 & 1 & 1 & 1 \\ 0 & 1 & 0 & 0 & 1 & 0 \\ 1 & 1 & 1 & 0 & 1 & 1 \end{bmatrix}$$

$$= \begin{bmatrix} 1 & 0 & 1 & 0 & 0 & 1 \\ 0 & 1 & 0 & 0 & 1 & 0 \\ 1 & 0 & 1 & 0 & 0 & 1 \\ 0 & 0 & 0 & 1 & 0 & 0 \\ 0 & 1 & 0 & 0 & 1 & 0 \\ 1 & 0 & 1 & 0 & 0 & 1 \end{bmatrix}$$

In the preceding matrix rows 1, 3, and 6 (and columns 1, 3, and 6) are equal and so are rows 2 and 5 (and columns 2 and 5). Hence according to the method described before, the models that represent these rows are strongly coupled. More specifically, in the current example there are two mutually exclusive SCCs; model1, model3, and model6 belong to the first SCC, and model2 and model5 to the second. Thus the system is now decomposed into the following: SCC_1 (model1, model3, model6), SCC_2 (model2, model5), and model4.

3. *System Scheduling*

Scheduling is the process of sequencing the models in a system for the purpose of executing them after eliminating or reducing the feedback loops. It was shown earlier that a computational system can have multiple feasible variable flow models in the presence of SCCs. We begin this section by first proposing the scheduling algorithm for coupled models. Further, the criteria for choosing the optimal variable flow model from the feasible ones are explained. Following that is the scheduling algorithm for noncoupled models.

Scheduling of coupled models. Presence of feedback loops makes it necessary to employ iterative methods for solving the SCCs. The more feedback loops, the higher the computational cost is. Feedback loops are formed when a model requires input from another model, which comes later in the execution sequence. Thus, reducing the feedback loops can reduce the time and computational cost for solving an SCC. Rearranging the models in the SCC is therefore crucial for reducing the feedback loops. The scheduling of coupled models is step 4 in the computational process modeler flow diagram shown in Fig. 1.

A genetic-algorithm-based approach for ordering complex design processes [9] is used for the rearrangement. Here we choose the number of feedback loops as the objective function to be minimized. (This objective function has been chosen after conducting extensive testing with different candidate objective functions on an aircraft conceptual design test case.) The GA approach has been chosen because, compared to other scheduling methods, it is independent of problem formulation, and therefore different objective functions can be formulated for different scheduling architectures.

The formula for calculating the feedback number is given in Eq. (10). In the equation, D is the DSM, which is populated based on the incidence matrix of an SCC. In the binary DSM, both rows and columns represent the models; 1 marked above the diagonal denotes a feedforward loop, and 1 below the diagonal denotes

	Model1	Model2	Model3	Model5	Model6
Model1	0	1	0	1	0
Model2	0	0	0	0	1
Model3	0	0	0	0	1
Model5	0	1	0	0	0
Model6	1	0	1	0	0

a) Number of feedback loops=1+1+1=3
Number of modified models=3

	Model1	Model2	Model3	Model5	Model6
Model1	0	1	0	1	0
Model2	0	0	0	0	1
Model3	1	0	0	0	1
Model5	0	1	0	0	0
Model6	0	0	1	0	0

b) Number of feedback loops=1+1+1=3
Number of modified models= 4

	Model1	Model2	Model3	Model5	Model6
Model1	0	0	1	0	1
Model2	1	0	0	1	0
Model3	0	0	0	0	1
Model5	0	1	0	0	0
Model6	0	1	0	0	0

c) Number of feedback loops=1+1+1=3
Number of modified models=3

	Model1	Model2	Model3	Model5	Model6
Model1	0	0	1	0	1
Model2	0	0	0	1	0
Model3	0	0	0	0	1
Model5	1	1	0	0	0
Model6	0	1	0	0	0

d) Number of feedback loops=1+1+1=3
Number of modified models= 2

Fig. 11 DSM representation of the incidence matrices of the SCCs.

a feedback loop.

$$nFdb = \sum_{i=2}^{n} \sum_{j=1}^{i-1} D(i, j) \tag{10}$$

Figures 11a–11d show the corresponding DSM representation of the four variable flow models from Figs. 9a–9d.

Each DSM in Fig. 11 is rearranged using genetic algorithm with the number of feedback loops as the objective function to be minimized. The final rearranged DSMs are shown in Figs. 12a–12d. It can be noted that the number of feedback loops of all of the DSMs is reduced as a result of the rearrangement.

It was shown earlier that there can be multiple feasible variable flow models for a system, in the presence of an SCC. To choose the optimal variable flow model that can lead to the shortest execution time, the number of feedback loops and the number of modified models (nMm) in the constituent SCCs of each variable flow model are taken into consideration. The flow model that has the lowest number of modified models in the constituent SCC is chosen as the optimal and is selected for further solving. If there are multiple variable flow models with the same number of modified models, then the one that has a smaller number of feedback loops is chosen as the optimum. The first criterion for selecting the optimal variable flow model (i.e., number of modified models) emerged after conducting a number of tests. These tests identified that the modified models add more to the computational cost compared to the feedback number. The higher the number of modified models, the higher is the computational cost. Nevertheless, the number of feedback loops still plays a significant role in rearranging the models of the SCC in each variable flow model, and so reducing this number significantly reduces the computational cost for the SCC.

Thus, from the variable flow models shown in Figs. 9a–9d, the one in Fig. 9d is considered as the optimal because it has the least number of modified model ($nMm = 2$). The execution sequence for the models in this variable flow model, which is obtained after scheduling, is shown in Fig. 12d. The selection

	Model1	Model5	Model2	Model6	Model3
Model1	0	1	1	0	0
Model5	0	0	1	0	0
Model2	0	0	0	1	0
Model6	1	0	0	0	1
Model3	0	0	0	1	0

a) Number of feedback loops=1+1=2
Number of modified models=3

	Model3	Model1	Model5	Model2	Model6
Model3	0	1	0	0	1
Model1	0	0	1	1	0
Model5	0	0	0	1	0
Model2	0	0	0	0	1
Model6	1	0	0	0	0

b) Number of feedback loops=1
Number of modified models=4

	Model1	Model3	Model6	Model2	Model5
Model1	0	1	1	0	0
Model3	0	0	1	0	0
Model6	0	0	0	1	0
Model2	1	0	0	0	1
Model5	0	0	0	1	0

c) Number of feedback loops=1+1=2
Number of modified models=3

	Model5	Model1	Model3	Model6	Model2
Model5	0	1	0	0	0
Model1	0	0	1	1	0
Model3	0	0	0	1	0
Model6	0	0	0	0	1
Model2	1	0	0	0	0

d) Number of feedback loops=1
Number of modified models=2

Fig. 12 Rearranged DSMs of the SCCs.

of optimal variable flow model is step 5 of the computational process modeler flow diagram shown in Fig. 1.

Scheduling of noncoupled models. The models that belong to an SCC (which are coupled) and that have already been arranged using the method described in the preceding subsection are considered as a single subsystem and are subsequently arranged together with the remaining noncoupled models.

The models and the SCCs that are to be scheduled are first populated in a DSM. By applying the graph theoretical algorithm [8] the DSM is rearranged to a lower-triangular matrix, thus ensuring all loops are feedforward. The scheduling of noncoupled models is step 8 of the computational process modeler flow diagram shown in Fig. 1.

The following example briefly explains the procedure. The system shown in Fig. 7a with X_7 as the chosen independent variable is considered here for demonstrating the sequential arrangement. In the incidence matrix models 1, 3, and 6 were identified as strongly coupled. The SCC, which contains these models along with the remaining noncoupled models, is represented in Fig. 13.

The global DSM arrangement for this system is

$$\text{DSM} = \begin{matrix} \text{model2} \\ \text{model4} \\ \text{model5} \\ \text{SCC} \end{matrix} \begin{bmatrix} 1 & 1 & 0 & 1 \\ 0 & 1 & 0 & 0 \\ 1 & 0 & 1 & 1 \\ 0 & 1 & 0 & 1 \end{bmatrix}$$

The rearranged DSM after applying the graph theoretical algorithm [8] is

$$\text{DSM} = \begin{matrix} \text{model4} \\ \text{SCC} \\ \text{model2} \\ \text{model5} \end{matrix} \begin{bmatrix} 1 & 0 & 0 & 0 \\ 1 & 1 & 0 & 0 \\ 1 & 1 & 1 & 0 \\ 1 & 0 & 1 & 1 \end{bmatrix}$$

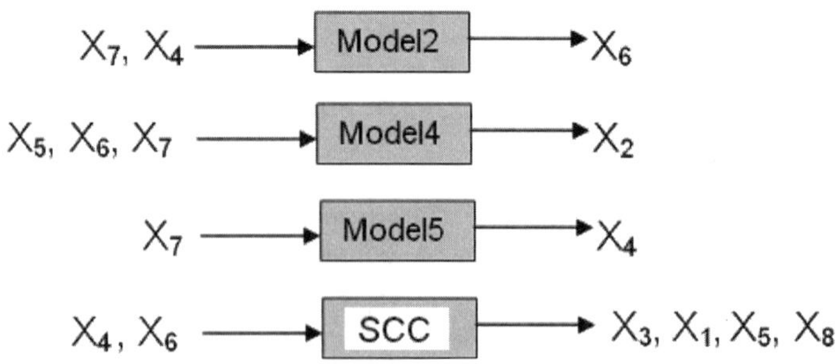

Fig. 13 Confined SCC and the remaining models.

The rearranged DSM is in a lower triangular form, which signifies that all the feedback loops are removed. Thus, the final sequential execution order for the system will therefore be model5 $\rightarrow$ model2 $\rightarrow$ SCC $\rightarrow$ model4.

III. Multi-Objective Optimization

The aim of conceptual design is not simply to synthesize an aircraft satisfying a set of requirements, but, rather, to design the best aircraft possible. The concept of "best" is obviously a relative one. When comparing different solutions, one can be considered better than another one only with respect to a specific comparison criterion or objective. In optimization terms the best solution for a particular set of requirements and a particular criterion is said to be an optimal solution. In fact, conceptual design can be seen as an optimization problem where the requirements coincide with a set of constraints to be satisfied and an objective will be a criterion chosen by the designer.

In real engineering design the number of criteria is usually greater than one. For instance, when designing an aircraft, the designer might wish to minimize the fuel capacity while maximizing the range of the aircraft and minimizing the life-cycle cost. Such problems are known as *multi-objective* or *multicriteria* optimization problems [13]. In general, the solution of a multi-objective optimization problem is nonunique; an infinite number of solutions exist, each satisfying a different compromise between the objectives.

When performing multi-objective optimization, the common approach is to define a set of preferences for each criterion. Such preferences represent a particular tradeoff or a compromise between the criteria. The methods making use of this approach are frequently referred to as *a priori articulation of preferences methods* [14]. The application of such methods presents two main problems: the first is because of the difficulty in defining the preferences by the designer, given an initial lack of knowledge, and secondly the difficulty in formulating correctly the preferences within the optimization problem.

An alternative possibility is that of generate first—choose later (GFCL) methods [15], or *a posteriori articulation of preferences methods* [14], in which a set of solutions is generated, each satisfying a particular combination of preferences or tradeoffs. It is then up to the designer to exercise his/her choice. Given the difficulties in defining effectively the preferences, the *a posteriori articulation of preferences* approach seems more appealing for application to conceptual design, also considering that, in the context of conceptual design, the calculation time for the analysis of a design point is several orders of magnitude faster than in the more detailed design phases.

As already mentioned, the solution of the multi-objective optimization problem is not unique. It is then natural to exclude from consideration any design solution that can be improved without deterioration of any objective and violation of the constraints—in multi-objective optimization terms, a solution that is dominated. This leads to the Pareto optimal solutions [13]. Mathematically, each Pareto point is a solution of the multi-objective optimization problem. In practice, however, the designer would select the ultimate solution among the Pareto set on the basis of additional (often subjective) requirements.

It follows then that the quality of the decisions depends on the availability of a sufficient number of well-distributed Pareto points ensuring good representation of the entire Pareto surface.

A. Multi-Objective Optimization Strategies and Methods

The general formulation of the multi-objective optimization problem is the following:

$$\min \boldsymbol{f}(\boldsymbol{x})$$

subject to K_i inequality constraints:

$$g_k(\boldsymbol{x}) \leq 0 \qquad k = 1, 2, \ldots, K_i \tag{11}$$

and P_e equality constraints:

$$h_p(\boldsymbol{x}) = 0 \qquad p = 1, 2, \ldots, P_e$$

where, given the design vector $\boldsymbol{x} \in [\boldsymbol{x}_L, \boldsymbol{x}_U] \subseteq \mathbb{R}^N$, each of the M objective variables $y_j = f_j(\boldsymbol{x})$ is to be minimized while satisfying the K_i inequality constraints and the P_e equality constraints.

The most common approach in multi-objective optimization is the *global criterion method* [14] in which all objective functions are combined to form a single scalar function AOF($\boldsymbol{x}$), also called *aggregate objective function* (AOF). Optimization problem (11) can then be reformulated as

$$\min \mathrm{AOF}(\boldsymbol{x}) = \mathrm{AOF}[f_1(\boldsymbol{x}), f_2(\boldsymbol{x}), \ldots, f_M(\boldsymbol{x})]$$

subject to K_i inequality constraints:

$$g_k(\boldsymbol{x}) \leq 0 \qquad k = 1, 2, \ldots, K_i \tag{12}$$

and P_e equality constraints:

$$h_p(\boldsymbol{x}) = 0 \qquad p = 1, 2, \ldots, P_e$$

Optimization algorithms based on the GFCL approach aim at determining a number of Pareto optimal solutions or points representative of the Pareto frontier, rather than generating a unique solution representative of a particular tradeoff between the objectives. An extensive review of GFCL methods is presented Marler and Arora [14].

Obtaining a well-distributed set of Pareto points is essential for the GFCL approach to be effective. Because the Pareto front cannot be obtained as a function of the design variables $\boldsymbol{x}$, but only a finite set of Pareto points can be obtained, it is fundamental that the Pareto points are well distributed on the full extent of the Pareto front, hence allowing a complete representation of the Pareto front in all of its regions. It has been demonstrated that conventional approaches, such as the weighted sum method [16], are not suitable for generating well-distributed Pareto points. As a result, new methods have been recently developed.

Shukla et al. [17] subdivide the methods for obtaining a representation of the Pareto frontier in two main groups: classical and nonclassical methods. The former are gradient-based, whereas the latter follow "some natural or physical principle" [17]. Shukla et al. also report an additional distinction between the two groups, namely, that the classical methods are mostly based on the common approach of reformulating the multi-objective optimization problem into a single-objective problem, as in the *global criterion method* (12), while the nonclassical methods deal directly with the original objective vector $\boldsymbol{y} = \boldsymbol{f}(\boldsymbol{x})$.

Rather than making a distinction between classical and nonclassical methods, the latter distinction between "reformulation" and "nonreformulation" methods is considered to be of greater interest for practical purposes. It is this characteristic, in fact, that defines how the multi-objective optimization problem is formulated.

As has been shown with the AOF, through reformulation the multi-objective optimization problem is transformed into a single objective optimization problem, Eq. (12). This implies that any conventional single objective optimization algorithm can be applied. Thus, with the AOF approach the problem of dealing with multiple objectives is managed through the reformulation. In contrast, when the multi-objective optimization problem is maintained in its original form, Eq. (11), it is the optimization algorithm itself that needs to be capable of handling multiple objectives. Thus the algorithm has to be specifically developed for such a purpose.

Because of this difference, algorithms making use of the reformulation approach require multiple optimizations to be performed. In fact, to obtain each of the Pareto points the formulation of the problem has to be modified, and an optimization has to be performed. Shukla et al. [17] refer to the principle followed by such methods as *one-at-a-time* strategy. Among these methods the ones considered to be the most interesting for producing well-distributed Pareto points [14, 17] are the normal boundary intersection (NBI) method [18], the physical-programming-based (PP-based) method [19], and the normal constraint (NC) method [20, 21].

Methods not making use of the reformulation approach follow a different principle, the *simultaneous* strategy [17]. The algorithms following this principle are mainly archive or population based and make use of the concept of dominance for dealing with the multiple objectives. Through application of this concept to an archive of points, a number of these can be compared, and those which are dominated can be eradicated. Such algorithms are then built specifically for making use of the dominance concept, and the optimizer is applied once for obtaining all of the Pareto points. Among these algorithms the ones considered most effective are the evolutionary multi-objective optimization (EMO) algorithms, among which one of the best known is Deb's nondominated sorting genetic algorithm, NSGA II [22].

Because of the characteristics of the conceptual design optimization problem one-at-a-time strategy algorithms are considered more interesting for application to conceptual design. This is because of a number of considerations. Firstly, as stated earlier, one-at-a-time strategy algorithms can be applied independently with respect to the optimizer used for obtaining each of the Pareto points. This allows the application of both evolutionary-based and gradient-based methods for solving each optimization. Secondly, given a constraint optimization problem for which the functions considered are generally differentiable, gradient-based methods are considered more effective [22, 23]. Gradient-based methods, in fact, make use of the gradients of the constraints in order to determine a search direction.

B. Methods for Generating a Well-Distributed Set of Pareto Points

A number of methods for obtaining an evenly distributed set of Pareto points have been developed in recent years. All of these are based on performing a subdivision of the criterion (objective) space in a set of domains. The optimization problem is then reformulated for each domain, for each of which a Pareto point is generated.

Das and Dennis [18, 24] were the first to provide a method for generating well-distributed Pareto points, the NBI method. Another method is the PP-based method developed by Messac and Mattson [19] as an extension of the *a priori articulation of preference* method known as the PP [25]. The new NC method [20, 21] developed recently also looks very promising.

All of the cited methods have a clear geometrical interpretation; they are all based on the well-known fact that a Pareto frontier belongs to the boundary of the feasible space towards the minima of the objective functions [13].

The NBI, NC, and PP-based methods all follow a similar approach for obtaining well-distributed Pareto points. The *M anchor points* $\boldsymbol{a}_i$ [26], which are the minima relative to each of the objective, are obtained first. Subsequently, a number of evenly distributed points belonging to the criterion space, the *utopia plane points* $\boldsymbol{p}$, are obtained as linear combinations of the *M anchor points* [20]. These are used as reference points allowing the reformulation of the optimization problem. Finally, for each utopia plane point $\boldsymbol{p}$ an optimization is executed in order to obtain a Pareto point. Each of these three steps is fundamental in order to obtain a complete representation of the Pareto frontier.

When dealing with multi-objective optimization problems, for which the number of objectives is greater than two, a *peripheral region* exists. The peripheral region is that region of the criterion space for which the orthogonal projection of the Pareto frontier on the *utopia plane* is external to the polygon spanned by the M anchor points $\boldsymbol{a}_i$ [18, 21]. In their work Das and Dennis [18] limit the importance of the peripheral region, stating that such region will be of no interest to the designer, whereas Messac and Mattson [21] are interested in obtaining a complete representation of the Pareto frontier.

Even though Das and Dennis believe that such points might be of no interest for the designer, it must be noted that the size of the peripheral region of the Pareto frontier is dependent on the positioning of the anchor points and can be significant as demonstrated by Fantini [27].

Messac and Mattson [21] proposed a two-step procedure for tackling the peripheral region problem. The first step allows increasing the size of the polygon, so that any vector perpendicular to the utopia plane and passing through any point belonging to the hypercube, enveloping the original polygon, intersects the expanded polygon. The second step is the *feasibility test*, which allows removing unnecessary utopia plane points. Through the implementation of the method and a number of tests, Fantini [27] demonstrated that the number of feasible utopia plane points obtained is sensitive to the relative positioning of the anchor points.

When performing multi-objective optimization, once a method for reformulating the optimization problem has been chosen it is applied to each of the utopia plane points. For each such point the optimization problem is reformulated accordingly, and an optimization algorithm is applied in order to obtain a Pareto point. Thus it is important that the adopted reformulation enables the effective application of the optimizer in order to determine well-distributed Pareto points. For allowing the optimizer to work effectively, it is important that the reformulation of the problem does not introduce any additional minima to the original optimization problem. For what concerns efficiency, it is important that the reformulation minimizes the number of optimizations to be performed in order to obtain the entire Pareto set. Fantini [27] reached the following conclusions with respect to the reformulation of the NBI, PP-based, and NC methods:

1) The only apparent drawback of the NBI method is the tightness of the equality constraints of the reformulated optimization problem. As shown by Fantini [27], the constraint tightness is such that the optimizer might fail to obtain a Pareto solution for some of the subproblems. The feasible region for the subproblem is reduced in fact to a line. Thus if there are ∞^M points in the global feasible space, there will be ∞^1 points for each of the subproblems. This reduces the effectiveness of any type of optimizer in obtaining a feasible solution.

2) For the PP-based method, the ideal sizes of the parallelepiped, as well as the value for the offsetting coefficients, are unknowns. Therefore, in order to determine well-distributed Pareto points, several optimizations need to be performed for each subproblem.

3) The NC method might fail to obtain Pareto solutions for more than two objectives when the anchor points are not "favorably" positioned. Furthermore, transforming $M - 1$ objectives in constraints, while minimizing only one objective, exposes the optimizer to the possibility of obtaining a solution that does not belong to a line orthogonal to the utopia plane and passing through the utopia plane point considered. This can occur in the case of the existence of a local minimum with respect to the single objective minimized.

The preceding considerations lead to the definition of a set of requirements that an effective method for generating well-distributed Pareto points should satisfy: 1) the reformulation should avoid the introduction of local minima, 2) the reformulation should minimize the possibility of the optimizer to fail, 3) a unique optimization should be performed for each of the subproblems minimizing the risk of failure, and 4) the method should be independent of the number of objectives.

The attempt [27] at improving the reformulation of the optimization problems, in order to remove the limitations associated with the existing methods, has followed three consecutive stages. The result of the first one is the modified PP-based method, developed through modification of the PP-based method [28, 29]. The second one has taken advantage of the experience gained from the development of the first one and has led to the development of the double hypercone boundary intersection (DHCBI) method [27, 30]. The third and last method developed is the NC+ method. The methods follow the tracks laid by the NBI, PP-based, and NC methods, combining the various approaches and the knowledge gained from them.

C. NC+ Method

The NC+ method is an improvement over the NC method. The formulation of the method is the following:

$$\min f_l(\boldsymbol{x})$$

subject to K_i inequality constraints:

$$g_k(\boldsymbol{x}) \leq 0 \qquad k = 1, 2, \ldots, K_i$$

P_e equality constraints:

$$h_p(\boldsymbol{x}) = 0 \qquad p = 1, 2, \ldots, P_e \tag{13}$$

subject to the additional $M - 1$ constraints:

$$\boldsymbol{v}_j(\boldsymbol{p}_i - \boldsymbol{f}) \leq 0 \qquad \forall j \in \{1, 2, \ldots, M\}, j \neq l$$

and subject to the additional constraints:

$$\boldsymbol{v}_j\left(\frac{\boldsymbol{f} - \boldsymbol{p}_i - n_c \boldsymbol{v}_l}{\|\boldsymbol{v}_l\|}\right) \leq 0$$

where $\boldsymbol{v}_j = (l_j/l_i)\boldsymbol{e}_l - \boldsymbol{e}_j$ for $j \neq l$, $\boldsymbol{v}_l = \boldsymbol{m}l\boldsymbol{e}_l/l_l - \boldsymbol{m}$, $\boldsymbol{e}_j\ \forall j$ are the base vectors of the coordinate system, $\boldsymbol{l}$ is the unit vector orthogonal to the utopia plane, n_c is a fraction of the Euclidean distance between two contiguous utopia plane points, and finally $\boldsymbol{m} \in \mathbb{R}^M$ is a vector such that $m_i = 1\ \forall i$ and $m_l = 0$.

The formulation is similar to the NC method, where $M - 1$ constraints are used for building $M - 1$ hyperplanes, which confine the solution to a region of the criterion space. Although in the NC method, these constraints are dependent on the anchor points; in the NC+ method they are built with respect to the coordinate system.

For the solution to belong to the line orthogonal to the utopia plane, passing through a particular utopia plane point, all hyperplanes need to be orthogonal to the utopia plane and have to intersect in the utopia plane point. In the NC+ method, the hyperplanes are determined with respect to the coordinate system in such a way that the orthogonality condition is always enforced [27].

To minimize the possibility of obtaining solutions not belonging to the line orthogonal to the utopia plane, passing through utopia plane point $\boldsymbol{p}_i$ an additional constraint is added. Constraint l is built in order to reduce the size of the feasible region, confining the solution in the proximity of the line passing through utopia plane point $\boldsymbol{p}_i$. As for the other $M - 1$ constraints, constraint l defines a

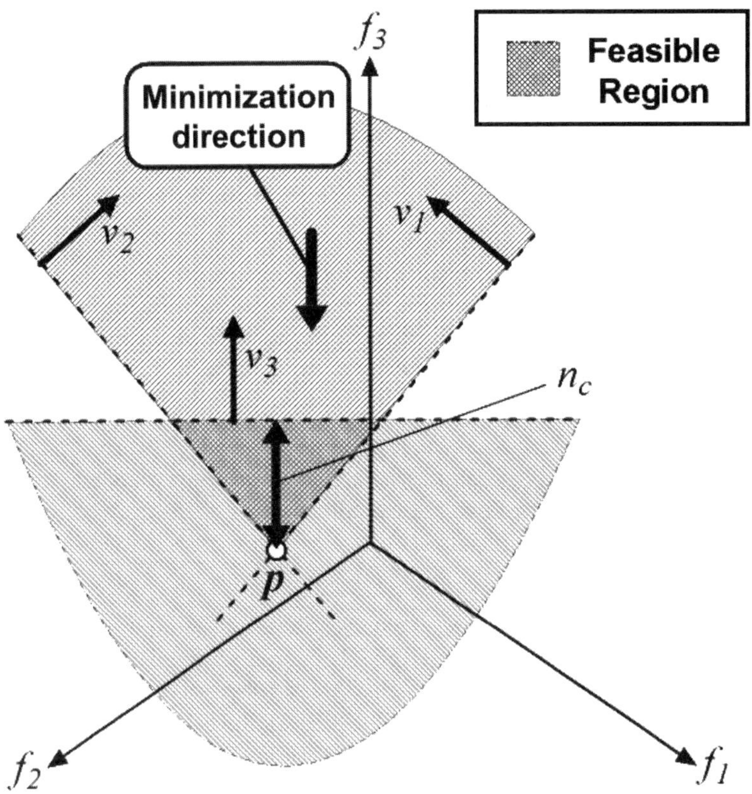

Fig. 14 Feasible region for utopia plane point *p* (viewed orthogonally with respect to the utopia plane).

hyperplane orthogonal to the utopia plane, but positioned at a distance n_c in the direction of vector $\boldsymbol{v}_l$.

Figure 14 depicts all of the three constraints for a three-objective optimization problem.

The positioning of the constraint at a distance n_c is simply obtained by adding $n_c\boldsymbol{v}_l/\|\boldsymbol{v}_l\|$ to utopia plane point $\boldsymbol{p}_i$ for constraint l, as also shown in Fig. 14. Then n_c can then be simply chosen as a fraction of the Euclidean distance between two contiguous utopia plane points.

D. Obtaining a Complete Representation of the Pareto Front

To obtain a complete representation of the Pareto frontier, it is necessary to obtain also those Pareto points that belong to the peripheral region. Here we propose an alternative approach for obtaining such points, with respect to the one proposed by Messac and Mattson [21]. Because the points belonging to the peripheral region have orthogonal projection to the utopia plane external to the polygon of vertices $\boldsymbol{a}_i$, a possible solution is that of generating the peripheral region utopia plane points $\boldsymbol{p}^+$ using the utopia plane points belonging to the edges of the polygon.

Let us consider a utopia plane point $\boldsymbol{p}^*$ belonging to the kth edge of the polygon. We define as edge vectors of the polygon, vectors

$$\boldsymbol{\nu}_i = \boldsymbol{a}_{i-1} - \boldsymbol{a}_i \qquad i = 1, \ldots, M-1 \quad (14)$$

A unit vector that is the outer normal to the edge considered can be obtained by means of the following linear combination of the edge vectors $\boldsymbol{\nu}_{i-1}$ and $\boldsymbol{\nu}_i$:

$$\boldsymbol{s}_i = \frac{\boldsymbol{\nu}_{i-1} + \beta_i \boldsymbol{\nu}_i}{\|\boldsymbol{\nu}_{i-1} + \beta_i \boldsymbol{\nu}_i\|}, \qquad \beta_i = \frac{-\boldsymbol{\nu}_{i-1}\boldsymbol{\nu}_i}{\boldsymbol{\nu}_i\boldsymbol{\nu}_i} \quad (15)$$

Therefore, following Eqs. (14) and (15), we can determine vector $\boldsymbol{s}_i$ parallel to the utopia plane and pointing towards the exterior of the polygon having as vertices the anchor points. For each of the utopia plane points $\boldsymbol{p}^*$ belonging to an edge, the *peripheral region utopia plane points* can be then obtained as

$$\boldsymbol{p}^+ = \boldsymbol{p}^* + q n_d \, \boldsymbol{s}_i \qquad q = 1, 2, \ldots \quad (16)$$

where n_d corresponds to the distance between two adjacent utopia plane points. Following this approach, each new peripheral region utopia plane point $\boldsymbol{p}^+$ belongs to the orthogonal projection of the peripheral region on the utopia plane and is generated from an initial utopia plane point $\boldsymbol{p}^*$, moving orthogonally to the edge of the polygon to which $\boldsymbol{p}^*$ belongs and parallel to the utopia plane. To obtain all Pareto points belonging to the peripheral region, for each of the edge utopia plane points $\boldsymbol{p}^*$, the step q in Eq. (16) is increased until the optimizer fails to obtain a solution.

The advantage of this approach over the modified PP-based method is that the problem concerning the rotation of the anchor points is removed. Furthermore, the approach is general and can also be applied to the NBI, PP-based, and NC methods.

E. Gradient-Based Global Multi-Objective Optimization

To solve the global optimization problem, one of the proposed methods is applied repeatedly to allow the identification of all local Pareto frontiers. Then, by evaluating the solutions obtained, the global Pareto set is easily determined. For doing this, it is sufficient to run one of the proposed methods as many times as the number of local minima obtained for each of the single objectives, each of which will correspond to a local anchor point. These are computed in a preprocessing phase of the algorithm. This gives the proposed approach the unique feature of generating multiple local Pareto fronts.

As discussed by Deb [31], we can define as local Pareto frontier a Pareto frontier within a particular neighborhood of the design space. The procedure has two main advantages: the first is that a more complete investigation of the design space is performed improving the chances of obtaining global Pareto solutions.

Second, it allows the DM to obtain and analyze a set of solutions that can be of interest from an engineering point of view, when accounting for additional considerations. Starting the sequence of optimizations from the utopia plane point closest to the ith local anchor point relative to objective $j\boldsymbol{a}_{ij}$ and moving toward the other $M-1$ global anchor points allows the obtaining of the local Pareto in the neighborhood of $\boldsymbol{a}_{ij}$. The detailed procedure for obtaining the local Pareto frontiers is described in Fantini [27] where the complete NC+ method is applied to a number of multi-objective global optimization problems in order to demonstrate its ability to generate the local and global Pareto frontiers.

F. NC+ Application Example

The Kursawe test case [32] is used to demonstrate the ability of the NC+ method to generate the multiple local and the global Pareto frontier for a global multi-objective optimization problem. The Kursawe test case has been developed in order to highlight the various difficulties that can be encountered when performing global multi-objective optimization. To the authors' knowledge, no attempt at solving problems of this level of complexity has ever been attempted with regard to the NBI, PP-based, and NC methods.

For the test case, parameter n_c was computed as $\frac{1}{4}$ of the distance between two adjacent utopia plane points. Also, 50 starting points were used in order to determine the multiple local minima needed in order to obtain the complete Pareto frontier described in the preceding section. Fifty utopia plane points were then generated for determining the Pareto fronts.

The weak Pareto solutions are included as part of the global Pareto solutions because there might be cases where the numeric roundoff error can make the distinction between global and local Pareto points difficult [27].

The formulation of test case Kursawe is the following:

Problem Kursawe:

$$\min[f_1(\boldsymbol{x}), f_2(\boldsymbol{x})]$$

where

$$f_1(\boldsymbol{x}) = \sum_{i=1}^{N} \left(-10e^{-0.2\sqrt{x_i^2+x_{i+1}^2}}\right)$$

and where

$$f_2(\boldsymbol{x}) = \sum_{i=1}^{N} \left[|x_i|^{0.8} + 5\sin(x_i)^3\right]$$

for

$$-5 \leq x_i \leq 5 \qquad i = 1, \ldots, N$$

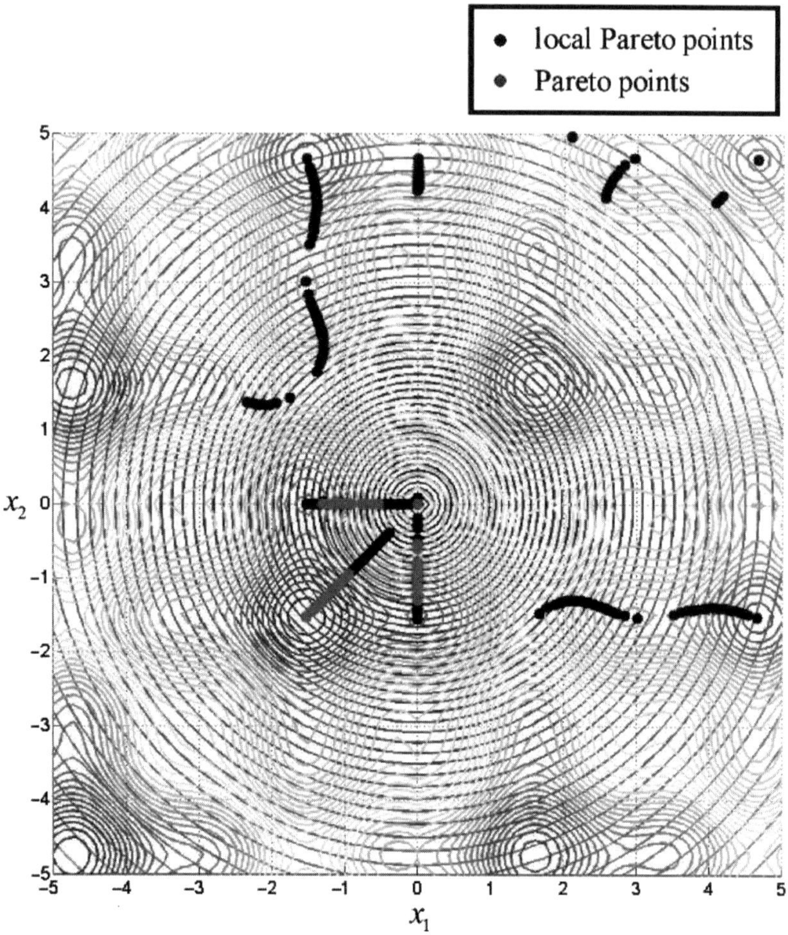

Fig. 15 Kursawe solution obtained with the NC+ method (design space).

The peculiarity of test case Kursawe is its symmetry, as shown in Fig. 15. Furthermore, the global Pareto front is also discontinuous.

A total of 33,345 analyses were required for obtaining the solution. Figure 15 shows the results obtained in the design space together with the overlapping contour lines of the two objective functions. Figure 16 shows the solution obtained in the criterion space.

As can be seen from the figures, the global Pareto front is composed of four distinct regions, one of which corresponds to the single point for which $\boldsymbol{x} = (0, 0)$, the minimum of objective function F_1.

The figures show that the NC+ method is capable of obtaining the complete global Pareto front, regardless of the symmetry problem.

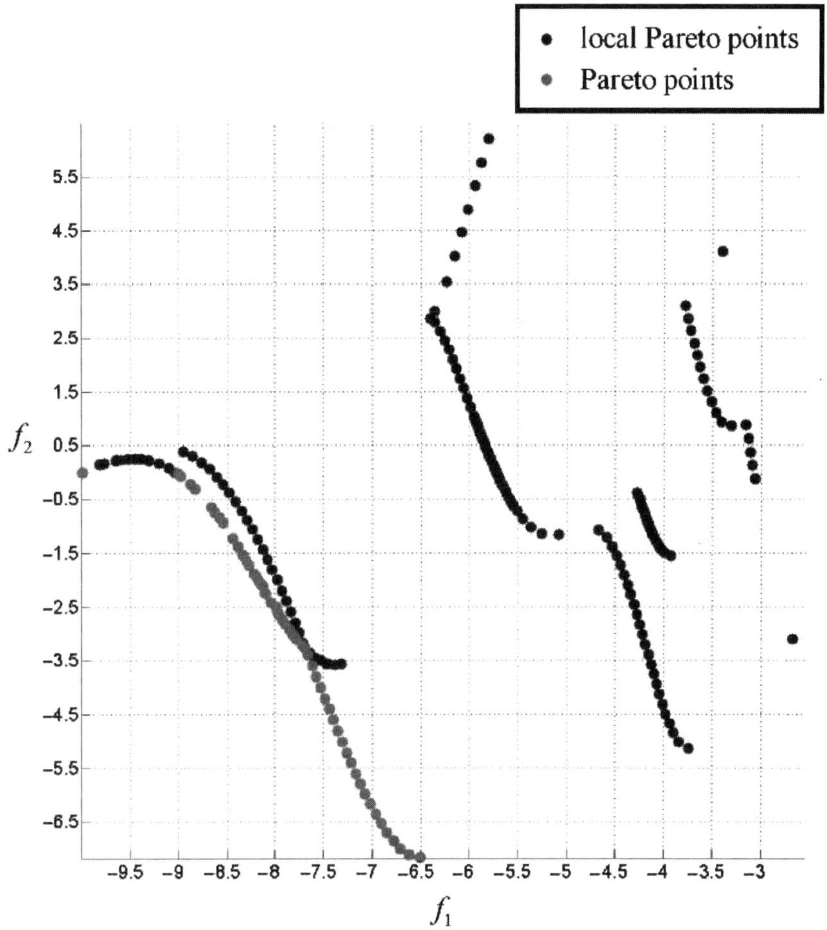

Fig. 16 Kursawe solution obtained with the NC+ method (criterion space).

The application of sensitivity analysis techniques for reducing the dimensionality of the optimization problem and hence its computational cost is presented in the next section.

IV. Sensitivity Analysis

A. Objectives of Sensitivity Analysis

Sensitivity analysis (SA) is the study of how changes in the outputs of a complex model represented as a black box (Fig. 17) can be apportioned qualitatively or quantitatively to variations in the different inputs. Absence of information or poor understanding of the internal mechanisms inside the model has led modelers to develop methods to understand how inputs and outputs are

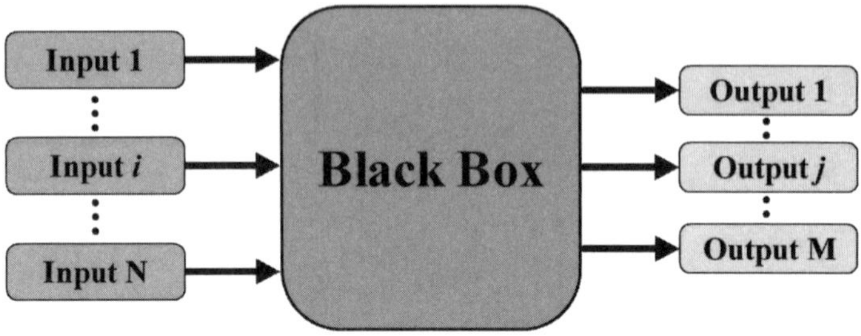

Fig. 17 Representation of a computer model: the black box.

related. These methods are based on different mathematical or statistical tools, depending on the problem and the information they provide.

Originally, SA was developed to deal simply with uncertainties in the input variables and model parameters, but it has been extended by providing an understanding of how the model response variables respond to changes in the inputs. Over the last few decades, a large number of SA methods have been developed to facilitate the decision-making process in the science and engineering fields for different purposes. In engineering, SA is mainly used for design reliability and robustness; in chemistry and the social sciences, for model validation and confirmation; and in economics and risk assessment, to determine the most influential parameters.

B. Sensitivity Analysis Procedure

Information on models used to describe a complex system cannot be obtained analytically because the internal mechanisms are not known. This is particularly true in MDO as complicated coupling between disciplines must be taken into account. In such a situation, only numerical results can be obtained. A way to get a better understanding of the model is to perform a sample-based sensitivity analysis.

A sample-based sensitivity analysis is a procedure in which the model is executed repeatedly for a set of input values. As described by Saltelli et al. [33] and illustrated in Fig. 18, the following steps can be identified in a sample-based sensitivity analysis:

1) Design of experiments: the question which the sensitivity analysis should answer is defined, and an appropriate strategy for sampling the input factors concerned with the analysis is decided.

2) Assign the probability distribution or range of variations to each input factor.

3) Generate a sample of input factors.

4) Execute the model to obtain the output for each input factor generated.

5) Perform sensitivity analysis to obtain the relevant information about the model.

When partial derivatives are used as sensitivity analysis methods for complex systems, they are often evaluated numerically by finite differences. In this case, a uniform distribution is assumed over the small interval selected for the evaluation.

Another important point is that when sample-based methods are used, both one-at-a-time (OAT) sensitivity analysis and multivariate sensitivity analysis can be performed. With sample-based techniques, input factors can be varied together to study interaction between sets of variables and their effect on the outcome of the model.

C. Variance-Based Methods

When implementation of sensitivity analysis is concerned, many different approaches can be followed, and a large number of techniques are available, which makes it difficult to select the appropriate method. We have chosen variance-based methods (VBM) because these are rigorous and theoretically sound approaches for global sensitivity calculation [34–39], exhibiting desirable properties for sensitivity analysis of complex models:

1) VBM are model independent. No assumption on linearity or additivity of the model is required to apply them. It is particularly interesting for MDO problems where models are usually very complex and highly nonlinear.

2) Very few assumptions on variables are necessary. They must be independent. In an MDO problem, variables are independent and could be varied on the whole design space.

3) VBM can deal with both discrete and continuous variables, which add to their flexibility.

4) VBM give a better insight into the design as input contributions are considered all together and not one at a time as in a traditional sensitivity analysis.

VBM provide quantitative information on the influence of each input factor to help the designer identify the most influential variables, on which the computational effort can be concentrated, and the nonsensitive variables, which can be discarded or frozen to a specific value.

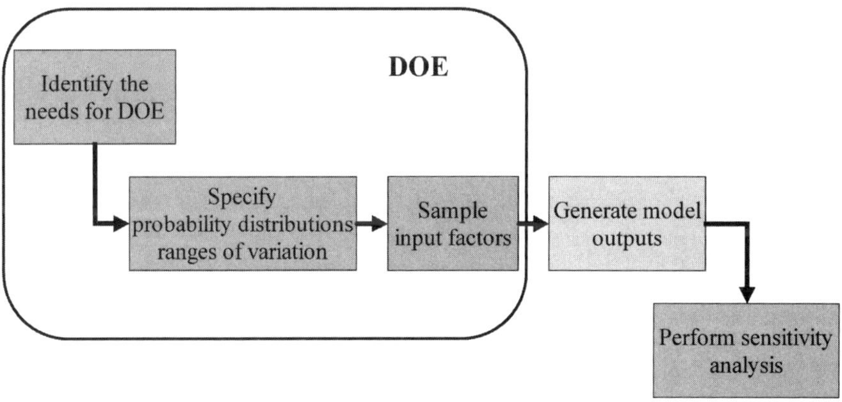

Fig. 18 Typical sensitivity analysis procedure.

Variance-based approaches for sensitivity analysis decompose the output variance into partial variances of increasing dimensionality:

$$V(Y) = \sum_i V_i + \sum_{i \neq j} V_{ij} + \cdots + V_{1,2,3,\ldots,k} \tag{17}$$

where

$$V_i = V\left[E\left(\frac{Y}{X_i} = x_i^*\right)\right] \tag{18}$$

$$V_{ij} = V\left[E\left(\frac{Y}{X_i} = x_i^*, X_j = x_j^*\right)\right] - V_i - V_j \tag{19}$$

and so forth.

In the decomposition of the variance, the term V_{ij} is the interaction effect between x_i and x_j. V_{ij} represents the part of the output variation caused by input parameters x_i and x_j, which cannot be explained by the sum of the first-order effects of parameter x_i and x_j. Similar considerations can be made for higher-order terms.

In the expression $V[E(Y/X_i = x_i^*)]$, the expectation is calculated over all variables X_j with $i \neq j$, and the variance is calculated over the values of X_i. $E(Y/X_i = x_i^*)$ is a function of X_i, and for an influential factor its value will vary a lot with X_i resulting in a large value for the variance $V[E(Y/X_i = x_i^*)]$. Therefore, the usefulness of $V[E(Y/X_i = x_i^*)]$ as a measure for sensitivity analysis is easy to grasp.

The signification of $V[E(Y/X_i = x_i^*)]$ is derived from the equation

$$V(Y) = V\left[E\left(\frac{Y}{X_i} = x_i^*\right)\right] + E\left[V\left(\frac{Y}{X_i} = x_i^*\right)\right] \tag{20}$$

which can be rewritten as

$$E\left[V\left(\frac{Y}{X_i} = x_i^*\right)\right] = V(Y) - V\left[E\left(\frac{Y}{X_i} = x_i^*\right)\right] \tag{21}$$

The term $E[V(Y/X_i = x_i^*)]$ is the expectation of the variance of Y when the input X_i is frozen. It results that $V[E(Y/X_i = x_i^*)]$ is the actual reduction of the output variance that one can expect if one fixes the input X_i to a particular value. The term V_{ij} represents the part of $V[E(Y/X_i = x_i^*, X_j = x_j^*)]$ that cannot be explained by the sum of the first-order terms V_i and V_j.

Let X_{-i} be the vector made up of all x_j with $j \neq i$, and let x_{-i}^* be a particular value of X_{-i}. $V[E(Y/X_{-i} = x_{-i}^*)]$ is the contribution to the output variance that is not caused by i. It means that the difference $V(Y) - V[E(Y/X_{-i} = x_{-i}^*)]$ is equal to the sum of all terms in the variance decomposition that include X_i. Two coefficients of main interest for sensitivity can be defined:

1) The main effect index

$$S_i^1 = \frac{V[E(Y/X_i = x_i^*)]}{V(Y)}$$

This gives the first-order contribution of X_i to the output response.

2) The total effect index

$$S_i^T = \frac{V(Y) - V[E(Y/X_{-i} = x_{-i}^*)]}{V(Y)}$$

This gives the total contribution of X_i to the output response.

The first term gives the direct contribution of each variable to the output and the second term gives the total contribution. For each input, the difference between its total effect and its main effect gives an indication of the importance the contribution to the output as a result of interactions with other inputs. Therefore, both the main effect indices and total effect indices are necessary to obtain information about the nonadditivity of the model and on the relative importance of variable interactions. Together, the main effect terms and total effect terms give a good description of the sensitivity of the model under investigation.

Two main methods, Fourier amplitude sensitivity test (FAST) [40, 41] and Sobol [42, 43], have been developed to compute the different terms of the variance decomposition and are described in the two next sections. Based on two different approaches, they enable the computation of both main and total effect indices and lead to the same predictions.

D. Reducing the Dimensionality of the Optimization Problem

In the context of MDO, we assume that the model, which evaluates the objectives and constraints of the optimization problem, can be represented as a black box. This stands for the fact that the designer is unaware of the complex internal mechanisms of the model, which relate objectives and constraints to the input variables. The methodology developed in this general case can then be readily implemented for a more specific purpose. Let us consider the multi-optimization problem described by Eq. (11). In the context of deterministic optimization, it is also assumed that all variables have uniform distribution defined by a minimum and maximum bound. In this approach, it is proposed to use VBM to evaluate the global sensitivity indices of each input with respect to all outputs. This allows the quantifying of the effect of the variations of the inputs on the outputs variance. Both main and total sensitivity indices are calculated. The discrepancy between these two indices represents the contribution to the output variance as a result of interaction effects among input variables. The main and total effects represent the minimum and maximum expected reduction of the output variance if the input is fixed to a specific value. Therefore, freezing an input variable with a negligible total sensitivity index will not affect the output variance.

It is proposed to remove from the original optimization formulation a variable with a negligible effect on all objectives and constraints [44]. When more than one variable is insignificant, the designer should consider these as a group of variables and see whether the sensitivity of the group is still below the significance threshold. Typically, this would imply performing another sensitivity analysis and resampling the design space. Such a procedure could be very computationally expensive. Instead, it is recommended to make sure that the sum of the group's total indices remains below a particular threshold value. Without loss of generality, let us assume that the first R variables are nonsignificant to all objectives and constraints. The problem can be reformulated as follows:

$$\min f(\boldsymbol{x}_{\text{red}})$$

subject to K_i inequality constraints:

$$g_k(\boldsymbol{x}_{\text{red}}) \leq 0 \qquad k = 1, 2, \ldots, K_i \tag{22}$$

and P_e equality constraints:

$$h_p(\boldsymbol{x}_{\text{red}}) = 0 \qquad p = 1, 2, \ldots, P_e$$

where $\boldsymbol{x}_{\text{red}} = (x_{R+1}, \ldots, x_N)$. All variables with negligible effect are fixed to a value $x_m = x_m^*$ for $m = 1, \ldots, R$.

The approach is presented in the case of a multicriteria optimization (see Sec. VI.C), but the same considerations can be made when a single objective is optimized.

V. Design Robustness

A. Robust Design Optimization

During the initial phases of product development, the complete problem frame is only approximately known; the adopted computer models have low-fidelity and assumptions based on previous experience have to be largely used. Such uncertain basis can cause, later on in the development cycle, the need for modifications to the particular (Pareto) solution that has been selected, as explained in the preceding sections, to respond to a specific tradeoff between the objectives. If the solution is sufficiently sensitive to perturbations of the design variables, such modifications can lead to a change in the predicted product attributes, which would be different with respect to the ones selected by the designer. One of the possible approaches to account for this issue consists in obtaining a *robust* Pareto front, in which, for each of the deterministic objectives, the specific measure of performance is traded off with its sensitivity to unforeseen variations.

The origins of robust design can be traced back to Taguchi, whose fundamental intuition was to understand that *quality*, interpreted as a minimization of the statistical variation of performance, has to be designed into the product, and not sought after only during the production phase [45]. His methodology, based on direct experimentation, was later extended to simulation-based design and gradually improved to exploit nonlinear constrained optimization techniques [46]. In its current version, the constrained robust optimization (RO) strategy can be thought of as made up of three main parts [47]. The first stage consists of identifying, qualifying, and quantifying the sources of uncertainty associated with the design input and the analysis modules. This is usually done by means of stochastic models. The second phase consists of propagating the uncertainty through the analysis system, to adequately model the probabilistic behavior of the objective functions and constraints. The obtained probabilistic quantities are hence used in the third stage of the process, during which the optimization is performed.

In RO, the probabilistic state can be defined in terms of expectation and variance of the deterministic objectives and constraints. A single-objective deterministic optimization problem turns then into a multi-objective robust problem if the two statistical moments are thought of as representing, two conflicting objectives such as a suitable average of the system performance and its sensitivity to unforeseen variations, respectively [48, 49]. Several approaches have been developed to adequately accommodate this issue, ranging from the weighted sum method to physical programming [46]. Alternatively, one among expectation and variance can be optimized and the other constrained [50, 51].

When the considered deterministic problem is multi-objective, multiple system performance metrics have to be traded off. A more complex robust design strategy might then be required to adequately consider objectives correlation structure [52].

In our case, the hypothesis for independence of the deterministic objectives is adopted. This leads to formulating, for example, a bi-objective problem as a four-objective one in which robustness is sought for by minimizing the objectives' variances, whereas expectations of the performance measures are optimized according to their physical meanings.

To maintain design feasibility with a specific level of confidence, given the prescribed input uncertainty, inequality constraints of the form $g_k(\boldsymbol{x}) \leq 0$ take the following form [53]:

$$\mu_{g_k}(\boldsymbol{x}) + t_{g_k}\sigma_{g_k}(\boldsymbol{x}) \leq 0 \tag{23}$$

The coefficient $t_{g_k} = \Phi^{-1}[P(g_k \leq 0)]$ guarantees a prescribed level of probability P of constraint satisfaction where Φ^{-1} is the inverse of the normal cumulative distribution function. This formulation is only approximate for nonnormal constraint functions, which however tend to normality, as insured by the central limit theorem, if the number of random design variables is large.

Recently, valid approaches to handle the challenging problem of equality constraints have been presented in literature, which are intuitively difficult to satisfy exactly if the function to be constrained is randomly varying [54]. Because they were not part of the original deterministic problem, to simplify the approach as much as possible, they are not considered as part of the presented RO formulation, which can then be formally expressed by extending Eq. (11) as follows:

$$\min_{\boldsymbol{\mu}_x} \boldsymbol{F} = \boldsymbol{\mu}_f(\boldsymbol{x}),\ \boldsymbol{\sigma}_f(\boldsymbol{x}) \tag{24}$$

subject to K_i inequality constraints:

$$G_k = \mu_{g_k}(\boldsymbol{x}) + t_{g_k}\sigma_{g_k}(\boldsymbol{x}) \leq 0 \qquad k = 1, 2, \ldots, K_i$$

with

$$\boldsymbol{x}_L + t_x\sigma_x \leq \mu_x \leq \boldsymbol{x}_U - t_x\sigma_x.$$

B. Uncertainty Propagation

The way in which mean and variance of each objective and constraint function are obtained starting from the knowledge of the uncertainty affecting $\boldsymbol{x}$ turns out to be crucial both for the efficiency and the accuracy of the whole RO approach.

If all of the variables are continuous, the first two moments of $y = f(\boldsymbol{x})$ are

$$\mu_f(\boldsymbol{x}) = \int_{-\infty}^{+\infty} f(\boldsymbol{\xi})p_X(\boldsymbol{\xi})\,\mathrm{d}\boldsymbol{\xi} \tag{25}$$

$$\sigma_y^2(\boldsymbol{x}) = \int_{-\infty}^{+\infty} \left\{f(\boldsymbol{\xi}) - \mu_f(\boldsymbol{x})\right\}^2 p_X(\boldsymbol{\xi})\,\mathrm{d}\boldsymbol{\xi} \tag{26}$$

where p_X is the joint probability density function corresponding to distributions modeling the uncertainty of the input variables and is not supposed to change with the design point $\boldsymbol{x}$. Because a closed-form solution of these integrals can be obtained only in a few cases of practical interest, uncertainty propagation is usually performed in an approximate fashion. Existing approaches include Monte Carlo methods [55, 56], Taylor-based method of moments [53, 57], surrogate models [58, 59], polynomial chaos expansion [51, 60], reliability-based techniques [61], and quadrature-based approach [62, 63]. The sigma-point (SP) approach [64] is well suited to the implemented optimization approach and can guarantee the required accuracy at an affordable computational cost. This method relies on a specific kind of reduced numerical quadrature, and gives mean and variance as follows:

$$\mu_{y_{\mathrm{SP}}} = W_0 f(\boldsymbol{x}_0) + \sum_{p=1}^{n} W_p\left[f(\boldsymbol{x}_{p+}) + f(\boldsymbol{x}_{p-})\right] \tag{27}$$

$$\begin{aligned}\sigma^2_{y_{\mathrm{SP}}} = \frac{1}{2}\sum_{p=1}^{n}\Big\{ & W_p\left[f(\boldsymbol{x}_{p+}) - f(\boldsymbol{x}_{p-})\right]^2 \\ & + \left(W_p - 2W_p^2\right)\left[f(\boldsymbol{x}_{p+}) + f(\boldsymbol{x}_{p-}) - 2f(\boldsymbol{x}_0)\right]^2\Big\}\end{aligned} \tag{28}$$

The weights are chosen as follows:

$$W_0 = \frac{h_{\text{sp}}^2 - n}{h_{\text{sp}}^2} \tag{29}$$

$$W_p = \frac{1}{2h_{\text{sp}}^2} \qquad \text{for } 1 \leq p \leq n \tag{30}$$

The sampling points are

$$\boldsymbol{x}_0 = \mu_x \tag{31}$$

$$\boldsymbol{x}_{p\pm} = \mu_x \pm h_p \sigma_{x_p} \boldsymbol{e}_p \tag{32}$$

where $\boldsymbol{e}_p$ is the pth column of the identity matrix of size n and h_p is equal to the square root of the kurtosis of the pth design variable distribution. The accuracy of the SP method, in particular for the mean estimate, is higher with respect to the largely adopted first-order Taylor-based method of moments (MM). However, it requires only $2n + 1$ function evaluations for each analysis, which is equal to the cost of linearization if function gradients are approximated by centered finite differences. When the function performing the system analysis is differentiable, this technique can be efficiently used in gradient-based optimization; the computational cost of a single optimization step in terms of function evaluations is $\propto n^2$ by using either the SP method or MM for the propagation phase, if the derivatives are obtained by finite differencing. If the source code of the analysis system is available and AD can be deployed, this cost decreases to $\propto n$ in both cases.

VI. Results

The results of the tests on the workflow manager and the coordinated application of the just-described treatments are presented in this section. The adopted test case is an Ultra Simplified Model of Aircraft (USMAC), which was provided by a major airframe manufacturer. USMAC can determine performance and sizing at conceptual design level for a short-to-medium-range commercial passenger aircraft. The test case contains 97 models and 125 variables.

A. Computational Process Modeler Results

Synthesis of an optimal computational plan has to be performed before any other treatment is applied to the test case. To test the capability of the computational process modeler for generating optimal computational plans, 23 variables (required in this case for the system to be well constrained) were chosen randomly as independent from the 125 variables set. Many such cases were considered in order to validate the computational process modeler. Here we explain the results generated by computational process modeler for two of those cases. Optimal computational plans were generated by the computational process modeler in both cases. The computational cost of each optimal computational plan was then compared with the cost of its corresponding nonoptimal plan.

The latter were generated during the computational process modeling and refer to the computational plans based on those variable flow models for the SCCs, which were not selected as optimal. The computational cost comparison was performed on the basis of the number of calls made to the models of the SCC during solving. The SCCs were solved by applying a fixed-point iteration method, whereas the modified models were resolved by applying the Gauss–Newton method [12].

1. *Case 1*

After decomposing the system, 13 out of the 97 models were identified as strongly connected. There were 12 variable flow models generated for the SCC. However, only four of these produced a converged solution. The nonconverged ones were those variable flow models that had a higher number of modified models and feedback numbers. Table 2 provides the details of the four converged models and also one of the nonconverged solutions for comparison.

Variable flow model 3 was chosen by the computational process modeler as the optimal one because it has the least number of modified models. It is shown in the table that the selected optimal flow model has the lowest computational cost for the SCC. From Table 2 it is also clear that when the number of modified models increases the computational cost for the SCC also increases. However, for variable flow model 4, even though the number of modified models is less than the one for the flow model 2, it has taken more calls to obtain a converged solution. This discrepancy was observed because the convergence of the SCC was not only depending on the number of modified models and the number of feedback loops, but also on other factors such as the starting (iteration) point for the unknown variables, mutual sensitivity of the switched (input with output) variables of the modified models, and possibly other factors which are yet to be discovered.

2. *Case 2*

In this case, two SCCs were identified after decomposing the system of 97 models. The first one had seven models, and the second one had six models. For the first SCC, two variable flow models were generated, and for the second one there were six variable flow models. Three variable flow models produced a converged solution. The results are summarized in Table 3.

Table 2 Details of computational process modeling and solving of SCC for case 1

Variable flow model	Number of feedback loops (*nFdb*)	Number of modified models (*nMm*)	Optimal flow model	Number of calls to the models in SCC	% Additional computational cost
1	3	6		117	95% more
2	5	11		158	163% more
3	6	3	■	60	Base
4	5	9		198	230% more
5	8	11		Not converged	—

Table 3 Details of computational process modeling and solving of SCCs for case 2

	Computational process modeling				Solving	
SCC	Variable flow model	*nFdb*	*nMm*	Optimal flow model	Number of calls to the models in SCC	% Additional computational cost
SCC1	1	1	4	■	110	Base
	2	1	5		110	Equal
SCC2	1	1	3		74	Base
	2	1	4		320	332% more
	3	1	2	■	86	16.2% more

The sixth column of Table 3 indicates that the flow model selected by the computational process modeler for SCC2 was not the optimum in terms of computational cost because variable flow model 1 made a fewer number of calls (74) compared to 86 calls for flow model3, which was chosen as the optimum. Nevertheless, these and many more extensive tests conducted on the process modeler demonstrated that the selections which it made were always among the best in terms of SCCs' convergence.

Figure 19 shows an example of a final computational sequence in an incidence matrix format (i.e., models in the rows and variables in the columns), generated by the computational process modeler for USMAC.

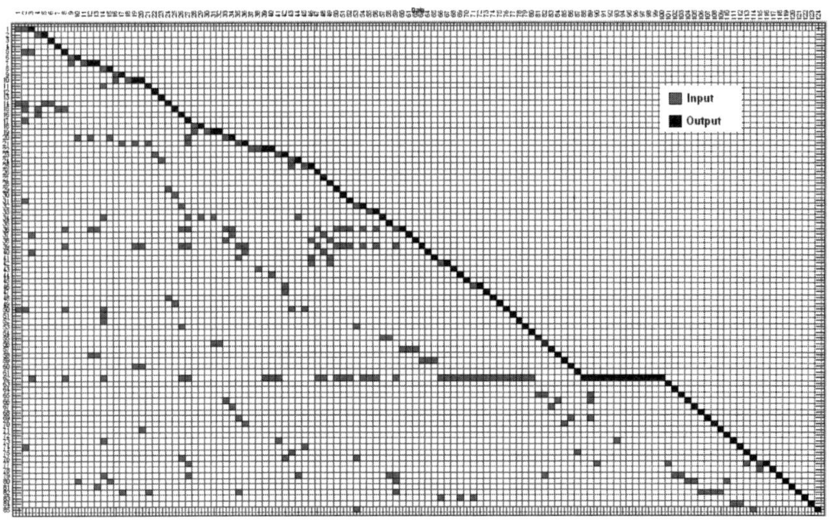

Fig. 19 Example of an USMAC final computational plan in a IM form for a particular set of input variables chosen by the designer.

Table 4 Input variables chosen for the application of subsequent treatments

Input variable	Unit
Npax (number passengers)	—
NpaxFront (number of passengers per row)	—
Naisle (number of aisles)	—
FNslst (engine thrust)	decaN
BPR (engine bypass ratio)	—
ne (number of engines)	—
Awing (wing area)	m^2
span (wing span)	m
ϕ (wing sweep angle)	deg
tuc (wing thickness to chord ratio)	—
RA (range)	NM
MTOW (maximum takeoff weight)	kg
Fuel (fuel quantity)	kg
altcrz (cruise altitude)	ft
Machcrz (cruise mach)	—
alto (takeoff altitude)	ft
altapp (approach altitude)	ft

B. Multi-Objective Optimization Results

The USMAC variables in Tables 4 and 5 were identified as the respective input and output of interest for the application of the multi-objective optimization treatment.

As can be seen from the input and output variables, the MTOW and the RA appear both in the inputs and the outputs, making the system nonhierarchically decomposable.

The multi-objective optimization problem is set as follows:

1) The independent variables considered were FNslst = [12,500;13,000] (decaN); Awing = [152,158] m^2; span = [30,38] m; ϕ = [28,32] deg; tuc = [0.07,0.1]; Fuel = [17,000;18,000] kg; and BPR = [6,7].

Table 5 Output variables chosen for the application of subsequent treatments

Output variable	Unit
MTOW (maximum takeoff weight)	kg
RA (range)	n mile
RA time (flight time)	h
tofl (takeoff field length)	m
vapp (approach speed)	kt
vzclb (climb rate)	ft/min
kfncth (cruise thrust coefficient)	—
Kff (wing fuselage fuel ratio)	—

2) The constant variables considered were Npax = 150; NpaxFront = 6; Naisle = 1; BPR = 6; ne = 2; altcrz = 35,000 ft; Machcrz = 0.82; altto = 0 ft; and altapp = 0 ft.

3) Objectives considered were range to be maximized and MTOW to be minimized.

4) Constraints considered were tofl $\leq$ 2000 m; vapp $\leq$ 120 kts; vzclb $\geq$ 500 ft/min; kfncth $\leq$ 1; and Kff $\geq$ 0.75.

The computational process modeler generated an optimal computational plan onto which the NC+ method was executed with the preceding settings. Five starting points were used in order to determine the multiple local minima. Forty utopia plane points were generated in order to determine the Pareto front. The results obtained are shown in Figs. 20–22. The values of the independent variables are shown in Fig. 21 against the Pareto solutions for MTOW from Fig. 20, while the values of the constraint variables are shown in Fig. 22.

It can be seen (Fig. 20) that when MTOW is at its minimum RA is also at its minimum and the aircraft is the smallest possible. In fact, Awing and span are at their minimum; ϕ, Fuel, and BPR are the smallest allowed, while the thrust FNslst is approximately 12,615 decaN (Fig. 21).

From Fig. 22 it can be seen that the active constraints are those relative to vapp, vzclb, kfncth, and Kff. Also, it can be noticed that the constraints on vapp and Kff are always active or very close to being active.

It also can be seen from the figures that for a slight increase of MTOW the span rapidly increases, thrust and FNsls rapidly decreases to its minimum,

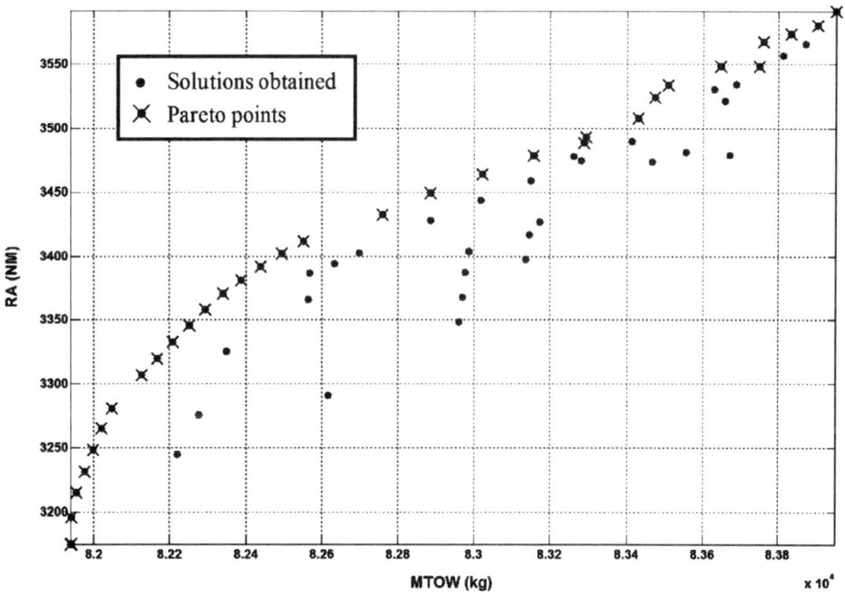

Fig. 20 USMAC test case Pareto front.

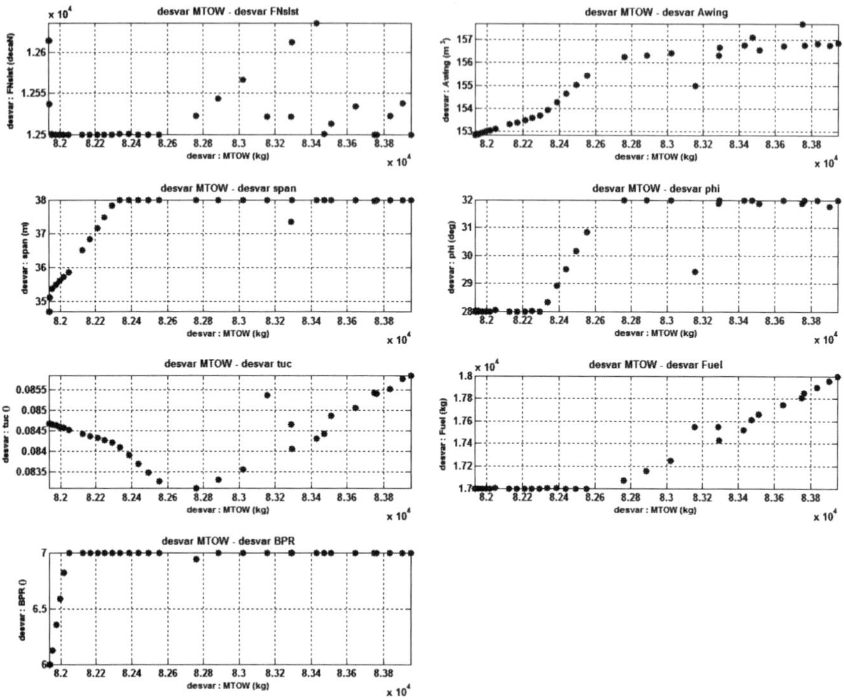

Fig. 21 MTOW—design variables.

while the constraint on kfncth becomes inactive. With a further increase of MTOW, RA and span rapidly increase to their maximum, which is maintained for all of the remaining Pareto points. From approximately 823,000 kg onwards, it is the increase in Awing, ϕ, and Fuel that drive the increase in MTOW and RA up to their maximum values.

An extension of the results visualization was carried out to illustrate how integrating design computation, parametric geometry, and configuration can aid the designer in choosing a particular (Pareto) solution. A snapshot of this off-line browsing tool is presented in Fig. 23. The tool allows for browsing each of the Pareto points while showing the changes to the geometry and indicating (in different color) which constrains are active for the particular point.

C. Sensitivity Analysis Results

The multi-objective optimization treatment can be computationally demanding. In such a case, identifying variables with negligible impact on the optimization outputs might allow the reduction of the dimensionality of the problem and hence the computational effort.

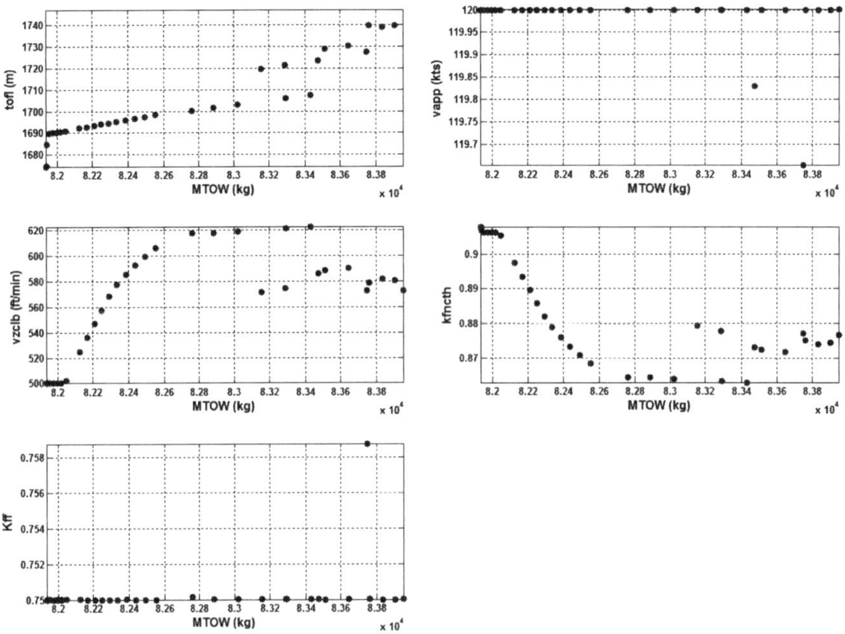

Fig. 22 MTOW—constraints.

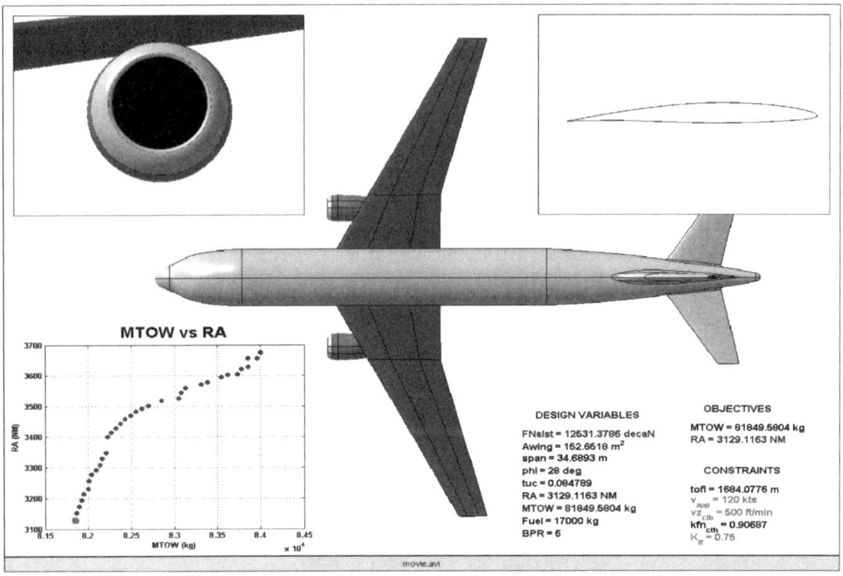

Fig. 23 Visualization of the Pareto frontier, including parametric geometry and constraints activation. (See also color figure section at the back of the book.)

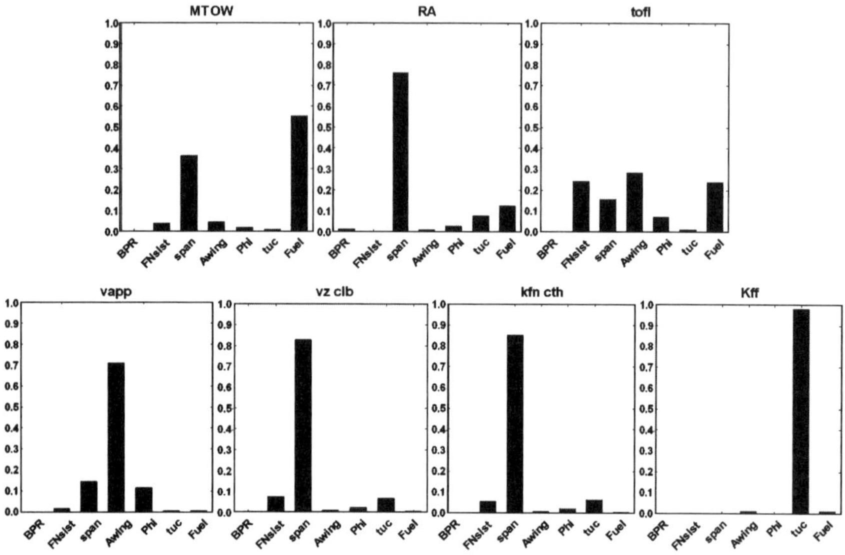

Fig. 24 USMAC global sensitivity analysis—VBM indices.

The global sensitivity indices of each output with respect to all inputs are obtained with VBM and are given in Fig. 24. For this particular test case and the specific range of variations, it appears that the contributions to the outputs variance as a result of interactions between inputs are very small. Therefore, one can conclude that contributions to the outputs are entirely caused by direct effects. It is appears that FNslst, Span, Wing, tuc, and Fuel are the most significant variables to the problem. Variable ϕ has a smaller effect but cannot be neglected as it affects tofl and vapp. The global indices for BPR for all optimization outputs are almost equal to zero, and therefore BPR appears to be negligible for this particular problem. The reduced optimization problem is derived by keeping all variables and fixing BPR to the mean value of its range of variation, that is, BPR = 6.5.

The original Pareto set and the one obtained with the reduced optimization problem are shown in Fig. 25. The two Pareto fronts are very similar in the criterion space, which indicates that a similar level of performance can be obtained by only considering variables with a real effect. Freezing BPR to a value resulted in a significant reduction of the computational effort: 16,627 function evaluations were needed to obtain the Pareto front compared to 65,332 when all variables are considered. Despite the encouraging initial results, more research will be needed to confirm these findings in general.

D. Robust Optimization Results

The robust counterpart of the deterministic optimization problem in Section B considers separately, without any *a priori* assumption on relative weights, mean, and variance for the two physical objectives. The problem to be solved is then a

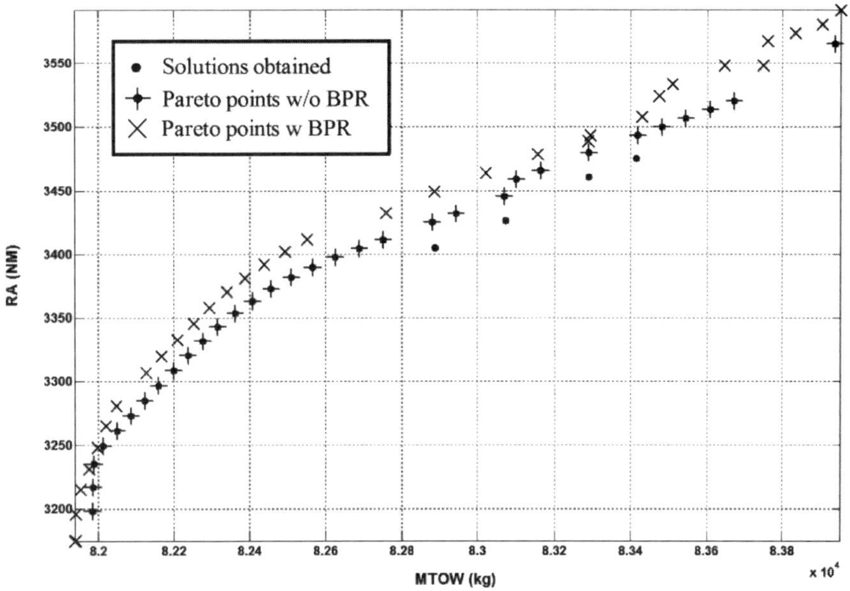

Fig. 25 USMAC Pareto front for full and reduced optimization problem.

four-objective optimization. The deterministic constraints are transformed into their robust counterpart by adopting Eq. (23), adopting $k = 1$ as weighting coefficient, to impose the robust feasibility with a probability of approximately 84%. The assumed uncertainties of input variables, in terms of standard deviation, are shown in Table 6.

It is useful to compare the deterministic with the obtained robust Pareto front by performing an *a posteriori* uncertainty analysis on the deterministic results in order to obtain mean and variance for each Pareto point. The mean of the objectives for the robust optimization is then superimposed on the same plot (see Fig. 26). The adopted representation is a bidimensional projection of a

Table 6 Uncertainties of input variables

Input variable	Standard deviation
FNslst (engine thrust)	100 decaN
BPR (engine bypass ratio)	0.2
Awing (wing area)	5 m^2
Span (wing span)	0.5 m
ϕ (wing sweep angle)	1 deg
Tuc (wing thickness to chord ratio)	0.02
Fuel (fuel quantity)	50 kg

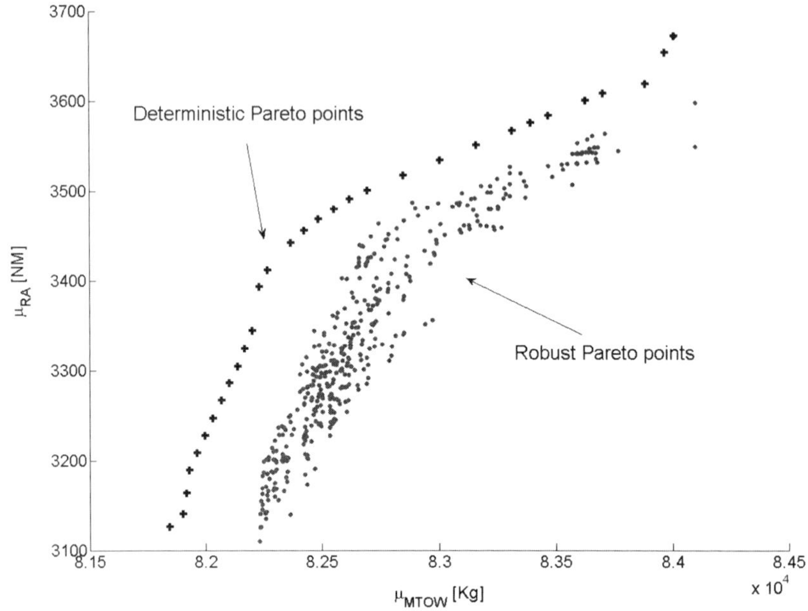

Fig. 26 Comparison of deterministic and robust Pareto fronts.

four-dimensional Pareto hypersurface. It has been judged to be appropriate for the problem at hand because the variation of standard deviation for both objectives is negligible with respect to their mean values. This is mainly because of the small input uncertainty considered. Thus the robust optimal solutions turn out to be dominated by the deterministic ones in terms of mean values mainly as a result of the imposed stricter constraints.

VII. Conclusions

Presented is a novel computational framework providing the capability for performing flexible design studies at preliminary design stage. It incorporates a workflow management device (WMD) and a number of advanced treatments, including multi-objective optimization, sensitivity analysis, and uncertainty management.

The WMD enables the designer to build, understand, manipulate, and share complex processes and studies. Results obtained after applying the WMD on various test cases showed a significant reduction of the iterations required for the convergence of the computational system. The tests also demonstrated the capabilities of the advanced treatments as follows:

1) The novel procedure for global multi-objective optimization has the unique ability to generate well-distributed Pareto points on both local and global Pareto fronts simultaneously.

2) The global sensitivity analysis procedure is able to identify input variables whose range of variation does not have significant effect on the objectives and constraints. It was demonstrated that fixing such variables can greatly reduce the computational time while retaining a satisfactory quality of the resulting Pareto front.

3) The novel derivative-free method for uncertainty propagation, which was proposed for enabling multi-objective robust optimization, delivers a higher accuracy compared to the one based on function linearization, without altering significantly the cost of the single optimization step.

This work demonstrated for the first time that such capabilities can be used in a coordinated way to enhance the efficiency of the computational process and the effectiveness of the decision making. Future work will concentrate on further integration of the treatments, which will allow their dynamic application as dictated by the computational process. Also, further integration with parametric geometry and configuration tools is planned in order to achieve the ultimate goal of this work, that is, to study unconventional new configurations with a higher level of detail and better risk assessment before proceeding to the next stages of the product development process.

Acknowledgments

The authors wish to thank our industrial partners for the fruitful discussions, constructive criticism, and the provision of test cases.

References

[1] Serrano, D., "Constraint Management in Conceptual Design," Ph.D. Dissertation, Dept. of Mechanical Engineering, MIT, Cambridge, MA, 1987.

[2] Bouchard, E. E., Kidwall, G. H., and Rogan, J. E., "The Application of Artificial Intelligence Technology to Aeronautical System Design," AIAA Paper 88-4426, Sept. 1988.

[3] Buckley, M. J., Fertig, K. W., and Smith, D. E., "Design Sheet: An Environment for Facilitating Flexible Trade Studies During Conceptual Design," AIAA Paper 92-1191, Feb. 1992.

[4] Cormen, T., Leiserson, C., and Rivest, R., *Introduction to Algorithms*, McGraw-Hill, New York, 1991, pp. 651–681.

[5] Ramaswamy, R., and Ulrich, K., "A Designer's Spreadsheet," *Proceedings of the Design Theory and Methodology Conference*, Vol. 53, ASME, New York, 1993, pp. 105–113.

[6] Rogers, J. L., "Tools and Techniques for Decomposing and Managing Complex Design Projects," *Journal of Aircraft*, Vol. 36, No. 1, 1999, pp. 266–274.

[7] Steward, D. V., "The Design Structure System: A Method for Managing the Design of Complex Systems," *IEEE Transactions on Engineering Management*, Vol. EM-28, 1981, pp. 71–74.

[8] Tang, D., Zheng, L., Zhizhong, L., Dongbo, L., and Zhang, S., "Re-Engineering of the Design Process for Concurrent Engineering," *Computers and Industrial Engineering*, Vol. 38, 2000, pp. 479–491.

[9] Rogers, J. L., "Reducing Design Cycle Time and Cost Through Process Resequencing," *Proceedings of the International Conference on Engineering Design* (ICED), Aug. 1997; also NASA Technical Paper 97-11iced-jlr http://www.tpub.com/content/nasa1997/NASA-97-11iced-jlr/index.htm [retrieved July 2009].

[10] Rogers, J. L., "A Knowledge-Based Tool for Multilevel Decomposition of a Complex Design Problem," NASA Technical Paper 2903, May 1989.

[11] Guenov, M. D., Balachandran (Libish), K., Tang, D., and Lockett, H., "Computational Design Process Modelling," *25th Congress of the International Council of the Aeronautical Sciences*, ICAS, Stockholm, Sweden, Sept. 2006.

[12] Denniz, J. E., Jr., and Robert, B., *Numerical Methods for Unconstrained Optimization and Nonlinear Equations*, Prentice Hall, Upper Saddle River, NJ, 1983, pp. 221–228.

[13] Miettinen, K. M., *Nonlinear Multiobjective Optimization*, Kluwer Academic, Norwell, MA, 1999, p. 5.

[14] Marler R. T., and Arora J. S., "Survey of Multi-Objective Optimization Methods for Engineering," *Structural and Multidisciplinary Optimization*, Vol. 26, No. 6, April 2004, pp. 369–395.

[15] Balling, R. J., and Richard., J., "Pareto Sets in Decision-Based Design," *Journal of Engineering Valuation and Cost Analysis*, Vol. 3, No. 2, 2000, pp. 189–198.

[16] Das, I., and Dennis, J. E., "A Closer Look at Drawbacks of Minimizing Weighted Sums of Objectives for Pareto Set Generation in Multicriteria Optimization Problems," *Structural Optimation*, Vol. 14, No. 1, Aug. 1997, pp. 63–69.

[17] Shukla, P. K., Deb, K., and Tiwari, S., "Comparing Classical Generating Methods with an Evolutionary Multi-Objective Optimization Method," *Evolutionary Multi-Criterion Optimization*, Vol. 3410/2005 in *Lecture Notes on Computer Science*, Springer Berlin/Heidelberg, 2005, pp. 311–325.

[18] Das, I., and Dennis, J. E., "Normal-Boundary Intersection: A New Method for Generating the Pareto Surface in Nonlinear Multicriteria Optimization Problems," *SIAM Journal of Optimization*, Vol. 8, No. 3, 1998, pp. 631–657.

[19] Messac, A., and Mattson, C. A., "Generating Well-Distributed Sets of Pareto Points for Engineering Design Using Physical Programming," *Optimization and Engineering*, Vol. 3, No. 4, Dec. 2002, pp. 431–450.

[20] Messac, A., Ismail-Yahaya, A., and Mattson, C. A., "The Normalized Normal Constraint Method for Generating the Pareto Frontier," *Structural and Multidisciplinary Optimization*, Vol. 25, No. 2, 2003, pp. 86–98.

[21] Messac, A., and Mattson, C., "Normal Constraint Method with Guarantee of Even Representation of Complete Pareto Frontier," *AIAA Journal*, Vol. 42, No. 10, 2004, pp. 2101–2111.

[22] Deb, K., Pratap, A., Agarwal, S., and Meyarivan, T., "A Fast and Elitist Multi-objective Genetic Algorithm: NSGA-II," *IEEE Transactions on Evolutionary Computation*, Vol. 6, No. 2, 2002, pp. 182–197.

[23] Frank, P. D., Booker, A. J., Caudel, T. P., and Healy M. J., "A Comparison of Optimization and Search Methods for Multidisciplinary Design," AIAA Paper 1992-4827, Sept. 1992.

[24] Das, I., "An Improved Technique for Choosing Parameters for Pareto Surface Generation Using Normal-Boundary Intersection." In *Short Paper Proceedings of the Third World Congress of Structural and Multidisciplinary Optimization WCSMO-3*, Springer, Berlin, Germany, 1999, pp. 411–413.

[25] Messac, A., "Physical Programming: Effective Optimization for Computational Design," *AIAA Journal*, Vol. 34, No. 1, 1996, pp. 149–158.

[26] Mattison, C. A., Mullur, A. A., and Messac, A., "Minimal Representation of Multiobjective Design Space Using Smart Pareto Filter," AIAA Paper 2002-5458, Sept. 2002.

[27] Fantini, P., "Effective Multiobjective MDO for Conceptual Design—An Aircraft Design Perspective," Ph.D. Dissertation, School of Engineering, Cranfield Univ., UK, 2007.

[28] Utyuzhnikov, S. V., Fantini, P., and Guenov, M. D., "Numerical Method for Generating the Entire Pareto Frontier in Multiobjective Optimization," *CD Proceedings of Evolutionary and Deterministic Methods for Design, Optimization and Control with Applications to Industrial and Societal Problems, EUROGEN 2005*, Munich, Germany, Sept. 12–14, 2005, FLM, TU Munich.

[29] Guenov, M. D., Utyuzhnikov, S. V., and Fantini, P., "Application of the Modified Physical Programming Method to Generating the Entire Pareto Frontier in Multiobjective Optimization," *CD Proceedings of Evolutionary and Deterministic Methods for Design, Optimization and Control with Applications to Industrial and Societal Problems, EUROGEN 2005*, Munich, Germany, Sept. 12–14, 2005, FLM, TU Munich.

[30] Fantini, P., Balachandran, L. K., and Guenov, M. D., Computational System for Multidisciplinary Optimization at Conceptual Design Stage, *International Journal for Simulation and Multidisciplinary Design Optimization*, Vol. 2, No. 3, 2008, pp. 177–185.

[31] Deb, K., "Multi-Objective Genetic Algorithms: Problem Difficulties and Construction of Test Problems," Dept. of Computer Science/LS11, Univ. of Dortmund, Technical Rept. CI-49/98, Dortmund, Germany, 1998.

[32] Kursawe, F., "A Variant of Evolution Strategies for Vector Optimization," *Parallel Problem Solving from Nature*, edited by H.-P. Schwefel and R. Manner, Vol. 496/1991 of *Lecture Notes in Computer Science*, Springer Berlin/Heidelberg, Germany, 1991, pp. 193–197.

[33] Saltelli, A., Chan, K., and Scott, M., *Sensitivity Analysis*, Wiley, New York, 2000, pp. 101–154.

[34] Saltelli, A., Tarantola, S., Campolongo, F., and Ratto, M., *Sensitivity Analysis in Practice: A Guide to Assessing Scientific Models*, Wiley, Chichester, UK, 2004, pp. 47–49.

[35] Saltelli, A., Andres, T. H., and Homma, T., "Sensitivity Analysis of Model Output: An Investigation of New Techniques," *Computational Statistics and Data Analysis*, Vol. 15, No. 2, 1993, pp. 211–238.

[36] Saltelli, A., Tarantola, S., and Chan, K. P. S., "A Quantitative Model-Independent Method for Global Sensitivity Analysis of Model Output," *Technometrics*, Vol. 41, No. 1, 1999, pp. 39–56.

[37] Chan, K., Saltelli, A., and Tarantola, S., "Sensitivity Analysis of Model Output: Variance-Based Methods Make the Difference," *Proceeding of 1997 Winter Simulation Conference (WSC'97)*, IEEE, Los Alamitos, CA, Dec. 1997, pp. 261–268.

[38] Homma, T., and Saltelli, A., "Importance Measures in Sensitivity Analysis of Non-Linear Models," *Reliability Engineering and System Safety*, Vol. 52, No. 1, 1996, pp. 1–17.

[39] Saisana, M., Saltelli, A., and Tarantola, S., "Uncertainty and Sensitivity Analysis Techniques as Tool for the Quality Assessment of Composite Indicators," *Journal of the Royal Statistical Society—A*, Vol. 168, No. 2, 2005, pp. 307–323.

[40] Cukier, R. I., Levine, H.B., and Shuler, K. E., "Non-Linear Sensitivity Analysis of Multi-Parameter Model Systems," *Journal of Computational Physics*, Vol. 26, No. 1, 1978, pp. 1–42.

[41] Saltelli, A., and Bolado, R., "An Alternative Way to Compute Fourier Amplitude Sensitivity Test (FAST)," *Computational Statistics and Data Analysis*, Vol. 26, No. 4, 1998, pp. 445–460.

[42] Sobol, I. M., "Sensitivity Estimates for Non-Linear Mathematical Models," *Mathematical Modelling and Computational Experiments*, Vol. 1, No. 4, 1993, pp. 407–414.

[43] Sobol, I. M., "Global Sensitivity Indices for Non-Linear Mathematical Models and Their Monte Carlo Estimates," *Mathematics and Computers in Simulation*, Vol. 55, Nos. 1–3, 2001, pp. 271–280.

[44] Maginot, J., Guenov, M. D., Fantini, F., and Padulo, M., "A Method for Assisting the Study of Pareto Solutions in Multi-Objective Optimization," AIAA Paper 2007-7792, Sept. 2007.

[45] Phadke, M. S., *Quality Engineering Using Robust Design*, Prentice Hall, Upper Saddle River, NJ, 1989, pp. 13–39.

[46] Park, G. J., Lee, T. H., and Hwang, K. H., "Robust Design: An Overview," *AIAA Journal*, Vol. 44, No. 1, 2006, pp. 181–191.

[47] Keane, A. J., and Nair, P. B., *Computational Approaches to Aerospace Design: The Pursuit of Excellence*, Wiley, New York, 2005, p. 329.

[48] Chen, W., and Allen, J., "A Procedure for Robust Design: Minimizing Variations Caused by Noise Factors and Control Factors," *Journal of Mechanical Design*, Vol. 118, No. 4, 1996, pp. 478–493.

[49] Das, I., "Robustness Optimization for Constrained Nonlinear Programming Problems," *Engineering Optimization*, Vol. 32, No. 5, 2000, pp. 585–618.

[50] Deb, K., and Gupta, H., "Introducing Robustness in Multiple-Objective Optimization," Kanpur Genetic Algorithms Lab., Indian Inst. of Technology, KanGAL Rept. 2004016, Kanpur, India, Oct. 2004.

[51] Molina-Cristobal, A., Parks, G. T., and Clarkson, P. J., "Finding Robust Solutions to Multi-objective Optimization Problems Using Polynomial Chaos," *Proceedings of the 6th ASMO UK/ISSMO Conference on Engineering Design Optimization*, 3–4 July 2006, Oxford, UK. http://citeseerx.ist.psu.edu/viewdoc/download?doi=10.1.1.61.5233&rep = rep1&type = pdf [retrieved July 2009].

[52] Murphy, T. E., Tsui, K. L., and Allen, K. J., "A Review of Robust Design Methods for Multiple Responses," *Research in Engineering Design*, Vol. 16, No. 3, Dec. 2005, pp. 118–132.

[53] Parkinson, A., Sorensen, C., and Pourhassan, N. A., "General Approach for Robust Optimal Design." *Journal of Mechanical Design*, Vol. 115, No. 1, 1993, pp. 74–80.

[54] Rangavajhala, S., Mullur, A., and Messac, A., "The Challenge of Equality Constraints in Robust Design Optimization: Examination and New Approach," *Structural and Multidisciplinary Optimization*, Vol. 34, No. 5, Nov. 2007, pp. 381–401.

[55] Halton, J. H., "A Retrospective and Prospective Survey of the Monte Carlo Method," *SIAM Review*, Vol. 12, No. 1, 1970, pp. 1–63.

[56] Helton, J. C., and Davis, F. J., "Latin Hypercube Sampling and the Propagation of Uncertainty in Analyses of Complex Systems," *Reliability Engineering System Safety*, Vol. 81, No. 1, 2003, pp. 23–69.

[57] Du, X., and Chen, W., "Efficient Uncertainty Analysis Methods for Multidisciplinary Robust Design," *AIAA Journal*, Vol. 40, No. 3, 2002, pp. 545–552.

[58] Jin, R., Du, X., and Chen, W., "The Use of Metamodeling Techniques for Optimization Under Uncertainty," *Structural and Multidisciplinary Optimization*, Vol. 25, No. 2, July 2003, pp. 99–116.

[59] Chen, W., Jin, R., and Sudjianto, A., "Analytical Uncertainty Propagation via Metamodels in Simulation-Based Design Under Uncertainty," AIAA Paper 2004-4356, Sept. 2004.

[60] Xiu, D., and Karniadakis, E. M., "Modeling Uncertainty in Flow Simulations via Generalized Polynomial Chaos," *Journal of Computational Physics*, Vol. 187, No. 1, May 2003, pp. 137–167.

[61] Youn, B. D., and Choi, K. K., "Selecting Probabilistic Approaches for Reliability-Based Design Optimization," *AIAA Journal*, Vol. 42, No. 1, 2004, pp. 124–131.

[62] Evans, D. H., "Statistical Tolerancing: The State of the Art, Part II," *Journal of Quality Technology*, Vol. 7, No. 1, 1975, pp. 1–12.

[63] Xu, H., and Rahman, S., "A Generalized Dimension-Reduction Method for Multidimensional Integration in Stochastic Mechanics," *International Journal for Numerical Methods in Engineering*, Vol. 60, No. 12, 2004, pp. 1992–2019.

[64] Padulo, M., Campobasso, M. S., and Guenov, M. D., "Comparative Analysis of Uncertainty Propagation Methods for Robust Engineering Design," *International Conference on Engineering Design ICED07*, 2007, http://www-legacy.aero.gla.ac.uk/Research/CFD/sergioc/PDF/iced_158_2007.pdf [retrieved July 2009].

Chapter 3

Helicopter Predesign Strategy: Design-to-Mass or Design-to-Cost?

Jos Stevens* and Jan-Floris Boer†

National Aerospace Laboratory (NLR), Amsterdam, The Netherlands

Cyrille Sevin‡

Eurocopter SaS, Marignane, France

and

Jos Vankan* and Wim Lammen§

National Aerospace Laboratory (NLR), Amsterdam, The Netherlands

Abbreviations

ATA = Air Transport Association
CAIV = cost as an independent variable
CER = cost-estimating relationship
DL = disk loading
DMC = direct maintenance cost
EMPRESS = energy method for power required estimates
FBW = fly by wire
FH = flight hour
GB = gear box
GSE = ground support equipment
HC = helicopter

*Senior Scientist.
†Senior R&D Manager.
‡Patent Engineer.
§R&D Engineer.

LCC = life-cycle cost
MCP = maximum continuous power
MDO = multidisciplinary design and optimization
MGB = main gear box
MMH = maintenance man-hour
MTOW = maximum takeoff weight
OEI = one engine inoperative
OGE = out-of-ground effect
RDTE = research, design, technology, and engineering
SAR = search and rescue
SFC = specific fuel consumption
SLL = service life limit
SPEAR = specification analysis of rotorcraft
TOP = takeoff power
VIVACE = value improvement through a virtual aeronautical collaborative enterprise

Nomenclature

b = number of rotor blades
c = rotor-blade chord
CdS = flat-plate drag area
C_T/σ = blade loading
M_{gross} = gross mass
R = rotor radius
T = main rotor thrust
V_{tip} = rotor tip speed
W = weight
ρ = air density
σ = rotor solidity

I. Introduction

TRADITIONALLY, a helicopter predesign is driven by flight and mission performance requirements. Other important requirements, such as costs, mass, and specific customer requirements, have not been treated in the same manner. Also a formalized decision process for the assessment of different design solutions by tradeoff analyses is often missing.

The need for cost-effective operations is becoming increasingly important. The new design goal would be to find the optimum helicopter design that not only fulfills the required performance requirements, but also satisfies the customer's requirements at the lowest possible cost. In contrast to fixed-wing operators, helicopter operators often use the same helicopter for a diversity of missions. The costs are influenced by the different mission characteristics (flight hours, flight profile, payload, etc.), but also by the maintenance policies applied, which in turn are affected by design choices (i.e., configuration, drive train architecture, chosen materials). To find an optimal compromise between the "driving" design parameters, a methodology that can find an optimal technical solution for the diversity in customer requirements is required. This requires the identification and evaluation of the cost impact of the driving parameters through the assessment of the sensitivity of the design to each of these parameters by means of tradeoff analyses. Such a methodology can also improve the efficiency of the helicopter design process by reducing the number of iterations during the subsequent design process.

A multidisciplinary design and optimization (MDO) case study has been defined, in which the following activities concerning the helicopter predesign have been performed: 1) evaluation of existing (predesign) methodologies/technologies and tools, 2) development and integration of a life-cycle cost (LCC) model in a predesign analysis tool, 3) identification of the cost driving parameters and performing the sensitivity analysis, and 4) development and implementation of a multidisciplinary design methodology to optimize the LCC.

The helicopter LCC model, which reflects the impact of both the major technical parameters and the major categories of customers and missions, has been developed by a major European helicopter manufacturer. A European aeronautical research institute has integrated the LCC model into an in-house developed rotorcraft specification analysis tool and has developed a helicopter sizing optimization methodology that enables a multimission design with LCC optimization.

The final objective was to include all enabling processes, models, and tools available for use in an aeronautical collaborative design environment, including the associated processes, models, and methods. This innovative environment will support the (pre-)design of a complete helicopter by providing optimal design information in an efficient way in the very early stage of the project.

II. Rotorcraft Analysis Tool

A proprietary predesign rotorcraft analysis tool SPEAR, or specification analysis of rotorcraft [1] has been developed. This computer program (Fig. 1) is able to estimate the (minimum) size and mass of a rotorcraft capable of fulfilling a specified set of operational requirements (flight performance requirements and mission tasks) for a given rotorcraft configuration. Valid solutions are those that comply with the flight performance requirements and for which the available fuel equals the required fuel to fulfill the most demanding mission task. The tool determines the rotorcraft gross mass, its main physical dimensions (like the rotor dimensions), the installed engine power, the fuel capacity, and the mass breakdown for the major vehicle components. The consequences of operational requirements on rotorcraft sizing can be analyzed, tradeoff studies can be performed, and the effects of technological developments on optimal rotorcraft mass and size can be assessed. The computer program uses the flight and mission performance calculation routines from the EMPRESS (energy method for power required estimates) code [2]. The tool contains a large amount of information on historical and current rotorcraft designs, such as major rotorcraft design relationships, major component characteristics, etc. Different kinds of graphical representations for the rotorcraft design results are included. The tool also includes the potential for LCC optimization or LCC tradeoff studies. The tool runs on Windows personal computers, thereby taking advantage of the Windows features.

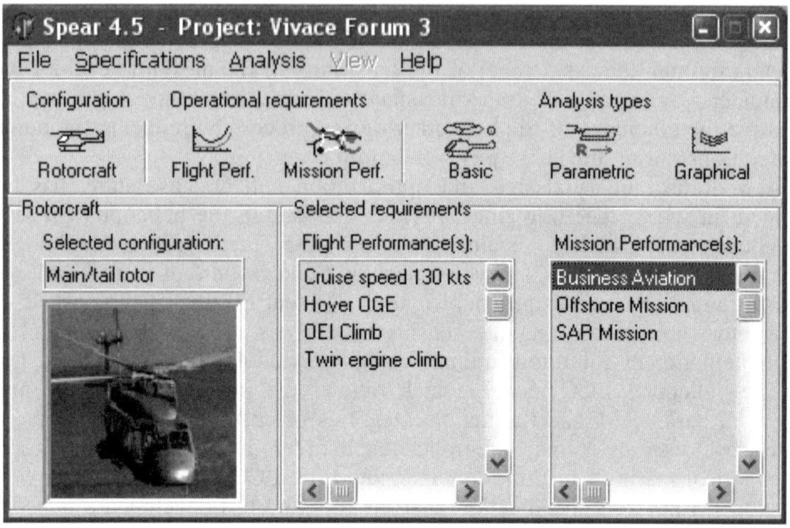

Fig. 1 Main window of the analysis tool.

III. Specification of Requirements

The rotorcraft designer has to specify a set of rotorcraft related requirements, which can be broken down into three parts: 1) rotorcraft configuration, describing the general layout plus some (aerodynamic) efficiency parameters; 2) flight performance, containing the data for the flight performance requirement(s) to be met; and 3) mission performance, containing the data for the mission profile(s) to be met. Each of the individual requirements is stored in the database, from which one or more can be selected for the analysis (Fig. 2).

A. Flight Performance Requirements

Each flight performance requirement is defined by an airspeed, ground-effect situation, atmospheric condition, number of engines operating, power setting, thrust and power margin, and a delta parasite drag area for any external equipment. Optionally a rotorcraft gross mass can be specified if the particular requirement has to be met at a specific gross mass.

B. Mission Performance Requirements

Each mission profile is specified by a number of mission segments, which are defined by a duration, airspeed, ground-effect situation, atmospheric condition, change of mass and/or drag as a result of (un)loading of payload, and engine power setting. The payload can be made up of persons (not the crew), cargo, weapons (in case of military use), specific mission equipment, or a mixture of these.

IV. Predesign Methodology

The methodology applied in the predesign analysis tool is largely based on [3]. The computer program will establish feasible rotorcraft dimensions that comply with the set of flight and mission performance requirements for the given rotorcraft configuration. Valid solutions are those that comply with the flight performance requirements and for which the available fuel equals the required fuel to fulfill the most demanding mission. The optimum solution is defined as the one that achieves these objectives at the lowest gross mass. (Traditionally helicopters are designed for lowest gross mass.) As suggested in [3], other criteria can be defined for the optimal solution, for example, the lowest LCC.

The actual sequence for the calculation of the various parameters is shown in Fig. 3. Comparing this predesign methodology to other MDO cases (e.g., aircraft wing MDO, see Chapters 5 and 6), the multidisciplinary calculations here are scheduled in sequential iterative loops. This combines the complete set of design calculations into one compact predesign tool. The calculation process runs efficiently by taking the main rotor disk loading as the driving variable. The rotor tip speed is fixed at a value compatible with the rotor technology state-of-the-art and, for example, with noise constraints. First the main rotor dimensions are determined, as these drive the other dimensions (like the rotorcraft length) and the flight performance. This is achieved by making an initial estimation for the gross mass M_{gross}.

Fig. 2 Available/selected flight and mission performance requirements.

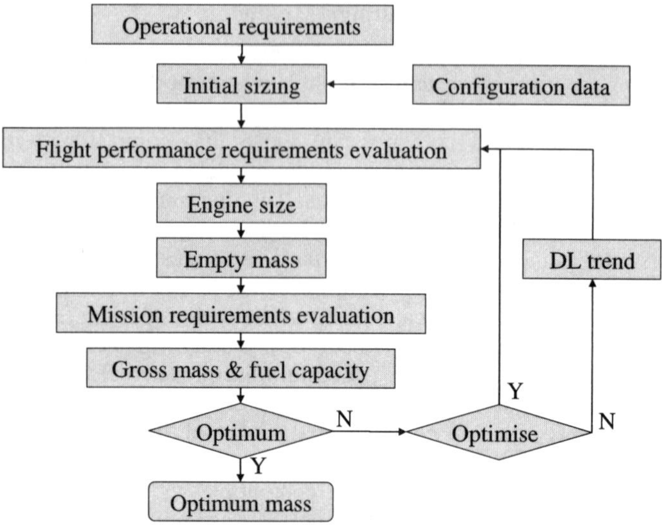

Fig. 3 Simplified flowchart for the calculation routine.

The rotor disk loading (DL) follows from historic data for disk loading vs gross mass:

$$\mathrm{DL} = 8.7188 * M_{\text{gross}}^{0.2264} - 23.685 \tag{1}$$

The main rotor radius R follows from the DL

$$R = \sqrt{\frac{M_{\text{gross}}}{\pi\,\mathrm{DL}}} \tag{2}$$

The maximum (limit) blade loading C_T/σ is a measure of the capability of the rotor blades to generate lift and depends on the rotor technology level. Maximum values for the main rotor-blade loading are specified in the input data. The main rotor thrust T is the total thrust required for the specific flight performance requirement and thus includes the gross mass and the download on the airframe. For each of the specified flight performance requirements, the rotor-blade solidity σ follows from the relationship between main rotor thrust, rotor radius, tip speed, and maximum blade loading:

$$\sigma = \frac{T}{\rho \pi R^2 V_{\text{tip}}^2 (C_T/\sigma)_{\max}} \tag{3}$$

The main rotor-blade chord c follows from the relationship between blade solidity, rotor radius, and number of rotor blades b:

$$c = \frac{\sigma \pi R}{b} \tag{4}$$

The highest value of the blade chord (for all performance requirement cases) is the valid one, as that value will give an acceptable blade loading value for all of those cases. Next the total power required is assessed. The most demanding flight performance requirement in terms of engine power defines the minimum engine power that is to be installed and thus defines the engine(s). At that point, an initial assumption for the fuel capacity is made, and the empty mass is assessed by estimating the masses of the major components, based principally on historic data. Next, the fuel required for actually fulfilling the various specified mission profiles is assessed. If the fuel mass needed to fulfill the most demanding mission appears to be different from the fuel mass available, the earlier assumptions for gross mass, fuel capacity, and disk loading are revised, and the calculation process is repeated. When the required and available fuel masses have been found to be equal, the process has converged to a valid design. Finally, the disk loading is varied with small steps, thereby no longer following the historic trend line. The calculation process is repeated in order to find the lowest gross mass at which the fuel criterion still holds, hence providing the optimum solution.

V. Analysis Features

The tool incorporates three analysis levels with an increasing amount of design options and associated capabilities.

A. Basic Analysis Level

The basic analysis (Fig. 4) determines the gross mass based on the selected rotorcraft configuration and operational requirements. For the limit blade loading C_T/σ in level flight (load factor equals one), several characteristic lines for various rotor designs are available in the database, and one of these designs has to be selected. For the engine there is a choice: either a hypothetical,

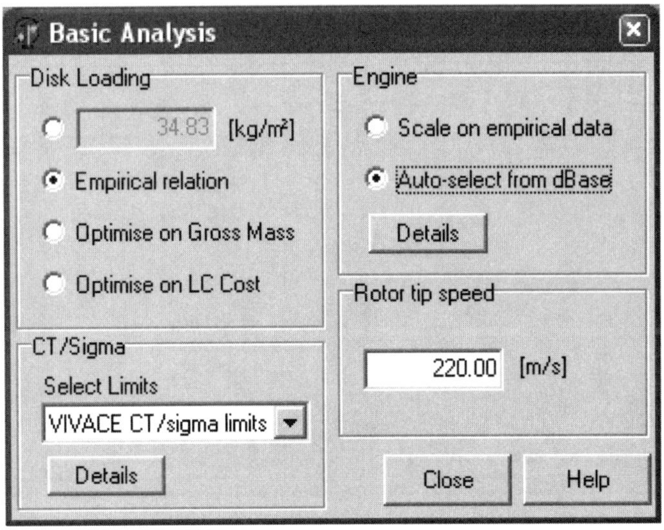

Fig. 4 Data input form for basic analysis.

fully compliant ("rubberized" on empirical data) engine or an existing engine can be automatically selected from the database. The value for the main rotor tip speed is defined by the designer. Historic data are used for fuselage parasite drag (assuming an "average" drag level), engine specific fuel consumption, and tail-rotor diameter.

B. Parametric Analysis Level

The parametric analysis (Fig. 5) provides more extensive options to further analyze the rotorcraft configuration. The effects of varying seven main rotor parameters can be analyzed: disk loading, blade loading, solidity, rotational speed, tip speed, diameter, and chord. Three of these parameters must be selected for the analysis. However, not every combination of three parameters is valid (e.g., disk loading and diameter cannot be selected at the same time). For some input data a selection can be made between a historic trend line value and a selected fixed value. This concerns the disk loading, the fuselage parasite drag area, the engine specific fuel consumption, the engine mass, and the tail-rotor diameter. In some cases the rotorcraft gross mass is limited to a maximum value, for example, because of the deck strength onboard a ship. For those cases the analysis process can be performed for a fixed gross mass (to be specified by the user), in which case the calculated rotorcraft gross mass is limited to the specified mass. It is possible, however, that the mission requirement(s) lead to a higher required mass than the one specified, implying that the design cannot fulfill all mission requirements.

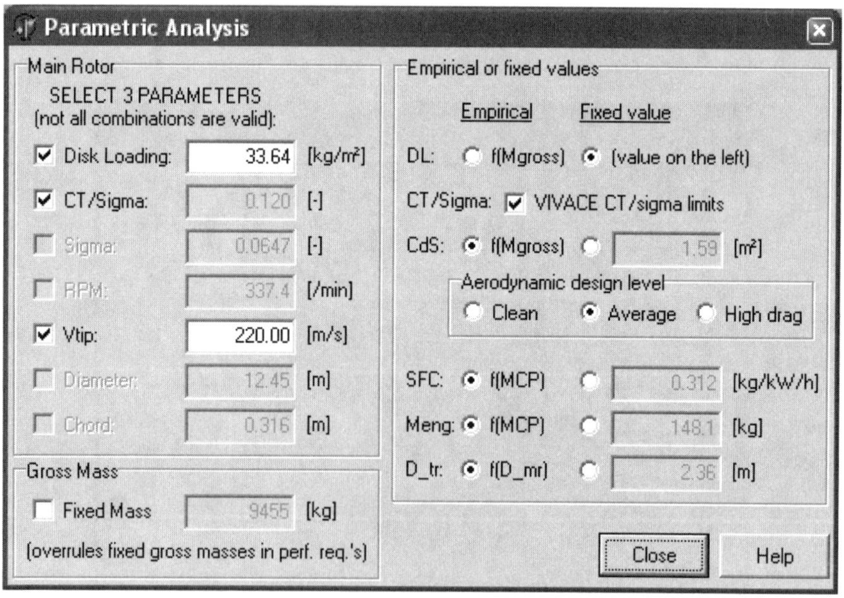

Fig. 5 Data input form for parametric analysis.

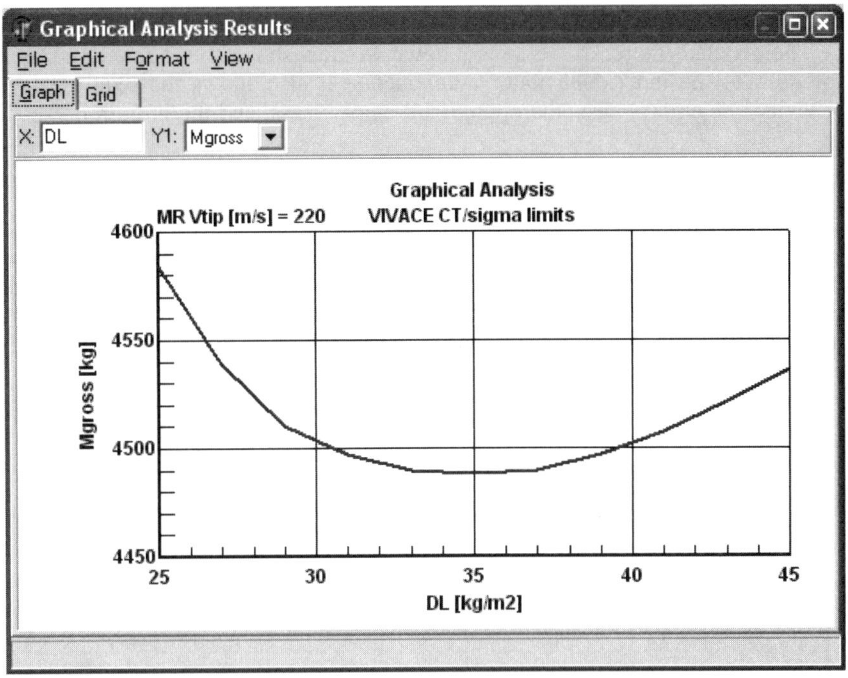

Fig. 6 Example output graph for graphical analysis.

C. Graphical Analysis Level

The graphical analysis presents the results in four types of graphs: 1) design chart (power required per kilogram of gross mass vs rotor disk loading); 2) parameter analysis chart (mass or power required vs one of the seven main rotor parameters); Fig. 6 presents the variation of rotorcraft gross mass vs the main rotor disk loading; 3) carpet plot (mass or power required vs two of the seven main rotor parameters); and 4) power curve (level flight power required vs airspeed for given values of gross mass, altitude, and temperature). Just like in the *basic* analysis level, historic data are used for fuselage parasite drag, specific fuel consumption, and tail-rotor diameter.

D. Analysis Results

The analysis results window (Fig. 7) provides an overview of the main results of the analysis. The seven main rotor design parameters, the figure of merit (for the isolated main rotor), the rotor mass, the tail-rotor dimensions, the engine data, the rotorcraft masses, the fuel capacity, and the parasite drag area are shown. In addition, it shows the names of the specific requirements (flight performance and mission profile) that have driven the main rotor design, the engine choice, and the fuel capacity. Additionally detailed breakdowns for the rotorcraft mass, the total LCC, and the mission result data can be shown.

E. Mass Breakdown

The detailed rotorcraft mass breakdown window shows the estimated masses for each of the individual major components. It also shows the required fuel capacity, the empty mass, the operational empty mass, and the gross mass.

F. Cost Breakdown

The input data for the detailed cost estimation process are specified in the cost input data window. A choice can be made whether the rotorcraft will be used (primarily) for civil or for military purposes. The rotorcraft acquisition cost ("the price") can be calculated either as a function of the gross mass or be based on the sum of RDTE (research, design, technology, and engineering) cost, production cost, and profit.

The calculated LCC (total operating costs for the number of acquired rotorcraft) is shown in the calculated cost results window (Fig. 8). The LCC is split

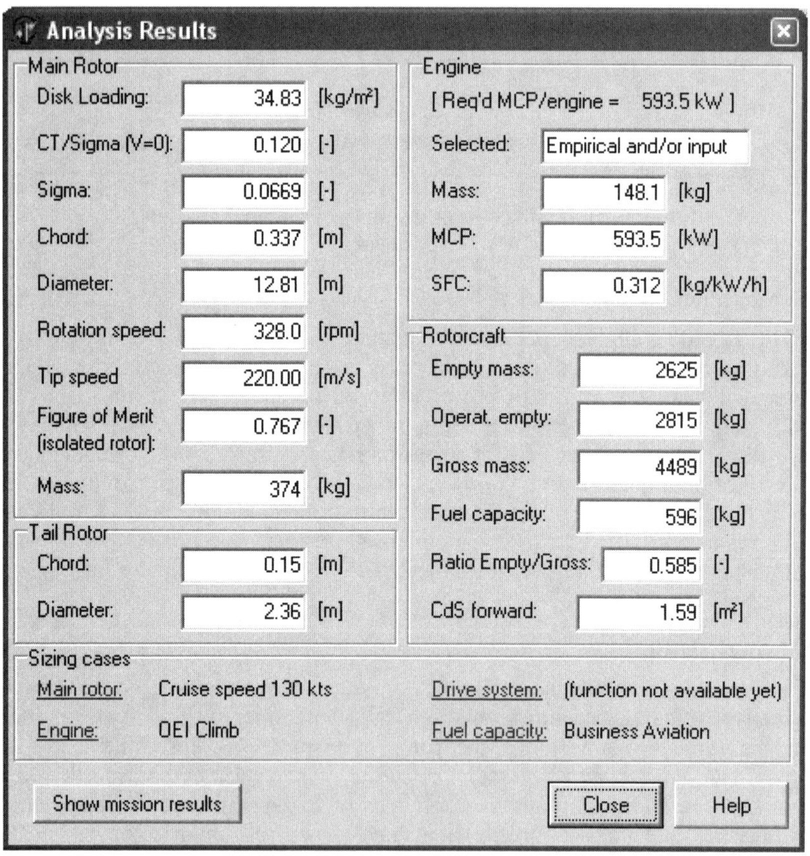

Fig. 7 Analysis results for VIVACE example requirements.

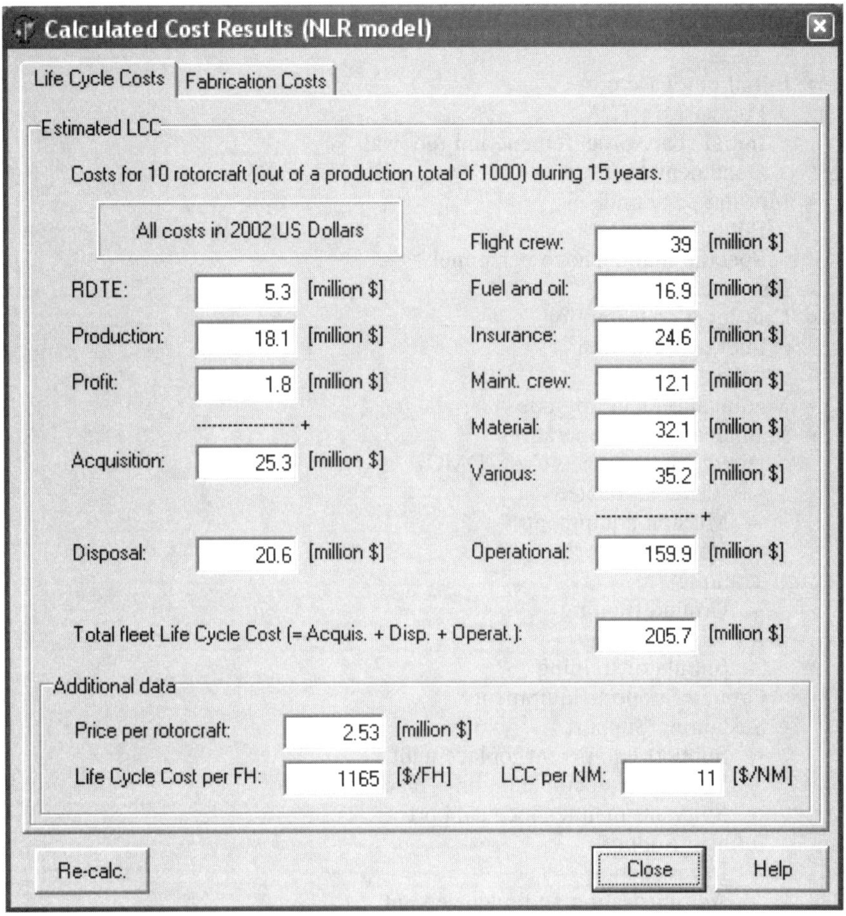

Fig. 8 Calculated cost results window.

into acquisition, disposal, operational, and fabrication costs (costs of making the individual major components). The estimated operating cost per flying hour and per nautical mile is also provided.

VI. Life-Cycle-Cost Model

A. Introduction

The cost model just mentioned is based on historical data and thus might not be valid for modern rotorcraft designs. Eurocopter has developed an LCC model, which is intended for use in the frame of predesign studies performed in house. The model, largely based on [4, 5], calculates global ownership costs for rotary-wing aircraft, either as costs per year, per helicopter, per flight hour, or per passenger. The detailed cost breakdown structure, as used in the LCC model, is shown here:

- Initial purchase cost
  - Purchase cost
  - Initial spare procurement and renewal
  - Documentation
- Mission personnel
  - Pilots salaries
  - Specific maintenance personnel
- Insurance
- Unit level consumption
  - Fuel consumption
  - Other
- Modification and upgrade
- Maintenance and operation
  - Direct Maintenance Cost (DMC)
    - Basic helicopter
    - Mission equipment
    - Consumables
  - Training
    - Ground training
    - Flight training
    - Simulator training
  - Ground Support Equipment
  - Sustaining Support
    - Support equipment replacement
    - Sustaining documentation replacement
    - Software maintenance support
  - Indirect Support
    - Infrastructure
    - Administration and management

In this model the decommissioning/disposal costs are not (yet) considered. No rules have been defined at the moment, and there is not enough return of experience to compute it sufficiently accurate in a general model. However, upon customer demand it can be included in a later phase.

B. Helicopter Breakdown

The major contributing parts in the cost analysis, the initial purchase cost (sale price) and the direct maintenance cost (DMC), are both determined by calculating their respective values for each of the major helicopter parts. The helicopter parts are chosen in accordance with the classical predesign ATA chapter breakdown (see Table 1).

C. Cost-Estimation Methods

The basic cost-estimation method for each item of the helicopter tree structure is made up of so-called cost-estimating relationships (CERs). CERs are

Table 1 Helicopter breakdown for LCC

Predesign breakdown	ATA chapter
Fuselage	52, 53, 55, 56
Landing gear	32
Main rotor blades	62-10
Main gear box (MGB)	63
Rotor hub	62-20
Tail rotor	64, 65
Electrical system	24
Avionics	22, 31, 34, 46
Flight controls	67
Hydraulics	29
Engine	71, 72, 76–77, 80
Fuel system	28
Furnishing and miscellaneous	21, 25, 26, 30, 33

mathematical expressions relating cost as the dependent variable to one or more independent cost drivers. These relations can be simple averages or percentages, or more complex equations that result from regression analyses and that connect the cost (the dependent variable) to the physical characteristics of the product (such as the mass, the output power, the percentage of a given material, etc.). Four cost-estimation methods can usually be discerned:

1) *Analogy:* comparing the system to a similar system with known cost and technical data.

2) *Parametric:* use of a database on similar elements to the item to be evaluated to generate a cost estimate based upon parameters representative of the performance characteristics of the item.

3) *Engineering:* bottom-up estimate from lowest subcomponents of a project (work breakdown structure).

4) *Extrapolation:* using information from the same system early in the project to estimate costs later in the project.

The parametric method has been selected here. This method can be used as soon as the technical specifications of the design are defined, when knowledge of cost, technical data, and hardware data is still limited. It consists of establishing a statistical correlation between the physical characteristics of the system (e.g., the weight, the volume, or the power) and its estimated cost.

D. Cost Drivers

For each major part of the helicopter, specific parameters—the so-called cost drivers—are used for the cost-estimation equation. These cost drivers are representative for the cost of that part of the helicopter. To permit the further design-to-cost optimization process using the rotorcraft predesign analysis tool, particular attention was paid to selecting these cost drivers as far as possible within the list of available parameters within that tool. The cost drivers are listed in Table 2. The columns "Cost driver 1 (purchase cost)" and "Cost driver 2 (DMC)" list the parameters most suitable for the calculation of purchase

Table 2 Technical data input for LCC

Predesign breakdown	Cost driver 1 (purchase cost)	Cost driver 2 (DMC)	Cost driver 3	Cost driver 4
Fuselage	Item mass, kg	—	% of composite	—
Landing gear	Item mass, kg	Skids or wheels	—	—
Main rotor blades	Item mass, kg	Item mass, kg	Technology factor (0 to 3)	—
MGB	TOP limit, kW	TOP limit, kW	Number of accessory power outputs	Number of reduction stages
Rotor hub	Centrifugal force, daN	Centrifugal force, daN	Number of blades	Rigid, Starflex©, or Spheriflex©
Tail rotor	Thrust max., daN	Thrust max., daN	Fenestron© or classic	—
Electrical system	Item mass, kg	Empty weight, kg	—	—
Avionics	Item mass, kg	Item mass, kg	—	—
Flight controls	Centrifugal force, daN	Centrifugal force, daN	Fly by wire or hydraulic	—
Hydraulics	MTOW, kg	Empty weight, kg	—	—
Engine	TOP limit, kW	TOP limit, kW	Reduction gear box or not	—
Fuel system	Number of tanks	Empty weight, kg	—	—

cost and DMC, respectively. The columns "Cost driver 3" and "Cost driver 4" represent other significant parameters, which are also used in both calculations, but have less influence.

E. Cost-Estimation Equations

Costs for all of the items from the predesign breakdown have been estimated using a mathematical method. The cost-estimation equations have been established and validated based upon the manufacturer's experience, using the cost drivers from Table 2. From experience, it is known that the cost of the hydraulic system, for sale price as well as for DMC, increases exponentially with the weight, whereby the rate of increase for each cost item is equal. The proposed laws are

$$\text{Sale Price} = a_1 W\_\text{MTOW}^{\alpha} \tag{5}$$

$$\text{DMC Price} = a_2 W\_\text{Empty}^{\alpha} \tag{6}$$

For electrical systems however, even though the increase is also exponential, the rate of increase for each cost item is different. The proposed laws are

$$\text{Sale Price} = b_1 W\_\text{Elec}^{\beta_1} \tag{7}$$

$$\text{DMC Price} = b_2 W\_\text{Empty}^{\beta_2} \tag{8}$$

The cost-estimation equations are obtained by extrapolation from a sample of former designs (known costs) to the new design and thus are based upon experience on existing designs. Therefore, the CERs cannot reflect the influence of new design factors on costs. A parametric model will be more reliable if the selected technical solutions for the new design are closer to those of the previous concepts.

F. Validation of the Cost Model

To validate the cost model, the results have been matched with a database containing data on the manufacturer's range of helicopter types. The mathematical relationships have been generated by data analysis (regression) and expert investigation. The model has been validated for both sale price (purchase cost) and DMC, with an achieved accuracy of 13% (see Table 3).

Table 3 Validation of the cost model over the manufacturer's product range

	EC120	AS350B3	EC130	AS365N3	EC155	AS332L1	AS332L2
Sale price (ref/calc)	1.03	1.13	1.06	1.02	0.93	1.00	1.02
DMC (ref/calc)	0.95	1.01	1.05	0.94	1.09	0.99	1.02

Fleet Data		
Number of H/C	5	[-]
Number of FH per year and per HC	600	[FH/year/HC]
Number of bases	1	[-]

Personnel and crew data		
Number of pilot per aircraft	2	[-]
MMH/FH	7.2	[h/FH]
Number of manager per base	2	[-]
Number of crew per aircraft	1.5	[-]
Annual cost for a Pilot	83333	[€/year]
Annual cost for a manager	62500	[€/year]
Annual cost for a technician	41667	[€/year]

Maintenance data		
Price for one GSE package	1166666	[€]
Documentation cost / Purchase price	1.5%	[%]
Spare parts rate / Purchase price	12%	[%]
Software support cost / Avionics Purchase price/year	3%	[%]
Ratio for GSE maintenance per year	2%	[%/year]
Ratio for Documentation upgrade per year	10%	[%/year]
Infrastructure cost per year per HC	14167	[€/year]
Major retrofit at mid life in % of the purchase price	10%	[%]
Upgrade per year in % of the purchase price	0.50%	[%]

Training data		
Ground training hours/year/base	200	h/year/base
Ground training cost per hour	125	€/h
Flight training hours per year per crew	10	h/year/crew
Flight training cost	6303	€/h/HC
Simulator training hours per year per crew	15	h/year/base
Hour cost of simulator	1667	€/h

General data		
Working hours / Day	8	[h/day]
Working days / Year	200	[days/year]
Insurance cost / Acquisition cost per year	7%	[%/Year]

Nb GSE package per HC	
Number of HC	Number of GSE packages
1	1
2	1.5
5	2
10	4
20	7
50	12

Fig. 9 Operational data input for LCC model.

Life Cycle Cost	Costs per HC in for the all life long	Costs per HC in /FH	Total Costs for the all fleet and for the all HC life long
Initial Purchase cost	17 300 000	1 115	86 500 000
Purchase cost (P. Cost)	15 000 000	1 000	75 000 000
Initial Spare procurement & renewal	2 000 000	100	10 000 000
Documentation	300 000	15	1 500 000
Mission personnel	11 000 000	650	55 000 000
Pilots salaries	7 500 000	400	37 500 000
Specific maintenance personnel	3500000	250	17 500 000
Insurance	40 000 000	2 222	200 000 000
Unit level consumption	2 141 603	119	10 708 017
Fuel consumption	2 087 603	116	10 438 017
Other	54 000	3	270 000
Maintenance and operation	38 075 000	2 115	188 750 000
Direct Maintenance Cost (DMC)	27 000 000	1 500	135 000 000
Basic HC	20 000 000	2 000	100 000 000
Mission equipment	6 000 000	400	30 000 000
Consummable	1 000 000	50	5 000 000
Training	6 150 000	500	23 750 000
Ground training	150 000	50	750 000
Flight training	4 000 000	300	20 000 000
Simulator training	2 000 000	100	3 000 000
GSE	400 000	100	2 000 000
Sustaining Support	3 500 000	100	19000000
Support Equipment Replacement	600 000	20	3 000 000
Substaining documentation replacement	900 000	50	6 000 000
Software Maintenance Support	2 000 000	100	10 000 000
Indirect Support	1 025 000	100	9 000 000
Infrastructure	425 000	30	6 000 000
Admistration & management	600 000	30	3 000 000
Modification & upgrade	6 000 000	200	30 000 000
LIFE CYCLE COST	114 516 603	6 421	570 958 017

Fig. 10 Data output for LCC model.

G. Mission Equipment

The mission system sheet in the LCC model contains various generic mission equipment items, which have an impact on the LCC of the helicopter through the following parameters: acquisition cost of the equipment; maintenance cost of the equipment; and life span [service life limit (SLL)] of the equipment. The selection of different mission equipment items is done by marking the relevant equipments in the mission system sheet.

H. Input and Output of the LCC Model

Figures 9 and 10 show a typical set of operational input data and output results.

VII. Life-Cycle-Cost Model in Predesign Analysis Tool

The LCC model already described has been integrated in the predesign rotorcraft analysis tool. The goal was to add the possibility to optimize the

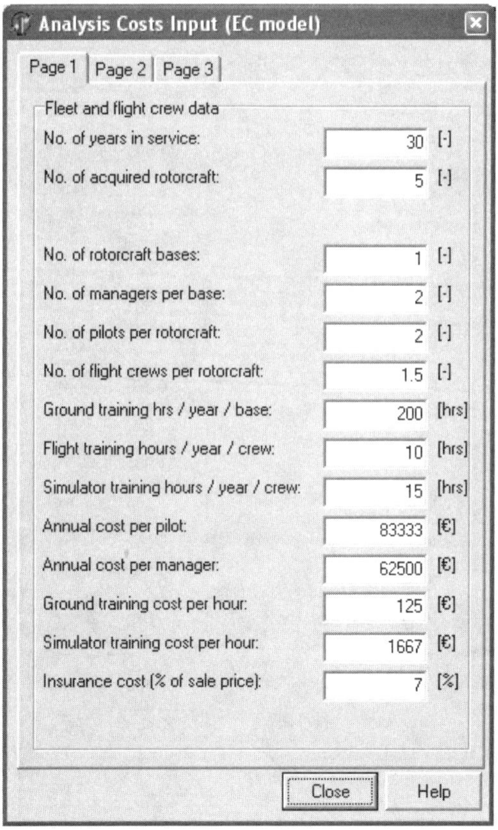

Fig. 11 Analysis costs input window.

design for minimum LCC. Dedicated windows have been added for the analysis costs input (Fig. 11) and the calculated cost results (Fig. 12). The calculated cost results data are shown on three data tab sheets:

1) The life-cycle-costs tab sheet shows the estimated total operating cost for the number of acquired rotorcraft during the stated period. The purchase cost is taken from the sale price tab sheet and the direct maintenance cost from the DMC tab sheet. Finally, the estimated operating cost per flying hour is provided.

2) The sale price tab sheet will show in detail the estimated costs of producing the individual major components. These add up to the sale price per rotorcraft.

3) The DMC tab sheet will show in detail the estimated direct maintenance costs per flight hour for the individual major components.

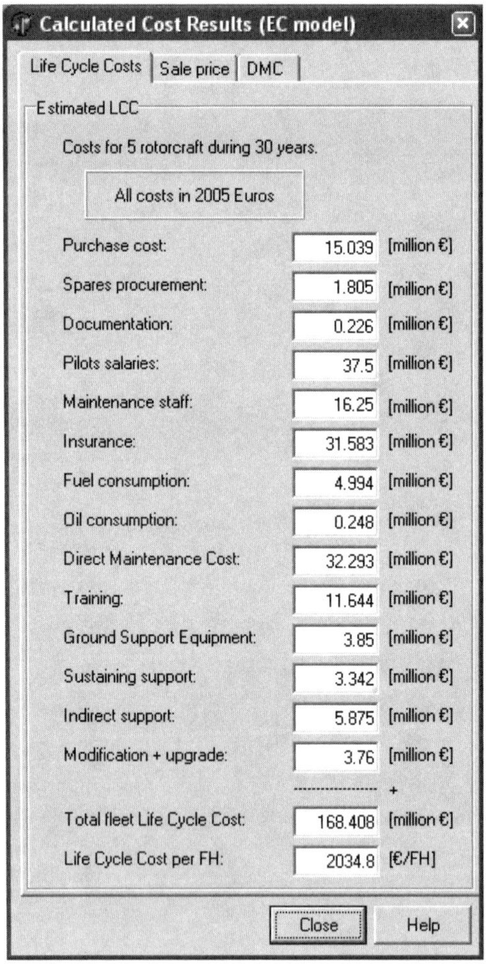

Fig. 12 Calculated cost results window.

The foregoing methodology allows the optimization of the design, either for minimum gross mass or for minimum LCC, but not in an automated way. Many design choices have to be made, such as the percentage of composites material in the structure, the complexity of the rotor system, and the number of fuel tanks. All of these will affect the results of the design process and might mutually affect each other as well. An optimization methodology has been developed to ease the (pre-)design process.

VIII. Optimization Methodology

A helicopter design optimization environment has been created by putting the combined analysis/LCC tool in an interactive MATLAB environment (see www.mathworks.com), which was achieved by compiling the model into a Windows dynamic link library (.dll) file. The functions in the .dll file are called with the appropriate design parameters as arguments and the values of the design objectives, that is, the helicopter mass and the helicopter LCC are returned. The functions and toolboxes in the environment, such as gradient-based algorithms [6], genetic algorithms [7, 8], and pattern search [7], can then be used for the evaluation and optimization of these helicopter design objectives.

The optimization of the helicopter design can be characterized as a mixed-integer programming problem, either single or multi-objective. Chapter 4 of this book gives a detailed description of optimization algorithms using continuous variables in general. The present chapter provides an extension of that approach by focusing on optimization using a combination of discrete and continuous variables. A specialized optimization algorithm ("fminconset") was applied, which combines a discrete branch-and-bound method [9] with the general-purpose nonlinear constrained optimization algorithm fmincon from the optimization toolbox [6]. The optimization algorithm treats the complete calculation of the helicopter design objective functions (gross mass and LCC) as a blackbox. Therefore, it is currently unavoidable for the algorithm to perform exhaustive evaluations, that is, to explore the whole discrete part of the design space. For the current tool and studies this still results in acceptable execution times in the order of a few hours on standard PCs. However, the concept of the branch-and-bound methods provides insight in the optimization problem and might contribute to future versions of the methodology.

The branch-and-bound algorithm (for single-objective optimization) is explained next, followed by the general concept of multi-objective optimization.

A. Branch-and-Bound Method

Branch-and-bound (BB) is a method applied in the area of combinatorial optimization, to find solutions of various types of integer optimization problems, for example, the traveling salesman problem [10] or other scheduling and assignment problems. The method is based on the idea that if one can predict that a specific branch of possible solutions has a better score than another branch, then the other branch can be excluded from search, thus saving computation time.

The optimization problem is posed here as finding the minimal value of a function $f(x)$ (e.g., the helicopter LCC), where x belongs to a set S of possible and acceptable designs (the design space). A branch-and-bound procedure consists of two steps being applied in a recursive way:

1) First, the *branching* step is performed. The set S of possible designs is divided into two or more smaller subsets $S_1, S_2, \ldots$, which cover S. Note that the minimum of $f(x)$ over S is equal to the minimum of the minima of $f(x)$ over each subset S_i. This way a tree structure is defined whose nodes represent the subsets of S (see Fig. 13).

2) Second, the *bounding* step is performed. Upper and lower bounds for the minimum value of $f(x)$ within a given subset S_i are computed.

If the lower bound of $f(x)$ over some subset of candidate designs S_c is greater than the upper bound of $f(x)$ over any other subset S_j, S_c can be safely discarded from the search. This can be implemented by maintaining a global variable m that records the minimum upper bound of all subsets examined so far. Any subset whose lower bound is greater than m can be discarded. Subsequently, the remaining subsets S_j are divided further into subsets $S_{j1}, S_{j2}, \ldots$ (see Fig. 13) to detect

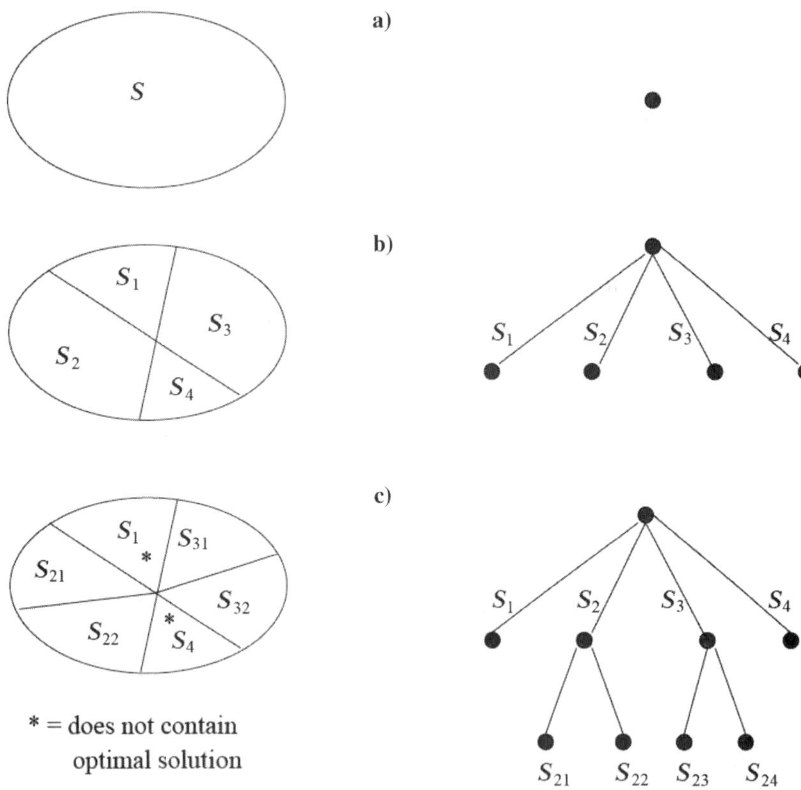

Fig. 13 Illustration of the tree structure created from a design set S with the branch-and-bound method. The set S of possible solutions is partitioned. The subsets S_1 and S_4 are not further subdivided in the tree structure because a bound function indicates that they do not contain the optimal solution (branch-and-bound method from [10]).

again possible subsets to be excluded from the search by means of the bounding technique. This process is repeated recursively until the set S of candidate solutions is reduced to a single element, or when the upper bound for set S matches the lower bound. Either way, any remaining element of S will be a minimum of the function $f(x)$.

The efficiency of the branch-and-bound method strongly depends on the effectiveness of the bounding algorithm used. For each particular problem a specific bounding technique should be designed. For this, detailed information about the behavior of the objective function (which is to be optimized) is necessary. The structure of our combined analysis and LCC tool allows for application of the branch-and-bound method in future cases, for example, if more design variables are to be explored.

B. Multi-Objective Optimization

A detailed description of the multi-objective optimization problem and the corresponding optimization search algorithms can be found in Chapter 6 of this book. The definitions and optimization approach are summarized here, using the same notation as in Chapter 6.

Multi-objective optimization, which can be considered as a generalization of single-objective optimization, deals with vector-valued objective functions $\boldsymbol{y} = \boldsymbol{f}(\boldsymbol{x})$, for example, both helicopter mass and LCC are minimized simultaneously. The definition of optimality in this case is nontrivial because of the indefiniteness of the relation among the multiple objective functions. To resolve such kind of optimization problems, consider the following definition, which is based on the concept of Pareto optimality [11]. According to this concept, an objective vector $\boldsymbol{y}^1$ is said to *dominate* any other objective vector $\boldsymbol{y}^2$ ($\boldsymbol{y}^1 \prec \boldsymbol{y}^2$) if the following two conditions hold: no component of $\boldsymbol{y}^1$ is greater than the corresponding component of $\boldsymbol{y}^2$; and at least one component of $\boldsymbol{y}^1$ is smaller than $\boldsymbol{y}^2$. Accordingly, it can be stated that a solution $\boldsymbol{x}^1$ is *better* than another solution $\boldsymbol{x}^2$, that is, $\boldsymbol{x}^1$ *dominates* $\boldsymbol{x}^2$ ($\boldsymbol{x}^1 \prec \boldsymbol{x}^2$), if $\boldsymbol{y}^1 = \boldsymbol{f}(\boldsymbol{x}^1)$ dominates $\boldsymbol{y}^2 = \boldsymbol{f}(\boldsymbol{x}^2)$. For example, let decision vector $\boldsymbol{x}^1 = -1$ and $\boldsymbol{x}^2 = 1$, and their corresponding objective vectors are $\boldsymbol{y}^1 = (1, 1)$ and $\boldsymbol{y}^2 = (9, 1)$, that is, in that case $\boldsymbol{y}^1$ dominates $\boldsymbol{y}^2$. Additionally, a solution vector $\boldsymbol{x}^u \in \boldsymbol{X}$ is said to be *Pareto optimal* if there exists no $\boldsymbol{x}^v \in \boldsymbol{X}$ for which $f(\boldsymbol{x}^v)$ dominates $f(\boldsymbol{x}^u)$. The set of (Pareto) optimal solutions in the decision space $\boldsymbol{X}$ is in general denoted as the *Pareto optimal set* $\boldsymbol{X}^* \subseteq \boldsymbol{X}$, and we denote its image in objective space as *Pareto front* $\boldsymbol{Y}^* = f(\boldsymbol{X}^*) \subseteq \boldsymbol{Y}$. One can derive the Pareto front of an objective space with calculated objective function results: $\{y^i = f(x^i) | x^i \in X\}$ by so-called *nondominated* sorting. The decision space $\boldsymbol{X}$ is divided in to classes of decision vectors that do not dominate each other. Then each class is given a *Pareto rank.* The decision vectors that are not dominated by any other decision vector get rank 1. Subsequently, the decision vectors that are dominated only by the decision vectors with rank 1 get rank 2, etc. The rank maps the multidimensional output vectors in the objective space to one single dimension. The optimization search algorithms that calculate the Pareto optimal set are concerned with the shape of this rank mapping. This concept fits in with the generalization of the single-objective problem.

As an example (see also Chapter 4), consider the following being the simple multi-objective optimization problem:

$$\min_{x} y_1 = \sin(x), \quad y_2 = \cos(x); \quad x \in [0, 2\pi]$$

The solution of this problem consists of the Pareto front for y_1, y_2 as indicated by the thick part of the lines in Fig. 14 ($x \in [\pi, 3\pi/2]$). There are many different algorithms to find the *Pareto optimal set*. Examples of methods, with the focus on so-called evolutionary algorithms, are described in Chapter 6 of this book.

C. Formulation of the Optimization Problem

The multi-objective optimization problem of helicopter design with respect to gross mass and LCC can be formulated as follows: minimize both the helicopter gross mass and total life-cycle cost as a function of the following 13 design parameters, which have been chosen from the parameters listed in Table 2 (possible values are indicated between parentheses): 1) percentage of composite material (mass) in the fuselage (0–100%); 2) complexity of the main rotor blades (1 = low complexity or metal blade, 2 = moderate complexity or hybrid blade, 3 = high complexity or full composite blade); 3) type of main rotor hub (0 = rigid, 1 = Starflex, 2 = Spheriflex); 4) type of flight control system (0 = mechanical, 1 = fly by wire, FBW); 5) type of tail rotor (0 = conventional,

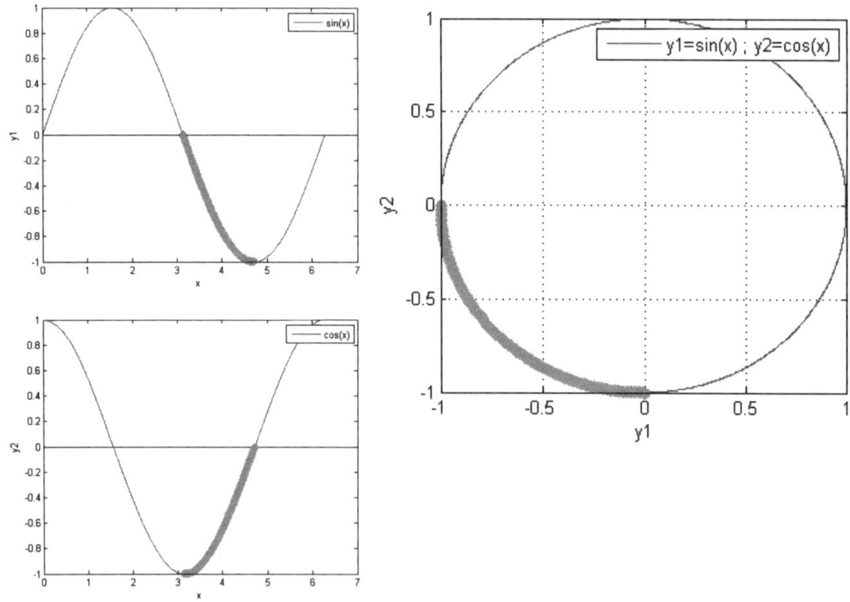

Fig. 14 Illustration of the solution of a general multi-objective optimization problem.

1 = Fenestron); 6) number of accessory gearboxes (1–5); 7) number of reduction steps in the main gearbox (2–5); 8) number of fuel tanks (1–5); 9) presence of an engine reduction gearbox (no, yes); 10) presence of a critical environment for avionics (no, yes); 11) fleet-wide number of business flights per year (0– ...); 12) fleet-wide number of offshore flights per year (0– ...); and 13) fleet-wide number of search/rescue flights per year (0– ...). Note that the first parameter is continuous, whereas all other parameters have discrete values. Parameters 11–13 express the numbers of missions that will be flown by the operator, expressed as total number of flights per year flown by its fleet.

D. Reference Design

For the illustration purposes of the MDO case study, a selection of certain parameter values has been made to limit the total number of potential combinations, that is, the search space. On the basis of engineering judgment and state-of-the-art technology, the following values are fixed (except the engine reduction gearbox) for the optimization calculations: 5) conventional tail rotor (lower mass and cost than Fenestron tail rotor); 6) one accessory gearbox; 7) two main gearbox reduction steps; 8) one fuel tank; 9) with and without engine reduction gearbox (its effect on mass and cost is not clear beforehand, as the reduction step is moved from the engine to the main gear box or v.v.); and 10) no critical environment for avionics (lower mass and cost).

For the mission combination a possible division of flights per year has been chosen (the effect of changing this division will be shown at a later stage): 11) 350 business flights per year, 12) 500 offshore flights per year, and 13) 150 search/rescue flights per year. The results presented here are based on a study with a reduced set of input parameters to illustrate the capabilities of the methodology only.

A reference helicopter design has been determined based on the aforementioned parameter choices, complemented with full metal construction, low-complexity rotor blades, Starflex rotor hub, and mechanical flight control system. The reference helicopter will have a calculated mass of about 3870 kg and total LCC of about 180 million Euros with engine reduction gearbox, or 3860 kg and 170 million Euros without engine reduction gearbox. As the removal of the engine reduction gearbox has a beneficial effect on mass (minor) and costs (major), it will no longer be used in the optimization strategy.

E. Optimization Evaluations

As described before, a mixed-integer programming algorithm can be applied to the design optimization problem as a whole. However, because it is expected that this algorithm will perform an exhaustive search, first a global evaluation of the effects of four design variables (the first four parameters just given) on the design objectives is performed, to gain insight in the design space. Because the LCC objective function is nonlinear, the global exploration of the design space also allows the regarding of LCC as an independent variable [cost as an independent variable (CAIV [12])], which is becoming more frequently used in military design cases and which might be desirable in future cases.

The first parameter (percentage of composite mass in the fuselage) is evaluated at 11 discrete values {0, 10, . . . , 100%}, and for the parameters 2 to 4 all possible values are evaluated. The resulting 198 evaluations of helicopter mass and LCC are given in Fig. 15.

From these results it is obvious that, to obtain a design that has minimum mass, a Starflex type main rotor hub must be used in combination with a high-complexity rotor blade and a FBW flight control system. However, for minimum LCC a mechanical flight control system should be selected. Also, to achieve minimum helicopter mass a high-percentage composite mass in the fuselage must be used, whereas the lowest LCC is achieved for a lower-percentage composite mass in the fuselage.

It is therefore decided that more detailed analyses are needed to find the best value for the percentage composite mass. Hence, separate optimizations are performed for the helicopter mass and LCC as a function of the percentage composite mass and the type of flight control system. In both of these minimizations, the optimal area, already indicated by the global evaluations, is zoomed in. The Starflex-type main rotor hub and a high blade complexity (i.e., full composite blades) are used. The mixed-integer programming algorithm fminconset, as mentioned before, was used for the minimization over the percentage composite mass and the type of flight control system. The results of these optimizations for helicopter mass and LCC are given in Figs. 16 (circle) and 17 (square).

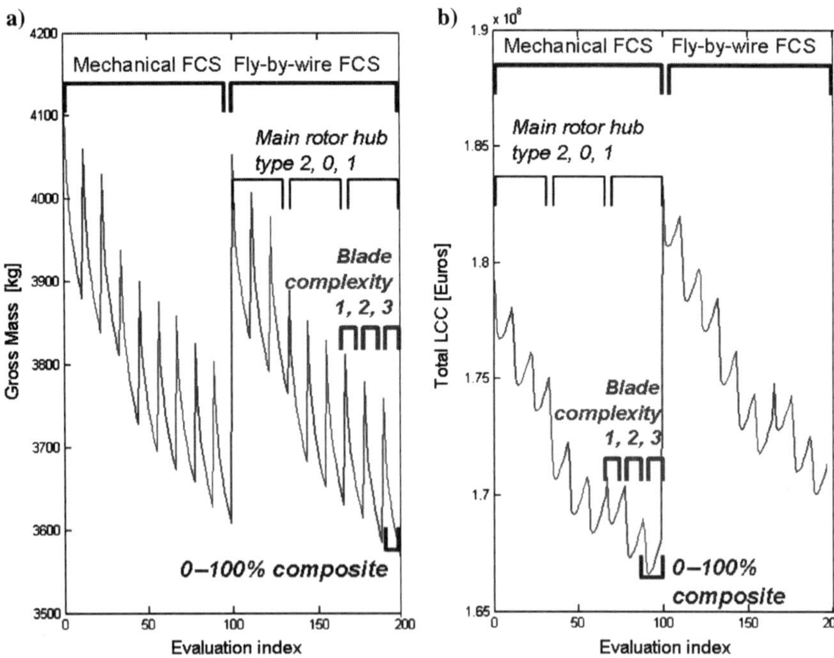

Fig. 15 Global evaluation of a) helicopter mass and b) LCC: their dependency on the four different design variables.

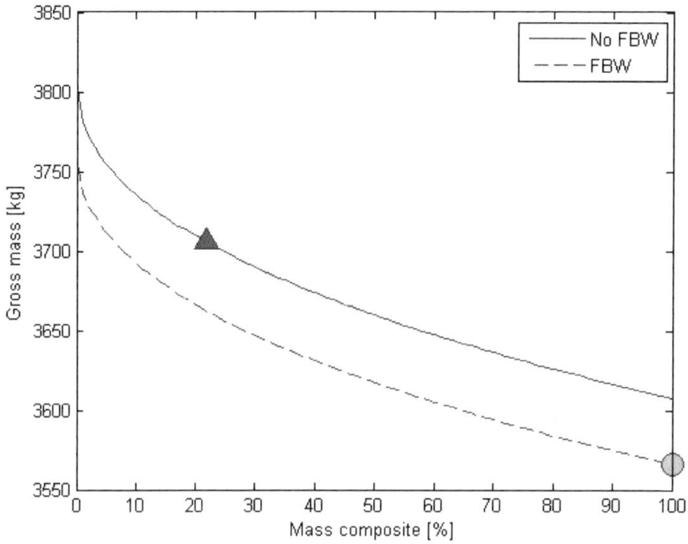

Fig. 16 Minimum helicopter mass (circle) found for 100% composite mass in the fuselage and a fly-by-wire flight control system.

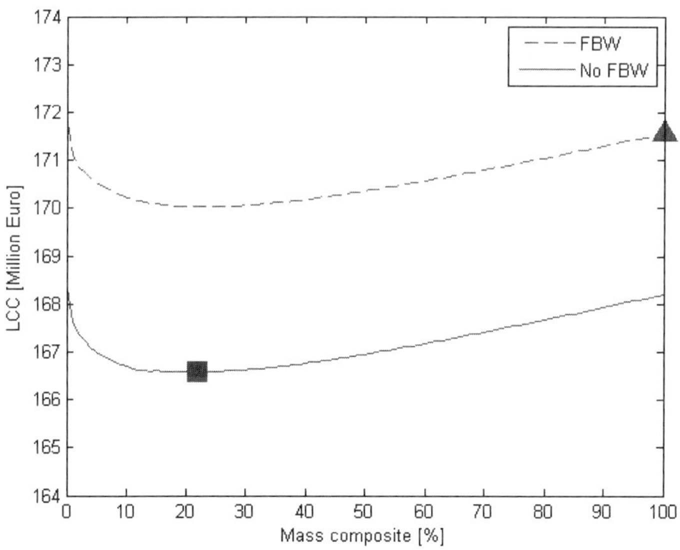

Fig. 17 Minimum helicopter LCC (square) found for 22% composite mass in the fuselage and a mechanical flight control system.

The triangle in both figures indicates the optimum design for the other objective. The triangles show that the design that is optimized for mass has a corresponding LCC value of about 172 million Euros, which is higher than the minimal LCC value of about 167 million Euros. At the same time the design that is optimized for LCC has a corresponding mass value of about 3705 kg, which is higher than the minimal mass value of about 3565 kg. Hence, these single-objective optimum design points provide poor values for the other design objective that is not optimized.

To efficiently take into account more than one design objective in the helicopter design optimization study, a multi-objective optimization approach can be used. Efficient algorithms for solving such multi-objective optimization problems are available (see Chapter 6 of this book or [8]). This approach is used for the helicopter mass and LCC objectives. A tradeoff between mass and LCC can be performed by plotting these objectives directly against each other. A switch is performed from the design space to the objective space. The key in this approach is that the compromise solutions for the best values for both objectives are pursued. Such optimization problem can be formulated as a Pareto optimal [11] design problem (see preceding section), having a set design points as the optimal solution, the Pareto optimal set (or Pareto front in the objective space). This Pareto optimal set is shown in the Fig. 18.

The result was found for the optimization of mass and LCC as a function of the percentage composite mass in the fuselage and the type of flight control system, just like the preceding single-objective optimizations. Also here, the Starflex-type main rotor hub and a high blade complexity (i.e., full composite blades) were used. Obviously, from this Pareto optimal set the optimum design points for mass or LCC can be easily selected. Also the tradeoff between mass

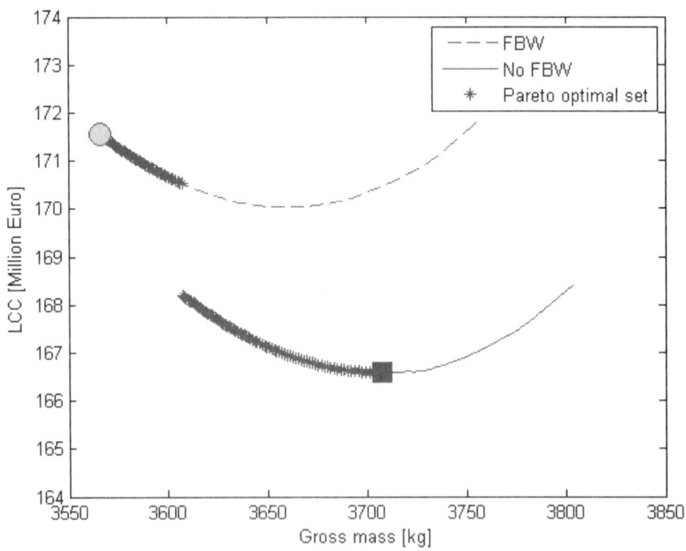

Fig. 18 Results of helicopter mass and LCC multi-objective optimization problem.

and LCC can be directly made. The results shown in the figure are given in the objective space, that is, the resulting LCC values plotted vs the mass values. The Pareto optimal set of helicopter designs is represented by the star symbols. In the figure also the single-objective optimum design points for mass (circle) and LCC (square) are indicated.

F. Comparison with Reference Helicopter Design

Figure 19 shows the reference helicopter design (star) together with some results from the preceding optimization strategy. In comparison to the reference design, the introduction of high-complexity blades and a FBW flight control system does significantly reduce the helicopter mass, but has almost no effect on the total LCC as a result of the higher acquisition cost being balanced by the lower maintenance effort (moving left in the graph).

As shown before, the removal of the engine reduction gear box (GB) has a small effect on the helicopter mass, but significantly reduces the total LCC as a result of a lower maintenance effort (moving down in the graph). A further reduction in total LCC can be achieved by replacing the FBW flight control system by a mechanical one, but then the helicopter mass will slightly increase again (moving to the bottom line in the graph).

From the different design points in the graph, it becomes clear that a helicopter design can be either optimized for lowest mass or for lowest total LCC; however, these designs will have a different configuration with respect to the systems used.

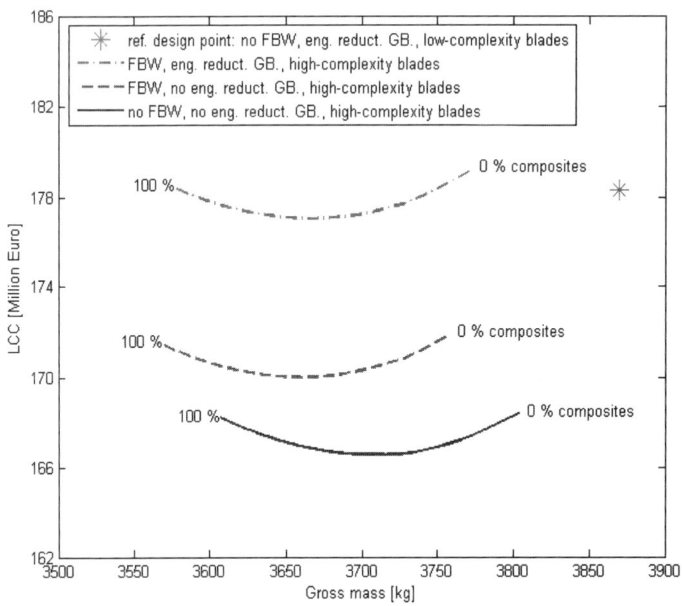

Fig. 19 Combined results of reference helicopter design and optimization results.

G. Design for Multiple Mission Combinations

In the preceding part the optimization process has concentrated on optimization of the combined mass and LCC design objectives. This has been done for a single helicopter operator with one specific mission combination (defined as 350 business flights, 500 off-shore flights, and 150 search/rescue flights per year), illustrating how this method can help operators gain insight into the consequences of their requirements. A helicopter manufacturer however is interested in multiple operators having multiple mission combinations. Therefore a next step in the optimization process is to optimize the LCC for these multiple mission combinations. This results in different LCC values for the helicopter design that is being used for different mission combinations during its life cycle.

As an illustration, a multi-objective optimization of LCC has been performed for two different mission combinations during the life cycle: combination 1 represents the LCC if 350 business, 500 off-shore, and 150 search/rescue flights per year would be flown during the life cycle, and combination 2 represents the LCC if 2000 business, 0 off-shore, and 0 search/rescue flights per year would be flown during the life cycle. The helicopter design is then optimized for both of these two mission combinations.

Figure 20 shows the optimum design point for combination 1 that was found in the previous mass-LCC optimization (square; helicopter design with 22% composite mass).

Additionally, the optimum design point for combination 2 is found (diamond; helicopter design with 45% composite mass). The line in the figure connects a series of design points, the so-called Pareto optimal set, which represent a compromise optimal helicopter designs for both combination 1 and combination 2. These design points are found for helicopter designs with the percentage composite mass increasing from 22 to 45%.

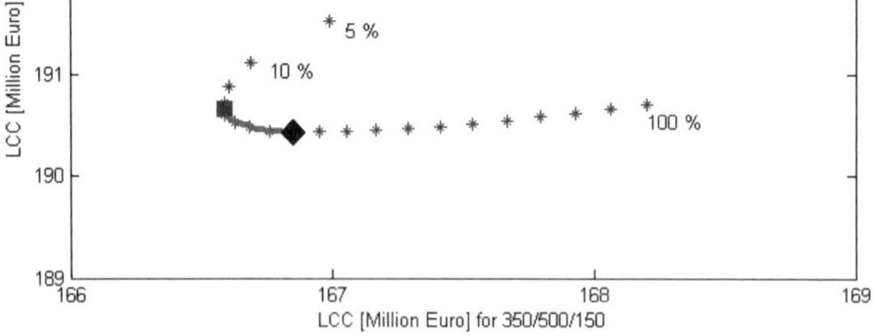

Fig. 20 Results of helicopter multi-objective optimization problem for mission combination 1 (horizontal axis) and combination 2 (vertical axis).

IX. Concluding Remarks

The helicopter predesign is normally driven by performance requirements, and traditionally the helicopter mass is considered the design optimization criterion. However, the need for cost-effective operations urges the manufacturers to design helicopters that reach the performance requirements, not only at a low mass, but (also) at the lowest possible operating costs. Therefore a LCC model that reflects the impact of both the major technical parameters and the major categories of customers and missions is needed.

In the case study the LCC model has been integrated into a rotorcraft predesign analysis tool. A helicopter design optimization environment for the evaluation and optimization of the helicopter design objectives has been created in an interactive environment. The optimization methodology applied in this study is based on the formulation of a generic optimization problem that allows for, among others, single- or multi-objective optimization problems, nonlinear constraints, and discrete variables.

The results of the optimization strategy have been compared with a reference helicopter design. From the resulting different design points, it becomes clear that a helicopter design can be either optimized for lowest mass or for lowest total LCC, resulting in different design choices. The optimization strategy gives a clear insight in what design choices contribute to a reduction in mass and/or a reduction in LCC. A tradeoff analysis can be performed using a Pareto optimal set of designs. The LCC calculations in the objective function are partly based on estimations of the future and therefore have a statistical uncertainty margin in absolute sense. However, it is expected that the relative uncertainty of these calculations is much smaller and that the same optimization strategy is applicable if new LCC estimates are performed in the future.

Because helicopter manufacturers are interested in multiple operators each having multiple mission combinations, an additional optimization study has been performed to optimize the LCC for these multiple mission combinations. This resulted in different LCC values for the helicopter design that is being used for different mission combinations during its life cycle. The calculation results show the Pareto optimal set of design points, which represents the set of compromise optimal helicopter designs. The optimal design point depends on the actual combination of the defined missions.

The discussed method and supporting tool can likewise be used by helicopter operators to select the optimal helicopter configuration for their combination of missions. It also allows the operators to assess the consequences of individual mission requirements in terms of the resulting helicopter configuration and total LCC. The Pareto approach and global exploration of the design space also allow the regarding of LCC as an independent variable, which is becoming more frequently used in military design cases and which might be desirable in future cases.

The resulting predesign strategy contributes to a reduced number of iteration loops in the preliminary design process and applying the Pareto technique summarizes the set of compromise optimal helicopter designs in an effective way, leading to a less time-consuming preliminary design phase; reduced development costs of future helicopter designs through the ability to better predict the LCC of the helicopter; reduced operational cost for the operators/owners of helicopters;

support to helicopter marketing by providing the LCC relationship for multimission combinations; and a better insight into the consequences of design choices.

Further research is necessary to improve and validate the models and enable useful optimization strategies for the development of cost-efficient multirole helicopters for multiple-operator-defined combinations of missions.

References

[1] Boer, J. F., and Stevens, J. M. G. F., "SPEAR: A Rotorcraft Specification Analysis Program," *Reference Guide and User Manual*, National Aerospace Laboratory (NLR), NLR-TR-2002-503, Amsterdam, The Netherlands, 2002.

[2] Stevens, J. M. G. F., "EMPRESS: A Rotorcraft Performance Calculation Program," *Reference Guide and User Manual*, National Aerospace Laboratory (NLR), NLR-TR-2001-312, Amsterdam, The Netherlands, 2001.

[3] "Engineering Design Handbook: Helicopter Engineering, Part One, Preliminary Design," U.S. Army Material Command, AMCP-706-201, 1974.

[4] Dhillon, B. S., *Life Cycle Costing, Techniques, Models and Applications*, Gordon and Breach, New York, 1989, pp. 46–81.

[5] Glade, M., *Life Cycle Cost Modelling: Maintenance Cost and Reliability Forecast and Their Application to Aeronautics*, Presse Univ. de Lyon, Lyon, France, 2005.

[6] MATLAB Optimization Toolbox, MATLAB, Natick, MA, http://www.mathworks.com/products/optimization [retrieved July 2007].

[7] MATLAB Genetic Algorithm and Direct Search Toolbox, MATLAB, Natick, MA, http://www.mathworks.com/products/gads [retrieved July 2007].

[8] Deb, K., Pratap, A., Agarwal, S., and Meyarivan, T., "A Fast and Elitist Multi-Objective Genetic Algorithm: NSGA-II," *IEEE Transaction on Evolutionary Computation*, Vol. 6, No. 2, 2002, pp. 181–197.

[9] Land, A. H., and Doig, A. G., "An Automatic Method of Solving Discrete Programming Problems," *Econometrica*, Vol. 28, No. 3, July 1960, pp. 497–520.

[10] Clausen, J., "Branch and Bound Algorithms – Principles and Examples," Dept. of Computer Science, Univ. of Copenhagen, Denmark, March 1999.

[11] Goldberg, D. E., *Genetic Algorithms in Search, Optimization and Machine Learning*, Addison Wesley Longman, Reading, MA, 1989, pp. 197–198.

[12] Rush, B. C., "Cost as an Independent Variable: Concepts and Risks," *Acquisition Review Quarterly*, Spring 1997, pp. 161–172.

Chapter 4

Aeroelasticity and Sizing in a Multilevel Modeling Approach for Preliminary Design

Wolf R. Krüger*
German Aerospace Center, Göttingen, Germany
Anja Wetzel†
German Aerospace Center, Braunschweig, Germany
Martin Spieck‡
German Aerospace Center, Göttingen, Germany
and
Jens Baaran†
German Aerospace Center, Braunschweig, Germany

Abbreviations

AME = aircraft model engine
BAM = basic aircraft model
BEA = beam element analysis
CFD = computational fluid dynamics
DI = disciplinary interface
DM = disciplinary model
DME = disciplinary model engine
FEA = finite element analysis
HALE = high-altitude long endurance

*Research Engineer, Institute of Aeroelasticity, Department Head.
†Research Engineer, Institute of Composite Structures and Adaptive Systems.
‡Research Engineer, Institute of Aeroelasticity.

LP = load point
MBS = multibody simulation
MM = mission module
MST = mission sequence template
PAM = physical aircraft model
RF = relaxation factor
SA = structural analysis

Nomenclature

A = cross section of equivalent beam
E = Young's modulus
G = shear modulus
I_T = polar torsional moment of inertia
it = torsional stiffness
ixx, izz, ixz = flexural stiffness
iyy = extensional stiffness
Mx, Mz = wing bending moments, local wing coordinates
R = strength of the isotropic material,
RF = relaxation factor
SF = safety factor
T = thickness
σ_V = von Mises stress

I. Motivation

IN the preliminary design phase, aircraft are described with a limited level of detail, the focus being on fast investigation of a large number of cases and quickly changing configurations and parameters. An evaluation chain at preliminary design usually consists of a number of different methods, including analytical equations (e.g., for range), frequency response (e.g., for handling qualities), and time simulations. It is a highly coupled, iterative process. In many cases the analysis environments for preliminary design are heterogeneous, incorporating methods and tools ranging from standard software to in-house solutions. Process chains are very much adapted to the specific needs of a group or a company, and interactions of analysis methods are performed manually or are hard coded in the process.

Some attempts have been made to generate a more coherent process chain for preliminary design; see Chapter 2. For an industrial use, one of the major motivations of re-examining the existing design loops is the need for modularity of the used analysis methods. In the same environment, the designer wants to be able to execute analysis modules with varying complexity, for example, to change a description of the structure from a mass or beam element type to a small finite element representation, without having to change the complete process.

As part of the Prelude case study, the Simulation Toolbox task was aimed at a modularization of the preliminary design loop. An outline of a modular design environment had been defined by the industrial partner at the outset of the project. As a representative example for an analysis element, a module for nonlinear time-domain simulation was selected to be implemented. The module is based on a multibody simulation (MBS) approach, which is well suited for the preliminary design task. This is because it is focused on models of a medium level of complexity, and also because it has inherent capabilities for fast parameter variation and for a simple coupling of engineering disciplines, especially rigid-body flight mechanics, structural dynamics, and aerodynamics.

In the majority of cases, dynamic aeroelastic deformations are neglected at an early aircraft design stage. However, as aircraft become increasingly more flexible (resulting from the application of new materials and structural optimization), the introduction of aircraft elasticity is expected to greatly increase the reliability of the analysis results even earlier in the design process than previously possible. In many cases, aeroelastic calculations, load calculation, and flight dynamics are treated in separate approaches and independent evaluation modules during the evaluation phase. An integrated modeling of flight mechanics and aeroelastic effects is essential for time-domain analysis. An example is the calculation of dynamic (maneuver) loads where an introduction of aeroelastic effects can significantly change the results. The sample application presented in this chapter has been taken from that field.

II. Basis Considerations for a Modular Framework for Preliminary Design

The toolbox framework presented in the following has three basic modeling layers: aircraft models, model engines and interfaces, and software tools. The software tools can range from simple analytical formulas over user-defined programms to commercial software packages. The tools can also be replaced by other "result-generating" approaches, for example, response surface methods.

A. Aircraft Models

Three levels of aircraft models are used, the basic aircraft model, the physical aircraft model, and the disciplinary model.

The basic aircraft model (BAM) represents the most basic, or generic, modeling level. The model consists of approximately 10–20 parameters that define the aircraft's most important characteristics. These basic parameters can be changed directly to perform tradeoff studies. These modifications to the model should propagate through the different modeling layers.

The physical aircraft model (PAM) contains all information of the detailed aircraft model: geometry, masses, propulsion, structural properties, etc. All data necessary for the disciplinary analyses can be derived from this model. The PAM must be scalable in respect to the basic parameter set of the BAM; that is, it has either to be parameterized directly, or PAM data must be modified every time a basic parameter of the BAM is altered. Additionally, the PAM will contain extensive information that will not, or only indirectly, be controlled by the BAM.

A disciplinary model (DM) is a generalized data set that contains all information necessary to perform all analyses and/or evaluations that have been assigned to this specific discipline. As an example, the aerodynamics DM contains the geometric data of the aircraft model in a generalized form. The input file for a specific aerodynamic analysis tool is then derived from the DM. A DM can consist of two parts: the standardized data set, which is compulsory, and an annex, which is reserved for any additional, tool-specific information. More information on such data conversion tools is provided in the chapter on engineering data management.

In this concept, a tool-specific input file is not considered as a disciplinary model. Modeling and data-handling strategies as well as syntax vary significantly from software tool to tool, but most of the underlying data do not change. These data are contained in the DM. Each software tool has a specific interface which then generates the input file from the DM data. An example can be taken from the definition of structures: structural codes require node positions, elements, material properties, etc.; this information is stored in the DM, either explicitly or in abstract form. The interfaces translate these data into the syntax of a proprietary finite element program like NASTRAN or ANSYS. Figure 1 shows an outline of the concept just described.

B. Model Engines and Interfaces

Corresponding to the three modeling layers, three so-called model engines are needed.

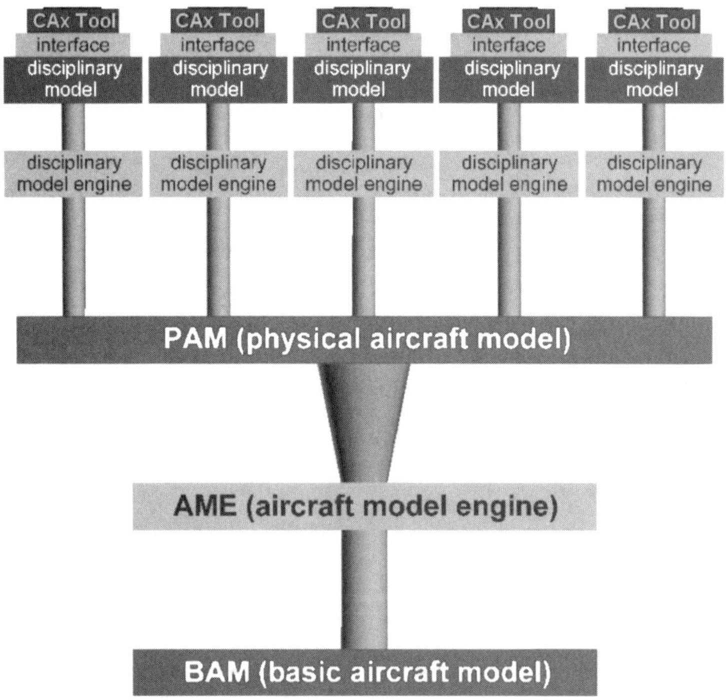

Fig. 1 Basic structure of simulation toolbox framework.

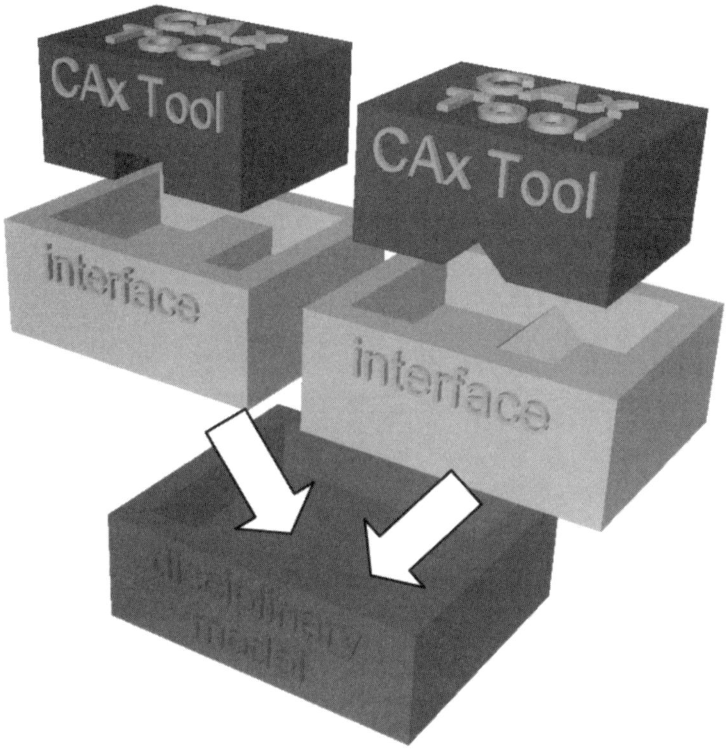

Fig. 2 Modular disciplinary analysis block.

The *aircraft model engine* (AME) assesses the basic model data (BAM) and generates/modifies the corresponding physical model. The AME should be equipped with a graphical user interface, which supports the modeling, especially the specification of PAM-specific data (modeling criteria, e.g., fidelity and model details that cannot be derived from BAM data).

The *disciplinary model engines* (DME) extract the required DM data from the physical model. They can be user controlled (for model setup) and/or process controlled (for model updating in interdisciplinary analyses and optimization).

Each analysis, simulation, or evaluation tool incorporates a *disciplinary interface* (DI). A DI assesses the disciplinary model data stored in the DM and converts it into an input that can be used by its corresponding tool; see Fig. 2.

C. Modular Disciplinary Analysis Block

A disciplinary analysis block is composed of the DM, the interface, and the analysis tool. This allows a very flexible choice of analysis tools (e.g., of different levels of fidelity).

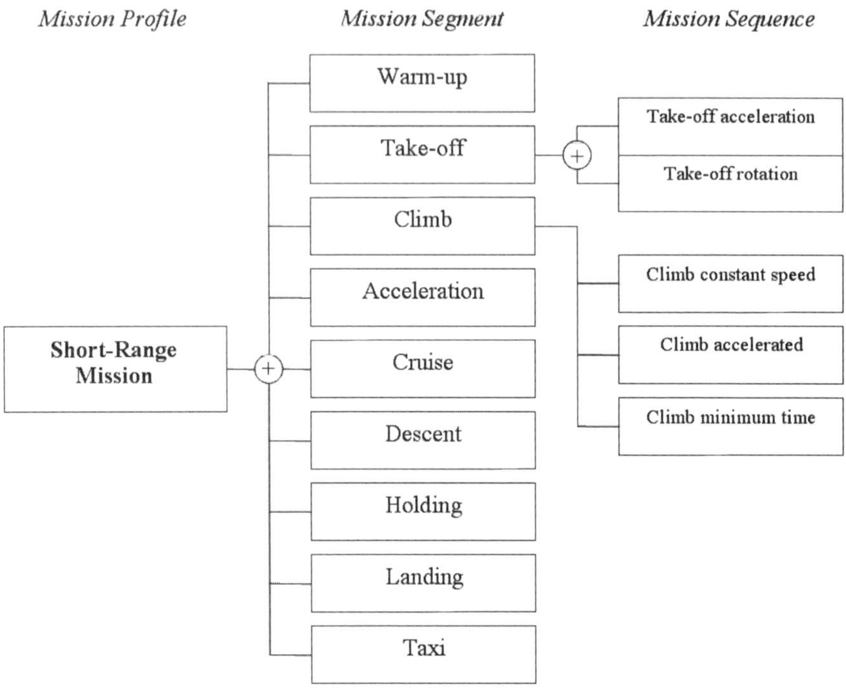

Fig. 3 Generalized mission composition.

D. Mission Module

Aircraft conceptual design is a highly coupled, iterative process. This process is controlled by the *mission module* (MM).

The process sequence that has to be performed for analysis of a mission is defined in a *mission sequence template* (MST). An MST can be imagined as a hard-wired controller that calls the framework components to generate the results. The mission is also referred to as the aircraft flight profile.

A mission profile can also be defined in a modular approach. The mission is decomposed into independent, elementary mission objects. A library containing many different kinds of such elementary mission segments can be formed. Any mission profile can then be constructed by arranging the corresponding mission sequences in an admissible order. New elementary mission sequences can be added when needed.

When decomposing a mission into elementary sequences, two things have to be carefully considered: the conceptual boundaries and the relationship of these elementary sequences, in order for the objects to be used for different mission profiles.

In the mission module hierarchy, three layers exist:

1) A *mission profile* usually describes an entire flight, for example, from warm-up prior to takeoff to taxiing to the gate after landing.

2) A *mission segment* is a fundamental part of a mission profile, for example, climb, cruise, descent, or landing.

3) A *mission sequence* is a subclass of a segment; it represents a logical part of a segment, and its definition is mainly influenced by technical reasons. As an example, a takeoff segment can be broken down to a takeoff run (accelerating aircraft with friction but no aerodynamics), rotation (aerodynamics and flight control), and initial climb (aerodynamics of the trimmed aircraft). Further subclassifications are possible if required.

A mission profile can then be composed by a selection of mission sequences, organized in mission segments (see Fig. 3).

Each mission segment is assigned to an MST, which specifies which tools have to be called in what order to analyze the respective segment.

A schematic design structure matrix of an MST is depicted in Fig. 4, whereas Fig. 5 shows how a mission segment template connects the analysis tools of the framework (at their DM socket) and thus controls the process flow for the analysis or simulation of the segment.

III. Simulation Toolbox—an Application of a Module for Multidisciplinary Design

The structural sizing of an aircraft wing at a preliminary design level, based on dynamic loads from a maneuver simulation, has been implemented as an example for a modular simulation environment. The disciplines involved are structural dynamics, flight mechanics, and aerodynamics. The involved tools include multibody dynamics for modeling and simulation of the flying aircraft and a beam element analysis tool for the sizing of the wing. Some preliminary results have

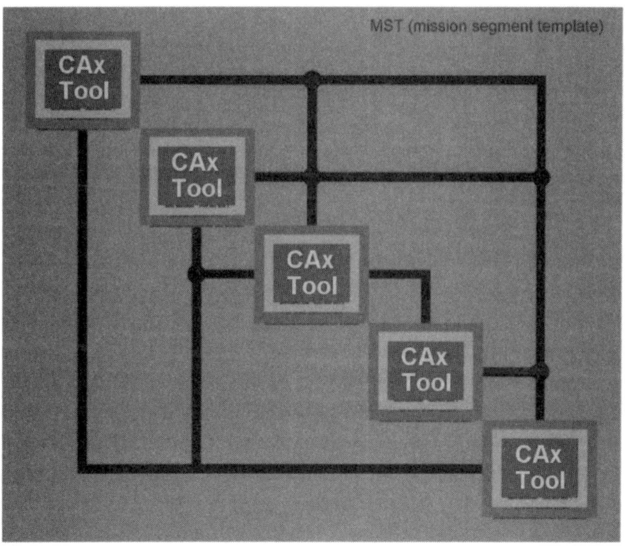

Fig. 4 Example of a mission segment template connecting the disciplinary tools.

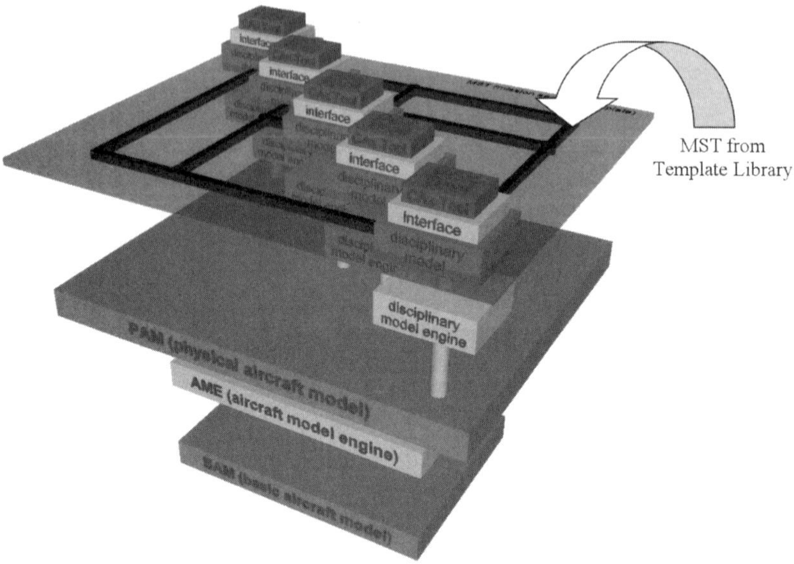

Fig. 5 Interplay of framework and mission module. (See also color figure section at the back of the book.)

been presented in [2]. In the following sections, the tools and methods along with the used models will be presented.

A. Multibody Simulation in Aircraft Design

1. Multibody Simulation

Multibody simulation (MBS) has been shown to be a valuable software tool for virtual aircraft design. In aeronautics, it is the state-of-the-art approach especially in the area of landing-gear design, ground maneuvers (takeoff, landing, taxiing, ground handling), and the layout of high-lift systems as well as in helicopter and tilt-rotor analysis. In addition, the medium level of complexity of typical multibody applications makes MBS a suitable tool for the application in aircraft preliminary design. Comprehensive simulation allows the analysis and evaluation of performance, structural loading, and dynamic behavior of the system, as well as the optimization of the design of the concept. It is becoming more and more important to perform these computations in complex, realistic scenarios, accounting adequately for aerodynamic effects on the flexible aircraft structure. This is an essential factor for such interdisciplinary simulations. The development of reliable aerodynamic models in addition to an existing interface for complex elastic structures has therefore been a major activity in the DLR Institute of Aeroelasticity during the past years [1], ranging from strip-theory and lifting-line-type models [2, 3] to interfaces with high-end computational fluid dynamics (CFD) tools [4, 5]. Aeroelastic analyses using MBS have been shown for example for loads calculation of civil [6] and military [7] aircraft. A major advantage of using multibody dynamics for aeroelastic applications is the straightforward introduction of flight mechanics into the aeroelastic simulation. Although the consideration of linear flight mechanics is a standard procedure both in aeroelastic stability analysis and for the calculations of maneuver loads, a full coupling between those disciplines has been described several times [8–10], but is not a common practice. The full advantage of using a complex multibody tool for that purpose becomes most evident for systems with large rotations such as combined aircraft/landing-gear analysis, including optimization [11], helicopters or tilt rotors, and for aircraft with large elastic deflections [3]. In the Prelude case study (see VIVACE Web site at http://www.vivaceproject.com), a multibody simulation model of a free-flying, maneuvering aircraft is modeled as an example for a modular simulation application. A similar approach has been taken in the German national MODYAS project [2]. The aim is to determine flight and landing loads on a model of medium complexity, which includes an elastic airframe, distributed aerodynamics, and realistic flight mechanics. Two test cases have been selected to test the approach and to analyze the difference between calculation of flight and ground loads on a rigid aircraft vs calculation on an elastic aircraft on a preliminary design level.

Multibody simulation has been selected as the method of choice because it is able to include all of the disciplines just mentioned in a straightforward manner. Furthermore, the simulation environment can be included in a larger design loop. The multibody simulation tool is described online at http://www.simpack.com. Some functionalities, for example, the trim module, have been implemented using a mathematical analysis program described online at http://www.scilab.org.

2. *Elastic Bodies in Multibody Simulation*

Two major approaches exist for the simulation of the dynamics of large mechanical models, finite element analysis (FEA) and multibody simulation (MBS). Finite element models are an established way to describe elastic systems, being used for static and dynamic analyses, with models accommodating a large number of degrees of freedom. Leaving the very time-consuming crash simulation aside, most dynamic applications of FEA use linear models with small deflections, neglecting large rigid-body motion. Results of dynamic calculations are often in the frequency domain, that is, natural frequencies and mode shapes that are input for stability analysis. Multibody codes are generally used for the simulation of complex dynamic systems with large, nonlinear motion in combination with reduced elastic models to describe small elastic deformations. Examples are road and wheel/rail vehicles, aircraft, and machines. Nonlinear forces can easily be described, and the analysis results are typically in the time domain. The resulting simulation models in MBS are usually considerably smaller than those of FEA approaches and are used not only for system evaluation but also as input for control design, real-time applications, and optimization. This makes multibody simulation the preferred solution for the analysis of a number of aeroelastic problems, especially for aircraft with large wing deflections and for helicopters.

There are two common approaches to representing elastic properties of elastic structures in multibody dynamics. The first approach, sometimes called the "chain-link" approach, is probably the older one, representing elastic beams as a combination of rigid bodies connected by torsion springs. The properties of those springs have to be derived from measurements or available analysis results. The second approach, being the standard one of the used multibody program, makes use of the modal representation of finite element based structures. In a preprocessing step, modal analysis of a finite element structure is performed, and the resulting model is included in the multibody simulation, taking coupling terms between elastic and rigid-body motion as well as geometric stiffening and small geometric nonlinearities into account. The advantages of the modal approach are that models can be easily derived from, often already available, finite element models in an arbitrary degree of complexity. Yet, the modal approach assumes small, linear deformation, whereas in the case of connected bodies no assumptions are made concerning the nature of the connecting force elements and the deflections. Another reason for choosing the chain-link formulation is that changes in properties of the elastic structure make it necessary for each parameter change to repeat all preprocessing steps for the modal approach, whereas for a chain-link approach parameter changes can be quickly made in the MBS model directly.

3. Multibody Model Setup and Model Data

Figure 6 shows the basic layout of an elastic aircraft using a multibody modeling approach. A generic four-engine transport aircraft has been selected as a reference configuration. The structure is discretized in bodies, connected by rotational springs to represent structural elasticity. Input data are the geometry, the discretized mass distribution (mass and local center of gravity with respect to the elastic axis), and the stiffness of the rotational springs connecting the bodies. The data were extrapolated from diagrams of existing aircraft; data were available for wing bending and torsion as well as for fuselage bending and torsion. A typical stiffness distribution for wing and fuselage bending is shown in Fig. 7. For the aerodynamic data, strip theory has been implemented for the first simulation test cases. This method assumes that the aerodynamic properties of the wings can be described in spanwise strips across the wings, not taking any interference between the sections into account. Two-dimensional aerodynamic equations are implemented, using a flat-plate assumption, that is, a linear approach with 2π as gradient for the dependency of lift from angle of attack. No initial wing twist, that is, built-in angle of attack, was assumed. The necessary wing reference area for lift calculation was estimated from three-side views of related aircraft. Control commands are introduced by changes of the local lift coefficients for the respective surfaces. For the landing approach, the local lift coefficients have been multiplied by a factor of 1.4 to account for the effect of high-lift devices. For the touchdown case, rigid-body landing-gear models representing the nonlinear gear dynamics [6, 12] have been added.

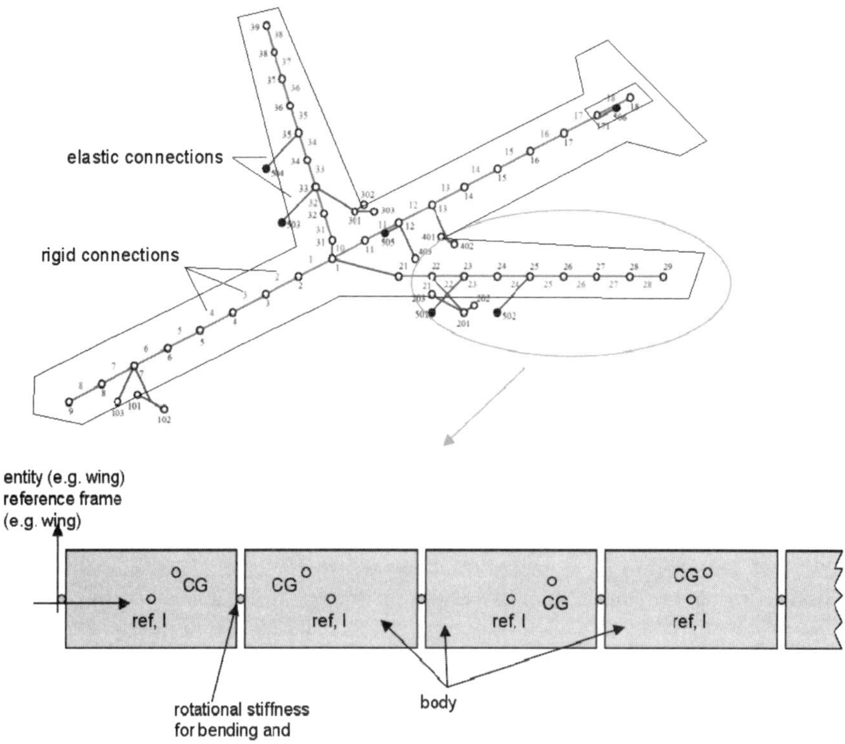

Fig. 6 Multibody model of aircraft and close-up of wing.

The collected data are coherent. Its accuracy is sufficient to be comparable to real aircraft, if not too close to a real four-engine configuration, and sufficient for the determination of trends and comparisons between rigid and elastic modeling approaches. Figure 8 shows a screen shot of the resulting model of the aircraft in the modeling environment for the touchdown case.

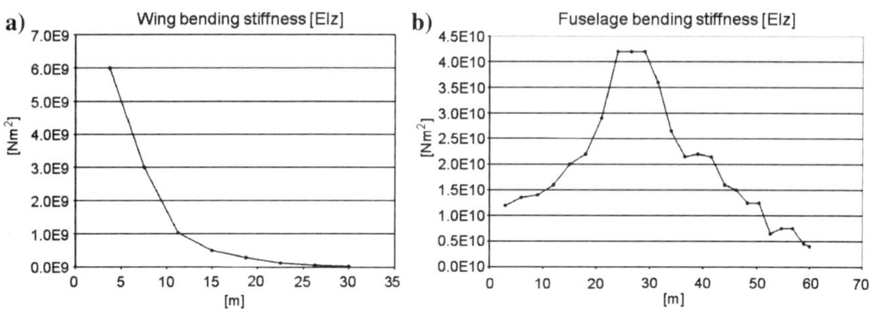

Fig. 7 Bending stiffness distribution of aircraft wing and fuselage.

B. Beam Element Analysis for Wing Structural Design

1. General Remarks

The tool used for the structural sizing is a beam element analysis (BEA) approach, based on the analysis of wing cross sections. For a given wing design it calculates distributed mass, stiffness, and, for a given load distribution, the resulting stresses and strains. For a given design, an equivalent beam is calculated. The tool can be used for wings with an open cross section or with one or more cells in closed cross sections.

The wing is described by a series of cross sections along its length; see Fig. 9.

The outer contour of the cross sections is a curved profile; inner walls are assumed to be flat. The use of fiber-enforced material is possible (though not yet validated). Figure 9 gives an overview of a cross section with a curved airfoil and four spars (a, c, f, i in Fig. 9) forming three cells. All walls of the cells including the so-called shells between the spars (b, d, e, g, h, j in Fig. 9) are assumed to be straight (without curvature). Shells and spars join at the so-called caps (1–8 in Fig. 9).

Nose edge located before the leftmost spar and the trailing edge located behind the rightmost spar are not considered to be part of the load-carrying structure and are therefore neglected. If they are to be taken into account, it is possible to define "dummy spars" at the leading or trailing edge of the airfoil. The loads (three forces and three moments) are acting at the quarter-chord of the profile.

Fig. 8 Multibody model of the aircraft at touchdown.

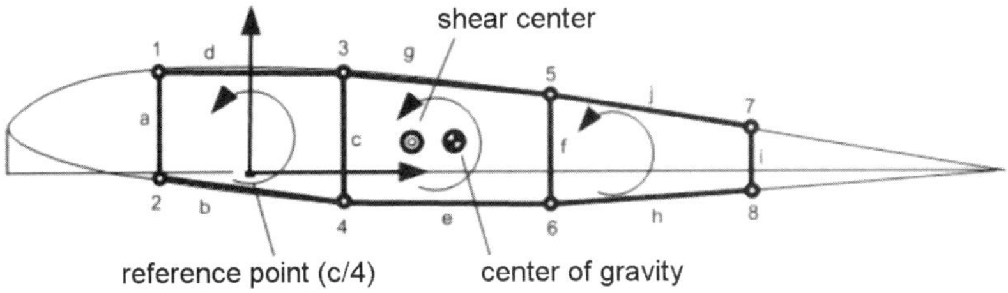

Fig. 9 Input data for beam analysis tool.

2. *Transfer Parameter and Interfaces*

Data structures using the XML definition are used for transferring the input and output parameter of the BEA tool. The main parameters are the material properties, the geometry of the airfoils, the configuration with spars and caps of the wing section, and the loads for each wing section.

In the multibody time simulation the aircraft is modeled by mass points connected by rotational stiffness for bending and torsion as shown in Fig. 6. Therefore, for each section the structural analysis tool needs to provide mass per length; the extensional stiffness, $iyy = \int E \, dA$; the flexural stiffness, $ixx = \int Ez^2 \, dA$; $ixz = \int Exz \, dA$; $izz = \int Ex^2 \, dA$ of all sections with the Young's modulus E and the cross section area A; the torsional stiffness $it = GI_T$ with the polar torsional moment of inertia I_T; and the shear modulus G.

All of the wing-section properties are calculated on the basis of dimensioned thicknesses of the spars and the airfoils. Therefore, a structural sizing routine is included into the BEA tool. The structural sizing routine is described more precisely in the following section.

3. *Methodology of Structural Sizing*

The structural sizing is based on a fully stressed design, which means that the calculated stresses are compared to maximal allowed stresses including safety factors (SF). For isotropic material the von Mises stress is used as comparison stress:

$$\sigma_V = \sqrt{\sigma_x^2 + \sigma_y^2 + \sigma_x \sigma_y + 3\tau_{xy}^2}$$

The following procedure is chosen for the structural sizing:

The loop over sizing iterations contains a loop over all sections and a loop over all load cases. The in-plane stresses are calculated based on the current geometry and assuming a two-dimensional stress state for all shells (lower and upper airfoil, spars) of a section, as illustrated in Fig. 10.

At the marked points in Fig. 10, the needed thickness

$$t_{\text{needed}} = SF \frac{\sigma_V}{R} t_{\text{current}}$$

is calculated, where R is the strength of the isotropic material and SF is a safety factor. Because the stress transfer occurs within the section, the thickness

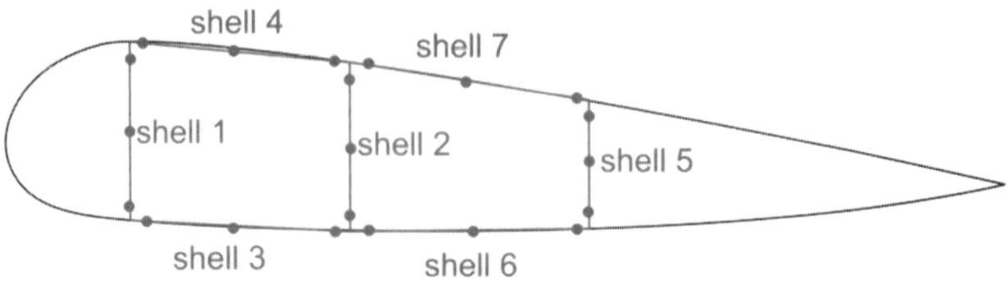

Fig. 10 Shells of a wing section.

increase or decrease is not done in one step. A relaxation factor RF is used to determine the new thickness:

$$t_{\text{new}} = t_{\text{current}} \pm \text{RF}|t_{\text{needed}} - t_{\text{current}}|$$

Subsequently, the maximum new thickness for the lower airfoil, the upper airfoil, and each of the spars are determined considering all load cases. Thereby, the thickness must not go below a minimum value. The iteration stops if the thickness change is less than 0.01 mm.

C. Sizing Loop Using the Multibody Simulation and a Beam Element Analysis Approach

The final product of the Simulation Toolbox in the Prelude case study (data online at http://www.vivaceproject.com) is a sizing loop for an aircraft wing using time simulation for load determination and a beam element tool for load analysis, coupled by a general sizing algorithm. The workflow is described in Fig. 11.

A first design of the wing (geometry, masses, stiffness) is introduced into the time simulation in which dynamic loads for the elastic aircraft and one or more given maneuvers are calculated. If the initial masses and stiffness are not known, they can be approximated by the structural analysis tool using the initial geometry data. The load envelope found in the multibody time simulation is exported to the structural analysis tool where the given design is checked for maximum stresses. According to the result, the wing design is altered, or, if the design meets the requirements, the loop is finished, and the results are exported.

IV. Example: Structural Sizing of a Transport Aircraft Wing

A. Simulation Case and Results

In this section a representative application of the sizing loop using dynamic loads from time simulation is presented. Results for a 2.5-g pull-up maneuver are shown here for a demonstration of the workflow. The time simulations are based on the multibody model. Results for a looped sizing process are given. Studies for a two-point landing with a vertical touchdown velocity of 3.05 m/s, obtained with the same modeling approach, but not including the sizing routine, can be found in [2].

The objective of the example is to provide simulation capabilities for conceptual design, not to analyze and evaluate any actual aircraft.

B. Pull-up Maneuver

As an example for the coupled simulation, a 2.5-g pull-up has been defined. The simulation has been performed both with a rigid and with an elastic aircraft. One goal of the simulations is to assess of the differences between the two approaches, most notably to see whether an elastic model has an influence on the prediction of the dynamic loads at the wing root. The comparison is a good way to show the differences between the two modeling approaches. The simulated aircraft has a weight of 180 t and is calculated starting with a steady flight from trimmed conditions: $vx = 200$ m/s, elevator setting = 10.2 deg, and angle of attack = 1.85 deg.

In the simulation, the elevator setting is changed from 10.2 deg such that an acceleration of the center of gravity of 2.5 g is reached. Because of the structural response, a higher elevator deflection has to be commanded for the elastic aircraft compared to the rigid one to achieve the same c.g. acceleration for both cases. Using the respective inputs, both rigid and elastic aircraft have roughly the same overall system response. Although the airframe dynamics have an influence

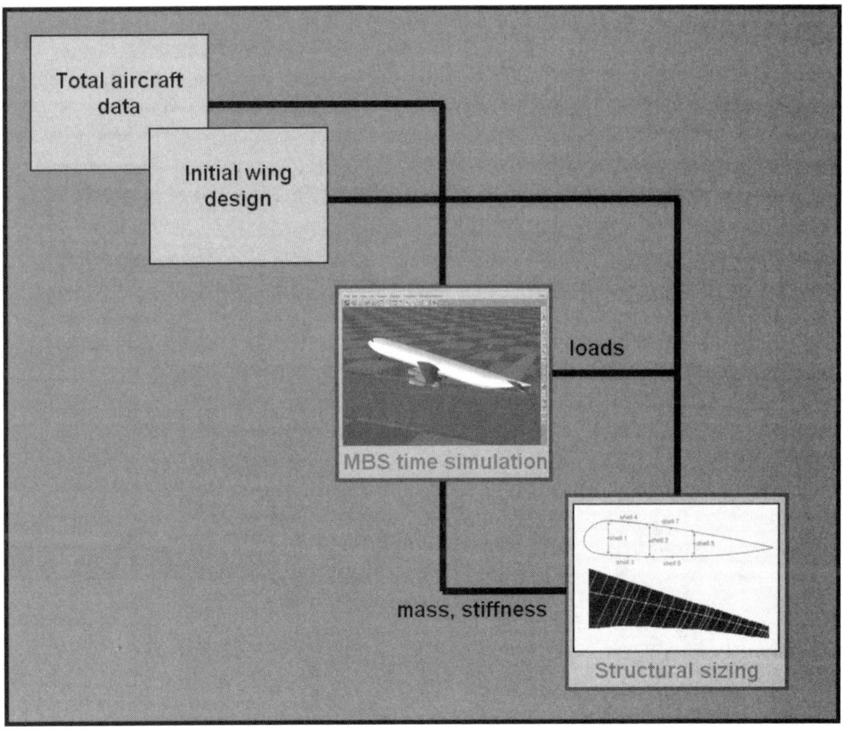

Fig. 11 Structural sizing loop including multibody simulation and structural tool.

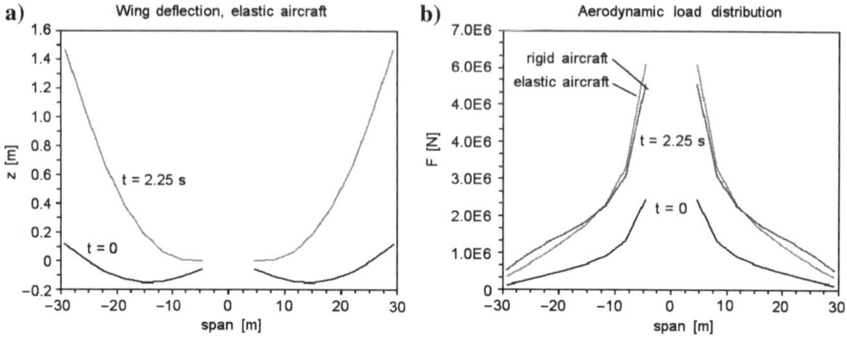

Fig. 12 Wing deflection and aerodynamic load distribution for rigid and elastic aircraft.

on the rigid-body flight mechanics, the differences are so small that they are not relevant for the current investigation. Wing deflection and aerodynamic load distribution are shown in Fig. 12. In Fig. 12a the initial wing deflection is the reference shape, that is, the flight shape in steady cruise (1 g, $t = 0$); the line for $t = 2.25$ s is the new deflection under the air load at 2.5 g. A rigid aircraft shows no deflection against the reference shape, of course. In Fig. 12b the lift distributions are compared, at cruise ($t = 0$, equal for rigid and elastic aircraft), and at 2.5 g. The elastic twist bends the wing upwards and, because of the wing sweep, introduces a downward tilt of the profile, effectively reducing the local angle of attack and thus the local air force on the outboard part of the wing. The effective aerodynamic center is therefore shifted inwards for an elastic wing when compared to a rigid wing. This effect leads to a significantly lower

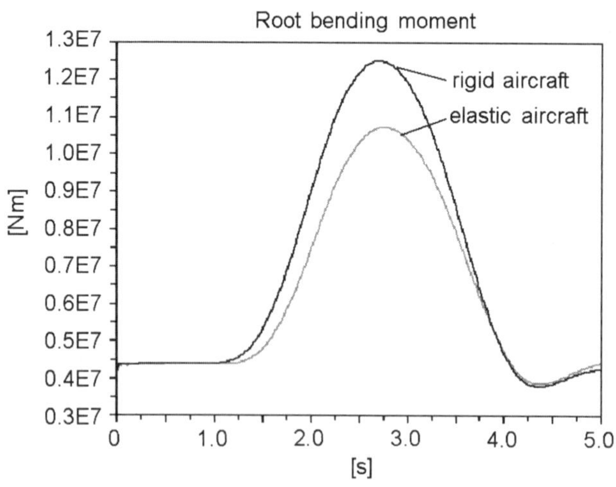

Fig. 13 Calculated root bending moment for rigid and elastic aircraft.

wing root bending moment of the elastic aircraft when compared to the rigid model; in the current setup the difference is approximately 20% (see Fig. 13). Thus, for the given flight conditions and control input, a reduced calculated root bending moment can be shown for an elastic aircraft against a rigid one. This fact indicates that for most cases a rigid simulation gives conservative load assumptions.

However, this assumption might not be valid for gust or control excitations in the frequency range of the wing or fuselage natural frequencies, where elastic models might exhibit higher responses. The landing simulation in [2] shows partially increasing loads at the cockpit station for reduced wing attachment flexibility as a result of a phase shift in elastic fuselage response. In another case for ground runs of aircraft, shown in Krüger [12], the rigid aircraft model underestimated the dynamic loads by a factor of three for a given configuration.

In summary, it can be stated that taking aircraft elasticity into account will give valuable insight even at an early design stage. The additional information more than compensates the additional effort needed for the implementation of elastic models and the respective simulation capabilities.

C. Example for Structural Sizing

1. Geometrical Input for the Wing Box Beam

The geometric data for the design of the wing was based on a three-side view of the whole aircraft and three cross sections (root, middle, outboard) of the wing box, typical for existing wing designs. Root and middle cross sections contain a middle spar.

The aerodynamic load for sizing was transferred from a preceding multibody simulation of the maneuver. The load was represented in eight points on a straight line along the wing. Theoretically, this line should coincide with the quarter-chord line. However, as the MBS geometry and the geometry used for the structural analysis tool originated from different sources, the quarter-chord lines of both layouts did not match completely. To make up for the difference, the x coordinates of the load points in the MBS model were shifted correspondingly.

The geometric data used are shown in Fig. 14.

The geometrical data have to be processed to produce the geometrical input for the structural analysis tool. For each multibody load point, one cross section is needed for the BEA tool.

To correctly model the beam, its cross sections must be aligned to be perpendicular to the beam axis, that is, the quarter-chord. As seen in Fig. 14, the middle and outboard profiles in a realistic aircraft more or less fulfill this requirement, while the root cross section lies in the x–z plane. Therefore this profile is rotated in the x–y plane around the intersection of the trailing edge with the fuselage by the angle of the quarter-chord, 30 deg.

Next, the cross sections corresponding to the multibody load points are interpolated between the root and middle cross sections (for the first three load points) and between the middle and the outboard cross section (for load points four to eight). The resulting cross sections can be seen in Fig. 15.

Next, profiles have to be defined. The origin of the z axis determines the z location of nose and trailing edge of the profile and is assumed to be in the

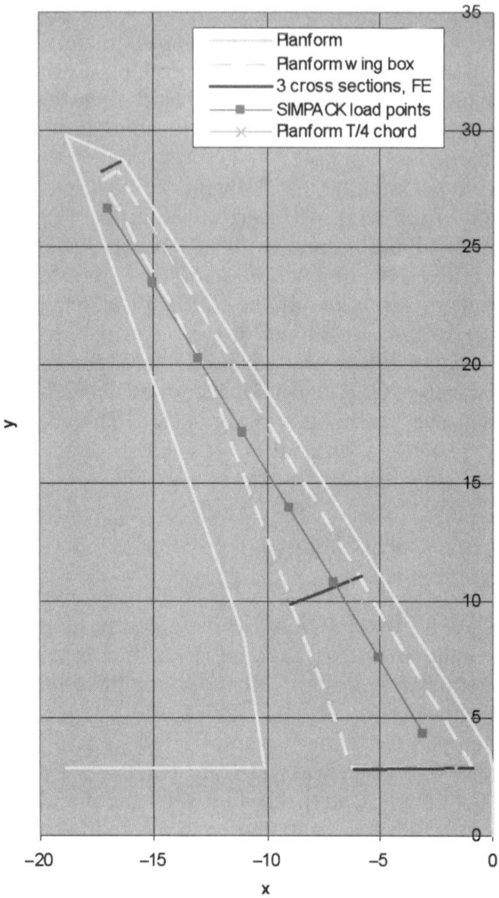

Fig. 14 Geometric data of the wing, *x*–*y* plane, MBS and BEA layouts. (See also color figure section at the back of the book.)

center of the wing box. The profiles must enclose the box beam cross sections in such a way that the quarter-chord point is at the correct location within the wing box because this is where external forces and moments are acting. In the BEA approach, the contour of all profiles is approximated by a polygon with linear interpolation between the vertices of the polygon. The location of the spars is identified by their x coordinate in the wing coordinate system. Therefore all of the caps (i.e., intersections of the spar and the profile) must be located on the profile. As a final step, the profile geometry is scaled to a length of one. Two example profiles for different cross sections are shown in Fig. 16.

The forces in the three spatial directions were given for the eight load points along the quarter-chord of the wing. Because the BEA wing coordinate system (y along quarter-chord) used for the determination of the structural parameters (stiffness, strength, stresses) is rotated relative to the coordinate system for the

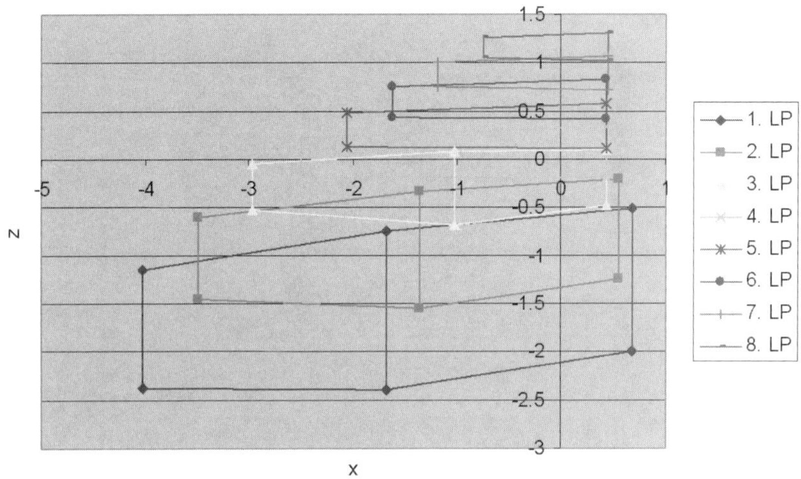

Fig. 15 Cross section for the eight load points (LP) in the wing coordinate system (*y* axis along quarter-chord). (See also color figure section at the back of the book.)

determination of the loads (global coordinate system with x pointing in flight direction) by an angle of −30 deg, the MBS loads have to be transformed. Additionally, the bending moments around the beam x and z axis have to be evaluated by taking the lever arms of the x and z forces into account. The torsional moment will be automatically taken care of by summing up the normal force from outboard towards the wing root and applying it at the quarter-chord in

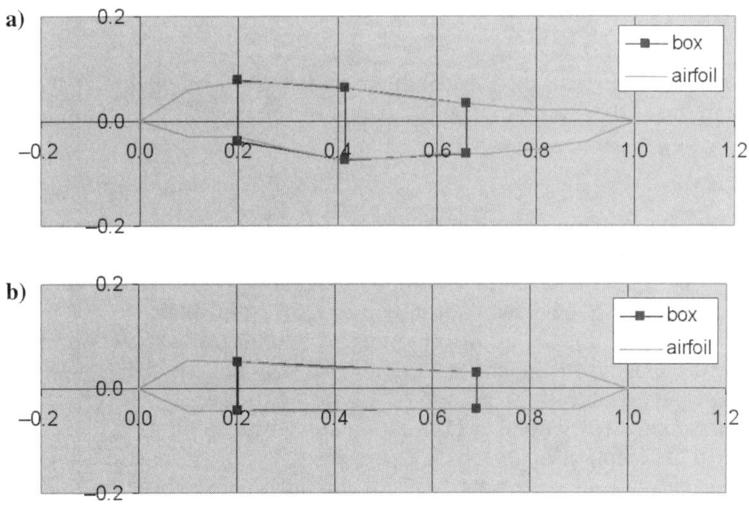

Fig. 16 Beam cross sections and profiles for a) cross section 2 and b) cross section 5.

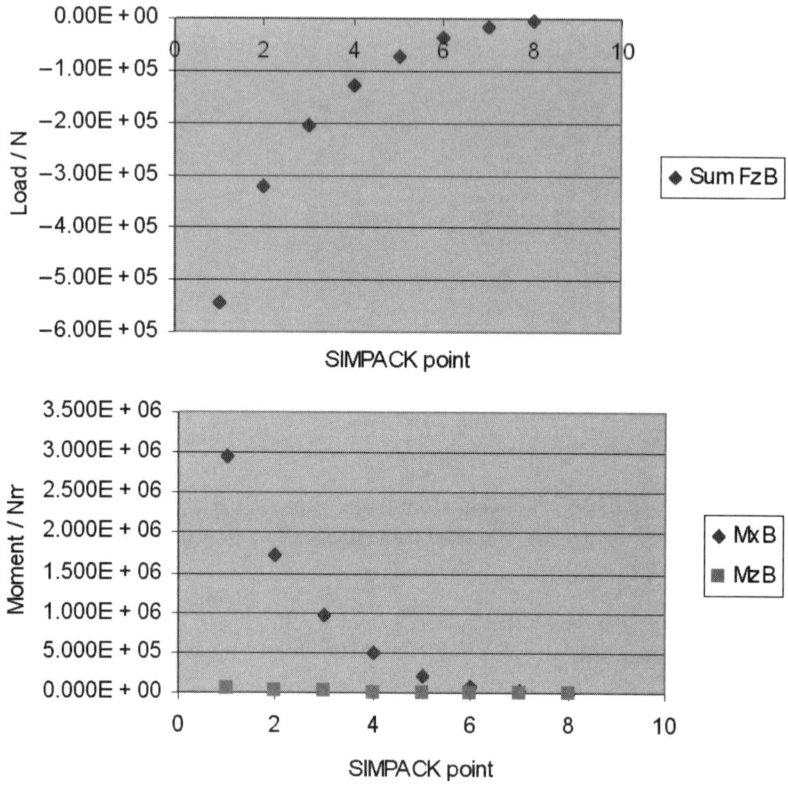

Fig. 17 Load distribution from root (point 1) to outboard (point 8) for transverse shear (*Qz*) and the two bending moments *Mx* and *Mz*.

each cross section. The load distribution from root to outboard for transverse shear and the two bending moments *Mx* and *Mz* are shown in Fig. 17.

2. *Sizing Results*

For the sizing process, the wing is assumed to be made of aluminum. At the beginning all thicknesses of the airfoils and spars are set to 1.0 mm, which is, at the same time, chosen as minimum thickness. The relaxation factor is chosen equal to 0.5. The safety factors are 1.8 for the spars, 2.2 for the upper airfoil, and 2.4 for the lower airfoil. After 34 iterations the sizing routine stops because the change of thickness between two consecutive sizing loops is less than the convergence threshold of 0.01 mm. For sections 1, 6, 7, and 8 the element thicknesses vs number of iteration are shown in Fig. 18.

With the exception of the thickness of spar 2 of section 1, the thicknesses converge within the first 10 iterations. The computational time of the 34 iterations is less than 15 s on a typical desktop computer. The maximum occurring thickness is 3.8 mm for the lower airfoil of section 3. For the outer sections 6, 7,

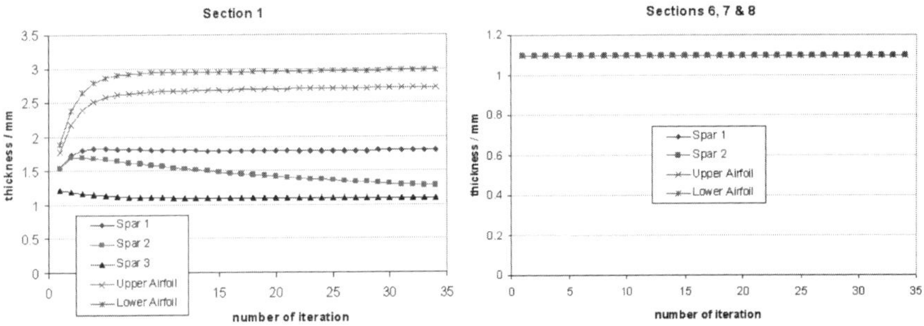

Fig. 18 Thicknesses vs number of iteration.

and 8 the thickness of all elements is restrained by the minimum thickness value instead of the fully stress design.

After all iterations the flexural stiffness *ixx*, *izz*, and *ixz*, the torsional stiffness *it*, and the extensional stiffness *iyy* are calculated for all eight sections. The stiffness distribution vs the section numbers are shown in Fig. 19.

The sizing algorithm is capable of treating several load cases at the same time, selecting per section the maximum load of all load cases applied.

V. Results of the Simulation and Sizing Process Loops

The process loops of time simulation of the flying aircraft and subsequent sizing of the wing were performed as follows: first, a sizing of the wing was performed on the basis of a static 2.5-g maneuver. With this elastic wing, a time simulation of a dynamic 2.5-g pull-up was performed, recording the time history of the loads on eight selected points along the wing. The worst case from those time steps was selected, the criterion being the maximum transitory root bending moment. Note that this operational point would eventually be only one of several dimensioning cases that have to be treated in parallel in the sizing procedure.

After the wing was sized based on the maximum dynamic load, the new wing design was transferred to a multibody formulation, included in the multibody

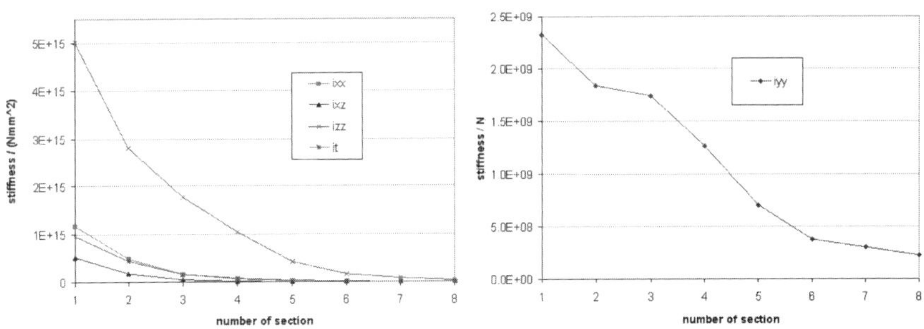

Fig. 19 Stiffness vs number of sections. (See also color figure section at the back of the book.)

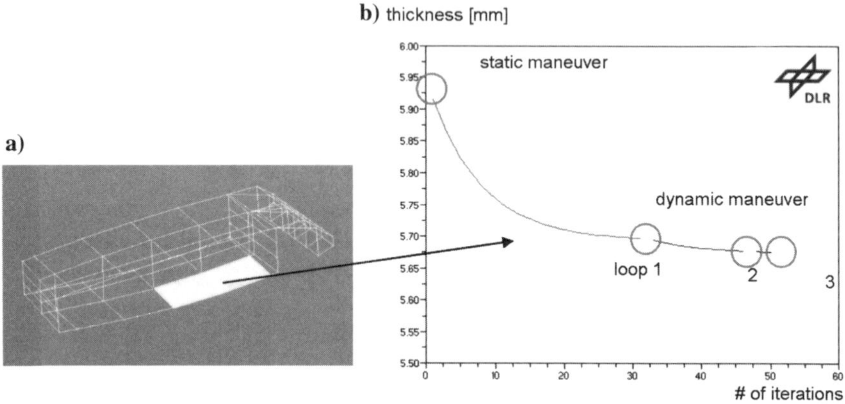

Fig. 20 Results of looped sizing based on time simulations with the new structural design.

model, and a new run of the dynamic simulation was performed. After that step, the sizing was repeated. After three times, convergence was observed.

Figure 20 shows the results for the coupled sizing loop based on the simulation presented in Section B and the sizing approach presented in Section C. The sizing is performed for the wall thickness of the wing beam. A corresponding segment from a finite element segment using shell elements is shown in Fig. 20a for illustration. The evolution of the calculated wall thickness is given in Fig. 20b.

The first value, called "iteration 0," is the calculated wall thickness for the static 2.5-*g* case. The value for the first iteration is the wall thickness calculated from the dynamic maneuver, simulated with the same wing model as the static case. Iterations two and three show thickness values for maneuvers from simulations with updated wing models, where the corresponding stiffness and mass data from the sizing procedures of the iteration steps have been fed back.

It can be seen that the necessary wall thickness at the selected point calculated for the dynamic case is approximately 5% smaller than for the static case. This indicates that for the selected dynamic maneuver maximum stress taken from single time intervals is distributed differently than for a static consideration. However, changes in mass and stiffness resulting from the new sizing induce little change in the dynamic properties of the wing. The changes in calculated sizing for the updated model are small, less than the overall modeling error.

In summary, it can be stated that the use of dynamic loads for structural sizing has the potential of delivering a less conservative structural design. The method gives important insight into the mechanical system. Depending on the maneuver, lower local stresses than from static considerations can be expected for a number of standard cases. However, the examples given in [2], especially the landing case, showed that for some examples higher local stresses might be possible for dynamic than for static considerations.

Another part of the original idea of the approach was to feed back an updated wing design from the sizing into the simulation loop to generate new loads from the updated design. However, the example shows that the change in the structural properties from using dynamic loads for sizing is negligible for the overall aircraft flight dynamics, if the original structural layout is already of good quality.

VI. Summary and Outlook

In the case study, a structural sizing at preliminary design level, using a time simulation loop based on multibody dynamics and a beam element analysis tool, was implemented and demonstrated for the dynamic maneuver of a transport aircraft. It was demonstrated that multibody simulation is an approach well suited for combined aeroelastic/flight mechanics analysis, and that beam element analysis fits in well with complexity. Structural elasticity is usually considered to be conservative, that is, to yield reduced calculated loads; however, this might not be true for a significant number of cases where structural dynamics play a dominant role. It was shown that aeroelastic calculations give valuable insight even at an early design stage. Time simulation, here shown for the first time and exemplarily for a coupled multibody-beam element model, will thus extend the common preliminary design approach based largely on handbook cases and rigid models.

The coupling of multibody dynamics and beam element analysis tool is an example of a modular simulation environment. The solution can be integrated in existing loops. It successfully combines commercial, in-house, and freely available software, being portable and running both on XP and Linux. The approach is well suited to other applications, for example, helicopters or high-altitude long-endurance aircraft design.

As the work has the status of a proof of concept, there are several options to improve the quality of the solution as well as the process. The next step for the multibody simulation application will be the implementation of an interface to a higher-order, unsteady aerodynamics method. For a preliminary design process, this could be an unsteady vortex-lattice method. Furthermore, additional maneuvers, including nonlinear dynamics (e.g., landing impact) have to be included into the design loop to gain a more realistic load envelope for sizing. Finally, the solution has to be introduced into operating design environments.

References

[1] Krüger, W. R., and Spieck, M., "Aeroelastic Effects in Multibody Dynamics," *Vehicle System Dynamics*, Vol. 41, No. 5, 2004, pp. 383–399.

[2] Krüger, W. R., "A Multi-Body Approach for Modelling Manoeuvring Aeroelastic Aircraft During Preliminary Design," *Proceedings of the IMechE, Part G: Journal of Aerospace Engineering*, Vol. 222, No. 6, 2008, pp. 887–894.

[3] Krüger, W. R., "Multibody Dynamics for the Coupling of Aeroelasticity and Flight Mechanics of Highly Flexible Structures," *Proceedings of the International Forum on Aeroelasticity and Structural Dynamics (IFASD), 2007 Congress*, Stockholm, Sweden, paper IF-107, sponsored by the CEAS (Council of European Aerospace Sciences), 17–20 June 2007.

[4] Krüger, W. R., Heinrich, R., and Spieck, M., "Fluid-Structure Coupling Using CFD and Multibody Simulation Methods," *Proceedings of the ICAS 2002 Congress*,

Toronto, paper ICAS 2002-296, sponsored by the International Council of the Aeronautical Sciences (ICAS), Stockholm, Sweden, 8–13 Sept. 2002.

[5] Arnold, J., Einarsson, G., and Schütte, A., "Multibody Simulation of an Aeroelastic Delta Wing in Roll Manoeuvres," *Proceedings of the ICAS 2006 Congress*, Hamburg, paper ICAS 2006-3.10.2, sponsored by the International Council of the Aeronautical Sciences (ICAS), Stockholm, Sweden, 3–8 Sept. 2006.

[6] Spieck, M., "Ground Dynamics of Flexible Aircraft in Consideration of Aerodynamic Effects," Ph.D. Dissertation, Technical Univ. of München, Munich, Germany, 2004.

[7] Krüger, W. R., and Spieck, M., "Multibody Simulation of Low Frequency Aeroelastic Wing Loads," *Functional and Mechanical Integration of Weapons with Land and Air Vehicles*, RTO Scientific Report RTO-MP-AVT-108, paper 11, NATO Research & Technology Organization, Paris, 2005.

[8] Waszak, M. R., Buttrill, C. S., and Schmidt, D. K., "Modeling and Model Simplification of Aeroelastic Vehicles: An Overview," NASA-TM-107691, Sept. 1992.

[9] Looye, G., "Integrated Flight Mechanics and Aeroelastic Aircraft Modeling Using Object-Oriented Modeling Techniques," AIAA Paper 99-4192, 1999.

[10] Meirovitch, L., and Tuzcu, I., "Integrated Approach to the Dynamics and Control of Maneuvering Flexible Aircraft," NASA/CR-2003-211748, June 2003.

[11] Cumnuantip, S., Krüger, W. R., and Spieck, M., "Multidisciplinary Design Optimization Process for Landing Gear System Conceptual Design of Blended Wing Body Aircraft," Deutscher Luft- und Raumfahrtkongress, paper DGLR-2004-026, sponsored by the DGLR (Deutsche Gesellschaft für Luft- und Raumfahrt), Dresden, Germany, 20–24 Sept. 2004.

[12] Krüger, W., "Integrated Design Process for the Development of Semi-Active Landing Gears for Transport Aircraft," *Mechanics of Structures and Machines*, Vol. 30, No. 4, 2002, pp. 493–526.

Chapter 5

Multidisciplinary Wing Design Optimization

Martin Laban* and Paul Arendsen*

National Aerospace Laboratory (NLR), Amsterdam, The Netherlands

Abbreviations

CFD = computational fluid dynamics
CG = center of gravity
FAR = federal aviation regulations
FE = finite element
FEM = finite element method
GSP = gas turbine simulation program
L/D = lift over drag
MDO = multidisciplinary optimization
MTOW = maximum takeoff weight
SOC = start of cruise
TO = takeoff
TOC = top of climb
VFR = visual flight rules

Nomenclature

a = angle of attack
C_D = drag coefficient
C_L = lift coefficient
C_p = pressure coefficient
D = drag
L = lift

*Senior Scientist, Aerospace Vehicles Division.

M = Mach number
Re = Reynolds number
T = temperature
T_2 = time to double
V_r = rotation speed
V_{zrc} = zero rate-of-climb speed
V_2 = takeoff safety speed

I. Introduction

WING design is inherently a multidisciplinary activity that includes disciplines such as aerodynamics, structural mechanics, propulsion, space allocation, manufacturing, cost, etc. Traditionally, wing design optimization relies on the knowledge and experience of the human experts involved. It is common that each of these experts focuses on his/her own discipline. The interaction between the disciplines involved is then effectuated by interaction between these experts on a regular base. Such interactions can comprise, for example, verbal communication and exchange of data files. Typical timescales of these design iteration cycles can be weeks/month. The success of modern aircraft testifies to the effectiveness of this way of working.

However, increasing requirements on aircraft performance to meet new challenges (e.g., the targets laid down in ACARE Vision 2020) justifies investigating innovative ways of working. The innovation of the approach described in this chapter is to be found in the strong coupling of (a selected set) of disciplines under control of a mathematically oriented optimization algorithm substituting the human-centered (verbal) interdiscipline communication links. This is not to say that the human experts are made redundant. Their role is still required in setting up the process, to monitor it while executing, to define the design space to be explored, and to judge the final results. The advantage of the proposed process is the automated exploration of a much larger part of the design space than possible with the traditional human-centered process.

To this end, aircraft analysis tools from various disciplines are selected to form a generic *multidisciplinary analyses* program tool suite applicable to aircraft wings. The wing geometry is described in a parametric way to allow coupling to the *global-level* optimizer driving the system. This parameterized design description is consistent for each involved discipline. Depending on the scope of the study, the design space of an aircraft wing shape can become large (ten to hundreds of design parameters). It is anticipated that any selected optimizer will typically be required to make hundreds to thousands of calls to the underlying analyses program suit. A key issue in this multidisciplinary challenge is to select appropriate tools for the underlying aircraft analysis that combine sufficient levels of fidelity with low computational requirements.

The current chapter will describe the generic multidisciplinary analyses program tool suite applicable to aircraft wings that has been continuously improved during the last decade. Applications comprise blended wing bodies [1], supersonic civil transport aircraft [2, 3], supersonic business jets [4], and low-fuel-burn open-rotor driven novel aircraft concepts. Results obtained during these applications will appear in the various illustrations included in this chapter. Although some of the described analysis tools are in-house developments, each of these can be substituted with equivalent tools for example, to be purchased from commercial suppliers. Chapter 6 entitled Meta-Modeling and Multi-Objective Optimization in Aeronautic MDO, is dedicated to the wrapping of the optimizer loop around the analysis tools described here and will elaborate on a specific application of this system.

II. Wing MDO Framework Architecture

A generic multidisciplinary analysis and optimization framework for aeronautical products has been developed by the authors in the course of various projects. Figure 1 introduces the main functional modules of this framework. Each of the individual modules will be explained and illustrated with actual applications in the remaining sections of this chapter. The modules are briefly introduced next:

1) A *geometry generation module* provides the parametric external (aerodynamic) and internal (structural) shape of the configuration. The wing planform is modeled as a double trapezium. The geometry generation module accepts external wing planform/airfoil parameters (i.e., span, chord, sweep, section-cord-to-thickness ratio, section twist, section shape parameters, etc.) through which the interface to the global-level optimizer is effectuated. The external shape, supporting (CFD-based) aerodynamic evaluations, takes aeroelastic deformations caused by aerodynamic and inertial loads into account. The internal wing shape, supporting (FEM-based) structural mechanics computations, represents

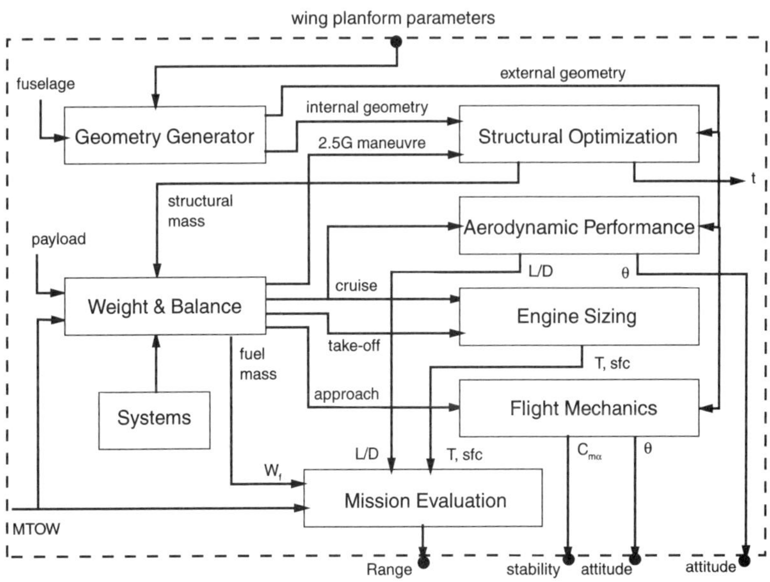

Fig. 1 Generic template used for multidisciplinary wing design optimization.

the wing covers, ribs, spars, and stringers. Emperical rules are used to arrive at a representative layout of the structural elements over the wing planform.

2) A *weight and balance module* keeps a record of all items contributing to the mass and center of gravity of the configuration. Fixed masses are assigned to the fuselage, payload, and tail planes. The mass of the wing primary structure is based on data delivered by the structural mechanics discipline. Secondary wing mass items (leading/trailing edges) are based on empirical rules. Mass of the propulsion system is modeled as a fraction of the engine takeoff thrust rating. Mass of the landing gear is modeled as a fraction of aircraft maximum takeoff weight (MTOW). With MTOW as global-level input to the system, available fuel mass is computed as the difference between MTOW and the sum of all other mass components.

3) A *structural optimization module* sizes the structural element thicknesses to arrive at a minimum weight wing-box structure. Structural element sizing is based on a finite element (FE) analysis implemented in Nastran, combined with analytical representations of details not covered by the FE representation. In this combination, Nastran is performing the actual optimization (SOL200) as well as the internal load-path calculations. The analytical representations are used to infer the allowable stress levels to be applied within the SOL200 optimization. The driving scenario for sizing of the structural elements is a combination of relevant load cases (e.g., 2.5-*g* pull-up, full aileron deflection, taxi loads, etc.). The surface pressure loads caused by these maneuvers are computed with CFD technology (e.g., lifting-line method, full-potential flow solver, Euler flow solver). Spline interpolation techniques are used to translate this information into load vectors at the structural nodes.

4) An *aerodynamic performance module* analyzes the aircraft lift-over-drag (L/D) performance. Cruise drag is obtained by means of a high(-er)-fidelity CFD analysis (e.g., full-potential/boundary layer or Euler/boundary layer) complemented with low(-er)-fidelity methods (e.g., lifting-line theory or empirical relations) to cover the complete flight envelope.

5) An *engine sizing module* sizes the propulsion system to meet the aircraft thrust requirements. Performance data decks of several engine concepts are modeled using NLR's Gas Turbine Simulation Program (GSP) [5]. This includes high(-er)-bypass-ratio turbofans suitable for transonic transport aircraft, various low(-er)-bypass-ratio turbojet concepts suitable for supersonic aircraft, and counter-rotating open-rotor propulsions systems. All engine data decks are "rubberized" by means of applying a scale factor on the engine thrust and fuel flow data. The engines are scaled to the actual airframe thrust requirements at takeoff and top of climb.

6) A *flight mechanics module* assesses the aircraft takeoff and landing performance as well as the longitudinal handling qualities. The aircraft low-speed characteristics form important global-level constraints and call for an analysis of the aircraft in the takeoff (10) and landing configuration. The aircraft field performance requirement is met by computing the required engine TO-thrust rating, which forms one of the inputs to the engine sizing module. For tail-less aircraft, the aircraft longitudinal handling qualities form potential global-level constraint values.

7) A *mission analysis module* collects the results from all contributing analysis disciplines and computes the aircraft mission range and mission fuel burn through a numerical integration of the aircraft trajectory. Mission range constitutes one of the global-level optimization constraints, which is fed back by the optimizer to close the loop with the wing planform variables as well as aircraft MTOW.

8) An *optimization module* (not shown in Fig. 1) closes the loop between the system outputs and inputs and drives the wing design towards the (constrained) optimum point in the design space. This topic will not be elaborated on in the current chapter. Chapter 6 will show an example of how response surface techniques in combination with genetic algorithms can be used to construct Pareto fronts.

The wing optimization strategy selected is that of a *multilevel* approach. On a *global level*, the multidisciplinary optimization objective is maximum overall aircraft performance subject to a selected set of constraints. Frequently applied choices of overall aircraft performance are, for example, minimum aircraft fuel burn for a fixed mission or minimum aircraft takeoff weight for a fixed mission. The global-level constraint set can incorporate variables such as mission range, approach speed, static stability margin, cruise pitch angle, approach pitch angle, etc. The global level comprises only those design parameters that directly impact all disciplines, typically a selected set of wing (planform/airfoil) geometry parameters complemented with variables such as, for example, MTOW and start-of-cruise altitude.

For every global-level iteration, several (monodiscipline) *local-level* suboptimizations might take place. For example, on the structural mechanics level groups of wing material element thicknesses can be optimized to arrive at a minimum weight structural design. The thermodynamic cycles of the engines can be optimized to arrive at minimum installed fuel consumption. The aircraft ascent and descent trajectories can be optimized to yield minimum lost range or minimum noise footprints. A decision on which local-level optimization to activate is dependent on the objective of the study. This typically needs the knowledge of the various human experts.

The coupling between the global and local levels is not dealt with in a formal mathematical way. Iterations of analyses are used to obtain convergence, while carefully tuned underrelaxation factors are applied to avoid oscillations of design solutions.

The exchange of information between the different modules is effectuated through reading/writing to a central product database, which records the most relevant information of each aircraft variant being analyzed. The MDO framework was initially developed as a set of programs linked together under control of a UNIX script. At a later stage, graphical user interfaces have been explored to better visualize the flow of information through the MDO framework [6].

The individual building blocks currently operational in the MDO framework, as introduced in Fig. 1, are described in more detail in Sections II–VIII. Also the data exchange and interdependencies between the disciplines, which is actually far more complex than presented here, will be highlighted.

III. Parametric Geometry Generation

The geometry generation module is responsible for defining the external (aerodynamic) and internal (structural) shape of the configuration. Although the scope of the investigation is limited to wing planform/airfoil modifications only, the complete aircraft geometry definition is still required to support global-level aircraft performance evaluations. To allow fast design space explorations, the wing geometry generation is set up in a parametric way. Geometry information is required at several instances during the multidisciplinary analysis:

1) The weight and balance module (Section IV.A) requires wing leading-/trailing-edge geometry information to compute the mass of these items.

2) The structural optimization module (Section V) requires the geometry of the wing-box structural elements.

3) The aerodynamic performance module (Section VI) requires the external aircraft shape to evaluate aircraft L/D performance.

A. External (Aerodynamic) Configuration Shape

The external shape of the configuration is computed facilitating the needs of CFD-based aerodynamic evaluations (Section VI). The wing planform is modeled as a double trapezium with parameterization as illustrated in Fig. 2. Nine parameters are required to define the wing planform. Additional parameters are used to define the wing-section thickness-to-chord ratio and the section twist.

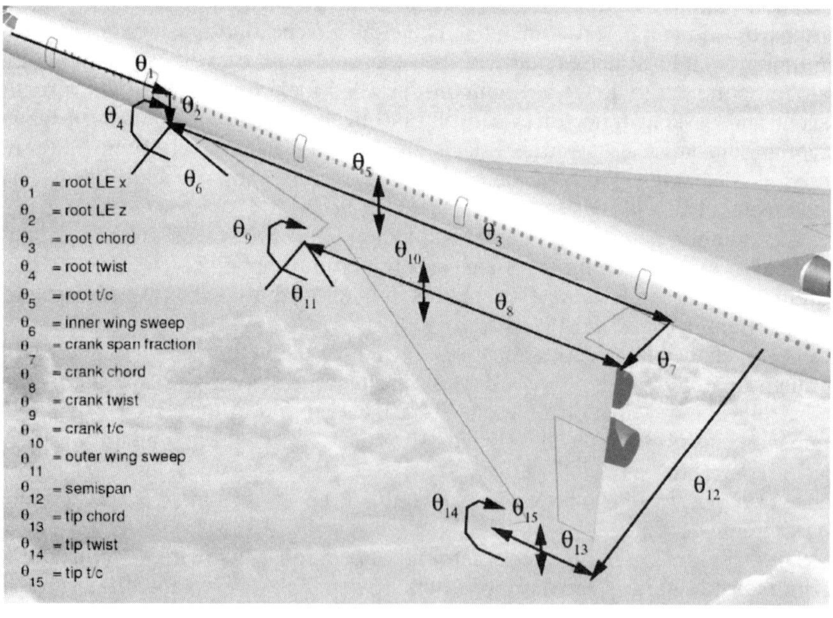

Fig. 2 Definition of 15 geometry parameters used during a supersonic transport aircraft wing planform optimization study [3, reprinted with permission of NLR].

The wing definition sections include root, crank, tip and, for certain applications, one or two additional outer-wing stations. Wing definition sections (airfoils) can either be frozen (to be read in as data files) or can be modeled parametric as well. Through this set of parameters, the geometry generation module interfaces to the global-level optimizer. The shapes of the fuselage and the tailplanes remain fixed and are read in as a priori defined data files. The engine nacelle geometry is also predefined, but is then scaled to follow thrust requirements delivered by the engine sizing module.

For delivery to the aerodynamics module, the surfaces of the individual aircraft components (fuselage, nacelles, wing, stabilizer, fin) are defined as structured surface meshes. The wing/fuselage intersection is computed, and part of the wing that falls inside the fuselage is removed. A rounded wing/fuselage fairing is added, and the wing tip is closed. The wing leading-/trailing-edge movables are deflected to the requested angles (to model, for examples, load alleviation devices). Aeroelastic wing deformation, obtained from the structural optimization module, is applied to transform the jig shape into a flight shape.

Different implementations of the geometry generation module have been developed for various applications. A dedicated Fortran program with some basic functionality has been in use by the authors, whereas others have built an implementation centering around the commercially available ICAD [7] software and investigated the use of CATIA-V5 with built-in parametric modeling support.

B. Internal (Structural) Wing-Box Shape

The internal shape of the wing box is computed facilitating the needs of FEM-based structural mechanics evaluations (Section V). The structural elements include spars, ribs, covers, and stringers. Figure 3 provides a typical example for a transport aircraft wing. In this case, the wing covers are supported by two spars and a number of ribs placed at 80-cm-span intervals. The spar and rib layout take the position of the landing-gear bays and the fuselage and engine attachment points into account. The structural topology needs to follow changes in wing planform/airfoils during the global-level optimization process. The wing structural topology layout is made by definition of a set of knowledge based rules. For multispar/multirib applications where the structural topology interacts with the landing-gear bay cutout (e.g., delta-wing planforms found on supersonic aircraft), this process can become quite complicated to automate. A structural design expert is required at this stage to formulate appropriate rules.

The wing panels are stiffened using hat-type stringers for the upper-wing covers, Z-type stringers on the lower-wing covers, and blade-type stringers on spars and ribs. To reduce structural modeling turnaround times, a number of stringers are lumped together to form "numerical" stringers rather than to model each stringer individually. Also, stringers follow the wing planform taper and do not run out on the front/aft spar. The number of *physical* stringers that are being represented by a single *numerical* stringer is a design variable under control of the structural optimization module.

The structural elements are represented by a set of structured surface meshes and are delivered to the structural optimization module for sizing of the structural elements.

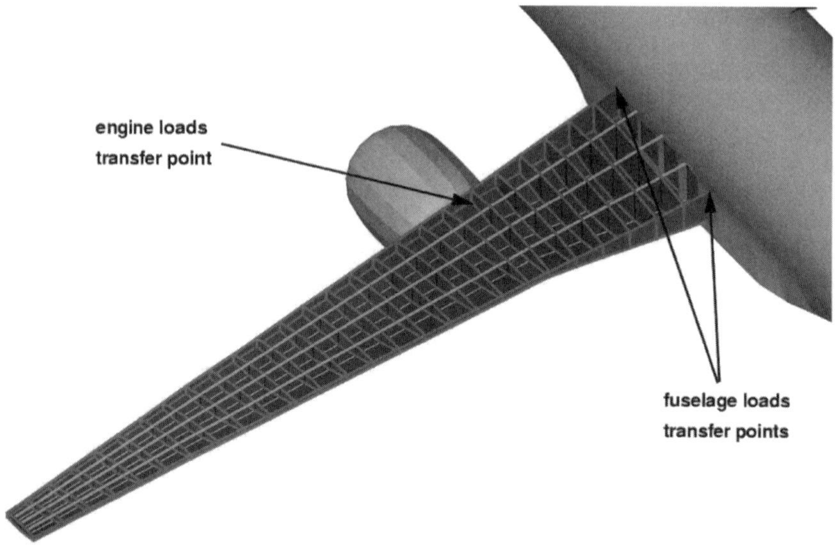

Fig. 3 Wing structural layout.

IV. Weight and Balance

The weight and balance module is responsible for keeping a record of all items contributing to the mass and center-of-gravity location of the aircraft. This information is required at several instances during the multidisciplinary analysis:

1) The structural optimization module (Section V) requires aircraft mass and mass breakdown information to infer wing loads.

2) The aerodynamic performance module (Section VI) requires aircraft mass information at start-of-cruise conditions to evaluate aircraft cruise L/D performance.

3) The engine sizing module (Section VII) requires aircraft mass information at takeoff and top-of-climb conditions to match the engine thrust with the airframe thrust requirements.

4) The flight mechanics module (Section VIII) requires center-of-gravity envelope data for handling-qualities analysis.

5) The mission analysis module (Section IX) requires fuel mass information to infer top-level aircraft performance data.

A. Mass Breakdown

Mass items are classified as follows:

1) *Fuselage and tailplanes:* Mass of these items are represented by a fixed mass, irrespective of wing geometry changes. In case mass information on these items is not available, estimates can be obtained from conceptual design handbooks.

2) *Payload:* Passengers and cargo are represented by a fixed mass per passenger.

3) *Wing-box structural items:* These include items such as covers, spars, ribs, and stringers. Wing-box mass is strongly linked to the wing geometry, which is variable during the optimization. The wing-box structural mass is delivered by the structural optimization module.

4) *Wing nonstructural items:* These include items not belonging to the primary wing-box structure, for example, leading-/trailing-edge movables and actuators, landing-gear brackets, and fuel systems. As the wing geometry varies during the optimization, the wing nonstructural items' mass will vary as well. Estimates for the masses of these items can be based on empirical "mass-to-area" relations found in conceptual design textbooks. The required movable (slats, flaps, spoilers, ailerons) areas are delivered by the geometry generation module.

5) *Landing gear:* Mass of the landing gear is modeled as fraction of MTOW. Conceptual design textbooks can supply empirical information for different gear layouts. MTOW is a variable under control of the global-level optimizer; hence, the landing-gear mass can change during the optimization process.

6) *Propulsion system:* Engine thrust requirement is a function of the wing planform, and hence engine mass varies during the optimization process. This information is delivered by the engine sizing module.

7) *Fuel:* Available fuel mass is computed as the difference between MTOW and the sum of all mass items just mentioned. Hence, available fuel mass is strongly linked to the wing geometry. The available wing volume is checked for storing all fuel at takeoff using the wing tanks exclusively. In case wing volume is a limiting factor, the available fuel is set equal to the wing volume, and the aircraft takeoff weight is reduced accordingly.

B. Load Cases Driving the Analysis Disciplines

The weight and balance module is assigned the task to assemble the individual mass components into critical *load cases* driving the individual analysis disciplines. For each load case, a full set of information comprising mass, center of gravity, flight condition, etc., is generated and written to the central database. These data present the driving scenario for the subsequent analyses:

1) *Scenario for structural wing-box optimization:* The driving scenario for sizing of the wing-box structural elements should ideally be chosen such that it comprises all load cases required to certify the airframe structure according to the FAR25 rules. Computing time restrictions might call for a less ambitious approach. If a single load case is selected, the $+2.5$-g pull-up maneuver is likely to be an interesting case already sizing large parts of the wing-box structure. The aircraft loading is then to be configured such that the wing structure experiences maximum bending moments, that is, MTOW, maximum payload, full fuel tanks.

2) *Scenario for aerodynamic performance assessment:* The driving scenario for the evaluation of the aircraft cruise L/D performance is the start-of-cruise

(SOC) Mach number and SOC altitude. Aircraft loading is maximum payload onboard and remaining fuel at SOC. Hence, the total aircraft mass is somewhat below MTOW. The fuel mass burnt between takeoff and SOC is estimated by running the mission analysis module (Section IX). It is assumed that a continuous ascent path will be flown during cruise such that the lift coefficient remains constant until reaching the end-of-cruise condition.

3) *Scenario for engine sizing:* One of the driving scenarios for matching the engine thrust with the airframe thrust requirements is the top-of-climb condition. This condition is identical, in terms of Mach number, altitude and aircraft mass, to the start-of-cruise condition. It differs in terms of the climb rate, which is now set to 500 ft/min.

4) *Scenario for handling-qualities evaluation:* One empty payload and one full payload scenario, combined with variable fuel mass, is defined to construct the outer bounds of the weight and balance envelope. This information drives the handling-qualities evaluation performed by the flight mechanics module (Section VIII).

V. Structural Wing-Box Optimization

Any change in the global-level wing shape parameters will impact the wing-box structural mass and wing-box stiffness properties. The structural optimization module is responsible for keeping track of this information. It is required at several instances during the multidisciplinary analysis:

1) The geometry generation module (Section III.A) requires wing-box deformation information to translate wing jig shapes into flight shapes.

2) The weight and balance module (Section IV.A) requires wing-box mass data to complement the aircraft mass breakdown information.

Wing-box mass is computed by a numerical volumetric integration of the material distribution over the wing-box structure. The geometry of the wing-box structural elements (i.e., covers, spars, ribs, and stringers) is delivered by the geometry generation module (Section III.B). At this stage, however, the individual wing-box structural elements have not yet been assigned structural properties (thicknesses, stringer topology, stringer pitch, etc.). These structural properties actually form *parameters* of a local-level structural optimization problem in which the *objective* is to arrive at minimal wing-box mass while the structure is sufficiently strong to withstand the various external load scenario's (Section IV.B). The strength requirements are actually formulated in terms of allowable stresses that form optimization *constraints*. Other requirements can also be imposed, for example, maximum deflection values, eigenfrequencies, or even flutter stability.

The simplest approach to infer wing-box mass is to use bending-beam theory. However, because of the computing power available today, *finite element method* (FEM)-based methods are now emerging in MDO schemes for this purpose. The FEM-based approach will need to operate on relatively coarse numerical discretizations to keep the problem turnaround time within reasonable limits. This section will detail how a structural optimization process can be set up centered around the commercially available Nastran code.

A. Structural Wing-Box Optimization Scheme

Figure 4 zooms in on the Structural Optimization Module introduced in Fig. 1 and shows how it is embedded in various loops interacting with modules from other disciplines:

1) The geometry generation module delivers the geometry of the wing-box structural elements (Section III.B).

2) The weight and balance module delivers load case scenario's (Section IV.B). In turn, the structural optimization module delivers wing-box mass information to the weight and balance module (Section IV.A).

3) The aerodynamics module delivers the surface-pressure information corresponding to the selected load cases (Section VI), for which the geometry generation module delivers the external shape (Section III.A), corrected for the wing deformation information delivered by the structural optimization module.

Inside the structural optimization module shown in Fig. 4, various submodules are identified: meshing the wing-box structural geometry, mapping of external loads (i.e., aerodynamic surface pressures and inertial loads) into force vectors on the structural element nodes, computing the structural panel allowable stresses and panel stringer layout by means of a local-level optimization process, assembling all information into a Nastran bulk data deck file, the Nastran-SOL101 based stress analysis, the Nastran-SOL200 based optimization, and the actual volumetric integration process to infer wing-box mass information.

The following dependencies between input and output data are identified in Fig. 4:

1) The definition of the structural sizing load cases requires aircraft mass breakdown information. At this stage, however, the wing-box mass is still unknown (this is the outcome of the structural sizing process); hence, the aircraft mass breakdown is still incomplete.

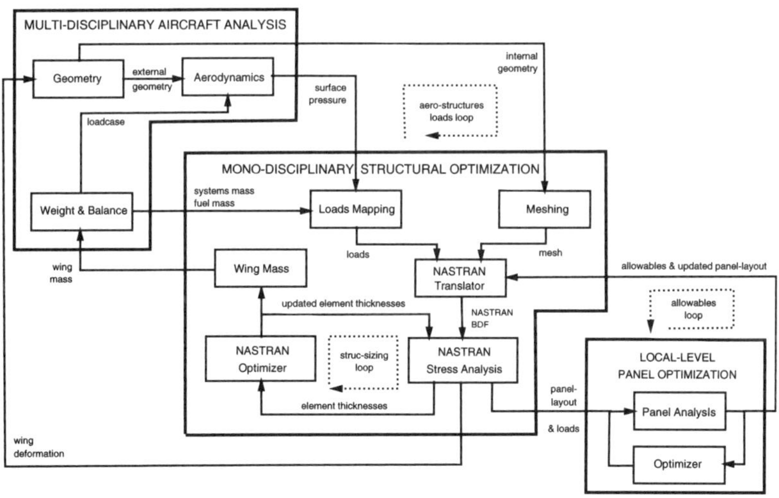

Fig. 4 Zooming in on the structural optimization module.

2) The wing structure deforms under loading. These deformations impact the aerodynamic loads. This holds especially for wings featuring high sweep angles where wing bending translates into wing twist deformation, which in turn affects the wing-span load distribution. Wing stiffness information is not available a priori as this is actually the outcome of the structural sizing process. This gives rise to an *aerostructures loads loop*.

3) The allowable internal stresses as well as the optimal stringer stiffened panel layout are a function of the individual structural panel length, width, and thickness distribution. However, the thickness distribution is unknown *a priori*. Actually this is the outcome of the structural sizing process. This gives rise to an *allowables loop*.

4) The Nastran SOL200 optimizer requires several calls to the Nastran stress analysis module.

A pragmatic approach is taken to solve the input/output dependencies. Loops are performed where needed. In case instabilities occur, appropriately chosen under-relaxation factors are applied to enforce stability of the scheme. This was found necessary in the aerostructures loads loop as well as the allowables loop where under-relaxation factors of respectively 0.33 and 0.50 are applied. An acceptable level of wing-box mass convergence is found after performing approximately five iterations through the aerostructures loads and allowables loops.

For slender wings, the load path will be more or less independent of the stiffness distribution; therefore, it is to a large extent statically determined. Such stress-driven designs are observed to converge rapidly. For stiffness-driven designs and/or more complex structures, the convergence of this pragmatic approach has been observed to be problematic. More formal MDO coordination schemes are envisaged as future extensions.

In the following sections, selected elements of the structural optimization module are discussed in more detail, and illustrative examples are provided.

B. Structural Wing-Box Model

The geometry module delivers the geometry of the individual wing structural items. Each structural item carries a nametag through which the structural item properties (e.g., item type, type of material, identification of connecting items) are passed on to the structural optimization module. The structural elements now need to be represented in Nastran's native format as a set of structural nodes ("GRID" entries in Nastran) and a connectivity table that groups these nodes into elements. Quadrilateral elements ("CQUAD4" elements in Nastran) are used to model the covers, spars, and ribs. Beam elements ("CBEAM" elements in Nastran) are used to model the stringers. The center of the beam elements are offset from the surface by half the stringer height. The stringer bending/torsional stiffness is modeled by specification of the beam element cross-sectional area and moments. The physial stringer pitch does not necessarily match the numerical stringer pitch. This is compensated by applying a scale factor to beam element area and moments being equal to the ratio of numerical vs physical stringer pitches. Each individual structural element carries a unique shell property definition ("PSHELL" and "PBEAM" entries in Nastran) defining the initial material thicknesses and contains a link to a material properties group (Nastran's "MAT1" entries). The current application defaults to Aluminum 2024-T3 for the lower covers, spars, and ribs and Aluminum 7075-T6 for the upper covers.

The wing box comprises a semispan model. Inboard of the wing/fuselage intersection, the wing box continues as a straight section until the symmetry plane. At this location, appropriate single-point constraints ("SPC" entries in Nastran) are applied to connect the two wing parts.

The structural elements are grouped into *design areas*. Each design area corresponds to a single uniformly distributed material thickness, which is eventually controlled by the Nastran SOL200 optimizer. Hence, the number of design areas correspond to the degrees of freedom in the structural optimization process. Initially, each individual wing cover, spar, and rib panel will be assigned its own design area. For certain applications, this approach was found to lead to instable design solutions where the optimizer is given too many degrees of freedom. In such cases, a number of design areas can be grouped and replaced by a single design area. This process requires judgement of the structural design engineer monitoring this process during the initial problem setup phase.

Figure 5 shows an example of a typical structural wing box model. It comprises a total of 606 quadrilaterals for the wing covers, spars, and ribs grouped into 201 design areas. Problem turnaround time for a single Nastran SOL101 stress analysis is * seconds.

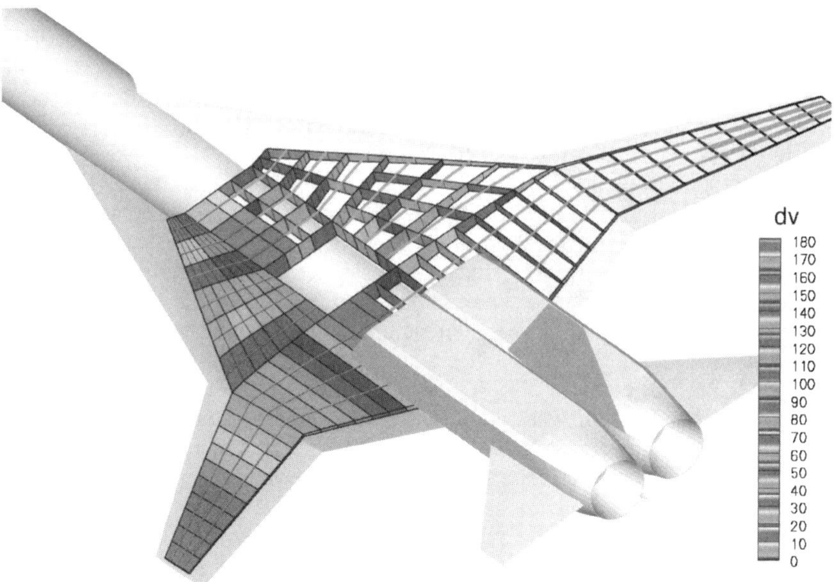

Fig. 5 Finite element mesh and design area distribution for the wing box of a supersonic business jet [4, reprinted with permission of NLR]. (See also color figure section at the back of the book.)

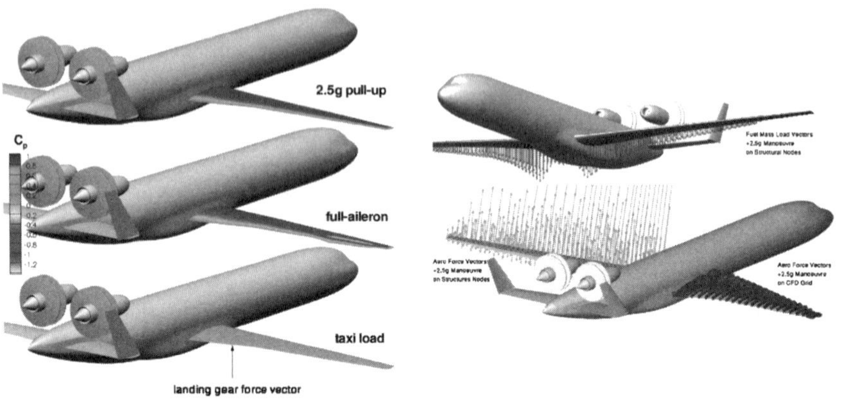

Fig. 6 Aerodynamic surface-pressure loads computed with CFD (left-hand side) and aero and inertial fuel load vectors mapping to the structural nodes (right-hand side) for an innovative low-fuel-burn aircraft concept. (See also color figure section at the back of the book.)

C. Wing Loads

The driving scenario for sizing of the wing-box structural elements is delivered by the weight and balance module (Section IV.B). The corresponding aerodynamic surface-pressure loads are provided by a dedicated flow solution for the selected load cases by a request made to the aerodynamics module (Section VI). The aerodynamic surface pressures are translated into elementary force vectors on the aerodynamic wing surface grid. These force vectors are then mapped, using spline interpolation techniques, to the structural node points on the aerodynamics/structures interface (the wing-box cover panels). The result is a set of Nastran "FORCE" entries.

In case the load-case definition includes grounds loads, a single vertical force vector is modeled to act at the structural node nearest to the main landing-gear attachment point again using a Nastran "FORCE" entry.

Wing fuel loads for the selected load cases act as hydrostatic loads on the wing-box lower-cover panels. The various wing tanks are filled in an assigned sequence to equipotential levels to reach the fuel volume prescribed by the load-case definition. Fuel is modeled as concentrated masses, using Nastran "CONM2" entries, connected to the wing lower-cover structural nodes.

All nonstructural items connected to the wing (landing gear, engines, leading- and trailing-edge movables) are delivered by the weight and balance module and are connected to the nearest structural node of the wing box, again using Nastran "CONM2" entries.

Mass items provide inertial relief loads, which will partly counter the aerodynamic surface-pressure loads. This requires definition of the acceleration caused by gravity, using Nastran's "GRAV" entry, together with the normal load factor, specified in Nastran's "LOAD" entry.

The wing box connects to the fuselage. A rib is positioned at this spanwise position. The fuselage loads enter the wing at this point. These forces are not specified explicitly. Instead, single-point constraints (Nastran's "SPC" entries) are defined at the wing/fuselage rib/spar intersection points, and Nastran will automatically place reaction forces needed for overall force and moment equilibrium at these points.

All external loads are assembled into a so-called "Nastran loads card." This information is appended to the Nastran bulk data deck file containing the structural model (Section V.B). At this point, sufficient information is available to run, for example, a Nastran SOL101 stress and deformation analysis.

Figure 6 shows an example of the wing-box sizing loads for innovative low-fuel-burn aircraft concepts featuring a high-aspect-ratio wing. The wing load cases selected in this particular example comprise a +2.5-*g* pull-up maneuver, a full-aileron deflection case, and a taxi load case. An inviscid full-potential solver is used to provide wing surface-pressure loads (Section VI.B).

D. Stress Allowables and Stringer Layout Optimization

The structural model, with a priori choices for the structural element thicknesses (Section V.B), together with the structural loads (Section V.C) would allow the computing of the actual stresses in the wing-box structure (e.g., through running Nastran SOL101). Pursuing a minimum weight fully stressed design calls for an assessment of the *actual* stresses vs the *allowable* stresses. The allowable stress levels, however, are a strong function of the structural panel dimensions (rib pitch and panel width), the stringer pitch and stringer cross-sectional characteristics, the panel material characteristics, and the panel running loads. Assuming a single typical value for the allowable stresses is one option (e.g., an allowable von Mises stress level of 250 MPa), but more accurate results can be obtained by analyzing each wing panel individually by means of a dedicated analytical panel analysis and optimization process interacting with the Nastran-based FEM analysis; see Fig. 7. This constitutes another local-level optimization loop within the wing-box structural optimization task.

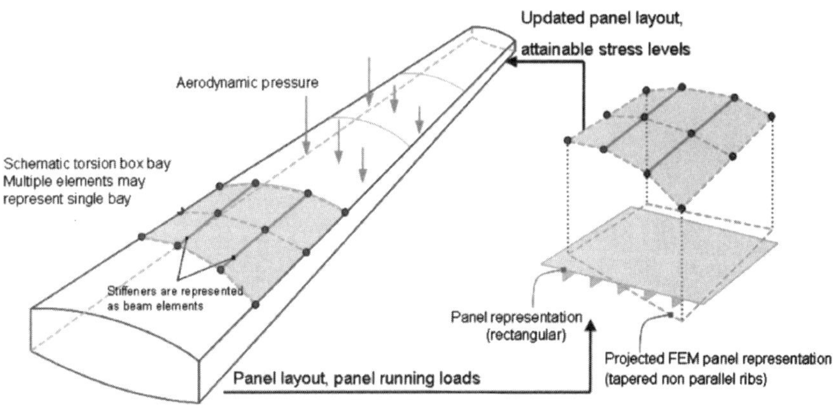

Fig. 7 Structural panel analysis and optimization loop delivers stress allowables and optimum stringer layout to the Nastran-based wing-box optimization task.

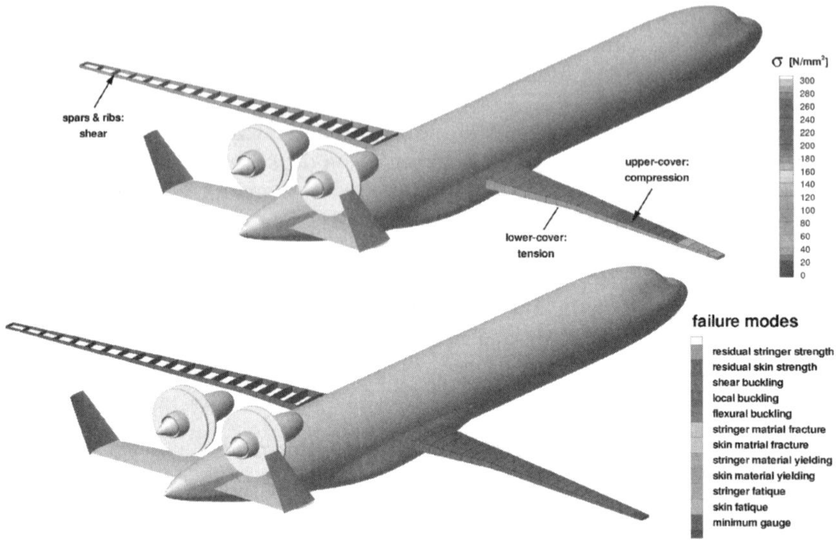

Fig. 8 Allowable stresses and critical panel failure modes for an innovative low-fuel-burn aircraft concept. (See also color figure section at the back of the book.)

Each individual part of the wing cover that runs between two successive wing ribs is modeled analytically as hat, blade, or Z-stiffened panel and is analyzed to infer the maximum allowable limit load stresses for fatigue loading (lower-wing panels) or buckling loading (upper-wing panels). In this process, the optimal values for the stringer pitch and stringer cross-sectional area are also determined.

Again this process is iterative as it requires information on the wing panel loading from the FEM analysis, while it provides input to the FEM analysis in terms of stress allowables and stringer layout.

Figure 8 shows a typical example of the critical wing-box panel failure modes and allowable stresses.

E. Structural Optimization

With the structural wing-box model, the wing loads, the allowable stresses, and the optimum stringer layout available, the actual structural optimization can be performed. Two options exist. Nastran SOL101 stress analysis can be linked to any external optimization algorithm. Alternatively, Nastran's internal optimization functionality, implemented in SOL200, can be used. This latter approach turns out to be effective.

Nastran SOL200 control runs through definition of design parameters. These design parameters ("DESVAR" entries in Nastran) are linked to the structural element properties (Nastran's "PSHELL" entries) through definition of an appropriate "design variable to property relation" ("DVPREL1" entries in Nastran). The optimization process is made responsive (through Nastran's "DRESP1" entries) to structural wing-box mass as well as the Von Mises stresses in the

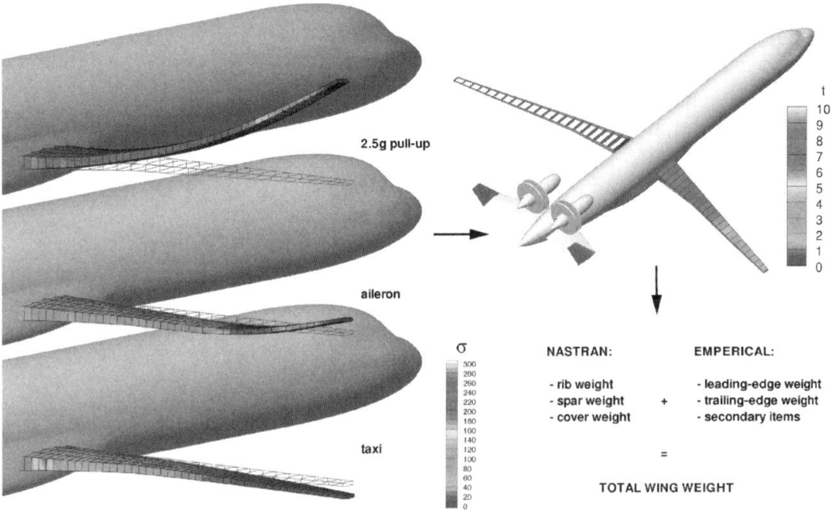

Fig. 9 Von Mises stresses and optimum material thickness distribution for an innovative low-fuel-burn aircraft concept. (See also color figure section at the back of the book.)

centers of the quadrilateral elements. Optimization constraints are defined ("DCONSTR" entries in Nastran) that link the allowable limit-load stresses (Section V.D) to the appropriate design responses. Additional constraints are formulated in terms of a 2-mm minimum element thickness. Further optimization directives on convergence criteria and number of cycles are specified.

With the SOL200 optimizations directives added, the bulk data deck file is now ready to be processed by Nastran. Convergence to a fully stressed minimum weight design is usually obtained after around 10 SOL101/SOL200 iterations. Problem turnaround time is typically three to five minutes. Figure 9 shows an example of the resulting Von Mises stress distribution and structural element material thickness distribution. Based on this information, the wing-box mass can be computed by means of a volumetric integral over the complete wing-box structure.

VI. Aerodynamics

Information concerning the aircraft aerodynamic characteristics, in terms of wing surface-pressure loading, maximum attainable lift coefficient, L/D performance and center of pressure information, is required at several instances during the multidisciplinary analysis:

1) The structural optimization module (Section V) requires the wing surface-pressure distribution during selected maneuvers to infer wing load information.

2) The engine sizing module (Section VII) requires aerodynamic drag information to infer top-of-climb thrust information.

3) The flight mechanics module (Section VIII) requires aerodynamic inputs, in terms of angle of attack vs lift, drag, and pitching-moment coefficients as well as

attainable maximum lift coefficient for aircraft field performance, stability, and control assessment.

4) The mission analysis module (Section IX) requires L/D performance information at multiple flight conditions.

5) The global-level optimizer uses cruise attitude (i.e., angle of attack) information as a constraint value.

Because of the computing power available today, CFD-based methods are now attractive means in MDO schemes to provide this information. However, as the various analyses call for aerodynamic evaluations at numerous Mach-C_L combinations, it would become prohibitively expensive to compute every entry in the Mach-C_L plane using high-fidelity CFD technology for every aircraft variant. As a compromise, CFD technology will be used for selected conditions only. Computationally less expensive methods will be used to complement this information for the off-design phases. The various CFD options and their range of applicability are reviewed in the following sections.

A. Solutions Based on the Euler Equations

Solutions based on the Euler equations are particularly useful for high-speed aircraft applications where shock waves and vortical flows dominate. For example, supersonic aircraft applications rely on this technology. For supersonic flows, the numerical discretizations can remain relatively coarse, the solutions generally converge rapidly, and boundary-layer interactions have second-order effects only such that flat-plate Reynolds corrections are sufficiently accurate.

CFD grid generation can be carried out employing block-structured grid technology [8]. This requires manual effort in meshing the baseline aircraft. Grid morphing techniques, based on volume spline techniques [9, 10], can be used to mesh subsequent aircraft variants fully automatically. Numerical flow inaccuracies as a result of grid anomalities will be consistent between aircraft variants and therefore will have little effect on the gradients in the design space. This allows the use of relatively coarse grids. As long as the grid topology and dimensions are retained, flow solutions of subsequent aircraft variants can restart from the converged flow solution of the baseline aircraft. Both measures are effective to reduce problem turnaround times to acceptable limits for MDO applications.

The Euler solutions will only capture the wave- and vortex-drag components. This needs to be complemented with the missing viscous drag component according to the Prandtl–Schlichting turbulent boundary-layer friction drag formula applied to all exposed wetted surfaces S_{wet} with characteristic length l of the aircraft,

$$C_{D_{\text{viscious}}} = \sum \frac{S_{\text{wet}}}{S_{\text{ref}}} \cdot C_F \tag{1}$$

$$C_F = \frac{1}{0.455} r_T (\log Re_l - 2.8 \log r_T)^{-2.58} \tag{2}$$

$$r_T = 1 + 0.18 M^2 \tag{3}$$

Fig. 10 Technique based on solving the Euler equations for cruise drag assessment of a supersonic business jet wing/body configuration [4, reprinted with permission of NLR]. (See also color figure section at the back of the book.)

Figure 10 shows an Euler solution example applied to a supersonic business jet configuration. Only the wing/body geometry is retained in this particular application. Viscous corrections apply to all wetted surfaces (i.e., including the tail planes and nacelles). The initial numerical grid is "hand made" by an aerodynamics expert using semi-automated techniques and comprises 57 individual blocks and 185,216 grid cells for a semispan configuration. Grid morphing to subsequent aircraft variants by means of volume spline techniques proceeds fully automatic and requires 0.5 minutes turnaround time (Pentium-4, 2 GHz). The flow solver (ENFLOW [8]) iterates towards the prescribed cruise lift coefficient in about 100 multigrid cycles in about 10 minutes to deliver cruise L/D information. The same setup is also used to provide wing surface-pressure loading to support the structural optimization process. In this case a coarser sublevel grid is used, which requires only 2 minutes turnaround time.

B. Solutions Based on the Full-Potential/Boundary-Layer Equations

Full-potential/boundary-layer solutions have proven to be very efficient in analyzing transonic flows. In case of weak shock waves and reasonably two-dimensional boundary-layer flow, conditions common for conventional transport aircraft wing with limited leading-edge sweep angles, their accuracy is almost comparable to the full Reynolds-averaged Navier–Stokes solutions.

Figure 6 shows an application of this technique to the innovative low-fuel-burn aircraft concept using the MATRICS-V code. Cruise drag is computed in 10 minutes turnaround time with inner-flow integral boundary-layer solutions interacting with the full-potential outer flow. The same technique is also used for loads evaluation (Fig. 6). In this case the flow is modeled inviscid, which reduces the problem turnaround time to about 3 minutes.

C. Solutions Based on the Linear Potential Equations

Solutions that make use of surface singularity distributions (panel codes) have been in use for decades. The main area of application is subsonic inviscid flow. Once the matrix of *influence coefficients* has been processed, solutions caused by different boundary conditions (because of change in incidence angle, control deflections, etc.) can be obtained at little additional computational cost. This allows effective computation of aircraft stability and control characteristics that can be used, for example, for aircraft handling-qualities assessment (Section VIII.B). This technique can also be used to an advantage when large amounts of structural load cases need to be computed.

D. Solutions Based on the Lifting-Line Equations

The classical lifting-line theory dates back to Prandtl in 1918. The method is based on a single lifting vortex to capture the circulation of a lift generating surface. The strength of this circulation varies along span. The rate of change in circulation is shed as a continuous sheet of trailing vorticity, which in turn causes induced velocities at the lifting-line location. This approach turns out to be very efficient in transforming two-dimensional airfoil aerodynamic coefficients into three-dimensional lifting surface data. The method can be applied in the subsonic as well as (low) transonic regiem. It applies to cleans wings as well as wings with deflected high-lift devices for which it can provide maximum attainable lift coefficient estimates. Applications are limited to relative slender wings. The required two-dimensional airfoil database can be set up with commercially available airfoil analysis codes like XFOIL, ISES, and MSES.

In the current context of wing planform optimization with fixed airfoil geometries, it does allow the assessment of the planform variation effect on global aircraft coefficients with very low computational cost. Complete polars for multiple Mach numbers can be produced in less than a minute problem turnaround time. This method is particulary suited to complement the relative expensive CFD data to cover the complete flight envelope range.

E. Solutions Based on Empirical Aerodynamic Equations

The literature provides various closed-form equations to predict lift [11] and drag data [12] of various aircraft configurations with low computational effort. These equations are well suited to complement the CFD-based L/D results for the off-design conditions.

VII. Engine Sizing

Any change in the global-level wing shape parameters will impact the aircraft thrust requirements. Data decks of several baseline engine concepts are computed using NLR's Gas Simulation Program GSP [5]. The engine data are tuned to the specific needs of the airframe by applying a scale factor to the engine geometry, thrust, fuel flow, and mass information. This information is required at several instances during the multidisciplinary analysis:

1) The geometry generation module (Section II) requires the engine scale factor to update the nacelle geometry. The external nacelle shape contributes to the airframe drag, and hence, affects global aircraft performance.

2) The weight and balance module (Section III) requires propulsion system mass information to complement the aircraft mass breakdown.

3) The mission analysis module (Section IX) uses a tuned engine data deck to compute mission fuel burn.

A. Engine Data Decks

In the coarse of various projects, data decks for different generic engine concepts have been computed using NLR's Gas Simulation Program GSP [5]. These comprise a classical turbofan engine representative for the popular CFM56 engine powering various aircraft in service today, a number of low-bypass-ratio engines suitable to drive supersonic transport and business jets cruising at Mach numbers between 1.3 and 2.0, and a turboshaft core engine combined with a counter-rotating open rotor suitable to drive a low-fuel-burn novel aircraft concept. The engine data deck file for the first two concepts comprises look-up tables with flight Mach number, altitude, turbine entry temperature as input and propulsive thrust, fuel flow, and overall efficiency as output. For the third concept, the core engine data deck is almost identical with the exception that the propulsive thrust is replaced by engine shaft power. This is complemented with a model of the open rotor, which takes the format of a conventional propeller chart, that is, the propeller advance ratio $J = V/nD$ and power coefficient

$$C_P = \frac{P}{\rho V^3 n^5}$$

as inputs and the thrust coefficient

$$C_T = \frac{T}{\rho V^2 n^4}$$

blade angle and propulsive efficiency as output. Figure 11 shows an example.

B. Engine Sizing Procedure

During the wing (planform) shape optimization process, the actual airframe thrust requirement will differ from the baseline engine thrust rating. For that purpose, the engine is rubberized by means of applying a scale factor on the engine thrust and fuel flow data, and the square root of this scale factor applies to the engine geometry (nacelle diameter and length). This scale factor is determined by comparing the actual airframe thrust requirements vs the baseline engine thrust rating at two points along the aircraft mission:

1) *Takeoff:* The required engine TO-thrust rating is computed by the flight mechanics module (Section VIII.B) based on a takeoff runway length constraint.

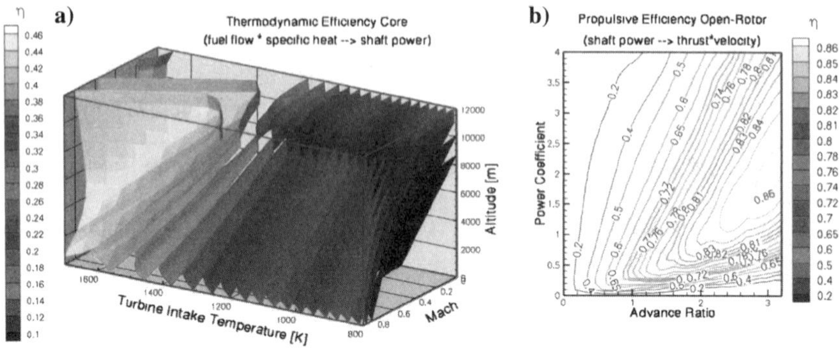

Fig. 11 Example of a data deck of a turboshaft engine driving a counter-rotating open rotor: a) the core engine thermodynamic efficiency and b) the rotor propulsive efficiency chart. (See also color figure section at the back of the book.)

2) *Top of climb:* This condition, in terms of Mach number, altitude, and aircraft mass, corresponds to the start-of-cruise condition. The required engine TOC-thrust rating corresponds to the airframe drag plus excess thrust to achieve a 500-ft/min climb rate.

The final engine scale factor is based on the maximum value of the two individual cases considered. This scale factor is used to translate the baseline data deck into a tuned engine data deck for the current aircraft after which it is ready to be used by the mission analysis module.

VIII. Flight Mechanics

The flight mechanics module is responsible for an assessment of the aircraft takeoff field performance as well as the longitudinal stability and controllability of the aircraft. This information is required at several instances during the multidisciplinary analysis:

1) The engine sizing module (Section VII) needs information on the engine takeoff thrust rating required to meet the aircraft takeoff runway length constraint.

2) The global-level optimizer uses a constraint value based on the actual center-of-gravity range vs an allowable center-of-gravity range.

A. Takeoff Field Performance

The driving scenario for takeoff distance computations is a continued takeoff at MTOW after engine failure at rotation (i.e., the decision speed V_1 and the rotation speed V_r are assumed to coincide). The takeoff distance is defined as the distance traveled until screen height (35 ft) and takeoff safety speed V_2 are reached. The analysis requires aerodynamic data on lift and drag (Section VI) and uses functionality implemented in the mission analysis flight-path integration module (Section IX). For supersonic delta-wing configurations, the minimum speed, or stall speed V_s, which plays a central role in takeoff and landing analysis, is substituted by the zero-rate-of-climb speed V_{zrc}. The zero-rate-of-climb speed is defined as that speed at which the aircraft (gear-up and out-of-ground effect) can maintain level flight with one engine inoperative and the other engine operating at maximum takeoff thrust rating. The rotation speed, liftoff speed, and takeoff safety speed are computed according to the FAR25 rules with respect to their relation to the minimum speed and required one engine-out second segment climb gradient $\gamma > 0.024$ (three or four engined configurations) or $\gamma > 0.030$ (two engined configurations). Additionally, an angle-of-attack limit ($\alpha < 12°$) at liftoff (tail scrape) is fulfilled by increasing rotation and liftoff speeds until this constraint is met.

The takeoff distance computation module is embedded in a local-level optimization loop in which the engine takeoff thrust rating is a variable. The system is run with the desired runway length constraint active to find the minimum required engine takeoff thrust rating. The minimum required engine takeoff thrust rating is one of the drivers of the engine sizing module described in Section VII.B.

B. Longitudinal Handling Qualities

Longitudinal handling qualities can become an issue strongly linked to the wing (planform) shapes for tail-less aircraft concepts. The blended wing body and the supersonic transport aircraft are representative examples in this respect. In such cases, additional longitudinal handling-qualities constraints from the flight mechanics discipline need to be incorporated at the global level.

The weight and balance module provides mass distribution information on the various aircraft components. This information is used to infer the *actual* bounds of the aircraft weight and balance envelope caused by different payload/fuel scenarios. At the same time, *tolerable* bounds, arising from a number of handling-qualities constraints, can be displayed in the same format. A global-level constraint value can be defined based on the distance between the actual vs tolerable center-of-gravity bounds.

Figure 12 illustrates this procedure applied during the blended-wing/body aircraft case where the following handling-qualities constraints were imposed: takeoff rotation pitch acceleration $dq/dt > 3\ \mathrm{deg/s^2}$, providing a forward c.g. limit; elevon trim at minimum speed $|\delta| < 20\ \mathrm{deg}$, providing both a forward and an aft c.g. limit; push-over pitch acceleration at minimum speed $dq/dt < -5.5\ \mathrm{deg/s^2}$, providing an aft c.g. limit; and aircraft longitudinal instability, formulated in terms of the "time to double" for flight-path distortions $T_2 < 0.8$ s, providing an aft c.g. limit.

Figure 12 identifies a rotation control power problem for full-payload takeoff of the blended-wing/body where the actual full-payload center of gravity violates the forward tolerable center-of-gravity bound.

IX. Mission Analysis

The main analyses modules in Fig. 1 (structural mechanics, aerodynamics, propulsion, flight mechanics) are centered around performing single-discipline analysis tasks. All of this information now needs to be synthesized into

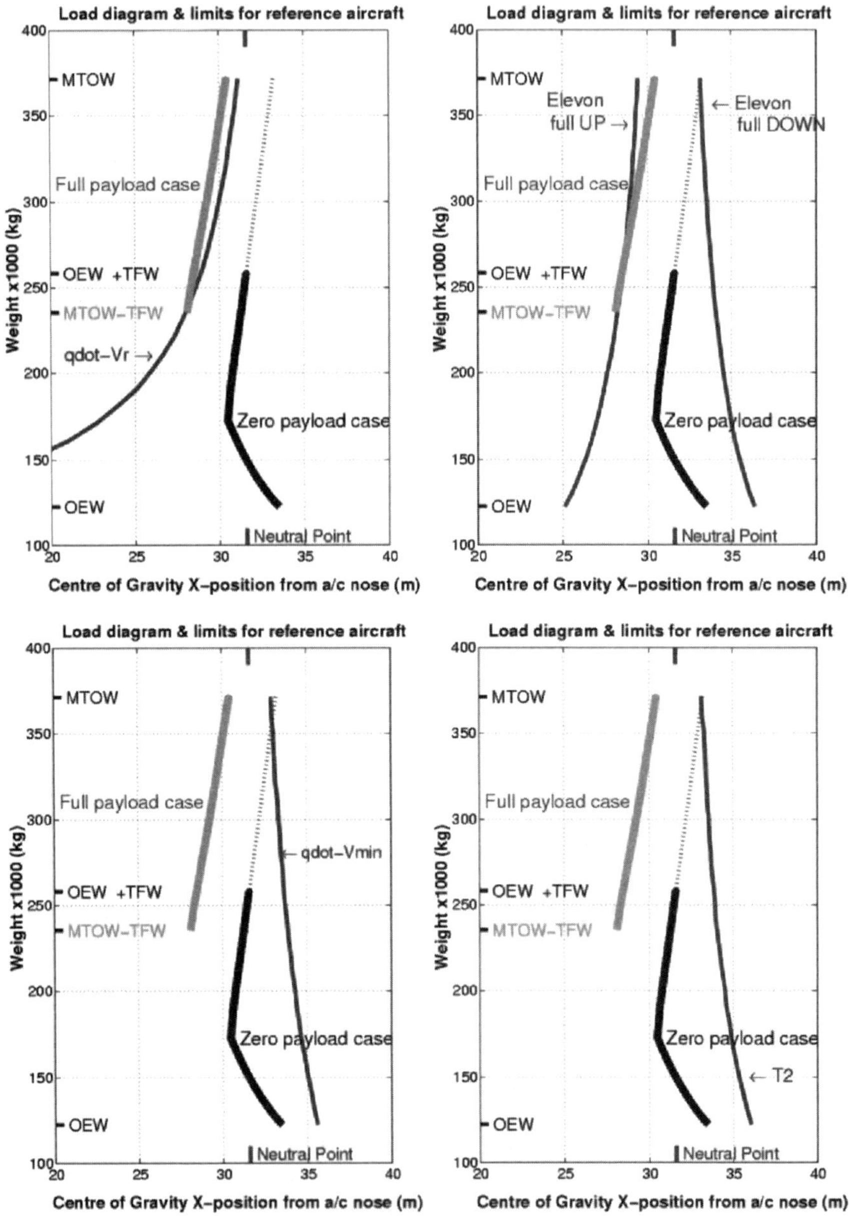

Fig. 12 Weight and balance check for the blended-wing/body configuration [1, reprinted with permission of NLR]. (See also color figure section at the back of the book.)

global-level aircraft performance data that can be used to close the loop with the wing design parameters. Global-level aircraft performance data comprise, for example, maximum aircraft mission range and mission fuel burn. The mission analysis module is assigned the task to combine mass, aerodynamic, and engine data to provide this information.

Two options arise for computing the aircraft range performance. The simplest option is to make use of the classical Breguet range equation, with optional *lost range* corrections (i.e., correcting for climb and descent flight phases, fuel reserves, etc.). A more elaborate, yet more precise, option is to perform a complete numerical integration of the aircraft trajectory from takeoff to landing.

A. Breguet Range Equation

Range according to the Breguet equation is formulated as

$$R = C \cdot sfc_{\text{cruise}} \cdot \text{L/D}_{\text{cruise}} \cdot \ell n\left(\frac{\text{MTOW}}{\text{MTOW} - m_{\text{fuel}}}\right)$$

which shows how the three disciplines of propulsion sfc_{cruise}, aerodynamics $(\text{L/D})_{\text{cruise}}$, and weights m_{fuel} combine into the overall aircraft range performance. Note that the effect of wing mass (-variation) is not directly seen in the equation but runs though the available fuel mass (-variation).

The use of the Breguet range equation option requires aerodynamic/engine data at cruise only. Therefore the aerodynamic analysis needs to be performed at a single point, which might be useful to reduce overall problem turnaround time. The drawback of this option is that the design will basically become, from an aerodynamic point of view, a *single-point design*, which is insensitive to off-design performance.

B. Mission Trajectory Integration

In the second option, mission range is obtained through a numerical integration of the aircraft trajectory. At each time instant, aerodynamic drag, engine thrust, and fuel flow are computed and used to propagate the aircraft state to the next discrete time step. This open-loop integration process is brought under control by a model of a flight-management/flight-control/auto-throttle system, which effectively steers the aircraft along a number of "waypoints" in the Mach vs altitude plane.

The mission range evaluation actually constitutes another local-level optimization process. The timing of the various events (e.g., initiation of descents) as well as ascent and descent trajectories are optimized in order to maximize range under the constraint of arriving at the alternate destination with the required fuel reserve.

Figure 13 provides an example of a supersonic business jet going through its mission. For this particular application, the mission profile has been adopted from the NBAA standard mission and modified to the needs of supersonic flight. The flight phases comprise takeoff, climb to Mach 0.95 @ 31,000 ft altitude, accelerated climb to start-of-cruise altitude/cruise Mach number, continuous ascent

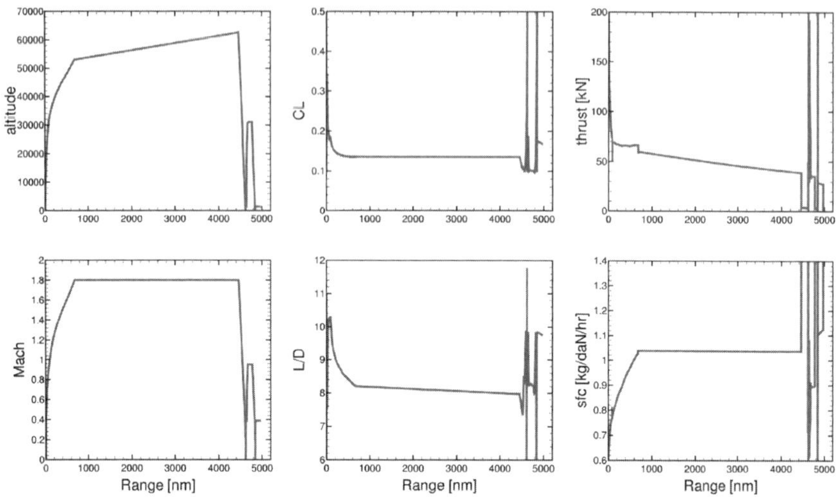

Fig. 13 Mission trajectory integration results for a supersonic business jet [4, reprinted with permission of NLR].

supersonic cruise, descent and approach, a missed landing at scheduled destination, 5 minutes holding at 250 KCAS/1,500 ft altitude, climb to 31,000 ft/Mach 0.95, 200 miles diversion flight, descent and approach to alternate destination, and land with fuel reserves sufficient for an additional 30 minutes visual flight rules loitering flight.

During cruise flight, the aircraft continues to climb such that the lift coefficient C_L remains constant. Also the dimensionless engine operation parameter T/δ, in which δ indicates the ratio of the actual pressure at altitude vs sea-level pressure, remains constant. Hence, a continuous ascent cruise flight allows both airframe and engine to operate at their optimum (design) point during an extended part of the mission. This procedure might deviate from operation practice. (Air traffic control might insist on the conventional step-climb procedure.) Yet, the rationale for this choice is that now only a single (costly) high-fidelity CFD analysis needs to be performed to capture the complete cruise flight leg of the mission.

The mission trajectory integration option holds the advantage that the range computation is relatively precise; the design takes off-design conditions into account. From an aerodynamic point of view, it becomes a *multipoint design* with a well-balanced mix of cruise vs off-design objectives. The drawback, obviously, is that it requires complete drag polar data from the aerodynamic module as well as complete engine performance data sets from the propulsion module to be available.

X. Conclusions

This chapter showed how the chain-of-analysis tools, typically required for multidisciplinary wing design optimization work, can be integrated into a framework. The framework was described to some level of detail and highlighted the setup of appropriate individual discipline analysis tools, the flow of information between disciplines, and the many interdependencies that exist between these disciplines.

The problem turnaround time constraint is a key factor in setting up this framework. The number of degrees of freedom in the geometrical shape of an aircraft wing can run from 10 to 20 (planform only) up to 100 (planform and airfoil shapes), which can lead to 100 through 10,000 analyses calls to find the (global?) optimum in the design space. It is for this reason that setting up a framework for wing design optimization is not a trivial task in which one simply sequences all available highest-fidelity analysis tools. Also the many data interdependencies do not allow a straightforward sequential ordering of the individual discipline analysis.

This chapter illustrates a practical multidisciplinary wing design optimization framework that meets the problem turnaround time requirements. The authors have successfully applied this framework to a number of different aircraft wing design optimization problems and obtained practical results.

References

[1] Laban, M., Arendsen, P., Rouwhorst, W. F. J. A., and Vankan, W. J., "A Computational Design Engine for Multi-Disciplinary Optimization with Application to Blended Wing Body Configuration," AIAA Paper 2002-5446, Sept. 2002.

[2] Herrmann, U., "CISAP: Cruise Speed Impact on Supersonic Aircraft Planform; a Project Overview," AIAA Paper, 2004-4539, Sept. 2004.

[3] Laban, M., "Multi-Disciplinary Analysis and Optimization of Supersonic Transport Aircraft Wing Planforms," AIAA Paper 2004-4542, Sept. 2004.

[4] Laban, M., and Herrmann, U., "Multi-Disciplinary Analysis and Optimization Applied to Supersonic Aircraft, Part 1: Analysis Tools," AIAA Paper 2007-1857, April 2007.

[5] Visser, W. P. J., and Broomhead, M. J., "GSP, A Generic Object Oriented Gas Turbine Simulation Environment," American Society of Mechanical Engineers, Paper 2000-GT-002, May 2000.

[6] Vankan, W. J., "A Spineware Based Computational Design Engine for Integrated Multi-Disciplinary Aircraft Design," AIAA Paper 2002-5445, Sept. 2002.

[7] La Rocca, G., Krakers, L., and van Tooren, M., "Development of an ICAD Generative Model for Blended Wing-Body Aircraft," AIAA Paper 2002-5447, Sept. 2002.

[8] Boerstoel, J. W., et al., "ENFLOW, A Full Functionality System of CFD Codes for Industrial Euler/Navier-Stokes Flow Computations," *2nd International Symposium on Aeronautical Science and Technology (SASTI'96)*, Jakarta, Indonesia, 24–27 June 1996.

[9] Spekreijse, S. P., and Kok, J. C., "Semi-Automatic Domain Decomposition Based on Potential Theory," *7th International Conference on Numerical Grid Generation in Computational Field Simulations*, Whistler, Canada, 25–28 Sept. 2000, http://handle.dtic.mil/100.2/ADA391985 [accessed July 2009].

[10] Spekreijse, S. P., Prananta, B. B., and Kok, J. C., "A Simple, Robust and Fast Algorithm to Compute Deformations of Multi-Block Structured Grids," *Proceedings of the*

8th International Conference on Numerical Grid Generation in Computational Field Simulations, Honolulu, Hawaii, 3–6 June 2002.

[11] Torenbeek, E., *Synthesis of Subsonic Airplane Design*, Delft Univ. Press, Delft, The Netherlands, 1982, pp. 467–485.

[12] Torenbeek, E., *Synthesis of Subsonic Airplane Design*, Delft Univ. Press, Delft, The Netherlands, 1982, pp. 487–524.

Chapter 6

Meta-Modeling and Multi-Objective Optimization in Aeronautic MDO

W. J. Vankan,* W. Lammen,† and R. Maas†

National Aerospace Laboratory (NLR), Amsterdam, The Netherlands

Abbreviations

AAE = average absolute error
ANN = artificial neural networks
BLUP = best linear unbiased predictor
CFD = computational fluid dynamics
DOE = design of experiments
FEM = finite element method
GA = genetic algorithm
kriging-cC = kriging-constant-cubic fit, kcc
kriging-cE = kriging-constant-exponential fit, kce
kriging-cG = kriging-constant-Gauss fit, kcg
kriging-lC = kriging-linear-cubic fit, klc
kriging-lE = kriging-linear-exponential fit, kle
kriging-lG = kriging-linear-Gauss fit, klg
LoD = lift over drag
MAE = maximum absolute error
MAPE = mean absolute percentage error
MARS = multivariate adaptive regression splines
MDA = multidisciplinary design analysis
MDO = multidisciplinary design and optimization
MNSGA = nondominated sorting genetic algorithm in MATLAB®

---

*Senior Scientist.

†Medior R&D Engineer.

MOO = multi-objective optimization
MTOW = maximum takeoff weight
NSGA = nondominated sorting genetic algorithm
poly0 = zeroth-order polynomial fit
poly1 = first-order polynomial fit
poly2 = second-order polynomial fit
poly3 = third-order polynomial fit
poly4 = fourth-order polynomial fit
poly5 = fifth-order polynomial fit
RBF = radial basis functions
RMSE = root-mean-square error
SBX = simulated binary crossover
SVR = support vector regression

I. Introduction—MDA, MDO, and MOO in Aeronautics

THE continuous development of the methodologies for aircraft design and analysis is aimed at achieving higher levels of detail in shorter analysis turn around cycles. Moreover, driven by ever-increasing technical and commercial requirements as a result of global competition, more detailed design analyses are being required and applied in earlier phases of the aircraft design process where there are still very many degrees of freedom and few restrictions on the design space [1]. The analyses traditionally used in the early phases of the aircraft design are mostly based on semi-empirical rules [2]. Although computationally efficient, these analyses often have a limited range of validity, accuracy, and flexibility. Therefore, these methods are gradually being replaced by the more generic geometry- and physics-based detailed design analysis methods that are generally applicable and potentially highly accurate [3]. However, these methods are mostly computationally expensive.

Also, the required design analyses in aircraft multidisciplinary design and optimization (MDO) need to comprise case-dependent sets of disciplines and might be difficult to combine into an integrated aircraft design system. Therefore such an integrated aircraft design system is usually developed for a specific range of design problems, such as optimization of blended-wing-body [4, 5] or transonic transport aircraft planform [6]. Moreover, such an integrated aircraft design system often requires specific software (e.g., particular analysis tools) and hardware (e.g., dedicated computer servers) and is therefore prone to operational issues such as temporal unavailability of servers or licenses. The computational cost of (some of) the analyses in the integrated aircraft design system is another issue to be handled, especially when used within automated search or optimization loops that typically might require many (e.g., thousands) design analysis evaluations.

To deal with the just-mentioned issues, various approximation and interpolation methods have been proposed. These methods have been shown to effectively deal with the issues [7–9] by providing compact, accurate, and computationally efficient representations of the considered properties of the underlying aircraft design (in optimization context also termed as design objectives or fitnesses). In this chapter the term *meta-models* will be used to identify the just-mentioned methods; in literature also alternative terms like *surrogate models* or *response surface models* are used. The key of the approach presented in this chapter lies in the decoupling of the (computationally expensive) aircraft design analyses from the automated search and optimization process. The search process now makes use of the compact and computationally efficient meta-model and allows for high flexibility for further investigations. The meta-model, in turn, is based on a set of results that are obtained with the computationally expensive integrated aircraft design analyses. For this purpose, effective sampling of the multidimensional design domain is achieved by design of experiments (DOE) methods. In the selected sample points the design objectives and constraints are evaluated by (parallel) computations with integrated aircraft design analysis systems.

Many different fitting methods are available for the creation of the meta-models (e.g., [7]), each with different advantages for different types of problems. In this chapter we consider a number of different fitting (interpolation and approximation) methods and compare the meta-model quality for these fitting methods on the basis of appropriately defined quality-of-fit criteria. A proprietary multidimensional and multimethod data fitting software tool (MultiFit) [10] is used to statistically analyze the data sets that result from the design evaluations and to generate meta-models using different fitting methods. The representativeness of the meta-models is investigated, and the most suitable meta-models are applied in the aircraft design process where several optimization algorithms are used to find the most promising aircraft designs.

Traditionally, optimization methods are applied to automatically search for design variations by which one property, expressed in an objective function, is optimized subject to certain constraints. However, in aircraft design it is desirable to take into account multiple product properties of interest and optimize those simultaneously. Typical properties in aircraft design that need to be optimized are, for example, weight, lift, drag, payload, range, etc. (e.g., [11–14]). Such simultaneous optimization can be expressed by a multi-objective optimization (MOO) problem formulation, which can be considered as a generalization of the traditional (single-objective) optimization problem [15]. Besides being more general, MOO problems are also more costly to solve (in terms of numbers of calculations, i.e., objective function evaluations) and can become quite complex in cases of high dimensional optimization problems. A multitude of methods is available for dealing with MOO problems [16–23], of which many have evolved quite recently and are still under significant development. Among these methods, the category of evolutionary algorithms has proven to be quite efficient for MOO problems. Some examples of evolutionary algorithms are genetic algorithms, evolutionary computing, evolution strategies, and evolutionary programming, of which some will be described in more detail in the following sections.

In this chapter the methodology for creating and assessing the meta-models is described. This methodology is based on a combination of various existing techniques for multidimensional data fitting and statistical assessment. It is aimed at the efficient application of many different fitting methods to multidimensional data sets, as such extending on the commonly used tools in this area, which are often limited in dimensionality or number of available fitting methods. The different fitting methods that are considered here are described in the next sections, as well as the way in which the quality-of-fit criteria are determined. Also, some state-of-the-art methods for multi-objective optimization are described. For illustration of the approach and its benefits, an aircraft design optimization case study is described. Here, the meta-modeling approach is applied to the multidisciplinary design and multi-objective optimization of aircraft wings. In this case study, the wing planform and the maximum takeoff weight of a generic transonic aircraft are optimized where the aircraft range and fuel consumption are the objective functions. More information on the multidisciplinary design analysis (MDA) system is provided in Chapter 5 (multidisciplinary wing design optimization).

II. Meta-Modeling

In aircraft design problems the aim is to improve or optimize the characteristics (design objectives such as performance, behavior, etc.) of the product by variation of its properties (design parameters such as shape, material, etc.). In general, the product's properties (denoted as a vector $\boldsymbol{x}$ in this chapter) and characteristics (denoted as a vector $\boldsymbol{y}$ in this chapter) are expressed as real-valued (continuous) quantities, and their interdependency (denoted as a vector function $\boldsymbol{f}$ in this chapter) is nonlinear [$\boldsymbol{y} = \boldsymbol{f}(\boldsymbol{x})$]. Evaluation of the function $\boldsymbol{f}$ is often costly (in terms of time and computer resources) and can involve (iterative) computational analyses (e.g., finite element or computational fluid dynamics). Because of these computational complications, it is desirable to retrieve efficiently, that is, using as few as possible function evaluations, the desired product's characteristics $\boldsymbol{y}$ in the considered design domain (i.e., for the set of allowable values of the product's properties $\boldsymbol{x}$). This can be achieved, for example, by direct optimization of $\boldsymbol{y}$ for $\boldsymbol{x}$ using efficient gradient-based optimization algorithms (e.g., [24]).

However, lack of accurate gradient or Jacobian information ($\mathrm{d}\boldsymbol{y}/\mathrm{d}\boldsymbol{x}$), limited robustness and reliability of the computational analyses, or convergence into local suboptima, can hamper the effectiveness of this approach. It is therefore beneficial to apply "gradient-free" global search methods, such as genetic algorithms and pattern search (e.g., [25]), besides the gradient-based optimization algorithms. The large number of evaluations of the objective function [$\boldsymbol{y} = \boldsymbol{f}(\boldsymbol{x})$] that are typically needed by these search methods do not allow for a high computational cost per evaluation. An efficient approximate representation [$\boldsymbol{y}^* = \boldsymbol{f}^*(\boldsymbol{x})$], also-called meta-model, of the design problem is therefore required.

A variety of fitting methods, such as polynomial regression, neural networks, and kriging models, is available for creating such meta-models [7] from sampled data sets ($\boldsymbol{x}^i$, $\boldsymbol{y}^i$) of the design problem. To achieve an optimal meta-model, the most suitable fitting method for the considered design problem should be applied. There exist various statistical verification and cross validation methods [25, 26] by which the quality (or "representativeness") of the different meta-models can be assessed and the most suitable method can be identified. These methods consider a (small) subset of the data set, called *verification points*, in which the error of the prediction ($\boldsymbol{y} - \boldsymbol{y}^*$) is evaluated for a fit that is made for the data set without the verification points. For user-friendly creation, assessment and comparison of fits with a wide range of multidimensional interpolation and approximation methods, a dedicated software tool [10] was developed, which is used in the further evaluation and optimization of the considered design problem.

The two most used model classes in the analyses in this chapter are the polynomial models and the kriging model, which are described next. Besides these two, several other interpolation and approximation methods are available [10], such as artificial neural networks (ANN) and radial basis functions (RBF) models, of which a brief description is also given next. Kriging and RBF models are interpolating models, that is, they are exact in the given data points, whereas polynomial models and ANN are approximating models.

A. Polynomials

One commonly used and relatively simple class of meta-models is the polynomial regression models, which are also often referred to as response surface models. The coefficients of the polynomial regression model are usually determined according to a least-squares procedure, that is, by minimizing the sum of the squared residuals (the difference between the values predicted by the polynomial model and values in the data set). Polynomial regression models provide a global representation of the data, that is, consist of one regression function for the whole considered domain. Polynomial functions of different orders can be used as regression functions. The well-known linear regression model is the first-order polynomial approximation to the data. Of course also a zeroth-order regression model of the data can be made, which is equal to the mean value of the data. Higher-order models can also be built. However, with higher-order polynomials the risk of so-called overfitting of the data increases. In this case the (high-order) polynomial follows the data so closely such that it does capture local oscillations in the data, but it does not properly capture the global behavior of the data, as such compromising the quality of the prediction for points not in the data set. Furthermore, the reliability of polynomial models for extrapolation [i.e., predicting output values for input values (slightly) outside the range of the available data] is in general very poor, in particular for high-order polynomial models.

To illustrate the creation of polynomial regression models, consider the following simple example, in which we assume a data set that consists of N input-output combinations (data points) $\{(\boldsymbol{x}^i, y^i)|i = 1, \ldots, N\}$, where the input is three dimensional $\boldsymbol{x}^i = (x_1^i, x_2^i, x_3^i)$ and the output y^i is one dimensional. The following second-order polynomial regression model can be fitted to the data:

$$\begin{aligned} y^* = {} & c_0 + c_1x_1 + c_2x_2 + c_3x_3 + c_4x_1x_2 + c_5x_1x_3 + c_6x_2x_3 \\ & + c_7x_1^2 + c_8x_2^2 + c_9x_3^2 \end{aligned} \tag{1}$$

This model y^* is linear in the coefficients c_j, and the optimal c_j are found by least squares:

$$\boldsymbol{c} = \begin{pmatrix} c_0 \\ \vdots \\ c_9 \end{pmatrix} = \left(A^TA\right)^{-1}A^T\boldsymbol{y}_{\text{data}}$$

where

$$A = \begin{pmatrix} 1 & x_1^1 & x_2^1 & x_3^1 & x_1^1 x_2^1 & x_1^1 x_3^1 & x_2^1 x_3^1 & x_1^{1^2} & x_2^{1^2} & x_3^{1^2} \\ \vdots & \vdots & \vdots & \vdots & \vdots & \vdots & \vdots & \vdots & \vdots & \vdots \\ 1 & x_1^N & x_2^N & x_3^N & x_1^N x_2^N & x_1^N x_3^N & x_2^N x_3^N & x_1^{N^2} & x_2^{N^2} & x_3^{N^2} \end{pmatrix} \tag{2}$$

and

$$\boldsymbol{y}_{\text{data}} = \begin{pmatrix} y_1 \\ \vdots \\ y_N \end{pmatrix}$$

where A is the so-called "design matrix" for the input data $\{\boldsymbol{x}^i\}$. For other input dimensions and other orders of the polynomial, the same procedure can be applied. A polynomial of order m in a k-dimensional input space has $\begin{pmatrix} m+k \\ k \end{pmatrix}$ coefficients, which is then also the minimum number of the input-output combinations $(\boldsymbol{x}^i, y^i)$ data set that are required for creating this polynomial model. In practice the minimum number of data points needed for an acceptable fit is about 1.5 the number of coefficients.

B. Kriging

Kriging models have in the past decade evolved as a new and quite effective meta-modeling technique. Originating from geostatistics, kriging models have found their way into modeling of data coming from computationally expensive simulations, often referred to as design and analysis of computer experiments (DACE). Lophaven and coworkers [27] have presented a widely used MATLAB® implementation of the kriging method (the MATLAB kriging toolbox DACE). Although the derivation of the kriging models is often fully statistical, a more deterministic explanation is given in this chapter following some of the notions given by Lophaven et al. [27].

The kriging model combines a global regression model and local refinements into one interpolation model of a given data set. As will be explained next, the global model is a generalized regression model on top of which a local model is built using a correlation function to couple points in the data set that are in the vicinity of each other. (Illustrated in Fig. 1.)

As explained in the preceding section, polynomial functions can be fitted to a data set using a least-squares approach. The underlying assumptions in the least-squares fit [Eq. (2)] is that the residuals between the fit and the data are equally important in all data points and no correlation exists between the residuals in two data points. In kriging models this correlation between the residuals is assumed to exist given by the correlation function g defined as

$$g(\boldsymbol{\theta}, \mathbf{x}, \tilde{\mathbf{x}}) = g[(\theta_1 \ldots \theta_k), (x_1 \ldots x_k), (\tilde{x}_1 \ldots \tilde{x}_k)] = \prod_{j=1}^{k} g_j(\theta_j, x_j, \tilde{x}_j)$$

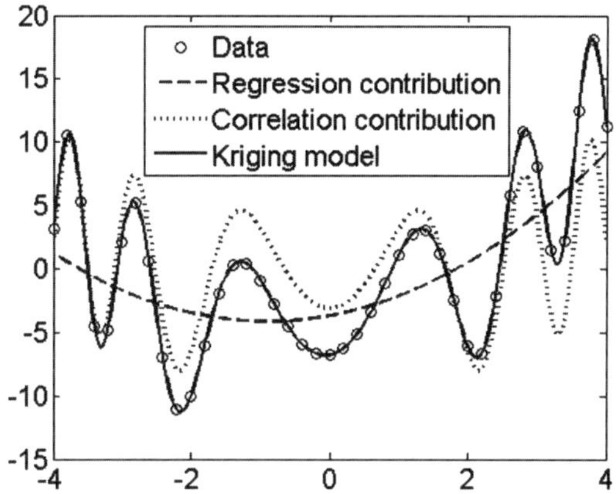

Fig. 1 Plot of a simple kriging approximation and the different contributions to the model. (See also color figure section at the back of the book.)

where θ is a scaling parameter for the correlation function and x and $\tilde{x}$ are two different input points, such that

$$g(\boldsymbol{\theta}, \mathbf{x}, \tilde{\mathbf{x}}) \geq 0 \tag{3}$$

and

$$g(\boldsymbol{\theta}, \mathbf{x}, \mathbf{x}) = 1$$

An example of such a function is a Gaussian correlation function:

$$g[(\theta_1, \ldots, \theta_k), (x_1, \ldots, x_k), (\tilde{x}_1, \ldots, \tilde{x}_k)] = \prod_{j=1}^{k} e^{-\theta_j (x_j - \tilde{x}_j)^2} \tag{4}$$

The regression part of the kriging model is now obtained by regression model to the data points with a low-order polynomial (up to second order) taking the correlation relations into account. The best fit to the data is obtained by the so-called generalized least-squares fit to the data:

$$\boldsymbol{c} = \begin{pmatrix} c_0 \\ \vdots \\ c_9 \end{pmatrix} = \left(A^T R^{-1} A\right)^{-1} A^T R^{-1} \mathbf{y}_{\text{data}} \tag{5}$$

where A again is the polynomial design matrix of x, as in Eq. (2) and $\mathbf{y}_{\text{data}}$ is the vector containing the output values of all data points. R is the $N \times N$ correlation matrix of the data points $\boldsymbol{x}^1, \ldots, \boldsymbol{x}^n$:

$$R = \begin{pmatrix} g(\boldsymbol{\theta}, \mathbf{x}^1, \mathbf{x}^1) & g(\boldsymbol{\theta}, \mathbf{x}^1, \mathbf{x}^2) & \cdots & g(\boldsymbol{\theta}, \mathbf{x}^1, \mathbf{x}^n) \\ g(\boldsymbol{\theta}, \mathbf{x}^2, \mathbf{x}^1) & g(\boldsymbol{\theta}, \mathbf{x}^2, \mathbf{x}^2) & \cdots & g(\boldsymbol{\theta}, \mathbf{x}^2, \mathbf{x}^n) \\ \vdots & \vdots & \ddots & \vdots \\ g(\boldsymbol{\theta}, \mathbf{x}^n, \mathbf{x}^1) & g(\boldsymbol{\theta}, \mathbf{x}^n, \mathbf{x}^2) & \cdots & g(\boldsymbol{\theta}, \mathbf{x}^n, \mathbf{x}^n) \end{pmatrix} \tag{6}$$

The kriging model consists of this regression model plus a correction model based on the correlation function such that the model is interpolating. For a quadratic regression function and a correlation function g, the kriging model can be written as (again for the three-dimensional example)

$$y^*(\mathbf{x}) = \begin{pmatrix} 1 & x_1 & x_2 & x_3 & x_1x_2 & x_1x_3 & x_2x_3 & x_1^2 & x_2^2 & x_3^2 \end{pmatrix} \begin{pmatrix} c_0 \\ \vdots \\ c_9 \end{pmatrix} + r(\mathbf{x})v(\mathbf{x}) \tag{7}$$

where $r(\boldsymbol{x}) = [g(\theta, \boldsymbol{x}, \boldsymbol{x}^1) \cdots g(\theta, \boldsymbol{x}, \boldsymbol{x}^n)]$ and $r(\boldsymbol{x})v(\boldsymbol{x})$ make up the local correction to the regression model. Here $v(\boldsymbol{x})$ is an n-dimensional vector that can be derived easily because the kriging method as explained here is an interpolating method in the known data points. Combining Eq. (7) for all of the known data points gives

$$\boldsymbol{y}_{\text{data}} = A\boldsymbol{c} + Rv(x) \quad \Rightarrow \quad v(x) = R^{-1}\left(\boldsymbol{y}_{\text{data}} - A\boldsymbol{c}\right) \tag{8}$$

Thus for an arbitrary point x the kriging approximation becomes

$$y(x)^* = a(x)\boldsymbol{c} + r(x)R^{-1}\left[\boldsymbol{y}_{\text{data}} - a(x)\boldsymbol{c}\right], \qquad \text{with}$$
$$\boldsymbol{c} = \begin{pmatrix} c_0 \\ \vdots \\ c_9 \end{pmatrix} = \left(A^T R^{-1} A\right)^{-1} A^T R^{-1} \boldsymbol{y}_{\text{data}} \tag{9}$$

where the first term on the right-hand side represents the global regression model and the second term represents the local correlation model.

In Lophaven et al. [27], three different regression models are considered: zeroth- (i.e., constant), first- (i.e., linear), and second- (i.e., quadratic) order polynomials. If the regression is restricted to the constant function, then one speaks of *ordinary kriging*, whereas kriging with higher-order polynomial regression is called *universal kriging*. Also several correlation functions can be applied. Besides the already mentioned Gaussian correlation function, also the exponential and the cubic spline correlation function are available [27].

For convenience, the formulations of the exponential and the cubic spline correlation function are also given here:

Exponential correlation function:

$$g[(\theta_1, \ldots, \theta_k), (x_1, \ldots, x_k), (\tilde{x}_1, \ldots, \tilde{x}_k)] = \prod_{j=1}^{k} e^{-\theta_j |x_j - \tilde{x}_j|} \tag{10}$$

Cubic spline correlation function:

$$g[(\theta_1, \ldots, \theta_k), (x_1, \ldots, x_k), (\tilde{x}_1, \ldots, \tilde{x}_k)] = \prod_{j=1}^{k} g_j(\theta_j |x_j - \tilde{x}_j|) \tag{11}$$

with

$$g_j(v) = \begin{cases} 1 - 15v^2 + 30v^3 & v \leq 0.2 \\ \frac{5}{4}(1 - v)^3 & 0.2 < v < 1 \\ 0 & v \geq 1 \end{cases}$$

The correlation functions have scale parameters θ_j. Although kriging models yield interpolating fits through the data points, independent of the values for the θ_j parameters, the shape of the model does depend on these θ_j parameters. If all θ_j are large, then the overall model will be (approximately) the regression model with local spikes around the available data points, whereas for small θ_j the model will interpolate more smoothly through the data points. In the creation of the kriging models, optimal values for the θ_j are determined based on minimization of the log-likelihood formulation [27]

$$\begin{aligned} \theta_{\text{opt}} &= \arg\min_{\theta} \left\{ -\frac{1}{2} k \log\Big[\big(\boldsymbol{y}_{\text{data}} - Ac\big)^T R^{-1} \big(\boldsymbol{y}_{\text{data}} - Ac\big) \Big] + \log[\det(R)] \right\} \\ &= \arg\min_{\theta} \left\{ \Big[\big(\boldsymbol{y}_{\text{data}} - Ac\big)^T R^{-1} \big(\boldsymbol{y}_{\text{data}} - Ac\big) \Big]^2 \det(R)^{\frac{1}{k}} \right\} \end{aligned} \tag{12}$$

As a closing remark, the kriging predictor has another interesting property. It can be proven that under certain restrictions a kriging model is a so-called best linear unbiased predictor (BLUP) in the data values y^i. This is generally the starting point of the explanation of the kriging models from a statistical point of view.

C. Artificial Neural Networks and Radial Basis Functions

In this section, ANN and RBF models will only be briefly described. Further information about those methods can be found in [28]. In this description only feed-forward ANNs with backpropagation learning functions are considered. The chosen architecture consists of input nodes, one hidden layer and an output layer. The hidden layer has a tanh (tan-sigmoid) activation function, while the output layer has a linear activation function. The architecture of this type of ANN is explained in the Fig. 2.

Because of the nonlinear activation functions of the nodes in the hidden layer, nonlinear input-output relations can be modeled. However building an ANN, that is, computing the optimal weights, requires a training for which in general many input-output combinations (i.e., data points) are needed. As a general rule of thumb, the minimum number of data points required is often computed as three times the number of connections in the network {i.e., $3[(n_{\text{in}}+1)n_h+(n_h+1)n_{\text{out}}]$}.

The RBF method considered here is an interpolating method on all data points. In this method so-called radial basis functions are centered on the input data points. Although the radial basis function can have (almost) any radial symmetric form, they are normally Gaussian functions:

$$f(\boldsymbol{x}, \boldsymbol{c}) = e^{-\theta\|\boldsymbol{x}-\boldsymbol{c}\|^2} \tag{13}$$

where $\boldsymbol{c}$ is the center of the radial basis function (Fig. 3), $\|\boldsymbol{x}-\boldsymbol{c}\|$ is the (Euclidean) distance from $\boldsymbol{x}$ to $\boldsymbol{c}$, and θ is a parameter (chosen in advance) that determines the width of the radial basis function.

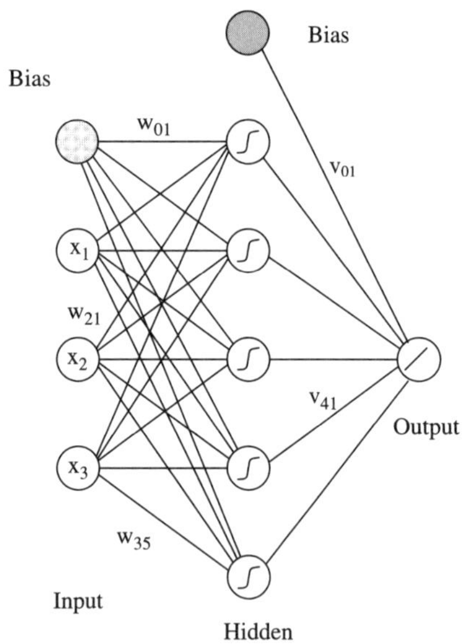

Fig. 2 Example of a feed-forward ANN. The inputs (including a constant bias) are multiplied by weights w_{ij} and summed at a hidden node after which the transfer function is applied. The resulting values from the hidden layer are multiplied by the weights v_{ij} and then summed to give the output of the network.

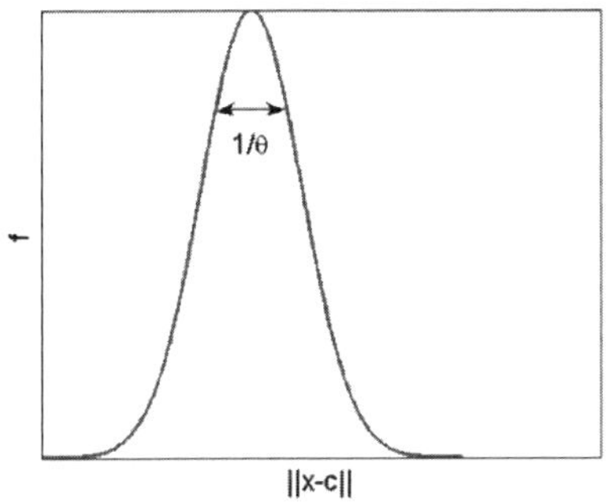

Fig. 3 Sketch of a radial basis function.

The predictive model based on the data pairs $\{(\boldsymbol{x}^1, y^1), \ldots, (\boldsymbol{x}^n, y^n)\}$ is then

$$y(\boldsymbol{x}) = \sum_{j=1}^{n} w_j e^{-\theta\|\boldsymbol{x}-\boldsymbol{x}^j\|^2} \quad \text{with}$$

$$\begin{pmatrix} w_1 \\ \vdots \\ w_n \end{pmatrix} = \begin{pmatrix} e^{-\theta\|\boldsymbol{x}^1-\boldsymbol{x}^1\|^2} & \ldots & e^{-\theta\|\boldsymbol{x}^1-\boldsymbol{x}^n\|^2} \\ \vdots & \ddots & \vdots \\ e^{-\theta\|\boldsymbol{x}^n-\boldsymbol{x}^1\|^2} & \ldots & e^{-\theta\|\boldsymbol{x}^n-\boldsymbol{x}^n\|^2} \end{pmatrix}^{-1} \begin{pmatrix} y^1 \\ \vdots \\ y^n \end{pmatrix} \tag{14}$$

D. Assessments of Meta-Models

Having so many different possibilities to build meta-models, tools to select the best model for the task at hand are needed. Some of the methods just presented (e.g., kriging models, RBF) are interpolating methods, which by definition are exact in the data points used in the modeling, while the approximating methods (polynomials, ANN) are not. However, it is not the exact value in the known points that is the most important, but the capability of the meta-model to generalize well over the total input space. For assessment of this capability, there are different methods available [10]: 1) validation by an independent data set that was not used in the creation of the model, 2) leave-one-out cross validation, and 3) P-fold cross validation.

In the first method the available data are split into two sets. With the first set the meta-model is built, while the approximation error on the second set is considered. A drawback of this method is that the comparison of the methods is only done in (normally a few) selected points, and thus this method has only local validity and is not very robust. On the other hand, the advantage is that the fitting and validation are only done once, which results in a better efficiency.

In the second method, the leave-one-out cross validation, each single data point is used in turn for validation, whereas a meta-model is built with the other

$N - 1$ data points. While now some global quality measure is obtained, the drawback of this method is that all considered fitting methods have to be applied and validated N times, which can be computationally expensive for large data sets.

The third method, the p-fold cross validation method, provides a global coverage of the estimated error while keeping computation time low. In this method the data set is split into p disjoint subsets of about equal size. While one subset is kept as a validation set, the models are built using the rest of the data. Notice that when $p = N$ this method is equal to the leave-one-out method. However, when N is large, a p-fold cross validation with a moderate p value can be a computationally efficient alternative to leave-one-out cross validation with regard to obtaining a global fit quality measure.

To assess the fit quality, several error metrics can be applied to the validation points, for example:

RMSE:

$$\sqrt{\frac{\sum_{i}^{N_{\text{val}}} (y_i - y_i^*)^2}{N_{\text{val}}}}$$

AAE:

$$\frac{\sum_{i}^{N_{\text{val}}} |y_i - y_i^*|}{N_{\text{val}}}$$

MAE:

$$\max_{i=\{1,\ldots,N_{\text{val}}\}} |y_i - y_i^*|$$

MAPE:

$$\frac{\sum_{i}^{N_{\text{val}}} \left|\frac{y_i - y_i^*}{y_i}\right|}{N_{\text{val}}} \cdot 100\%$$

The first three measures are absolute measures, quantifying different misfit properties of the model. The fourth error measure is a relative error measure. In case of leave-one-out or p-fold cross validation, the predictions of the validation points by all the different fits are taken together, and then the error formulas are applied.

E. Illustration of Fitting Methods

To conclude this section, some examples of fits and error measures on a simple sine function are presented. The one-dimensional data set consists of 11 data points $(x_k, y_k) \in \{[\frac{2\pi}{10}k, \sin(\frac{2\pi}{10}k)] | k = 0, \ldots, 10\}$. To these data points, four

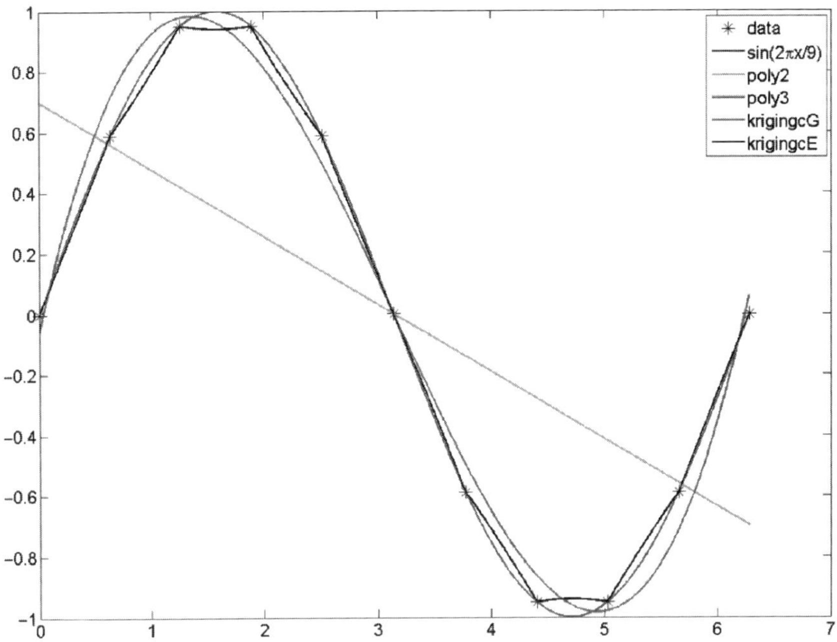

Fig. 4 Several models fitted to a period of the sine function. (See also color figure section at the back of the book.)

models are fitted: a second-order polynomial (poly2), a third-order polynomial (poly3), a kriging model with constant regression function and Gaussian correlation function (krigingcG), and a kriging model with constant regression function and exponential correlation function (krigingcE).

The results are plotted in Fig. 4. The krigingcG model gives the best representation of the sine function, while the poly2 model based on this data set approximates the underlying sine function very poorly. To assess the quality of the different fits, 629 data points are generated with the sine function $\{(k/100), \sin(k/100)|k = 0, \ldots, 628\}$. These 629 data points are then used as validation points, and the four described error measures are computed for the four different models (Table 1). In Fig. 4 the results are presented, again indicating, according to each of the four error measures, that krigingcG is the best model in this case.

Table 1 Fit quality for various fitting functions

	poly2	poly3	krigingcG	krigingcE
RMSE	0.466995	0.079639	1.25E-06	0.030739
AAE	0.417659	0.071183	6.06E-07	0.025641
MAE	0.699474	0.124509	4.17E-06	0.059566
MAPE	184.6193	18.9017	0.159321	4.942767

III. Optimization—MDA, MDO, and MOO in Aeronautics

Product design mostly aims for improvement (or optimization when possible) of one or more product properties. Traditionally, optimization methods (single objective) are applied to automatically search for design variations by which one property, expressed in an objective function, is optimized subject to certain constraints. However, it would be sensible to take into account all of the product properties of interest and optimize those simultaneously. Typical properties of interest in aircraft design are, for example, weight, lift, drag, payload, range, etc. (e.g., [10–14]). Such simultaneous optimization can be performed by MOO methods, which can be considered as a generalization of the traditional (single-objective) optimization approach [15]. Besides being more general, MOO methods are also more costly (in terms of calculations, i.e., objective function evaluations) and can become quite complex in cases of high-dimensional optimization problems.

A multitude of methods is available for dealing with MOO problems [16–23], of which many have evolved quite recently and are still under significant development. Besides the traditional gradient-based optimization algorithms, also another category of algorithms is considered, which can be characterized as evolutionary algorithms. Evolutionary algorithms can be considered as computer-based problem solving systems that use computational models of some of the known mechanisms of evolution as key elements in their design and implementation [29]. Some examples of different types of evolutionary algorithms are genetic algorithms, evolutionary computing, evolution strategies, and evolutionary programming.

This section provides a description of the general multi-objective optimization problem formulation and an overview of several state-of-the-art algorithms for solving these multi-objective optimization problems.

A. Multi-Objective Optimization

Multi-objective optimization, as opposed to single-objective optimization, considers the optimization of vector-valued objective functions instead of single scalar-valued objective functions. In the case of single-objective optimization, comparing solutions in single-objective space is trivial: we simply say (in case of a minimization problem) that a solution $\boldsymbol{x}^1 \in \boldsymbol{X}$ is ***better*** than another solution $\boldsymbol{x}^2 \in \boldsymbol{X}$ if $y^1 < y^2$, where $y^1 = f(\boldsymbol{x}^1)$ and $y^2 = f(\boldsymbol{x}^2)$.

As an illustration, consider the solution of the simple single-objective optimization problem:

$$\min_{x} y = \sin(x); \quad x \in [0, 2\pi]$$

with the solution $\boldsymbol{x}^1 = 3\pi/2$ with objective value $y^1 = -1$ (Fig. 5).

Note: For single-objective optimization problems, there can exist several optimal solutions in the decision space $\boldsymbol{X}$. These solutions can be either ***local*** or ***global***, where a local minimum represents a minimum in a (small) subspace of the considered decision space $\boldsymbol{X}$, and a global minimum represents the overall minimum in the whole considered decision space $\boldsymbol{X}$.

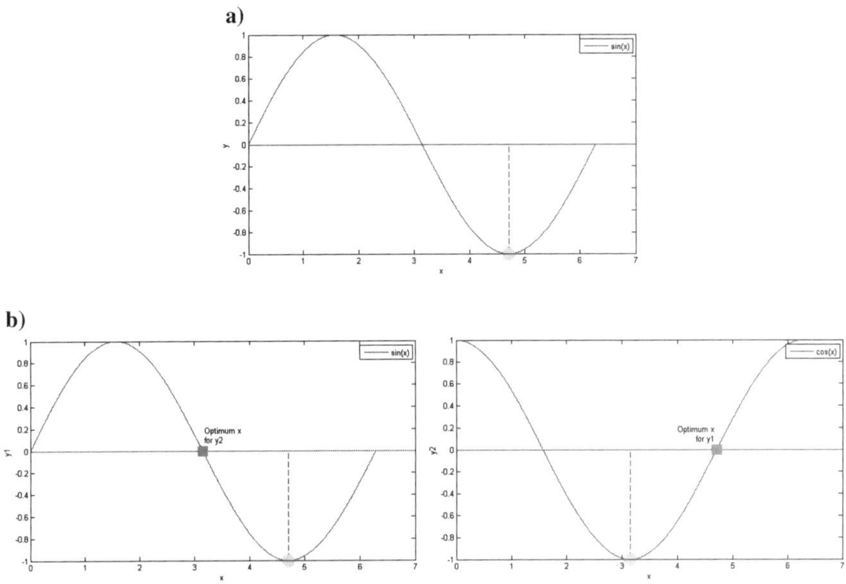

Fig. 5 (a) Optimum (dot at $x = 4.71$) of the simple single-objective function $y = \sin(x)$. (b) Example of two objective functions in a simple multi-objective optimization problem.

Multi-objective optimization, which can be considered as a generalization of single-objective optimization, deals with vector-valued objective functions $\boldsymbol{y} = \boldsymbol{f}(\boldsymbol{x})$. The definition of optimality in this case is nontrivial because of the indefiniteness of the relation among the multiple-objective functions. Hence, comparing two solutions of the vector-valued objective function, for example, $\boldsymbol{y}^1$ and $\boldsymbol{y}^2$ in the points $\boldsymbol{x}^1$ and $\boldsymbol{x}^2$, is not straightforward. For example, in the simple case where we have two objective functions, $\boldsymbol{y} = (y_1, y_2)$, both to be minimized, then the first objective function y_1 might be ***lower*** in $\boldsymbol{x}^1$ than in $\boldsymbol{x}^2$, while the second objective function y_2 is ***greater*** in $\boldsymbol{x}^1$ than in $\boldsymbol{x}^2$. Alternatively, in another point $\boldsymbol{x}^3$, y_2 might be lower than in another point $\boldsymbol{x}^4$, while y_1 is not.

As an illustration, consider the following simple multi-objective optimization problem:

$$\min_x \; y_1 = \sin(x), \quad y_2 = \cos(x); \quad x \in [0, 2\pi]$$

The optimum points of each of these two objective functions separately are indicated in the Fig. 5b (dots at y_1 and $y_2 = -1$); the values of the one function in the optimum of the other function are also indicated (dots at y_1 and $y_2 = 0$).

Obviously the optima of the two objective functions do not coincide, so that there is no clear unique solution to the multi-objective optimization problem.

To resolve such multi-objective optimization problems, consider the following definition, which is based on the concept of Pareto optimality [30]

(Fig. 6a). According to this concept, an objective vector $\boldsymbol{y}^1$ is said to *dominate* any other objective vector $\boldsymbol{y}^2$ ($\boldsymbol{y}^1 \prec \boldsymbol{y}^2$) if the following two conditions hold: 1) no component of $\boldsymbol{y}^1$ is greater than the corresponding component of $\boldsymbol{y}^2$, and 2) at least one component of $\boldsymbol{y}^1$ is smaller than $\boldsymbol{y}^2$. Accordingly, we can say that a solution $\boldsymbol{x}^1$ is *better* than another solution $\boldsymbol{x}^2$, that is, $\boldsymbol{x}^1$ *dominates* $\boldsymbol{x}^2$ ($\boldsymbol{x}^1 \prec \boldsymbol{x}^2$), if $\boldsymbol{y}^1 = \boldsymbol{f}(\boldsymbol{x}^1)$ dominates $\boldsymbol{y}^2 = \boldsymbol{f}(\boldsymbol{x}^2)$. For example, let decision vector $\boldsymbol{x}^1 = -1$ and $\boldsymbol{x}^2 = 1$, and their corresponding objective vectors are $\boldsymbol{y}^1 = (1, 1)$ and $\boldsymbol{y}^2 = (9, 1)$, that is, in that case $\boldsymbol{y}^1$ dominates $\boldsymbol{y}^2$. Additionally, a solution vector $\boldsymbol{x}^u \in \boldsymbol{X}$ is said to be *Pareto optimal* if there exists no $\boldsymbol{x}^v \in \boldsymbol{X}$ for which $f(\boldsymbol{x}^v)$ dominates

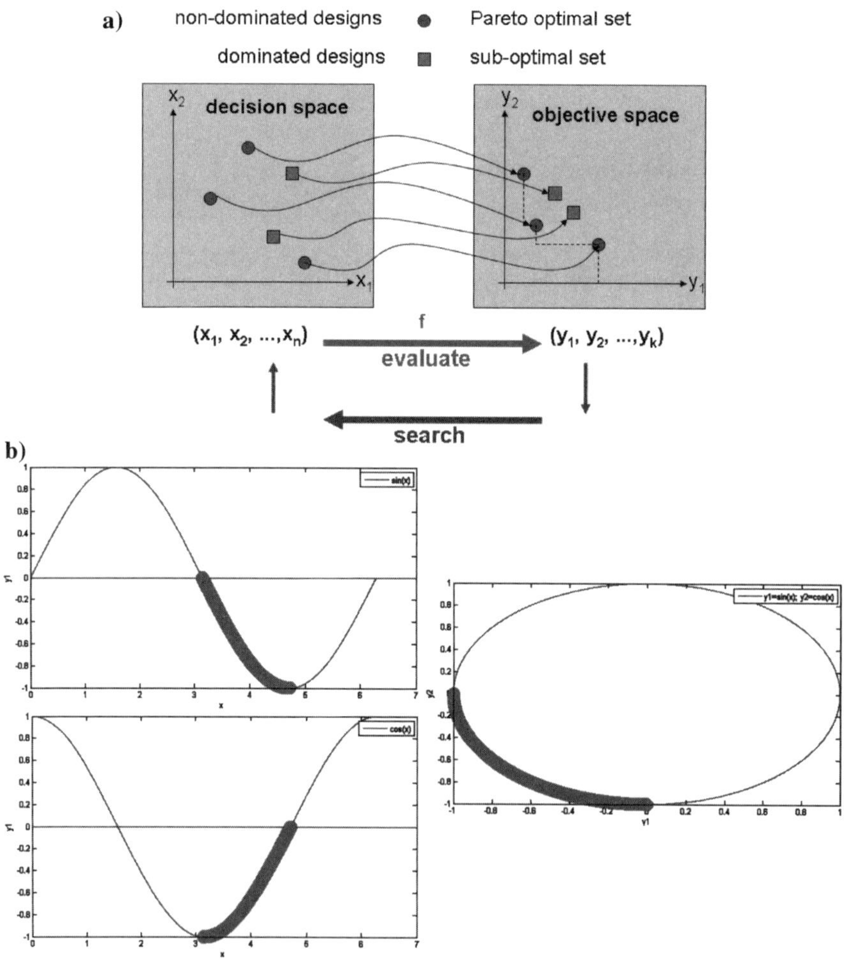

Fig. 6 (a) Illustration of the solution of a general multi-objective optimization problem. (b) Illustration of the solution of a general multi-objective optimization problem.

$f(\boldsymbol{x}^u)$. The set of (Pareto) optimal solutions in the decision space $\boldsymbol{X}$ is in general denoted as the *Pareto optimal set* $\boldsymbol{X}^* \subseteq \boldsymbol{X}$, and we denote its image in objective space as *Pareto front* $\boldsymbol{Y}^* = f(\boldsymbol{X}^*) \subseteq \boldsymbol{Y}$. This concept of Pareto optimality is illustrated in Fig. 6a.

The simple multi-objective optimization problem just presented is now handled according to the concept of Pareto optimality. The solution then consists of the Pareto front for y_1, y_2 as indicated by the thick lines in Fig. 6b ($x \in [\pi, 3\pi/2]$).

In summary, the solution of an MOO problem is adequately defined by the *Pareto optimal* definition, which can be considered as a multidimensional generalization of a single-objective optimum.

There exist many different algorithms to find the *Pareto optimal set*. However, in accordance with the No Free Lunch Theorems [31], there is no best algorithm for all classes of optimization problems. Therefore different methods, which give different performance for different classes of optimization problems, will be briefly presented and used in this study. The focus will be on the so-called evolutionary algorithms. These algorithms have developed strongly in the last decade and have shown to be very effective for multi-objective optimization problems. Also, various commercial implementations of evolutionary algorithms for multi-objective optimization problems are currently available, for example, in MATLAB [32].

B. Evolutionary Algorithms and MOO

Engineering optimization problems involve a number of characteristics due to which these problems are difficult to be solved by classical numerical optimization algorithms [17]. Some of these characteristics are existence of multiple conflicting objectives, existence of multiple optimums (local and global), existence of nonlinear constraints, nonsmooth or even noncontinuous functions, and stochastic and uncertainties in functions describing the optimization problem.

The term *evolutionary algorithm* stands for a class of stochastic optimization methods that mimic nature's evolution. Therefore these algorithms borrow some terminology from natural evolution. For instance, we call solution vectors *individuals*, a set of individuals is called *population*, and the objective function is called *fitness*. Roughly speaking, a general stochastic search algorithm consists of three parts: 1) a working memory that contains the currently considered solution candidates, 2) a selection module, and 3) a variation module as depicted in Fig. 7 [15] while *mating selection* mimics the competition for reproduction and *environmental selection* mimics the "survival of the fittest" among living beings: the other principle, *variation*, imitates the natural capability of creating "new" and "improved" living beings by means of so-called *recombination* and *mutation* operators [15].

An evolutionary algorithm is characterized by three features [15]: 1) a set of solution candidates is maintained (memory, Fig. 7), 2) a mating selection process is performed on this set (selection, Fig. 7), and 3) several solutions can be combined in terms of recombination to generate new solutions (variation, Fig. 7). Several evolutionary algorithm methodologies such as *genetic algorithms*, *evolutionary programming*, and *evolution strategies* have been proposed [15]. In this chapter, we look mainly at genetic algorithms (GAs).

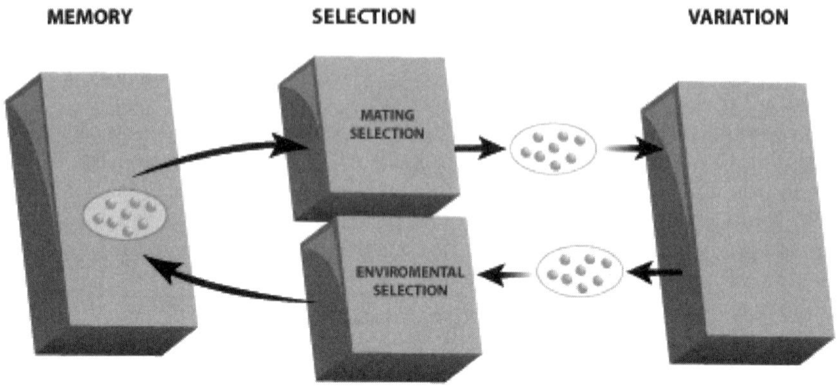

Fig. 7 Components of a general stochastic search algorithm as identified in [15].

In general, the evolutionary algorithm process is as follows. Say, at the first *generation* we have a population consisting of randomly created individuals. All of these individuals have values for their objective functions (or fitness). The evolutionary algorithm performs a *selection* process to select the most appropriate individuals (individuals that have the best fitness values). All of the best individuals found are kept in memory for the second generation (see Fig. 7). The question now is how to produce new individuals that will be better than the best individuals found in the first generation. Here, the natural evolution principles for variation are adopted in the algorithm. First, we consider *recombination*. Recombination is the process where from two selected individuals, called *parents*, certain parts are recombined to form new individuals, called *children*. Hence, a new pair of individuals is created by a recombination function ψ_{recomb} according to $(\boldsymbol{x}^{new1}, \boldsymbol{x}^{new2}) = \psi_{recomb}(\boldsymbol{x}^1, \boldsymbol{x}^2)$. The second process is called *mutation*. Mutation is applied in order to enforce genetic diversity from one generation to the next. In many evolutionary algorithms mutation is applied with a certain probability to the individuals (children) created by the recombination process. The mutation process creates one new individual by just changing small parts of a selected individual. Hence, a new individual is created by a mutation function ψ_{mut} according to $\boldsymbol{x}^{new} = \psi_{mut}(\boldsymbol{x})$. These two steps of selection and variation by recombination and mutation are iteratively repeated until a converged optimum solution is found, or until a predefined maximum number of iterations (or generations) has been achieved.

There exist many multi-objective evolutionary algorithms (MOEAs), which have recently been proposed in the literature; some of which are considered here more closely: the nondominated sorting genetic algorithm II (NSGA-II) [18], the ε-dominated multi-objective evolutionary algorithms, ε-MOEA [19], and the ε-dominated nondominated sorting genetic algorithm ε-NSGA [33]. These algorithms will be investigated in more detail because these algorithms have the most promising properties for aircraft design optimization.

In both NSGA-II and ε-MOEA algorithms, similar recombination and mutation operators are applied. They use *simulated binary crossover* [34] for recombination and *polynomial mutation*.

With simulated binary crossover (SBX) the children $[\boldsymbol{x}^{(1,t+1)}, \boldsymbol{x}^{(2,t+1)}] = \psi_{recomb}[\boldsymbol{x}^{(1,t)}, \boldsymbol{x}^{(2,t)}]$ of two parents of generation t are computed with the following equations:

$$\begin{aligned} x_i^{(1,t+1)} &= 0.5\big[\big(x_i^{(1,t)} + x_i^{(2,t)}\big) - \beta_{1,i}\big(x_i^{(2,t)} - x_i^{(1,t)}\big)\big] \\ x_i^{(2,t+1)} &= 0.5\big[\big(x_i^{(1,t)} + x_i^{(2,t)}\big) - \beta_{2,i}\big(x_i^{(2,t)} - x_i^{(1,t)}\big)\big] \end{aligned} \tag{15}$$

and

$$\gamma_i = \begin{cases} 1 + 2\dfrac{x_i^{(1,t)} - x_i^L}{x_i^{(2,t)} - x_i^{(1,t)}} & \text{for child 1} \\ 1 + 2\dfrac{x_i^U - x_i^{(2,t)}}{x_i^{(2,t)} - x_i^{(1,t)}} & \text{for child 2} \end{cases} \tag{16}$$

where x_i^U and x_i^L are the lower and upper bounds on the decision variables x.

The non-negative user-defined parameter, *distribution index for crossover* η_c controls the shape of the probability distribution of the spread of the children. A large value of η_c gives a higher probability for creating near-parent solutions and a small value of η_c allows distant solutions to be selected as offspring.

$$\beta_{\text{qi}} = \begin{cases} (u_i\alpha_i)^{\frac{1}{\eta_c+1}} & \text{if} \quad u_i \le \dfrac{1}{\alpha_i} \\ \left(\dfrac{1}{(2 - u_i\alpha_i)}\right)^{\frac{1}{\eta_c+1}} & \text{otherwise} \end{cases} \tag{17}$$

with u_i a random number between 0 and 1, and

$$\alpha_i = 2 - \gamma_i^{-(\eta_c+1)} \tag{18}$$

From Eqs. (17) and (18), one can see that the first child will lie between lower bound and midpoint between first and second parents, while the second child will lie between midpoint between first and second parents and upper bound. This implies that if two parents are far away, a usual condition for the initial population, almost any value of offspring can be achieved. When the solutions tend to converge, the parents are close to each other, and distant solutions are not allowed, thereby focusing the search to a narrow region. Thus, SBX helps in exploring the search space at the initial generation, while it exploits the acquired knowledge at later stages.

In polynomial mutation [15] the following steps are involved in mutating a solution $x_i^{(1,t)}$ to obtain the offspring $x_i^{(1,t+1)}$. Let u_i be a random number between 0 and 1.

$$\bar{\delta}_i = \begin{cases} (2u_i)^{\frac{1}{\eta_m+1}} - 1 & \text{if} \quad u_i < 0.5 \\ -[2(1-u_i)]^{\frac{1}{\eta_m+1}} & \text{otherwise} \end{cases} \tag{19}$$

$$x_i^{(1,t+1)} = x_i^{(1,t)} + (x_i^U - x_i^L)\bar{\delta}_i \tag{20}$$

The shape of the probability distribution of the distance between the mutation and its original is directly controlled by the user-defined parameter, *distribution index for mutation* η_m. A large value of η_m gives a higher probability for creating near-parent solutions, and a small value of η_m allows distant solutions to be selected as offspring.

C. NSGA-II

The NSGA-II procedure [18] for finding multiple Pareto optimal solutions for multi-objective optimization problems has the following three main features: 1) it is based on Pareto dominance sorting, 2) elitism by preservation of best individuals, 3) diversity-preserving mechanism by crowding distance sorting (distance here represents the Euclidian distance between an individual and its closest neighbor). These processes are further explained next.

Initially, NSGA-II randomly creates a population containing N individuals. This population, $P(0)$, is sorted according to their rank values to get $P(1)$. The sorting is based on the order of dominance of the individuals. An individual that is not dominated by any other individuals gets a Pareto rank 1. Subsequently, the individuals with rank 2 are the ones that are dominated only by the individuals with rank 1, etc. After sorting, the offspring population $Q(1)$ [i.e., children of $P(1)$], is created by selecting parents from population $P(1)$ and applying the genetic recombination and mutation operators [33, 34], as just described. Thereafter, the two populations [$P(1)$, $Q(1)$] are combined together to form the new population R_1 of size $2N$. Because this combined population R_1 is used, the best individuals of both the parents and children can be preserved, yielding a so-called *elitism* mechanism.

Then again, a nondominated sorting is applied to the entire population R_1. The new population is then filled by individuals of subsequent ranks (i.e., nondominated fronts F_1, F_2, etc.; see Fig. 8) starting with rank 1, followed by rank 2, etc.

Because the overall population size of R_1 is $2N$, not all fronts can be accommodated in the N slots (i.e., individuals) available in the new population. All fronts that are not accommodated at all are simply deleted. The front that can be accommodated only partly is treated separately. In Fig. 8 we see that not all individuals from front F_3 are included in the next generation. NSGA-II deals with this as follows. Per objective function, all individuals in the front F_3 are sorted according to their objective function values in ascending order. The minimum and maximum are selected and given infinity distance. These minimum and maximum will be put first on the list. This is to ensure that boundary points are always selected. Secondly, the distances for the rest of individuals

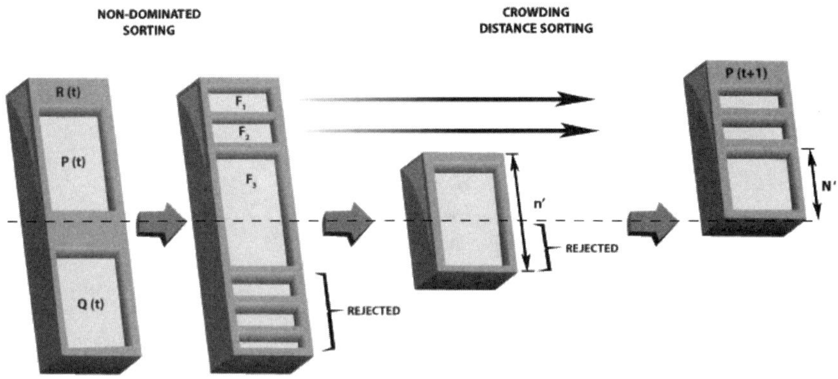

Fig. 8 NSGA-II procedure as described in [18].

are calculated by means of *cuboids*, formed by using the nearest neighbors of an individual as the vertices. After the rest of individuals' distances are calculated, they are sorted in descending order to form the newly ordered front F_3. In Fig. 8, this is illustrated as the small block right next to F_3. The individuals with smallest distance are rejected, and the rest are included into the next generation *P*(2). This process is called *crowding distance sorting*, and it is intended to maintain a good distribution of the individuals over the full Pareto front (i.e., to avoid convergence into 1 or a few Pareto optimal points).

After all *N* individuals in *P*(2) are obtained, the offspring population *Q*(2) is created. These populations are again combined to form R_2 containing 2*N* individuals. Then the same procedure is applied for every generation until a maximum number of generations, specified by the user is reached.

D. ε-MOEA

ε-MOEA is an evolutionary algorithm based on the ε-dominance concept introduced in Deb et al. [19]. The properties of this algorithm are as follows:

1) It is a *steady-state* MOEA. This means that it maintains *spread* (i.e., distribution of individuals over the Pareto front) while attempting to *converge* to the true Pareto optimal,

2) It emphasizes nondominated solutions by using usual dominance concept,

3) It maintains the diversity in the archive by allowing only one solution to be present in each preassigned hyperbox of size ε on the Pareto-optimal front (ε-dominance),

4) It uses an elitism principle by maintaining all of the best individuals in an archive throughout the computation.

In ε-MOEA, the objective space is divided into a number of grid cells (or hyperboxes), as illustrated on the left of Fig. 9. The diversity of the population is maintained by ensuring that there is only one solution in a grid cell (black dots). There are two co-evolving populations [19]: an evolutionary algorithm population *P*(*t*) and an archive population *E*(*t*), as shown on the right of Fig. 9, where *t* is the iteration counter and the archive population is intended to hold

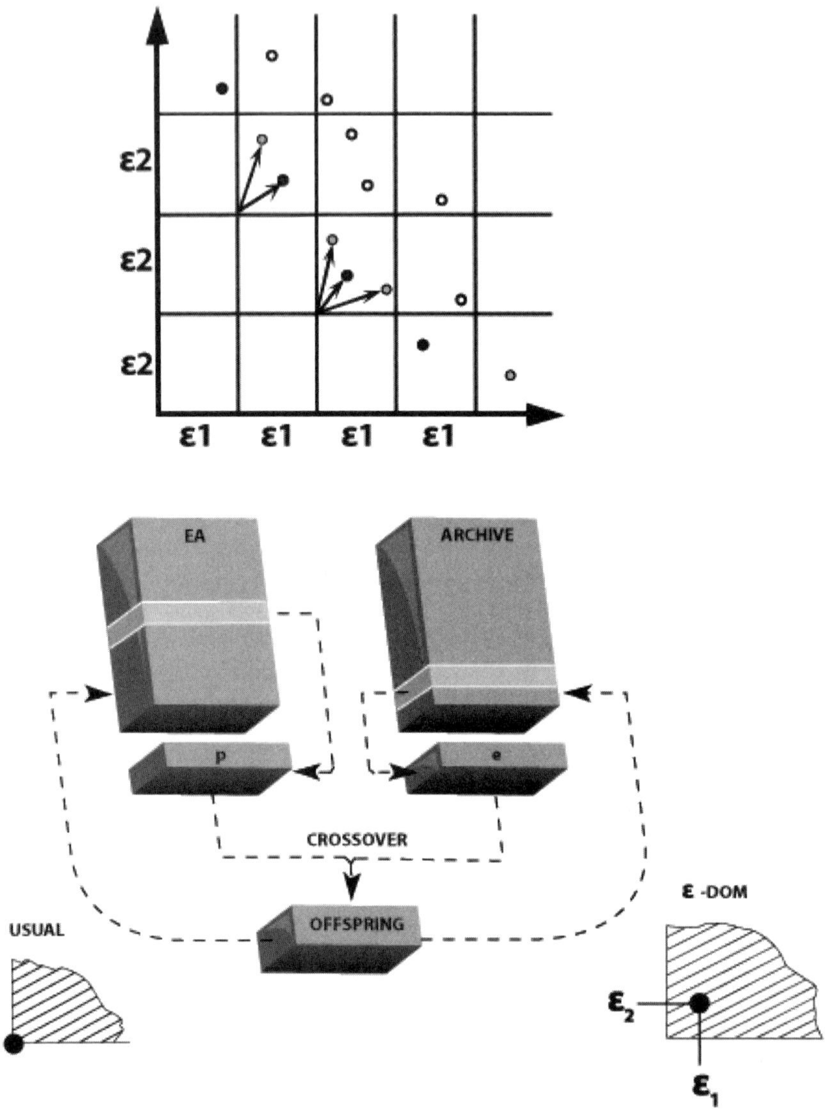

Fig. 9 ε-MOEA objective space (left) and evolution procedure.

the best individuals that have been found until iteration *t*. ε-MOEA begins with a (randomly generated) initial population *P*(0). The archive population *E*(0) is assigned with the ε-nondominated individuals of *P*(0). The ε-dominance concept is shown on the top of Fig. 9 and also in Fig. 10.

Thereafter, two solutions, one from each *P*(0) and *E*(0), are chosen for mating. To choose a solution from *P*(0), two population members from *P*(0) are picked up

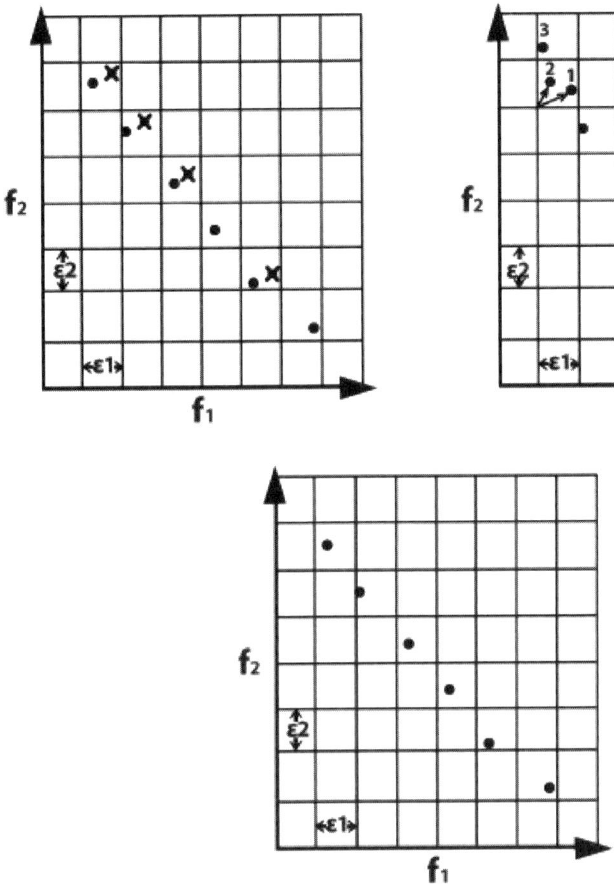

Fig. 10 Illustration of ε-dominance concept.

at random, and a domination check (in the "usual" sense, shown as dots on the left of Fig. 9 for minimization of objectives) is made. If one solution dominates the other, the former is chosen. Otherwise, it indicates that these two solutions are nondominated to each other, and simply one of them is chosen at random. Let us denote the chosen solution by p. To choose a solution e from $E(0)$, several strategies involving a certain relationship with the chosen p can be made [19]. In Deb et al. [19], for example, they randomly pick a solution from $E(0)$. Then solution p and e are mated to create λ offspring solutions, c_i ($i = 1, 2, \ldots, \lambda$). In the present study, $\lambda = 1$ is always used as also advised by Deb et al. [18].

The offspring solution c_i is compared to the archive members by ε-nondominated checking, based on ε-dominance concept [19]. The solutions are first compared on the basis of the Pareto dominance of the grid cells that they reside in. Subsequently, multiple solutions within one grid cell are compared on the basis of the usual Pareto dominance of these individuals. If the offspring dominates one or more archive members, the offspring replaces one of them (chosen at random). Every solution in the archive is assigned an identification array $\boldsymbol{B}$, which indicates the fitness score of a grid cell and which size equals to the total number of objectives as follows:

$$B_i(\mathbf{f}) = \left\lfloor \frac{(f_i - f_i^{\min})}{\varepsilon_i} \right\rfloor \tag{21}$$

where $\lfloor\ \rfloor$ represents the truncation operation to a whole number, $f_i^{\min}$ is the minimum possible value of the ith objective of all archive members, and ε_i is the allowable tolerance in the ith objective. This ε_i value is the same as the ε used in the ε-dominance definition [19]. The identification arrays are calculated for each archive member a and for the offspring c_i. If the $\boldsymbol{B}_a$ of any archive member a dominates that offspring c_i, then c_i is not accepted because it means that this offspring c_i is ε-dominated by archive. On the other hand, if $\boldsymbol{B}_{c_i}$ of the offspring dominates $\boldsymbol{B}_a$ of any archive member a, the archive member is replaced by the offspring. This is why individuals 3 and 4 in Fig. 10 are not accepted as an archive member. Although in the usual Pareto dominance sense they are nondominated solutions, they are dominated solutions in ε-dominance sence because their identification arrays are dominated by their neighbors. The grid cell of individuals 1 and 2 dominates the grid cell of individual 3.

If both the offspring and the archive are ε-nondominated, which means neither of the preceding two cases occur, another two operations are performed. If the offspring belongs to an unoccupied grid cell (different identification array), then offspring c_i is added into the archive. If it is in the same grid cell as an archive member (having same identification array), a check for the usual nondomination is conducted. It might appear that the offspring and the archive member are nondominated like individuals 1 and 2 in Fig. 10. If this is the case, then the one which is closer to the $\boldsymbol{B}$ vector in terms of the Euclidean distance is chosen (in this case individual 2), and the other (individual 1) is not kept in the archive.

The decision whether an offspring will replace any population member P can be made using different strategies. One of these is to compare each offspring with all population members. If the offspring dominates one or more population members, the offspring replaces one of them (chosen at random). On the other hand, if any population member dominates the offspring, it is not accepted. If both the offspring and the population members are nondominated, the offspring replaces a randomly chosen population member so that the evolutionary algorithm population size remains unchanged.

The preceding procedure is continued for a specified number of iterations, and the final archive members are considered as the obtained ε-nondominated solutions.

E. Improvements of ε-MOEA: ε-NSGA

There are some weaknesses of ε-MOEA: 1) the absence of *extreme* solutions, that is, the solutions with high- and low-objective values, on the Pareto-optimal front; and 2) for very many and expensive function evaluations, it turns out to be not efficient. The absence of extreme solutions occurs as a consequence of the hyperboxes. Although this issue is not always of critical importance, because the extreme solutions region is not always the most interesting design region, this issue can be alleviated by decreasing the epsilon values.

Because ε-MOEA calculates the objective function values for each individual separately, it means that we cannot take the advantage of vectorized calculation. Especially if objective function calculation is expensive, this is not efficient. Therefore we combine ε-MOEA and NSGAII (we name it ε-NSGAII) as shown in Fig. 11. Although such a combination has been proposed by Kollat and Reed [33], we use a different method in combining these two algorithms. In the proposed algorithm, a dynamic population size is used in comparison to the archive size. We found that this method is good to explore additional regions of the search space in the early generations. But in the later stage, as the archive size increases the population size is getting bigger and bigger until the archive size is stable. At this point, we found that for expensive function calculations, this method is inefficient.

According to the preceding reasons, we use fixed population size so that in the later stage the algorithm can still do the calculations efficiently. To explore additional regions of the search space in the early generation, we randomly generate new individuals as long as the archive size does not exceed half the population size. We found that this archive size limit of "half the population size" performs best for a population of up to 100 individuals.

The ε-NSGAII (Fig. 11) uses the NSGAII algorithm for generating new offspring individuals. After it finds new better individuals, ε-dominance sorting is applied, and the results are stored in the archive. New individuals to evolve in the next generation are randomly selected from both the archive and the new better individuals from the current generation. Here, we use 50% population from the current generation, and another 50% population is taken from the archive. If the archive size is less than 50% population, it generates new individuals randomly.

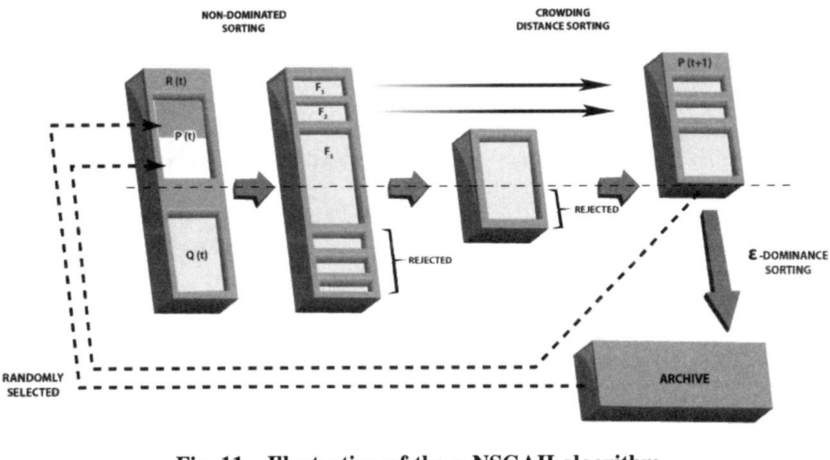

Fig. 11 Illustration of the ε-NSGAII algorithm.

IV. Case Study: Multi-Objective Optimization of Transonic Wing Design

A. Introduction of Case Study

This section applies the meta-models and optimization algorithms that were described in the preceding sections, in an example aircraft wing optimization study.

The case study considered in this chapter describes an investigation of aircraft range and fuel efficiency. Fuel efficiency here represents the distance flown per unit of fuel per unit of payload. The investigation concerns the optimization of transonic aircraft wings in the preliminary design phase. The design analyses in this investigation make use of a MDA system that evaluates the aircraft characteristics as a function of a set of design parameters [35]. The evaluations comprise, among others, wing structural sizing and optimization using finite element method (FEM) analyses, and cruise lift-over-drag performance using computational-fluid-dynamics (CFD) analyses. The design parameters include geometric wing planform parameters such as span, chord, sweep, as well as "aircraft operational parameters" such as maximum takeoff weight (MTOW) and cruise altitude. Given the design parameters inputs, the wing MDA system predicts the corresponding aircraft characteristics in terms of, among others, weight breakdown information, maximum range, and fuel consumption. More details of this MDA system are given in the chapter on wing multidisciplinary analysis.

To effectively handle the different (and possibly conflicting) design objectives, multi-objective Pareto front [30] optimization algorithms as already described are used in the presented aircraft wing design investigation. The multi-objective Pareto front results directly provide the design information on which further tradeoff considerations of the different objectives for the wing design can be based.

To limit the number of computationally expensive evaluations with the MDA system, the multi-objective optimization iterations have been decoupled from the MDA evaluations, according to the meta-modeling approach just described (see also [36]). The meta-modeling approach allows for computationally efficient exploration of the aircraft characteristics in a predefined design domain. The different meta-modeling methods, such as polynomial regression, kriging models, and neural networks, are used, and their predictive accuracy is carefully checked and compared in order to achieve the best representation. Obviously, the results of the optimization depend on the accuracy of the meta-models used and therefore also require careful assessment and validation, as is shown in the present aircraft design optimization study.

B. Aircraft Multidisciplinary Design Analysis

The MDA system described in the previous chapter is used in an aircraft wing design optimization study. Aircraft designs are pursued that have optimal performance for both range and fuel efficiency.

From the many results that come out of the MDA simulations, different variables can be selected as relevant objective or constraint functions in aircraft design optimization studies. In the present wing design optimization study, we look for optimal overall range and fuel efficiency, and hence we take into account the Breguet range and the total fuel consumption as the aircraft wing optimization objectives, which are both computed with the MDA system. The Breguet range R_B represents the actual distance traveled, taking into account the engine fuel efficiency during cruise c_{fs}, the aerodynamic lift-over-drag performance $L_{\text{oD-cruise}}$, and the actual amount of fuel consumed W_{fc} as follows:

$$R_B = \frac{v_{\text{cruise}}}{c_{\text{fs}}} \times L_{\text{oD-cruise}} \times \ln\left(\frac{W_{\text{MTO}}}{W_{\text{MTO}} - W_{\text{fc}}}\right) \tag{22}$$

The fuel weight W_{fc} depends on the aircraft structural weight, which is computed in the MDA through a structural optimization for the given aircraft configuration. The aircraft fuel efficiency η_f can be evaluated as a combination of range and actual fuel consumption and is calculated by

$$\eta_f = R_B \Big/ \left(\frac{W_{\text{fc}}}{n_{\text{pax}}}\right) \tag{23}$$

and is expressed in km/(l/person). These values can be easily compared to other fuel efficiency numbers as for example published for cars ($\eta_f \sim 14$ for single person driving a middle class car). From the many possible design parameters that are used in the MDA system, we selected four illustrative ones as the independent variables for the wing optimization: wing semispan, outer-wing leading-edge sweep angle, wing chords, and aircraft MTOW (Fig. 12). The three wing chords (at root, crank, and tip) are reduced to a single parameter, the wing chord scale factor, which linearly scales all three

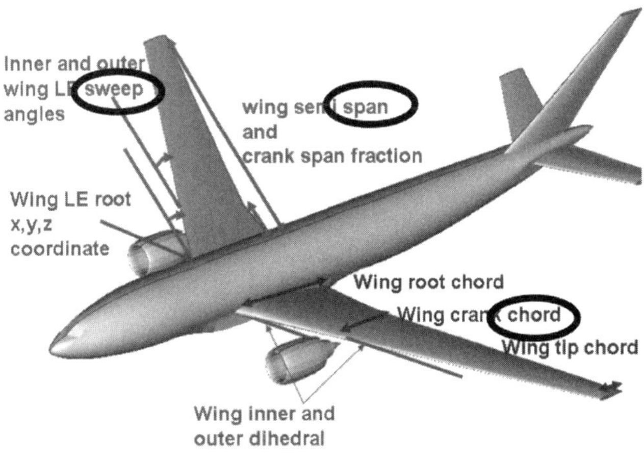

Fig. 12 Aircraft wing design parameters.

chords equally. All other design parameters of the MDA system are equal to their values for the reference aircraft and remain unchanged in this study.

The resulting multi-objective optimization problem for aircraft range and fuel efficiency can be formulated as follows:

$$\max_{(s_p, s_w, c_h, W_{\mathrm{MTO}})} (R_{\mathrm{Bcorr}}, \eta_f) \tag{24}$$

C. Meta-Models

To create the meta-models, first a suitable sample of the aircraft behavior in the considered design domain is pursued. This is achieved by a limited number of evaluations with the MDA simulation system in certain selected design points. These design points are defined according to a sequence of fractional factorial (i.e., fractions of full-factorial) sets of samples of the four-dimensional design space (i.e., parameter space of the design parameters: wing semispan, outer-wing sweep angle, wing chord, and aircraft MTOW). The semispan is varied between 29 and 32 m. The outer-wing sweep angle is varied between 21 and 39 deg. The wing chords at three stations (wing root, crank, and tip) are equally varied by one single chord scale factor, which is varied between 1.000 and 1.075. MTOW is varied between 150,000 and 280,000 kg.

In total, 99 design points are created in this parameter space and are evaluated with the full MDA simulation system, yielding (among many other data available in the integrated design model) the values for range and fuel consumption in these design points. As a quick preliminary design assessment, these range and fuel values are ordered according to a basic Pareto ranking procedure [30], as described in Section 4.3, in order to obtain a first indication of the interesting design regions. In this ranking procedure, the best (or nondominated) design points, that is, those points having the best values for range and fuel consumption, are assigned Pareto rank 1, the set of second best points are assigned Pareto rank 2, and so forth until all design points have been assigned a rank value.

The resulting rank values for these 99 design points, and their distribution in the objective space and their parameter values, are given in Fig. 13. The resulting data set with the values of the design parameters and of the range and fuel objectives in these 99 design points is then used to create the meta-models. The meta-models shall approximate as accurate as possible the objectives in each point of the parameter space.

A number of different polynomial functions (polyn in Tables 2 and 3), kriging models (kriging-xy in Tables 2 and 3), neural networks (ann in Tables 2 and 3), and radial basis functions (rbf in Tables 2 and 3) are applied [37], and the best-fit functions among these are determined. These best-fit functions are found through various cross-validation assessments on the data set, such that these functions' predictions of the design objectives (range, fuel efficiency) have the smallest residuals. Four different cross-validation assessments are performed by selecting different sets of validation points.

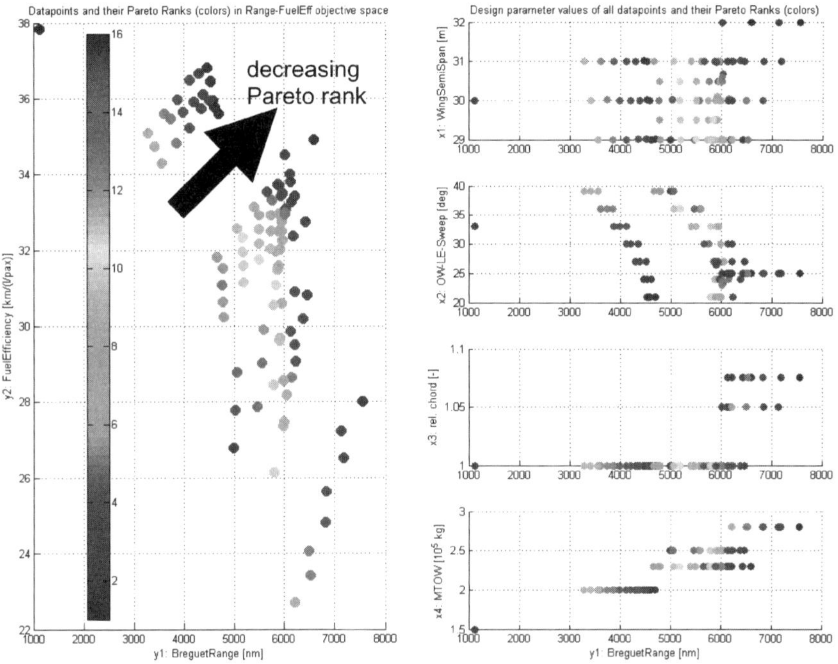

Fig. 13 Range and fuel results in the 99 design points in objective space (left) and in parameter space (right), scaled in grey by their Pareto rank. (The rank 1 points (black dots) have the highest range and highest fuel efficiency values.) (See also color figure section at the back of the book.)

Table 2 Range data: Accuracies of the different fit functions (identified in left column) for the different cross-validation

	RMSE				MAPE
Fit function	99/9	99/20	99/1/20	99/1/99	99/1/99/%
poly0	1824.8	1450.2	1464.0	993.2	18.5785
poly1	789.0	720.6	541.0	401.6	6.7994
poly2	739.3	509.2	460.8	234.1	3.7504
kriging-cG	1386.0	1155.3	886.3	400.3	4.2159
kriging-cE	1297.2	730.4	913.8	414.1	4.2473
kriging-cC	1025.6	722.3	814.8	367.0	3.8202
kriging-lG	**608.7**	**519.3**	**301.7**	**138.6**	**1.7258**
kriging-lE	567.6	418.8	465.5	210.1	2.2546
kriging-lC	600.9	440.5	411.0	186.8	2.2124
ann	1175.3	1053.7	957.3	859.6	12.8121
rbf	784.1	5130.0	205.0	99.7	1.1252

Table 3 Fuel-efficiency data: Accuracies of the different fit functions for the different cross-validation assessments

	RMSE				MAPE
Fit function	99/9	99/20	99/1/20	99/1/99	99/1/99/%
poly0	4.648	4.182	3.909	3.259	8.4636
poly1	1.984	1.499	1.368	0.995	2.3676
poly2	0.722	0.544	0.258	0.264	0.6371
kriging-cG	1.746	1.422	0.830	0.421	0.6577
kriging-cE	**2.435**	**1.289**	**0.947**	**0.430**	**0.3894**
kriging-cC	2.103	2.251	1.198	0.576	0.7798
kriging-lG	1.590	1.358	0.939	0.443	0.4836
kriging-lE	1.692	1.378	1.187	0.539	0.4136
kriging-lC	1.778	1.404	1.305	0.607	0.6339
ann	1.886	1.393	0.672	1.179	3.7600
rbf	6.990	66.977	4.740	2.140	1.1210

In a first cross-validation assessment the nine rank-one data points, that is, those data points having the best (lowest) Pareto rank values for range and fuel efficiency (dark dots in Fig. 13), are used as validation points, and the remaining 90 for building the model. The resulting rms values indicate that the kriging-linear-Exponential (kle) [27] and second-order polynomial (poly2) fit functions provide the best fits for range and fuel efficiency, respectively (99/9 column in Tables 2 and 3). However, this assessment represents the accuracy of the fits in only a local region around the rank-one data points, for fits where only design value points have been used that yield objective of rank two or worse. To obtain a more global accuracy assessment, we include some more validation points by adding the 11 Pareto rank-two data points to the validation set (99/20 column in Tables 2 and 3). Because this validation set is rather large (20 out of 99 points), the validation fits are made on relatively small data sets (79 points), and thus will differ significantly from the "full" fits made on the complete data set (99 points), and again the best points are actually removed from the fitting stage. Therefore, we also evaluate the rms residuals from a leave-1-out experiment [28] of this validation set (99/1/20 column in Tables 2 and 3). In this leave-1-out experiment, subsequently each point of the validation set is separated from the data set, a fit is made on the remaining 98 points, the residual in the validation point is evaluated, and the rms of the 20 residuals is calculated. Finally, as a real global accuracy assessment, we also performed a leave-1-out experiment on the complete data set (99/1/99 column in Tables 2 and 3). As an additional indication of the relative accuracy of the fits, we also include the MAPE of the global leave-1-out residuals (99/1/99/% column in Tables 2 and 3).

For the different cross-validation assessments we find reasonably consistent accuracies for most fit functions (Tables 2 and 3). The best rms residual found in each assessment is marked by the shaded cell. For the range data (Table 2), the rbf fit provides the best results for the leave-1-out experiments, but very poor fit quality according to the 99/20 experiments, and is therefore not selected as best fit for range.

Based on the results of each of the five assessments performed, and in particular on the global accuracy as measured by the leave-1-out experiments (Table 2, columns 99/1/99 and 99/1/99/%), it is concluded that the best fit for range is found by the kriging-linear-Gauss (klg) fit function. For fuel efficiency the poly2 fit performs quite well (Table 3), but its global accuracy as measured by the leave-1-out experiment (column 99/1/99/%) is worse than for some of the kriging fits. In addition, poly2 provides a least-square regression (noninterpolating) fit on the data, whereas the kriging models provide exactly interpolating fits on the data. Because the data represent results of deterministic computer simulations, it is concluded that the best fit for fuel efficiency is found by the kriging-constant-exponential (kce) fit function.

Regarding Tables 2 and 3, the values given are the rms of the residuals (or prediction errors) in the validation points. The thick-bordered cell marks a poor fit quality. The bolded cells mark the best overall fit method.

D. Design Optimization

A Pareto front optimization of the aircraft's range and fuel efficiency is performed using a multi-objective genetic algorithm (based on ε-NSGA-II as described in the preceding section), where the best fits for range and fuel efficiency are used as objective functions. In this optimization a population size of 99 individuals is used, where the 99 design points from the data set are used as the initial generation. The bounds of the search domain for the optimization are set to the minimum and maximum values of the design parameters of the 99 design points. In a first run three generations, so about 300 objective functions evaluations, are performed with the genetic algorithm. The resulting population is indicated by the circles in Fig. 14 in order to give some illustration of the convergence history of the genetic algorithm. Then this resulting population is used as the initial population for an extensive run of about 100 generations with the genetic algorithm. The total number of objective function evaluations in this extensive optimization is about 10.000, and takes about 20-s computational time on a standard PC (Pentium-4, 2.8 GHz). The resulting Pareto front solution (diamonds in Fig. 14) provides a set of clearly improved designs, as compared to the initial set of designs in the data set (small dots).

The behavior of the aircraft in the parameter space around the Pareto optimal design points (which were predicted on the basis of the meta-models) was further explored and interpreted by aircraft design experts. One candidate optimal design point was selected (Fig. 15) and accurately evaluated by the MDA simulation system. The results from this evaluation are given in Fig. 15. The parameter values of the candidate optimal design point are: span, 32.5; sweep, 25.1; chord, 1.08; MTOW, 285,000. From the MDA analysis the range is 7594.6, and the fuel efficiency is 27.8; from the meta-model the range is 7761.9, and the fuel efficiency is 28.4.

When considering these results more closely, we can conclude from the MDA analysis results that this point is an additional Pareto optimal design point

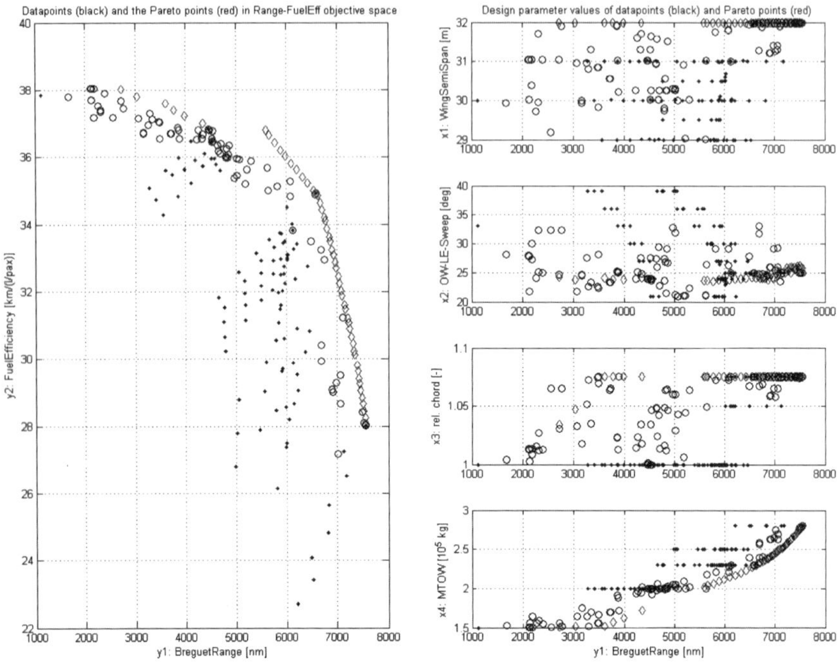

Fig. 14 Design points of data set (·), population after 3 generations (○, ●), and Pareto front after 100 additional generations (◇, ◆) for maximum range vs maximum fuel efficiency found with the kriging-linear-Gauss and kriging-constant-exponential meta models, respectively, for range and fuel efficiency. Results presented in objective space (left) and in the range-parameter subspaces (right) for each of the four design parameters. (See also color figure section at the back of the book.)

(Fig. 16). The meta-models predicted somewhat overestimated values for range and fuel efficiency for this point.

Furthermore, this new design point provides a valuable additional point for the data set on which the meta-models are created, and hence the meta-models can be further improved and used again in the multi-objective optimization. Therefore the meta-models for range and fuel efficiency were regenerated using the same kriging models as before (kcg for range and kle for fuel efficiency). In this optimization the 100 design points from the new data set are used as the initial generation, and the bounds of the search domain are set to the minimum and maximum values of the design parameters of the 100 design points.

The resulting improved Pareto front (shifted diamonds in Fig. 16) found with these improved meta-models provides a slight improvement compared to the Pareto front (diamonds in Fig. 16) found with the previous meta-models, as is shown in Fig. 16. The Pareto front again helps to further guide the computationally expensive full MDA evaluations to the most interesting designs for the team of expert designers.

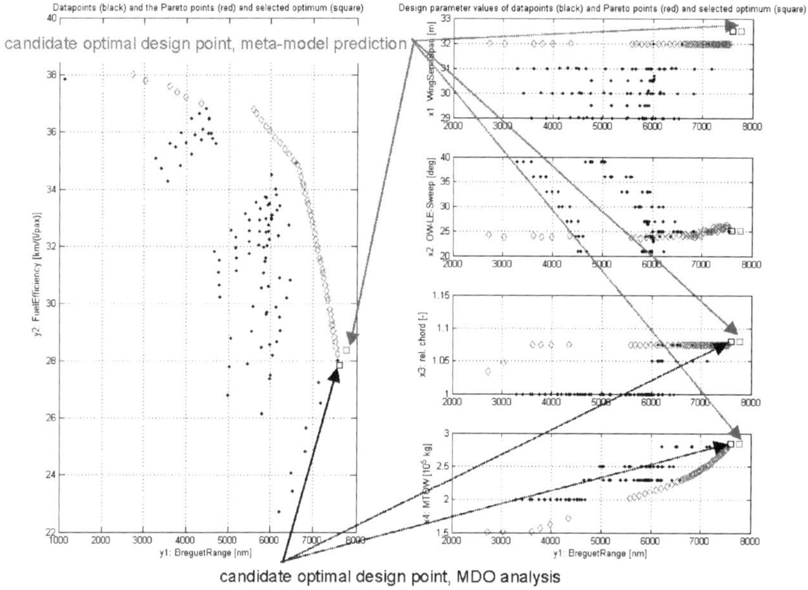

Fig. 15 Pareto front found with initial meta-model (◆, ◇), data set (·), and MDA analysis and meta-model predictions for candidate optimal design point (□). (See also color figure section at the back of the book.)

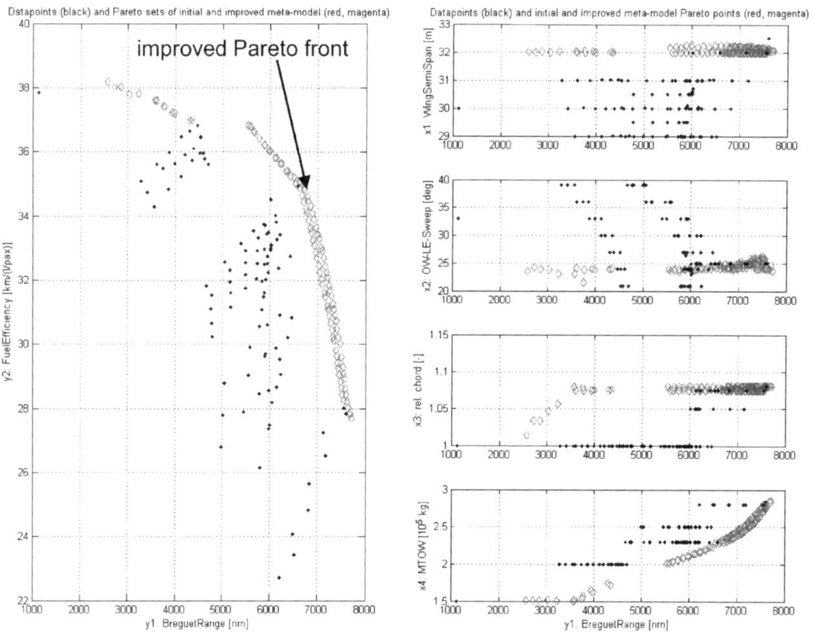

Fig. 16 Pareto fronts found with the initial meta-models (◇, ◆) and the improved meta-models (shifted diamonds). (See also color figure section at the back of the book.)

V. Conclusions

The combination of advanced meta-models and multi-objective optimization algorithms for aircraft design presented in this chapter is flexible and applicable to a variety of design problems. A key benefit of this approach is that large numbers of interesting (Pareto optimal) design points can be found relatively quickly and easily at the cost of only few computationally expensive analyses, while a reasonable level of the accuracy is maintained. Representation of the results in the design parameter space as well as in the objective space provides valuable information for design decisions, where involvement of design specialists is required.

However, for high-dimensional design problems the visualization, assessment, and selection of the most interesting design points require special attention. The same applies to the accuracy of the objective function values as predicted by the fits. Several ways to deal with the accuracy aspect were demonstrated:

1) Use as much as possible information that is available, for example, proper DOE, as many data points as possible, *a priori* knowledge of the underlying functions

2) Use different fitting methods and determine the best fit; carefully define appropriate validity domains for the fits (e.g., avoid extrapolation).

3) Try to account for fitting errors in the fit prediction by incorporating residual estimations.

The ε-NSGA-II algorithm has proven an efficient algorithm for calculating the Pareto optimal design points of the multi-objective design optimization, requiring only 20 s for this optimization run on a standard pc, which is very quick as compared to about 30-min computation time for a single MDA evaluation. The Pareto points based on the meta-model still need to be verified by the full MDA and validated by expert judgment. It was shown in the case study that, although the meta-models somewhat overpredicted the Pareto optimal points, the verified values are still on the Pareto front. From this verification the meta-model can be further improved, and this contributes to further refinement of the optimization process.

The optimization approach followed here was also used in the engine design case (chapter TBD), so is applicable to a larger set of design problems than only wing design.

References

[1] Pardessus, T., "Concurrent Engineering Development and Practices for Aircraft Design at Airbus," *Proceedings of the ICAS Conference 2004*, paper 2004-7.7.1, (Yokohama, Japan), International Council of the Aeronautical Sciences (ICAS), Stockholm, Sweden, Optimage Ltd., Edinburgh, UK, 2004.

[2] Torenbeek, E., *Synthesis of Subsonic Airplane Design*, Delft Univ. Press, Kluwer Academic Publishers, Delft, The Netherlands, 1982, pp. 215–262.

[3] Sobieszczanski-Sobieski, J., and Haftka, R. T., "Multidisciplinary Aerospace Design Optimization: Survey of Recent Developments," *Structural and Multidisciplinary Optimization*, Vol. 14, No. 1, 1997, pp. 1–23.

[4] Laban, M., Arendsen, P., Rouwhorst, W. F. J. A., and Vankan, W. J., "A Computational Design Engine for Multi-Disciplinary Optimization with Application to a Blended Wing Body Configuration," AIAA Paper 2002-5446, Sept. 2002.

[5] Ko, A., Leifsson, L. T., Mason, W. H., Schetz, J. A., Grossman, B., and Haftka, R. T., "MDO of a Blended-Wing-Body Transport Aircraft with Distributed Propulsion," AIAA Paper 2003-6732, Nov. 2003.

[6] Kesseler, E., Laban, M., and Vankan, W. J., "Multidisciplinary Wing Optimization," VIVACE Forum-1, Warwick, U.K., 20–21 Sept. 2006, Session 4, CD-ROM; http://www.nlr.nl/id~2608/l~en.pdf [retrieved July 2009].

[7] Simpson, T. W., Peplinski, J., Koch, P. N., and Allen, J. K., "Metamodels for Computer-Based Engineering Design: Survey and Recommendations," *Engineering with Computers*, Vol. 17, No. 2, 2001, pp. 129–150.

[8] Zink, P. S., DeLaurentis, D. A., Hale, M. A., Volovoi, V. V., Schrage, D. P., Craig, J. I., Fulton, R. E., Mistree, F., Mavris, D. N., Chen, W., Röhl, P. J., Lewis, K. E., Koch, P. N., Cesnik, C. E. S., and Simpson, T. W., "New Approaches to High Speed Civil Transport Multidisciplinary Design and Optimization," *Proceedings of the 2000 IEEE Aerospace Conference*, 2000, http://ideal.mech.northwestern.edu/pdf/IEEE-351-99.pdf [retrieved July 2009].

[9] DeLaurentis, D. A., Mavris, D. N., and Schrage, D. P., "System Synthesis in Preliminary Aircraft Design Using Statistical Methods," *Proceedings of the 20th ICAS Congress* (Sorrento, Italy), International Council of the Aeronautical Sciences (ICAS), Stockholm, Sweden, 1996, pp. 866–878.

[10] Vankan, W. J., Lammen, W. F., Kos, J., and Maas, R., "Complementary Approximate Modeling in Matlab and Modelica," *Proceedings of the EUROSIM 2004 Conference*, Paris, France, Sept. 2004, sponsored by EUROSIM (Federation of European Simulation Societies).

[11] Obayashi, S., Sasaki, D., Takeguchi, Y., and Hirose, N., "Multiobjective Evolutionary Computation for Supersonic Wing-Shape Optimization," *IEEE Transactions on Evolutionary Computation*, Vol. 4, No. 2, July 2000, pp. 182–187.

[12] Oyama, A., "Multidisciplinary Optimization of Transonic Wing Design Based on Evolutionary Algorithms Coupled with CFD SOLVER," *Proceedings of European Congress on Computational Methods in Applied Sciences and Engineering (ECCOMAS)*, Sept. 2000, http://flab.eng.isas.ac.jp/member/oyama/papers/eccomas2000.pdf [retrieved July 2009].

[13] Oyama, A., and Obayashi, S., "Multidisciplinary Wing Design Optimization Using Multiobjective Evolutionary Algorithm," *PPSN/SAB Workshop on Multiobjective Problem Solving from Nature (MPSN)* (Paris, France), 2000.

[14] Oyama, A., Obayashi, S., Nakahashi, K., and Hirose, N., "Coding by Taguchi Method for Evolutionary Algorithms Applied to Aerodynamic Optimization," *Proceedings of the Fourth ECCOMAS Computational Fluid Dynamics Conference* edited by K. D. Papailliou, D. Tsahalis, J. Périaux, and D. Knörzer, (Athens, Greece), Wiley, New York, Vol. 2, 1998, pp. 196–203.

[15] Zitzler, E., Laumanns, M., and Bleuler, S., "A Tutorial on Evolutionary Multiobjective Optimization," *Metaheuristics for Multiobjective Optimization*, Vol. 535 of *Lecture Notes in Economics and Mathematical Systems* edited by X. Gandibleux, et al., Springer-Verlag, Berlin, Germany, 2004, pp. 3–38.

[16] Deb, K., *Multiobjective Optimization Using Evolutionary Algorithms*, Wiley, Chichester, U.K., 2001, pp. 77–300.

[17] Deb, K., "Single and Multiobjective Optimization Using Evolutionary Computation," KanGAL, Technical Report 2004003, Feb. 2004.
[18] Deb, K., Agrawal, S., Pratap, A., and Meyarivan, T., "A Fast and Elitist Multiobjective Genetic Algorithm: NSGA-II," *IEEE Transactions on Evolutionary Computation*, Vol. 6, No. 2, 2002, pp. 182–197.
[19] Deb, K., Mohan, M., and Mishra, S., "A Fast Multiobjective Evolutionary Algorithm for Finding Well-Spread Pareto-Optimal Solutions," KanGAL, Technical Report 2003002, Feb. 2003.
[20] Fonseca, C. M., and Flemming, P. J., "Genetic Algorithms for Multiobjective Optimization: Formulation, Discussion and Generalization," *Proceedings of the Fifth International Conference on Genetic Algorithms*, edited by S. Forrest, Morgan Kauffman, San Mateo, CA, 1993, pp. 416–423.
[21] Knowles, J. D., and Corne, D. W., "Approximating the Non-Dominated Front Using the Pareto Archived Evolution Strategy," *Evolutionary Computation*, Vol. 8, No. 2, 2000, pp. 149–172.
[22] Zitzler, E., Laumanns, M., and Thiele, L., "SPEA2: Improving the Strength Pareto Evolutionary Algorithm for Multiobjective Optimization," *Evolutionary Methods for Design Optimization and Control with Applications to Industrial Problems*, Eurogen 2001, International Center for Numerical Methods in Engineering (CIMNE), 2002, pp. 95–100.
[23] Zitzler, E., and Thiele, L., "Multiobjective Evolutionary Algorithms: A Comparative Case Study and the Strength Pareto Approach," *IEEE Transactions on Evolutionary Computation*, Vol. 3, No. 4, 1999, pp. 257–271.
[24] Optimization Toolbox, The MathWorks, Natick, MA, http://www.mathworks.com/products/optimization [retrieved Aug. 11, 2009].
[25] Kleijnen, J. P. C., and Sargent, R. G., "A Methodology for the Fitting and Validation of Metamodels in Simulation," *European Journal of Operational Research*, Vol. 120, No. 1, 2000, pp. 14–29.
[26] Bradley, E., "Estimating the Error Rate of a Prediction Rule: Improvement on Cross-Validation," *Journal of the American Statistical Association*, Vol. 78, No. 382, 1983, pp. 316–331.
[27] Lophaven, S. N., Nielsen, H. B., and Søndergaard, J., "DACE—A Matlab Kriging Toolbox, Version 2.0," Informatics and Mathematical Modelling (IMM), Rept. IMM-REP-2002-12, Technical University of Denmark (DTU), 2002.
[28] Bishop, C. M., *Neural Networks for Pattern Recognition*, Clarendon Press, Oxford, U.K., 1997, pp. 170–180.
[29] Bäck, T., Hammel, U., and Schwefel, H. P., "Evolutionary Computation: Comments on the History and Current State," *IEEE Transactions on Evolutionary Computation*, Vol. 1, No. 1, 1997, pp. 3–17.
[30] Goldberg, D. E., *Genetic Algorithms in Search, Optimization, and Machine Learning*, Addison Wesley Longman, Reading, MA, 1989, pp. 197–198.
[31] Wolpert, D. H., and Macready, W. G., "No Free Lunch Theorems for Optimization," *IEEE Transactions on Evolutionary Computation*, Vol. 1, No. 1, 1997, pp. 67–82.
[32] Genetic Algorithm and Direct Search Toolbox, The MathWorks, Natick, MA, http://www.mathworks.com/products/gads [retrieved Aug. 11, 2009].
[33] Kollat, J. B., and Reed, P. M., "The Value of Online Adaptive Search: A Performance Comparison of NSGAII, ε-NSGAII and εMOEA," *Evolutionary Multi-Criterion Optimization: Third International Conference, EMO 2005*, in series *Lecture Notes*

in Computer Science, edited by C. A. Coello, et al., Vol. 3410, Springer, 2005, pp. 386–398.

[34] Deb, K., and Agrawal, R. B., "Simulated Binary Crossover for Continuous Search Space," *Complex Systems*, Vol. 9, 1995, pp. 115–148.

[35] Kesseler, E., and Vankan, W. J., "Multidisciplinary Design Analysis and Multi-Objective Optimization Applied to Aircraft Wing," *WSEAS Transactions on Systems and Control*, Vol. 1, No. 2, Dec. 2006, sponsored by WSEAS (World Scientific and Engineering Academy and Society), pp. 221–227.

[36] Vankan, W. J., Kesseler, E., and Baalbergen, E. H., "Distributed Collaborative and Multi-Disciplinary Design of Civil Aircraft Wings," *PDT Europe 2006*, 16–18 Oct. 2006, Toulouse, France, sponsored by Product Data Technology Europe, pp. 107–116.

[37] Vankan, W. J., Maas, R., and Laban, M., "Fitting Fitness in Aircraft Design," *ICAS 2006 Conference*, paper 2006-1.11.3, Hamburg, Germany, Sept. 2006, sponsored by the International Council of the Aeronautical Sciences (ICAS), Stockholm, Sweden.

Chapter 7

Fuselage Structure Optimization

Stéphane Grihon*

AIRBUS, Toulouse, France

Manuel Samuelides† and Antoine Merval‡

ONERA, Toulouse, France

Alain Remouchamps,§ Michael Bruyneel,¶ and Benoit Colson¶

SAMTECH, Liège, Belgium

and

Klaus Hertel**

AIRBUS, Bremen, Germany

Abbreviations

CFRP = carbon-fiber-reinforced plastic
COTS = commercial-off-the-shelf
DoE = design of experiments
GFEM = global finite element model
GUI = graphical user interface
HPC = high-performance computing
LHC = latin hypercube
MDO = multidisciplinary design optimization

*Expert in Structure and Multidisciplinary Optimization.
†Senior Researcher in Applied Mathematics.
‡Ph.D. Student in Structure Optimization.
§Leader of Optimization R&D.
¶Doctor Engineer in Optimization.
**Senior Engineer in Structure Optimization.

MoE = mixture of experts
MSE = mean squared error
NLFEA = nonlinear finite element analysis
NN = neural network
PB = postbuckling margin
PCL = PATRAN command language
PHP = hypertext preprocessor
RF = reserve factor
STIFFOPT = stiffened panel optimization

Nomenclature

aft = attached flange thickness
afw = attached flange width
D_{next} = catalog value and next in catalog
DC, D_{fast} = diameter of fasteners: continuous value
E_{skin}, E_{stringer}, E_{frame} = Young modulus of skin, stringer, frame
F_{XSR} = longitudinal force in the superstringer
fft = free flange thickness
ffw = free flange width
I = quadratic moment of the fuselage section
$IL(\lambda)$ = updated internal loads
IL^* = fixed internal loads
M_Y = bending moment in the fuselage section
N_{xx}, N_{yy}, N_{xy} = fluxes
PG = global optimization problem
P_L = local optimization problem
PLL = local optimization problem
p = stringer pitch
pt = pad thickness
pw = pad width
RF_{fatXRaid}, RF_{fatXSkin}, RF_{fatYSkin} = fatigue reserve factors
RF_{RS} = reserve factor for residual strength
RF_{stat} = reserve factor for static strength
rp = rivet pitch
S_{skin}, S_{stringer}, S_{frame} = section areas of skin, stringer, frame
sh = stringer height
st = skin thickness
$t_{angle}^{SKIN,i}$ = skin thickness variables (composite panel use case)
$t_{angle}^{STRINGER,i}$ = stringer thickness variables (composite panel use case)
$v = z - z_G$ = vertical distance with respect to the center of gravity
$W(\lambda)$ = weight function
wh = Web height
wt = Web thickness
λ = sizing design variables

λ_f = loading factor for buckling and nonlinear analysis
ν = Poisson ratio
σ = stress
σ_{all} = stress allowable
σ_{max} = maximum stress
σ_I, σ_{II}, σ_{VM} = minor, major principle and Von Mises stresses
σ_1, σ_2 = values based on experience

I. Introduction

A. Preamble

THE application of numerical optimization to engineering design was introduced several decades ago starting with structural optimization [1]. Since then, structural optimization has always been at the root of multidisciplinary optimization and today is a major component of multidisciplinary design optimization (MDO). The algorithms developed for structural optimization are fairly generic and therefore can be very useful to MDO, especially those regarding multilevel optimization as discussed in this chapter.

B. Context

The aeronautical industry is a business governed by increasing customer demands for product performance, cost, and time-to-market reduction. Numerical optimization, in turn, is a key technology to achieve these goals as a result of the successful marriage of mathematical theories and associated algorithms that guarantee an optimum performance and advanced software that facilitates the automation and integration of the design life cycle and thus ensures reduction of lead times [2–4].

In this context the trends for business globalization together with advances in massively parallel IT solutions and the World Wide Web encourage structural design to adopt distributed computing concepts [5, 6]. This is the notion of extended enterprise, discussed in more detail in Chapter 10. To fulfil these requirements, a modular multistep process for optimization of structural covers has been implemented. The basic idea was to fully map the manual sizing process.

C. Sizing Process

Because aircraft structures are made of thin shells working essentially as membranes, buckling strength must be taken in consideration. This is also the reason why longitudinal beams called stiffeners are added to the shell [7]. Such an arrangement is known as the stiffened panel concept, which is generally used for airframe covers (see Fig. 1).

Moreover, for local analysis needs, each stiffened panel is decomposed into panel bays also called superstiffeners, as depicted in Fig. 2. Hence, the superstiffener appears as the elementary pattern for the handmade stress and sizing process of airframe covers.

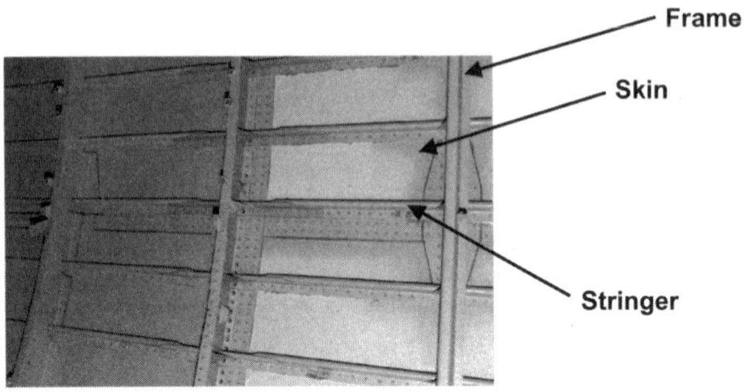

Fig. 1 Stiffened panel concept.

Of course, isolating a superstiffener to analyze the stability of covers assumes that the buckling/postbuckling is localized: this is generally a good approximation if a suitable transverse stiffening is provided (e.g., orbital frames for a fuselage structure).

The advantage of such an approach is that the sizing process is decomposed down to a very fine level of granularity allowing a very flexible implementation of optimization methods.

Thus, the superstiffeners are generally computed with standard stress tools, which are analytically based, including engineering methods relying on handbook formulas, test data banks, and experience.

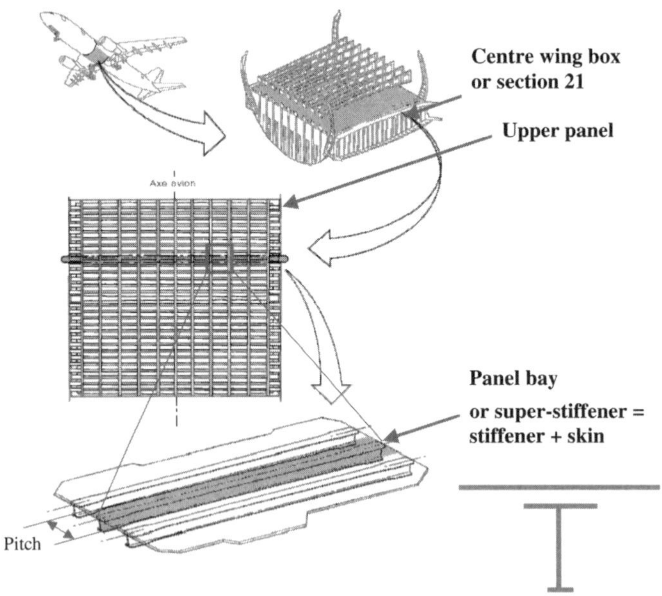

Fig. 2 Airframe structure decomposition.

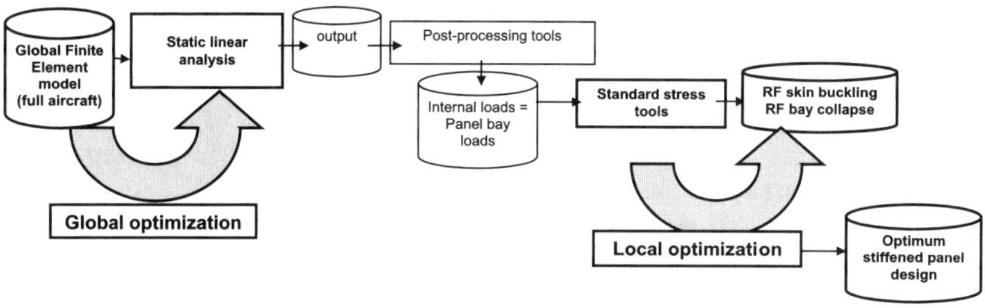

Fig. 3 Design and stress process/global-local analysis and optimization link.

The local analysis uses internal loads extracted from a global finite element model (GFEM) after a static linear analysis based on external loads. Specific postprocessing tools have to be used in order to compute a loading adapted to the superstiffener model; see Fig. 3.

The local analysis delivers reserve factors (RFs). A reserve factor is a value greater than one if the structure is feasible. For example, if σ_{all} is the failure stress for a given material, the reserve factor is the ratio $\sigma_{\text{all}}/\sigma$, where σ is the current stress value.

For buckling/postbuckling, the RFs include the RF for skin buckling and the RF for superstringer collapse (column collapse). Hence, the current design and stress process for a fuselage structure is a bilevel one with global and local analyses [8, 9].

D. Selection of an Optimization Process

In the same way, the optimization can be viewed as a local or a global optimization. Local optimization integrates local analysis and is based on fixed internal loads. The local optimization problem solved P_L is the following one:

$$(P_L)\begin{cases} \text{Min } W(\lambda) \\ \text{RF}(\lambda, IL^*) \geq 1 \end{cases}$$

Global optimization integrates global analysis and is based on updated internal loads. The global optimization problem solved is the following one:

$$(P_G)\begin{cases} \text{Min } W(\lambda) \\ \text{RF}[\lambda, IL(\lambda)] \geq 1 \end{cases}$$

To be properly solved, this formulation requires the computation of internal load sensitivities and the combination with RF sensitivities (chain ruling). This is feasible but complex. This process has been implemented since this research work at AIRBUS [10], but it is complex and heavy and cannot yet be used for a full structure sizing.

The position of this research work was more to reflect the manual sizing process, while keeping simplicity and modularity to make easier distributed computing approaches, knowing that it is not fully optimum. This manual sizing process consists of sizing each superstringer independently with fixed internal loads. The corresponding optimization scheme is then local optimization. However, this local optimization is done with fixed internal loads, while it is clear that changing the sizing also changes the stiffness, hence the distribution of internal loads. The internal load redistribution is generally considered further in the design when a new GFEM is built. No systematic convergence is searched for.

Thus, the optimization and quality of optimization results are only guaranteed after longer-term loops (FE and load loops of an aircraft project) have been performed. The purpose here is to go one step further and to iterate the internal load updates up to convergence.

Hence the algorithm used is the following one:

Step 1: Update the GFEM with new sizing properties λ_k.
Step 2: Compute new internal loads IL_k.
Step 3: Solve PL with $IL^* = IL_k$: gives λ_{k+1}.
Step 4: $\lambda_k \leftarrow \lambda_{k+1}$.
Step 5: Go to Step 1 while significant variation of $\underset{\cdot}{\lambda}$ and/or IL.

With this approach, the first objective is then to integrate this sizing process, that is, to integrate the superstiffener optimization capability in the global design and stress process. Starting from superstiffener optimization, as the main brick, a capability is built for full panel optimization, considering internal load redistribution based on GFEM updates and static re-analyses.

The rest of the chapter is organized as follows:

In Section II, the case study used in this work is introduced.

In Section III, the implementation of a local optimization session for a typical superstiffener is presented and then demonstrated on a set of superstiffeners.

In Section IV, the integration of the STIFFOPT software is described.

In Section V, a tight coupling integration is presented. The demonstration is first made on a set of panels. Then a full fuselage barrel is optimized, using a PC cluster.

In Section VI, a loose coupling scenario is explored based on design curves built with local optimizations and neural networks.

In Section VII, some complements to the sizing process are discussed.

In section VIII, a refinement of the local optimization process is proposed based on nonlinear finite element analysis of the superstiffener.

II. Presentation of the Test Case

The test case used for all method comparisons and validations consists of a fuselage barrel of a very large civil transport aircraft. This barrel is located between the left cargo door and the wing, more exactly between frame 38 and frame 46, as depicted in Fig. 4.

There are eight frames in orbital direction and 146 stiffeners in longitudinal direction. The zone contains consequently 1168 stiffeners. Fifty load cases were used for the sizing optimization, but an envelope was made (commented in Section VI).

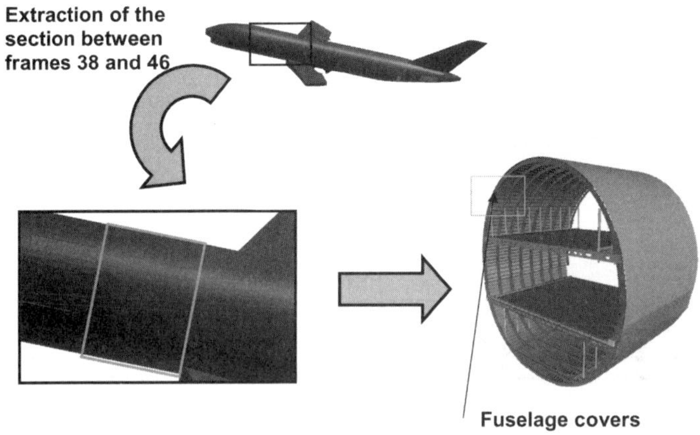

Fig. 4 Fuselage barrel test case.

III. Local Optimization

The purpose of this section is to present and demonstrate the main brick of the optimization process: the local superstringer optimization.

A. Definition

As described before, the purpose of the local optimization is to solve the following problem:

$$(P_L)\begin{cases} \text{Min } W(\lambda) \\ \text{RF}(\lambda,\, IL^*) \geq 1 \end{cases}$$

The exact instantiation for this local optimization problem and a more physical description are presented next.

1. Design Principle

The design principle selected for the use case is the following one: 1) material: metallic (isotropic aluminum); 2) stringer section: Z-shape (see Fig. 5).

2. Stress Hypotheses and Criteria

Specific computations are considered to compute internal loads from the direct output of the GFEM (NASTRAN static linear analysis—SOL101) [11]. Specific stress allowables are considered to take into account fatigue and damage tolerance.

For stability a skill tool is used: ASSIST [12]. ASSIST implements engineering formulas to analyze buckling/postbuckling of stiffened panels based on a superstringer pattern. The postbuckling behavior is considered up to collapse considering also material plasticity.

3. Objective Function

As the targeted fuselage section use case has a uniform frame pitch and material, optimizing the weight is equivalent to optimising the cross-sectional

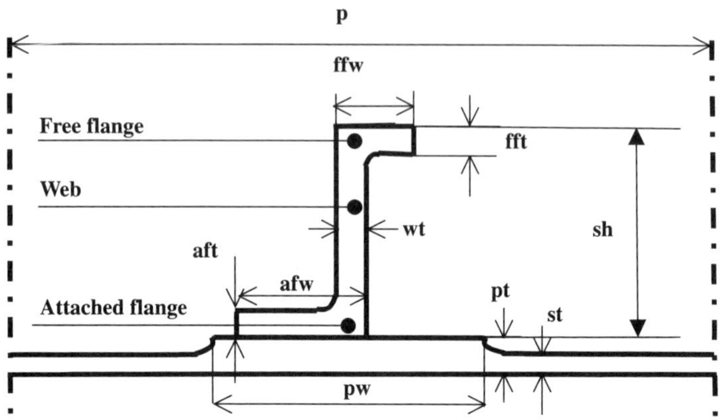

Fig. 5 Design variables for a Z-shaped stringer profile.

area of each superstringer. For this reason the objective function is the cross-sectional area of the superstringer.

4. *Design Variables*

The six following variables are considered within the given bounds (all sizes in millimeters): stringer height, $25 < sh < 55$; Web stringer thickness, $1.6 < wt < 4$; free flange width, $8 < ffw < 26$; free flange thickness, $2 < fft < 4$; attached flanged thickness, $1.6 < aft < 4$; and skin thickness, $1.6 < st < 8$.

The superstiffener section area is computed with these variables. The attached flange width and pad thickness values are linked with skin thickness. Pitch has not been selected as a design variable. It is possible to give some indications of optimum pitch values based on a local optimization. However it should be mixed with a parameterized GFEM to be able to change the pitch. This will be considered in a future work.

5. *Design Constraints*

A number of the six design constraints allow a geometry-conveying designer experience to avoid discontinuity in superstiffener geometry. The range of each constraint is set by designer experience and must be respected at the end of the local optimization process.

Three aspect ratios (width/height-to-thickness ratios) are used to control the geometry (refer to Fig. 5 for the definition of the various sizes):

$$3 < \frac{afw}{aft} < 20$$

$$3 < \frac{sh - fft - aft}{wt} < 20$$

$$3 < \frac{ffw}{fft} < 10$$

Two thickness ratios between each stringer dimensions are

$$\frac{aft}{wt} > 1.3$$

$$\frac{aft}{st} > 1.3$$

Attached flange width must be two times greater than free flange width in order to avoid a too long free flange:

$$\frac{afw}{ffw} > 2$$

Inequality constraints linked to fastener installation and pad are considered as active, that is, satisfied as equalities. This is based on experience from previous stiffened panel optimization studies and allows the reduction of the number of design variables by explicit relations. The width of the attached flange is given by the free edge distance and the minimum distance to the web. The attached flange width is thus linked to diameter and web thickness (4 is the fillet radius).

$$afw = 2 * D_{\text{fast}} + \frac{D_{\text{next}}}{2} + wt + 4$$

D_{fast} and D_{next} are the fastener diameter and the following diameter in a given catalog:

$$D_{\text{fast}} \quad \text{and} \quad D_{\text{next}} \in \{4;\ 4.8;\ 5.6;\ 6.4;\ 8\}$$

Pad width is linked to attached flange width:

$$pw = 2 * \left(3 + \frac{afw}{2}\right)$$

Pad thickness is linked to skin thickness only:

$$pt = 1.4 * st$$

The fastener pitch is taken as 4.5 times the diameter:

$$rp = 4.5 * D_{\text{fast}}$$

D_{fast} and D_{next} are discrete values computed using DC. DC is the continuous version of the fastener diameter. $DC = 0.5 * (aft + 1.4 * st)$ following the rule of the total fastened thickness. Then DC is rounded off to D_{fast} using the following staired function:

$$\begin{aligned} D_{\text{fast}} &= 4 + 0.8 * f(DC-4) + 0.8 * f(DC-4.8) \\ &\quad + 0.8 * f(DC-5.6) + 1.6 * f(DC-6.4) \end{aligned}$$

where f is a function defined $f(x) = [1 + \text{sign}(x)]/2$.

6. *Physical Constraints*

Buckling.
If $st \leq 3$, $PB \geq 60$.
If $3 \leq st \leq 5$, $PB \geq 80$.
If $st \geq 5$, $PB \geq 100$.

These three conditions build a margin policy for buckling on set: depending on the skin thickness, the skin is authorized to buckle at a certain percentage PB of limit loads.

To implement this step rule, the same principle as for diameter is used:

$$(\mathrm{PB})_{\min} = 60 + 20 * f(st{-}3) + 20 * f(st{-}5) \qquad \text{with} \quad f(x) = \frac{[1 + \mathrm{sign}(x)]}{2}$$

The reserve factor for collapse is computed by ASSIST and incorporates the effect of local stringer buckling.

Static strength.

$$RF_{\mathrm{stat}} = \frac{\sigma_{\mathrm{all}}}{\sigma_{\max}}, \qquad \sigma_{\max} = \max(\sigma_I, \sigma_{II}, \sigma_{\mathrm{VM}})$$

Residual strength.

$$RF_{\mathrm{RS}} = \sigma_1 * \left(\frac{E_{\mathrm{frame}} * S_{\mathrm{frame}}}{E_{\mathrm{frame}} * S_{\mathrm{frame}} + E_{\mathrm{skin}} * S_{\mathrm{skin}}}\right) + \sigma_2 \Big/ \frac{N_{yy}\mathrm{tenue}}{\mathrm{Skin}_{\mathrm{thickness}}}$$

This allowable was built by fatigue specialists and is based on tests for various panels and gives the stress allowable as a function of the orbital stiffening ratio.

Crack initiation in the stringer.

$$RF_{\mathrm{fatXStringer}} = \sigma_{\mathrm{all}} \Big/ \frac{F_{\mathrm{XSRfat}} * E_{\mathrm{Stringer}}}{E_{\mathrm{Stringer}} * S_{\mathrm{Stringer}} + E_{\mathrm{Skin}} * S_{\mathrm{Skin}}} - \frac{F_{\mathrm{Poisson}}}{S_{\mathrm{Stringer}}}$$

Crack growth in the skin.

$$RF_{\mathrm{fatXSkin}} = \sigma_{\mathrm{all}} \Big/ \frac{F_{\mathrm{XSRfat}} * E_{\mathrm{Skin}}}{E_{\mathrm{Stringer}} * S_{\mathrm{Stringer}} + E_{\mathrm{Skin}} * S_{\mathrm{Skin}}} + \frac{F_{\mathrm{Poisson}}}{S_{\mathrm{Skin}}}$$

Crack growth in the stringer.

$$RF_{\mathrm{fatYSkin}} = \frac{\sigma_{\mathrm{all}}}{N_{\mathrm{YYfat}}/st}$$

Crack initiation in the skin does not appear because it is covered by crack initiation in the stiffener.

When there is a stress in longitudinal direction, the skin tends also to be deformed in the orbital direction as a result of the Poisson effect:

$$F_{\text{Poisson}} = \frac{\nu}{st} * \frac{E_{\text{Stringer}} * S_{\text{Stringer}} * S_{\text{Skin}}}{E_{\text{Stringer}} * S_{\text{Stringer}} + E_{\text{Skin}} * S_{\text{Skin}}} * N_{yy}$$

All of these formulas take into account a local redistribution of internal loads between the skin and stringer. This is necessary because stringer and skin dimensions vary along the optimization.

Depending on the criterion considered, specific load cases are used: static ultimate loads for strength (limit loads for residual strength) and reference fatigue load cases for crack initiation and propagation.

The optimization problem finally consists in minimizing the superstringer cross-sectional area based on the just-mentioned design variables and taking into account design variable bounds (box constraints), (explicit) links between design variables, implicit links between design variables—design constraints, and structure feasibility constraints, where RF $\geq$ 1, also called physical constraints. The mathematical program is easy to write (based on this summary) and is not shown here for brevity.

B. Implementation

The purpose is to build a local optimization process embedding constraints based on simple equations and constraints based on the output of complex program (skill tools) and especially the ASSIST stability analysis tool.

Rather than writing the interface with a mathematical library like NAG [13] or NASTRAN, it was decided to use a software framework dedicated to integration of external software for easy and fast integration and to benefit from a graphic user interface. BOSS Quattro from SAMTECH company was used for that purpose [2–4].

Integrating an external tool and identifying design variables and results to be constrained or optimized is rather simple with an optimization framework as depicted in Fig. 6. Moreover, BOSS Quattro's library of algorithms was found quite satisfactory for solving the complex optimization problem defined before, considering also the nonlinear behavior of the responses as well as the singularities introduced (stair functions for diameter for example) [14, 15].

A practical implementation is worth noticing: as the framework interfaces external applications through files and as sensitivities are computed with finite differences (small perturbations: see Fig. 7 for illustration), it was found to be preferable to compute sensitivities externally (Table 1).

The reason for that was not at the least due to the benefit, which can be gained from the ASSIST implementation. That is, a list of geometries (as well as a list of load cases) can be given as input, and analyses are iterated inside the tool without re-initialization and additional input/output times. This approach gives much smaller computational times than doing computations separately.

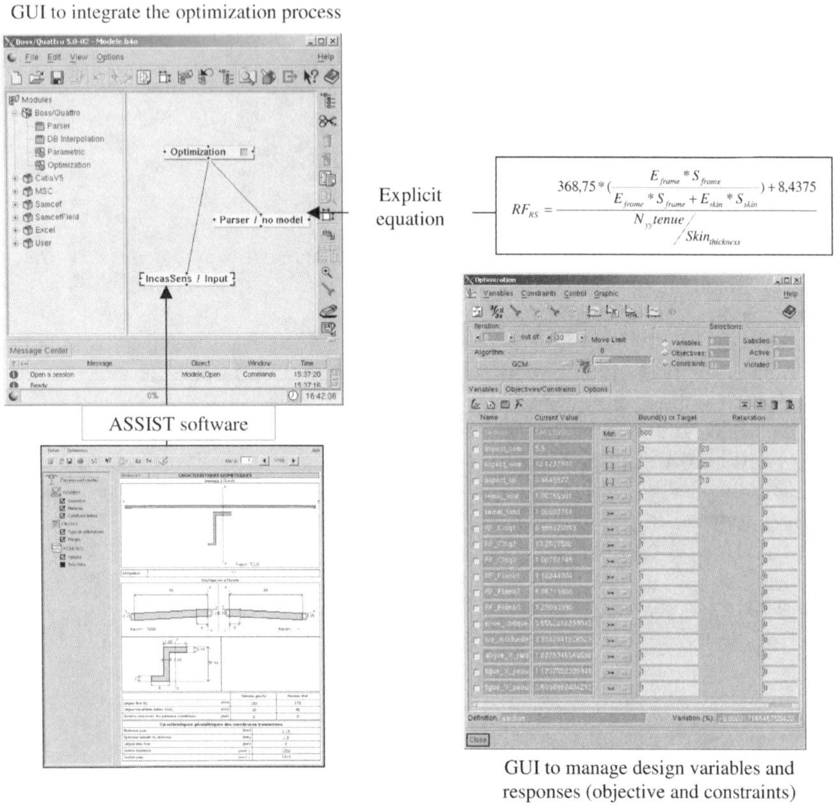

Fig. 6 Local optimization integration.

Considering that the relative step for sensitivities is $h = 10^{-3}$, Table 1 is given as input to ASSIST. This approach is appealing because it is in accord with the zero-order algorithms, which calculate one population at each iteration.

The time savings estimated for such an implementation above are about 60%: an iteration now takes 15 s compared with 40 s before parametric use. The main

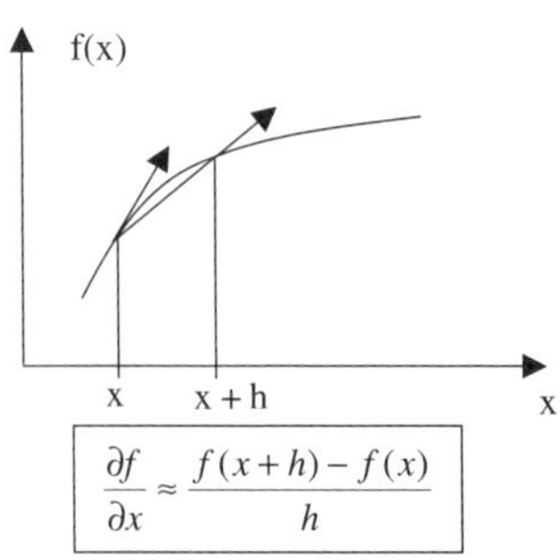

Fig. 7 Principle of finite difference sensitivity analysis.

Table 1 Geometries to be computed for current analysis and computation of sensitivities

sh	*wt*	*ffw*	*fft*	*aft*	*st*
$sh^*(1+10^{-3})$	*wt*	*ffw*	*fft*	*aft*	*st*
sh	$wt^*(1+10^{-3})$	*ffw*	*fft*	*aft*	*st*
sh	*wt*	$ffw^*(1+10^{-3})$	*fft*	*aft*	*st*
sh	*wt*	*ffw*	$fft^*(1+10^{-3})$	*aft*	*st*
sh	*wt*	*ffw*	*fft*	$aft^*(1+10^{-3})$	*st*
sh	*wt*	*ffw*	*fft*	*aft*	$st^*(1+10^{-3})$

reason for this lower time is the reduction of file input/output obtained through externalization of the finite difference calculations as illustrated in the Fig. 8.

C. Demonstration on a Set of Superstringers

1. *Test Case*

To check and demonstrate local optimization, a set of five superstringers was selected across the barrel in the frame bay between frame 42 and frame 43. Figure 9 illustrates the location of these superstringers.

In an actual detailed design all 72 super stringers will be calculated. This is to be doubled to consider the other half-barrel. Then top and lower stringers (S0 and S73), lying in the symmetry plan, have to be added. A total of 146 superstringers is thus to be considered for the full frame bay.

2. *Optimum Section Areas*

Figure 10 shows the section areas found for the superstringers at convergence. They are consistent with the loading of the superstringers: a high compression force results in a superstringer with a large section area. For example, stringer 1 supports 106116 N as compression force while only 23514 N for stringer 26 (Table 2).

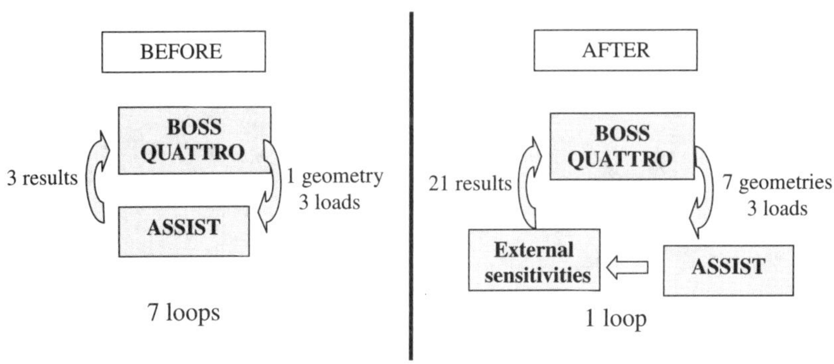

Fig. 8 Principle for implementation of sensitivity analysis.

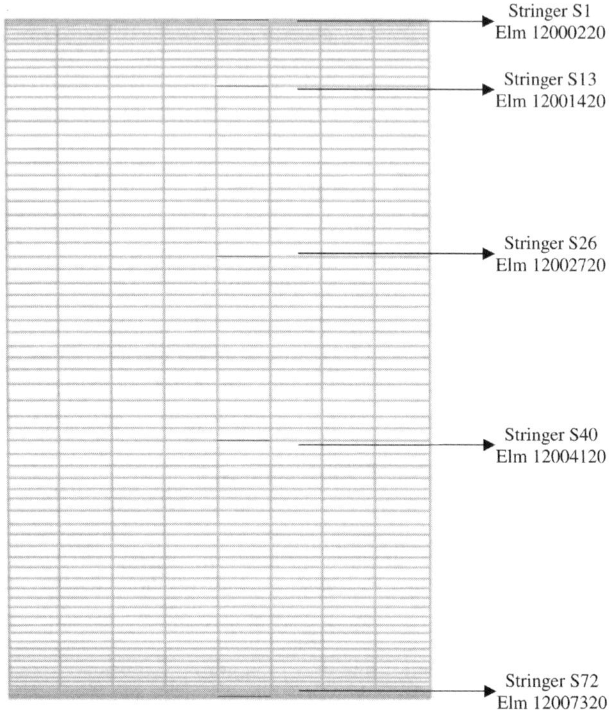

Fig. 9 Principle for implementation of sensitivity analysis.

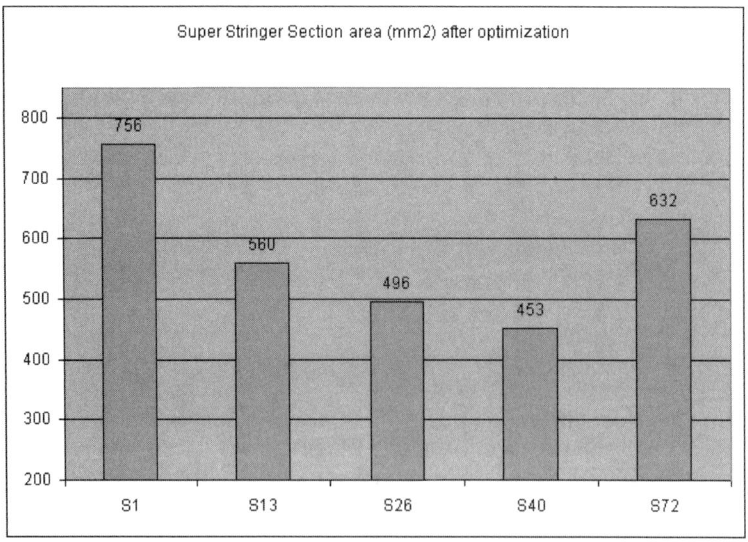

Fig. 10 Superstringer section areas after optimization.

Table 2 Superstringer loading[a]

Stresses	S1	S13	S26	S40	S72
Superstringer max compression force, N	−106,116	−46,561	−23,514	−20,499	−116,670
Static superstninger force, N	184,267	84,989	21,222	35,666	39,273
Fatigue superstringer force, N	58,563	27,943	21,294	14,733	6,289

[a]The superstringer S1 has a larger section than S72 but a lower compression load because the fatigue load is more important.

3. *Convergence History*

The convergence is difficult because of the nonlinearity of ASSIST: sometimes the responses are known to be nondifferentiable. This is the reason why the convergence histories appear chaotic and sometimes quite long.

Nevertheless, the optimization histories show a convergence in less than 40 iterations. Also looking at convergence plots in Fig. 11, it seems that the convergence can be truncated at about 30 iterations. This is a good lesson learned: as several cycles linked to internal load updates are performed, it is possible to truncate the convergence in order to have smaller computational times. This principle is applied further to the STIFFOPT process discussed next.

4. *Stringer Profiles*

Web height is high when a superstiffener with large section (heavily loaded) must be designed (stringer S1 and S72). Indeed, web height is known as the main driver for the column collapse RF.

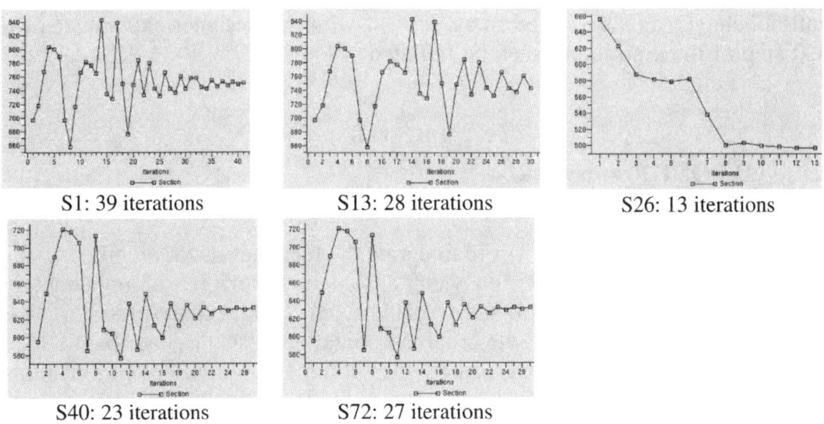

S1: 39 iterations; S13: 28 iterations; S26: 13 iterations; S40: 23 iterations; S72: 27 iterations

Stringer number	S1	S13	S26	S40	S72
Iteration number	39	28	13	23	27

Fig. 11 Optimization histories.

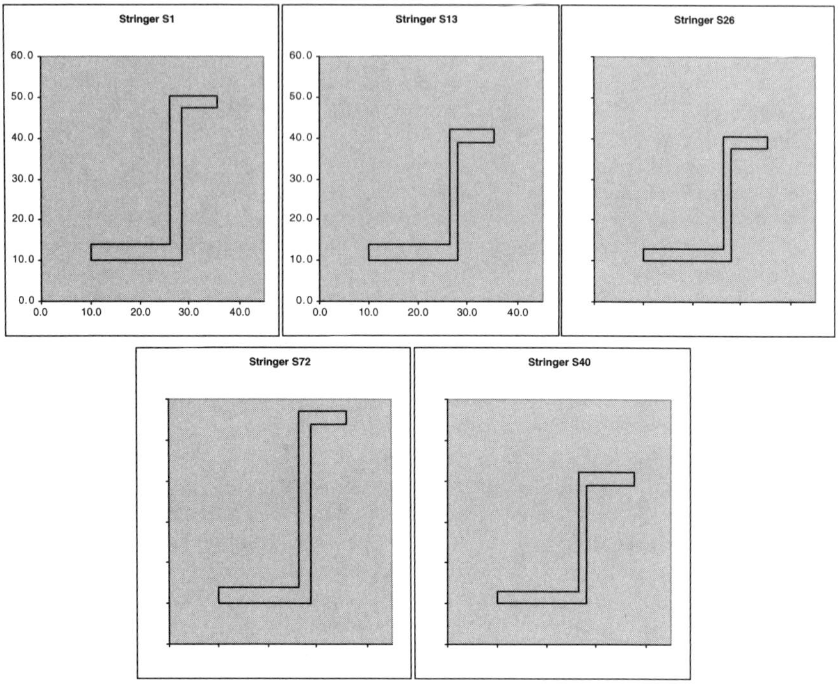

Fig. 12 Optimum stringer profiles.

Free flange and attached flange thickness are always larger than web thickness, see Fig. 12. Free flange width is less than attached flange width because of a specific design constraint added. This demonstration shows that the local optimization behavior is correct and in agreement with the designer experience. The next step of integration can then be initiated.

IV. Integration

A. STIFFOPT Framework

1. STIFFOPT Principle

The principle of STIFFOPT is to integrate the local optimization process validated and demonstrated earlier in a wider environment in order to implement the algorithm presented in the introduction. This supposes a mechanism to be able to initialize the sizing, prepare all local optimization sessions, launch all local optimizations, collect all sizing results after completion of these optimizations, update GFEM properties, run a new linear static analysis with this updated GFEM, postprocess the internal loads including a load envelope, and launch a new optimization based on the previous sizing results.

As the GFEM is compliant with the NASTRAN format and as part of the process requires updating the GFEM properties, PATRAN was considered as necessary in the process. Moreover, PATRAN offers a graphical user interface (GUI) and authorizes customization thanks to its internal command language PCL [16]. PATRAN can also be used to launch NASTRAN analysis such as SOL101 linear static analysis.

PATRAN can also be adapted to launch other processes such as local optimizations. Finally the integration is based on three commercial-off-the-shelf (COTS) tools: as shown in Fig. 13. It also integrates the Airbus in-house tool ASSIST.

Moreover PATRAN preprocessing capabilities allow defining the region of interest: the set of superstringers to be optimized. It is also possible to define groups of superstringers having the same sizing properties to simplify and reduce the size of the problem. This is further described next.

2. *STIFFOPT GUI*

The STIFFOPT process is displayed in a vertical menu of the PATRAN main window. Figure 14 presents this menu and the associated steps of the process.

Remark: Every second loop in Fig. 14 necessarily begins with new geometry updating in order to guarantee the consistency of the finite element results with the current definition of stringer bays.

Step 1: Group definition. Superstiffener groups are set during this step. Properties and materials are selected for each group. A group can be either created, modified, or deleted. The "modify" option enables the user to change the current design as he wishes and without any specific interaction with the process (except

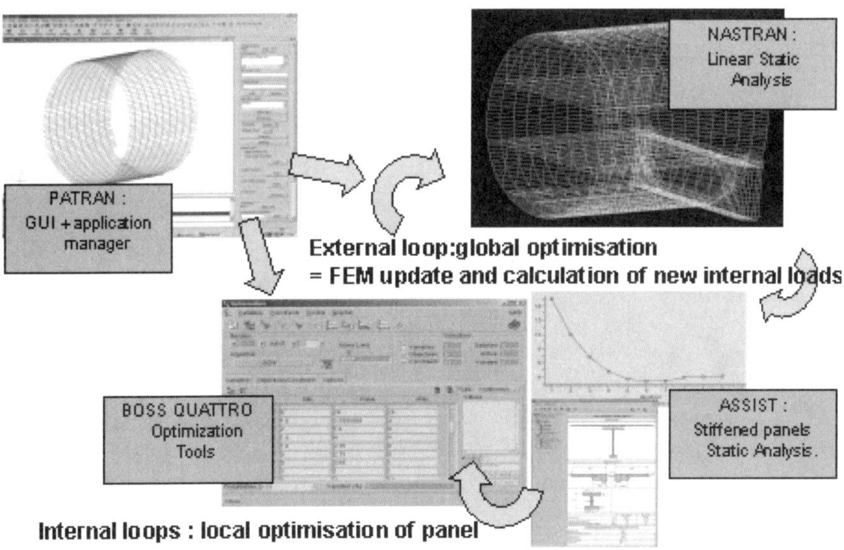

Fig. 13 STIFFOPT principle and COTS used. (See also color figure section at the back of the book.)

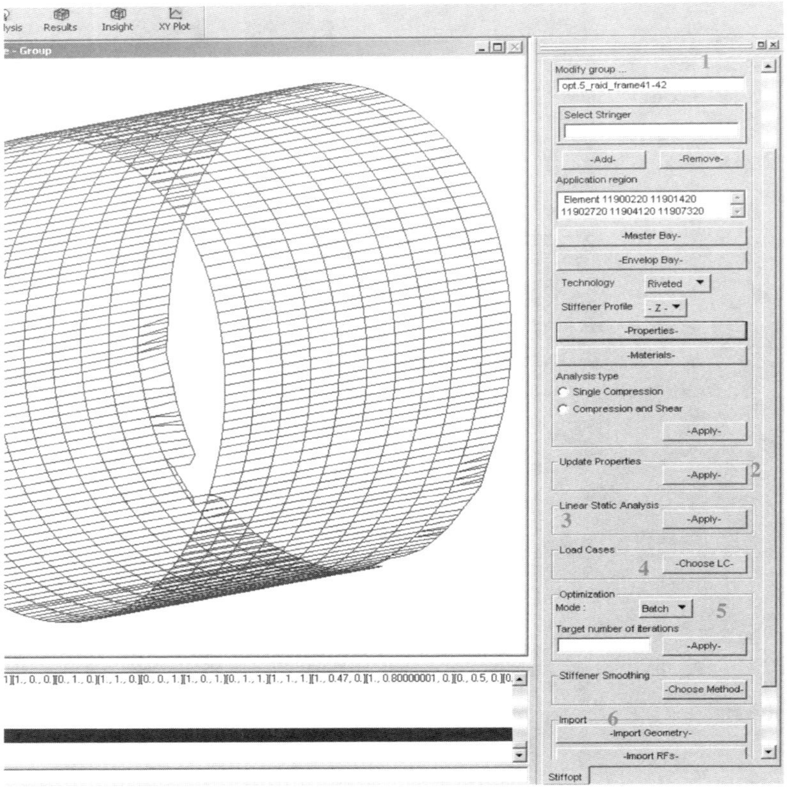

1 - Users create stiffener groups with properties for the first loop → **Group Definition**

2 - GFEM is updated with stringer geometry since the 2nd loop is done → **Update Properties**

3 - Internal load redistribution is computed using the new GFEM → **Linear Static Analysis**

4 - Users choose static and fatigue load cases → **Load cases**

5 - Envelope computing is made with new internal load redistribution and optimization is launched → **Optimisation**

6 - Results are imported in order to set new geometry in stiffener group → **Import**

Fig. 14 STIFFOPT vertical menu and successive steps.

a slower convergence if the design is changed too much with respect to the final optimum results). This option can be very important in the final stage to smooth the design in order to achieve a continuous variation of stringer dimensions (constant height for example). Some tools have been developed to help this manual process.

Master bay and envelope bay methods are optional. They allow grouping superstringers and reducing the number of optimizations to be performed for computational time reasons:

1) The "master bay" approach consists of attributing the sizing obtained for one superstringer to all superstringers of a group.

2) The "envelope bay" approach consists of optimizing a theoretical superstringer with an envelope of internal loads computed over the full group. Then the sizing of this theoretical superstringer is attributed to all superstringers of the group.

All superstringers are optimized independently in the default option.

Technology and stringer shape are then chosen. A riveted stringer with a "Z" shape is the only choice for the time being, but "I" and "J" shapes will be set in future versions [17]. Optimization sessions are already available from past applications.

Figure 15 illustrates the three ways to input properties: manually by writing in the listbox "value" with possibility to copy one sheet to the others for repetitive properties, using directly ASSIST GUI by clicking on the "Assist" button to get a graphical reference for properties, or importing an existing ASSIST file by clicking on the "Import" button for data already available. Of course, it is possible to copy-paste the information for one stringer to the other ones to limit the input effort. Materials are then to be chosen from a given list for skins, stringers, and frames.

The last step is to choose ASSIST computation type. "Single compression" is chosen if the study zone is not significantly loaded with shear. "Compression and shear" is the default choice.

Steps 2–4: Update properties, linear static analysis and load cases. "Update Properties" (see Fig. 16) allows the setting of new properties of stringer in the GFEM. The GFEM is updated with stringer and skin section area computed with new properties (web height, free flange thickness, etc). Stringer properties come from results of the last optimization loop or from properties setting.

Because the GFEM geometry has been modified, a linear static analysis SOL 101 must be launched [11]. This analysis gives new load redistribution of internal loads. Analysis steps are as follows as shown in Fig. 17: a *.blk file is created when users click on "Apply," an *.xdb is obtained after running NASTRAN, and an *.xdb file is attached by users on the GFEM. "Load Case" button opens a window where users choose static and fatigue load cases. These selections are used for linear static analysis and for envelope computing.

Step 5: Envelope computing and optimization. Before launching optimization, three files must be created for each stringer and put in a repository: a BOSS QUATTRO session file *.b4o, an Assist file *.ses, and a neutral file *.in.

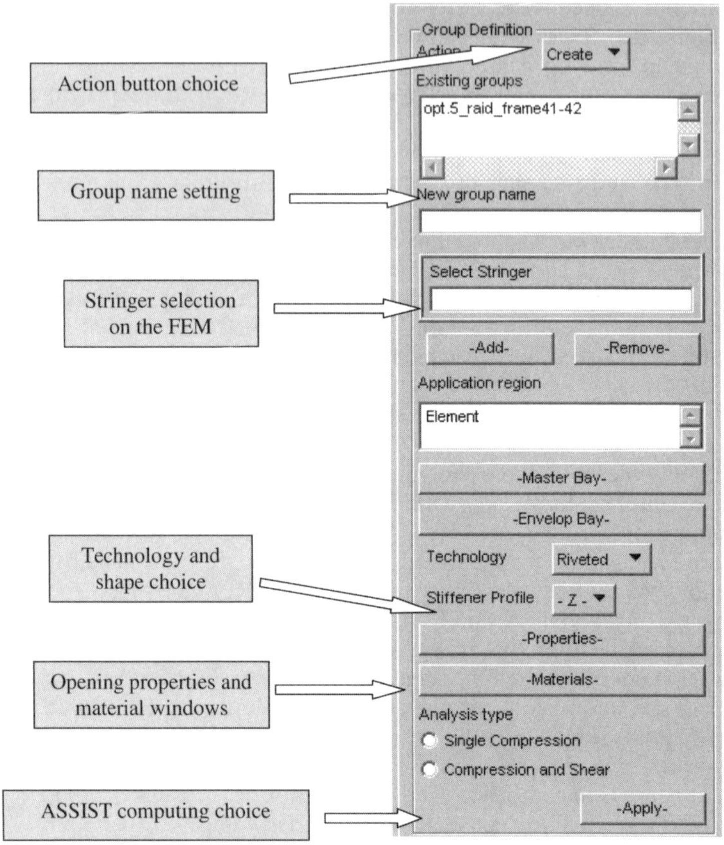

Fig. 15 STIFFOPT: definition of superstringer groups.

Figure 18 shows how these files are set in a repository tree composed of groups and stringers.

The files are created by PATRAN command language (PCL) using an envelope approach for internal loads in order to obtain sizing constraints values. Envelope computing for each superstringer is consequently done using a new load redistribution of internal loads and chosen load cases.

The longitudinal force in the stringer F_x and three fluxes in the skin (N_{xx}, N_{yy}, and N_{xy}) are taken from linear static analysis results for each stringer bay. The structural responses (see Section III) are computed using these four results for each stringer during a loop on selected load cases: maximum bay compression force (sum of the force in the stringer and adjacent skin fluxes multiplied by half-pitches) and associated shear (direct stress output), maximum shear and associated compression force, compression force and associated shear for maximum principle stress, maximum force for static load cases, flux N_{yy} for static load cases, maximum force for fatigue load cases, maximum force for fatigue load

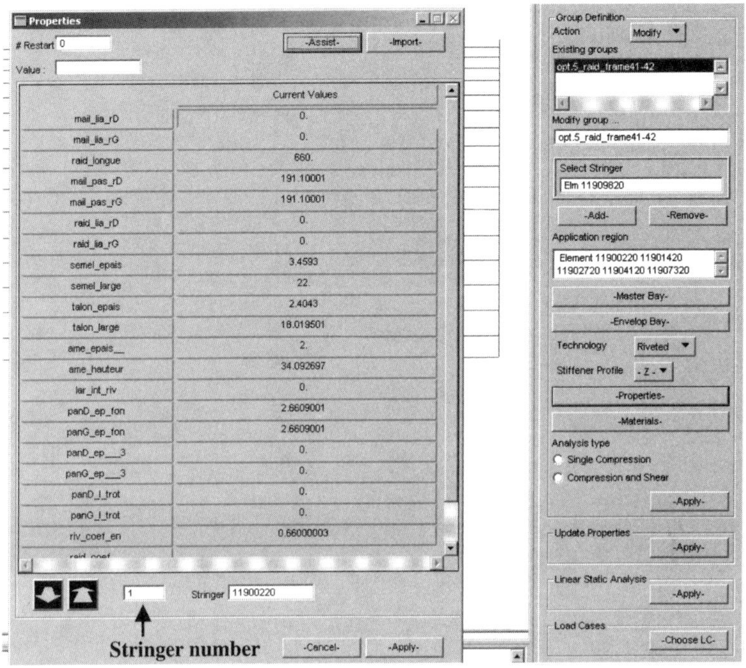

Fig. 16 Input of properties.

cases centered on the skin (sum of flux in the skin multiplied by pitch and half-forces in adjacent stringers), and flux N_{yy} on the skin for fatigue load cases.

To launch the optimization, the user can then choose "batch" or "interactive" modes. For batch mode, optimization session files are set, and the optimization is launched. For interactive mode, optimization session files are only set, and the optimization is not launched. The maximum number of iterations can be manually set if the users wish to accept an incomplete convergence. Finally, users click on "Apply" button (Fig. 19), and optimization is launched.

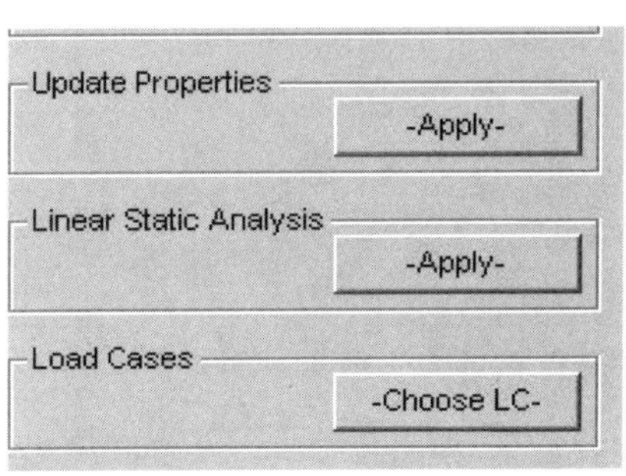

Fig. 17 Management of global analysis.

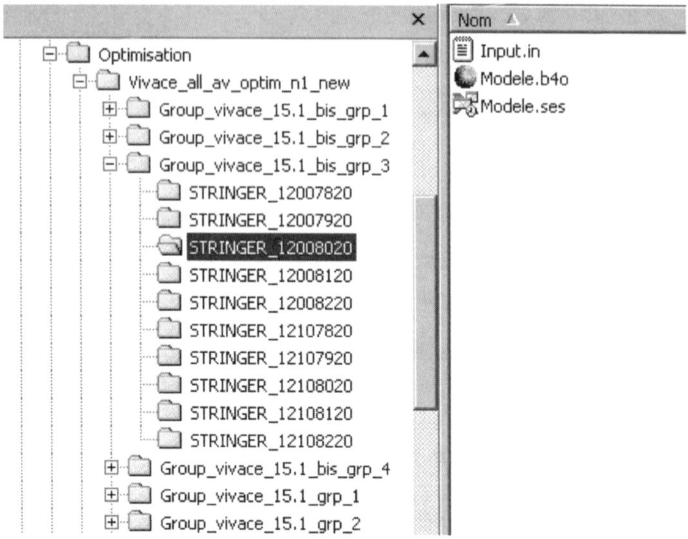

Fig. 18 Management of optimization data.

Step 6: Import. At the end of the optimization, three actions can be performed (Fig. 20): 1) setting new geometry in stringer properties → import geometry, 2) show buckling reserve factor in an array → import RFs, and 3) show mass for each stringer in an array → import mass. When the user clicks on "Import Geometry," each stringer repository is cleared, and only the final files are preserved (modele_end.b4o and input.in for example). Each stringer geometry is set in a file named modele.txt. This file is read, and new properties values are set in each properties stringer array.

B. Demonstration on a Single Panel

To validate the convergence of the global optimization process with the update of internal loads, a preliminary test case was performed on a limited zone. (Four stringer bays were optimized in panel 15.7.) A fast convergence of sizing was observed. Then a more extensive test case was solved, as explained next.

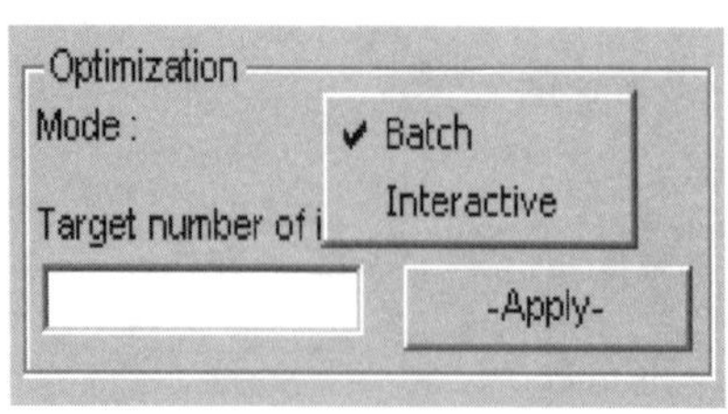

Fig. 19 Management of the optimization process.

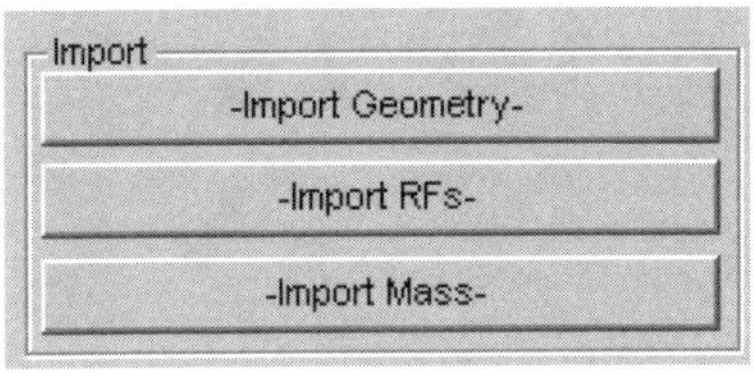

Fig. 20 Management of optimization results.

The optimization was performed for 72 stringer bays located in an area between frames 42 and 46 and stringers 0 to 21 for the panels 15.1 and 15.2 as depicted in Fig. 21. The initial sizing of the stringers is the same and is consistent with the finite element properties:

1) Initial stringer section area = 114.24 mm^2 (same for the GFEM).

2) Initial stringer bay section area = 661.24 mm^2 (for the GFEM stringer bay = 650 mm^2).

For the set of optimized stringer bays, a fast convergence has once again been observed. The dimensions are stabilized after three global iterations. As Fig. 22 shows, some convergence oscillations can be observed around limit values, and the convergence is nonmonotonic as already observed for local optimization test cases.

Stringer bay section areas are increased in the upper part of the fuselage (Str 1: final stringer section = 163.5 mm^2; final stringer bay section = 713.6 mm^2) and decreased in the lateral part (Str 15: final stringer section = 135.8 mm^2;

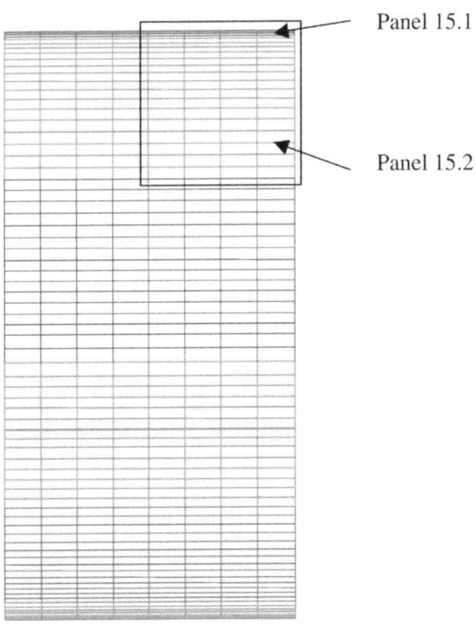

Fig. 21 Management of optimization results.

		init	1optim	2optim	3optim	4optim	5optim
Str 1	aft	4	2,618	2,62	2,618	2,618	2,618
	afw	10	23,782	23,798	23,782	23,782	23,782
	fft	1,6	1,817	1,929	1,818	1,819	1,876
	ffw	24	18,874	19,188	16,151	15,042	15,197
	wt__	1,6	2,182	2,198	2,182	2,182	2,182
	wh	22,4	31,972	32,966	32,727	32,727	32,727
	st	2,75	2,724	2,695	2,711	2,689	2,702
	tt	3,5	3,814	3,773	3,795	3,764	3,782
	pw	14,5	14,51	14,52	14,53	14,51	14,51
	S Str	114,24	166,318238	171,82368	163,034108	161,032988	162,181162
	S StrBay	650,365	702,418138	702,22124	712,336148	705,828488	709,604762
Str 10	aft	4	2,618	2,618	2,618	2,618	2,618
	afw	10	23,782	23,782	23,782	23,782	23,782
	fft	1,6	1,821	1,818	1,818	1,818	1,818
	ffw	24	15,901	9,841	9,839	9,79	9,827
	wt__	1,6	2,182	2,182	2,181	2,181	2,182
	wh	22,4	25	28,161	28,149	27,707	28,038
	st	2,75	2,275	2,295	2,295	2,266	2,288
	tt	3,5	3,185	3,213	3,213	3,172	3,203
	pw	14,5	14,51	14,51	14,51	14,51	14,51
	S Str	114,24	145,766997	141,599516	141,541547	140,488463	141,305678
	S StrBay	650,365	593,496097	593,264696	606,526907	599,586583	604,866978
Str 20	aft	4	2,618	2,618	2,618	2,618	2,618
	afw	10	23,782	23,782	23,782	23,782	23,782
	fft	1,6	1,818	1,818	1,818	1,818	1,818
	ffw	24	9,476	9,472	9,472	9,544	9,472
	wt__	1,6	2,182	2,182	2,182	2,182	2,182
	wh	22,4	25	25	25	25,598	25
	st	2,75	2,1	2,09	2,09	2,166	2,1
	tt	3,5	2,941	2,926	2,927	3,032	2,941
	pw	14,5	14,51	14,51	14,51	14,51	14,51

Fig. 22 Variation of four stringer bay dimensions: Str 1, 10, and 20, Fr 43–44.

final stringer bay section = 597.9 mm^2). This evolution is given as an example for three stringer bays in Fig. 22.

The variation along the iterations of maximum compression forces is surveyed in order to evaluate the internal load redistribution. In addition to the stringers whose dimensions are controlled (Str 1, 5, 10, 15, and 20), gauge stringers are used to investigate the variations of internal loads far from the optimized region. These are stringers 21, 31, and 41 for the interframes 43–44 and 44–45.

The variation of internal loads for the optimized stringer bays 1, 10, and 20 and for the gauge stringer bays 0, 21, 31, and 41 is reported in Fig. 22.

Some dimensions rapidly reach a common final value for all stringers. This is a result of the activation of several design constraints linking these dimensions as described in Section III.A. This shows that these constraints have probably to be relaxed and/or improved to get real optimum sections for the stringers.

Figure 23 shows the following:

1) There is a large variation of internal loads after the first optimization then a smaller variation showing that the optimization process is rapidly stabilized. Beyond the second optimization there is no real variation of dimensions; hence, internal load variations have not been estimated.

2) There is a significant reduction of internal loads (absolute value) in the upper part of the fuselage where the stringer bay sections are increased.

		init	1 optim	2 optim
Str 0	Fx StrBay	-99964	-98286	-98062
Str 1	Fx StrBay	-98365	-96630	-96753
	Fx Str	-23784	-23808	-24611
Str 10	Fx StrBay	-57888	-59126	-59008
	Fx Str	-11515	-14730	-14256
Str 20	Fx StrBay	-37747	-37179	-37159
	Fx Str	-7936	-9158	-9147
Str 21	Fx StrBay	-33730	-32961	-33049
Str 31	Fx StrBay	-12970	-12908	-12922
Str 41	Fx StrBay	23946	23956	23952

Fig. 23 Evolution of internal loads for three optimized stringer bays (Str 1, 10, and 20) and for four gauge stringer bays (Str 0, 21, 31, 41); Fr 43–44.

3) It also has an increase of internal loads in the lateral part where the stringer bay section is decreased.

4) A little variation of internal loads in the gauge stringer bays is shown.

Attempts to clearly explain the impact of sizing variations on the internal loads are difficult because there are both section increases and decreases in the region. Considering the beam theory explained in more detail in Chapter 4, the distribution of axial loads in the fuselage section is governed by the equation:

$$F_{\mathrm{XSR}} = \sigma \cdot S = \frac{M_Y}{I} \cdot v \cdot S$$

So the evolution of axial forces is a combination of global variations, (I, z_G) variations, and local section variations. Moreover, the same analysis should be made in the y axis for a lateral bending moment included in the set of sizing load cases (lateral gust). Hence, accurate computations should be done to justify clearly the behavior of internal loads.

Altogether, this first case of validation is encouraging giving a rapid convergence to the STIFFOPT process but must be now widely extended. Indeed the area optimized is still limited, and thus the impact of sizing changes on the distribution of internal loads is also rather limited. To make a more convincing test case, it is necessary to work on a full fuselage section. This is the purpose of the next section.

V. Tight Coupling

The tight coupling process is investigated in this section. The tight coupling consists of direct calls to the stress tool ASSIST inside the local optimizations. The tight coupling is defined in opposition with the loose coupling described in Section VI. The loose coupling process consists of replacing the local optimizations based on ASSIST with the use of surrogate models. These surrogate models are artificial neural networks built from intensive local optimizations performed a priori, before using STIFFOPT.

A. High-Performance-Computing (HPC) Implementation

1. *Why HPC Is Necessary and Easy to Implement*

Computational times are very heavy for the optimization of a full use case. Despite some potential time savings arising from external sensitivity analysis in the local optimizations, reduction of load cases based on an envelope, and optional use of master bay and envelope bay methods, superstringer per superstringer optimization can take a very long time. If STIFFOPT optimizes the zone between frame 38 and 46, 1168 local optimization session files must be launched (1168 stands for eight interframes times $2 * 72 + 2 = 146$ stringers per interframe: top and lower specific stringers have not been optimized in the symmetry plan). Because one iteration lasts 15 s and about 15 iterations are necessary to reach convergence, one stringer optimization lasts about 4 min. An optimization with 1168 stringers will last about 80 h.

Nevertheless, one large advantage of the approach selected is its inherent parallelism: each superstringer optimization is independent from the other ones. Hence, a natural and direct way of reducing computational times is to perform each optimization on a separate processor. Of course, 1168 processors are not necessarily available. So, at least, what can be done is to group computations in equal size sets and send each set to a separate processor.

2. *Computational Means*

A demonstration of such a process was made based on standard desk PCs. A set of 30 PCs was used overnight to parallelize computations. Figure 24 shows how the network is composed of 30 workstations Dell Precision that are used each day. The computation grid is consequently available only during the night. The technical characteristics are the following: worksations Dell Precision 360, processors PIV 2.5–3.2 GHz, 512 Mo and 2Go of DDR SDRAM, [80–120] Go disk (IDE & SATA), and videoboard Quadro4 and Quadro Fx.

3. *PC Grid Principle*

The grid computing method is composed of four steps as illustrated on Fig. 25:

1) Boss Quattro files are created on a master workstation.
2) The master workstation dispatches one file for each slave workstation.
3) Optimization is launched on slave workstation.
4) Result files are imported to master workstation after optimization.

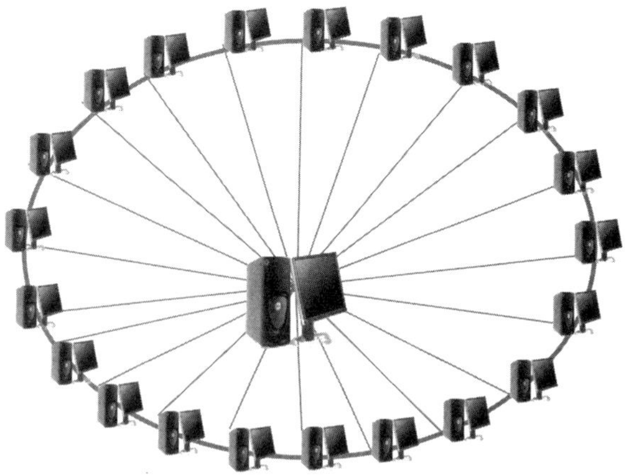

Fig. 24 PC grid principle illustration.

The theoretical speed-up factor is consequently the number of PCs. The practical speed-up factor is very close to the theoretical speed-up factor. If grid computing is used with 30 workstations, computation time is reduced by a factor closed to 30: 3 h vs 80 h as shown in Fig. 26.

4. Management of the Parallel Process

A GUI in PHP: Hypertext Preprocessor (PHP) programming language has been created to launch the computation grid [18, 19]. The steps are 1) process initialization, 2) census of available PC, 3) census of session file available, and 4) process launch. This program is launched with the master workstation as shown on Fig. 27.

With such an HPC approach, we can now imagine a full fuselage optimization. We have, however, focused our demonstration on the fuselage barrel test case, which is now possible to optimise in a "reasonable" time.

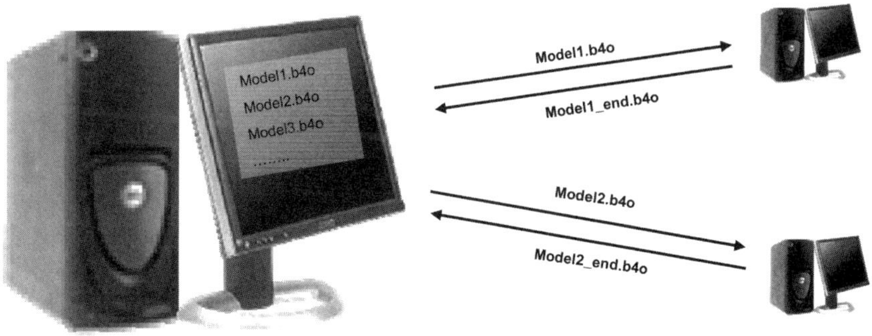

Fig. 25 Grid computing principle.

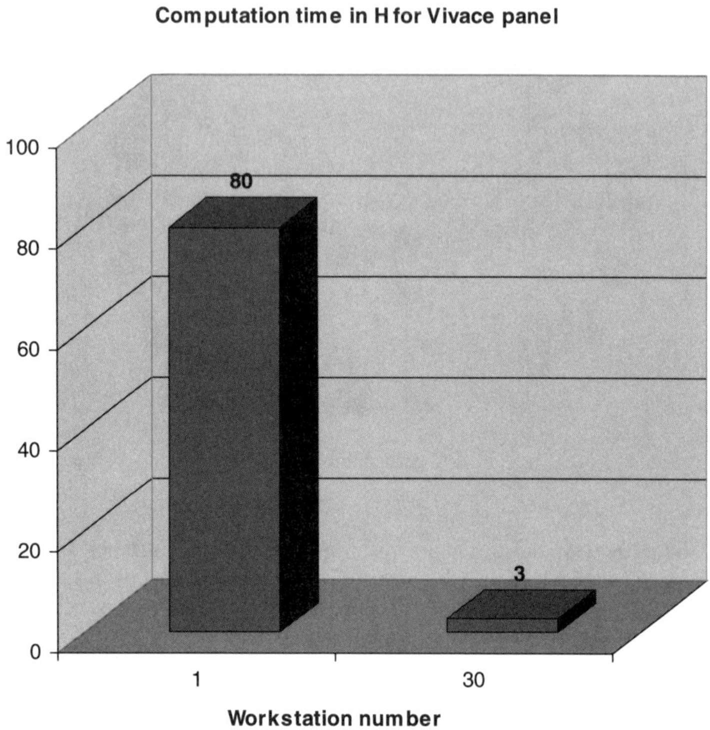

Fig. 26 Computation time with grid computing.

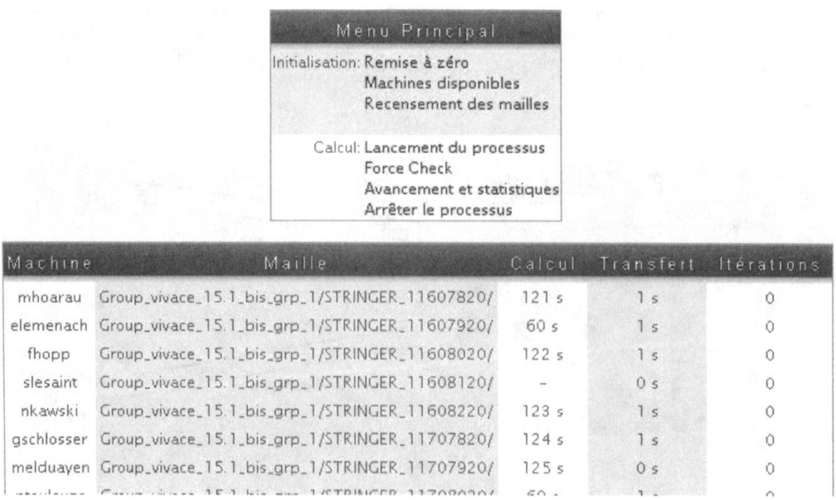

Fig. 27 Management of the parallel process.

Table 3 Topology of groups

Panels	Group 1	Group 2	Group 3	Group 4
15.1	15.1_grp_1	15.1_grp_2	15.1_grp_3	15.1_grp_4
15.1_bis	15.1_bis_bis_grp_1	15.1_bis_bis_grp_2	15.1_bis_bis_grp_3	15.1_bis_bis_grp_4
15.2	15.2_grp_1	15.2_grp_2	15.2_grp_3	15.2_grp_4
15.2 bis	15.2_bis_bis_grp_1	15.2_bis_bis_grp_2	15.2_bis_bis_grp_3	15.2_bis_bis_grp_4
15.3	15.3_grp_1	15.3_grp_2	15.3_grp_3	15.3_grp_4
15.3 bis	15.3_bis_bis_grp_1	15.3_bis_bis_grp_2	15.3_bis_bis_grp_3	15.3_bis_bis_grp_4
15.34	15.34_grp_1	15.34_grp_2	15.34_grp_3	15.34_grp_4
15.34 bis	15.34_bis_bis_grp_1	15.34_bis_bis_grp_2	15.34_bis_bis_grp_3	15.34_bis_bis_grp_4
15.42	15.42_grp_1	15.42_grp_2	15.42_grp_3	15.42_grp_4
15.42 bis	15.42_bis_bis_grp_1	15.42_bis_bis_grp_2	15.42_bis_bis_grp_3	15.42_bis_bis_grp_4
15.46	15.46_grp_1	15.46_grp_2	15.46_grp_3	15.46_grp_4
15.46 bis	15.46_bis_bis_grp_1	15.46_bis_bis_grp_2	15.46_bis_bis_grp_3	15.46_bis_bis_grp_4

B. Demonstration for a Full Barrel

Forty-eight groups of superstiffeners were set. But in each group the optimization has been performed superstringer per superstringer. Twelve panels exist in the section (six on each side of the fuselage), and each panel is divided into four groups: group 1, frame 38–40; group 2, frame 40–42; group 3, frame 42–44; and group 4, frame 44–46. Table 3 gives the 48 group names. They follow the physical panelization as illustrated on Fig. 21. The results provided in Fig. 28 show four optimization loops were computed because the convergence is fast. Indeed, the convergence is reached after loop 2 when the gap percentage is only 0.23%.

Results are analyzed panel per panel and not per group. There are 12 panels. Figures 29 and 30 show gap percentages between each optimization for each group. The panel convergence is not homogenous. Two convergence cases exist:

1) Case 1—nine panels: Because the initial constraints are not active, the initial mass is too great, and mass decreases with optimization. The gap percentage between optimization is always negative and reaches zero.

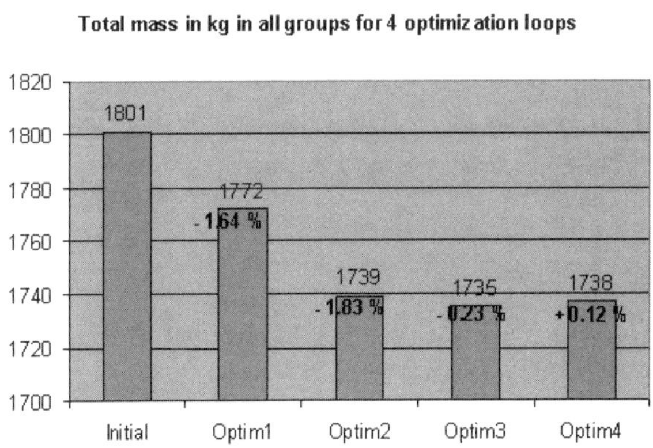

Fig. 28 Convergence history of the full section mass.

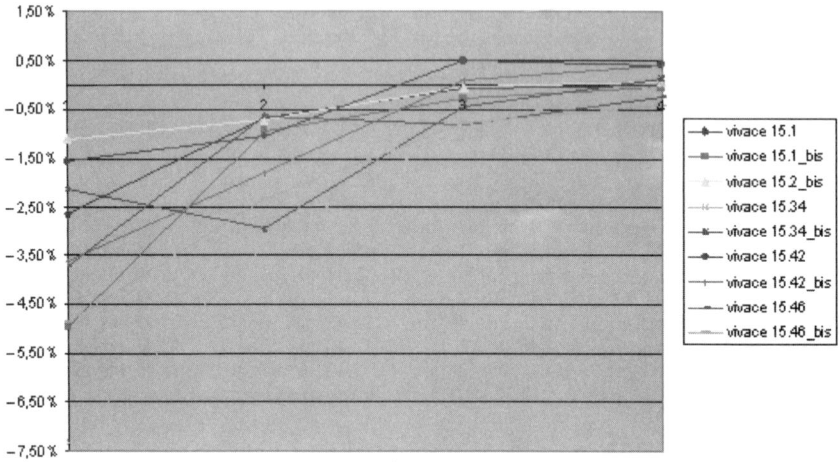

Fig. 29 Convergence of case 1. (See also color figure section at the back of the book.)

2) Case 2—three panels: Because some initial constraints are active or violated, mass increases after optimization 1 to permit constraints to be allowable. Then the problem is the same as case 1 for optimization 2. The gap percent is first positive and then negative. Finally, it reaches zero. That means the stringer is optimal.

Figure 31 shows the mass of each panel for the four optimization loops.

The mass of panel 15.2, 15.3, and 15.3_bis increases after the first loop. The panels that are in the middle top of the fuselage are close to the optimum. Indeed, initial mass and mass after the last loop is almost the same.

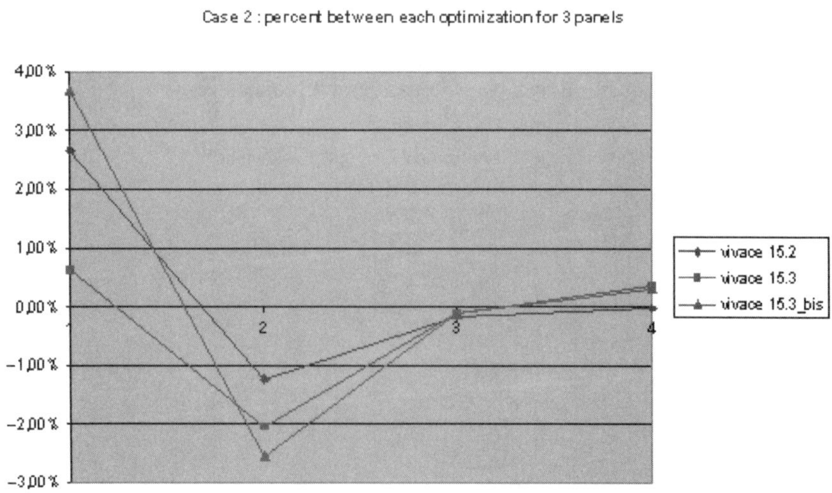

Fig. 30 Convergence of case 2. (See also color figure section at the back of the book.)

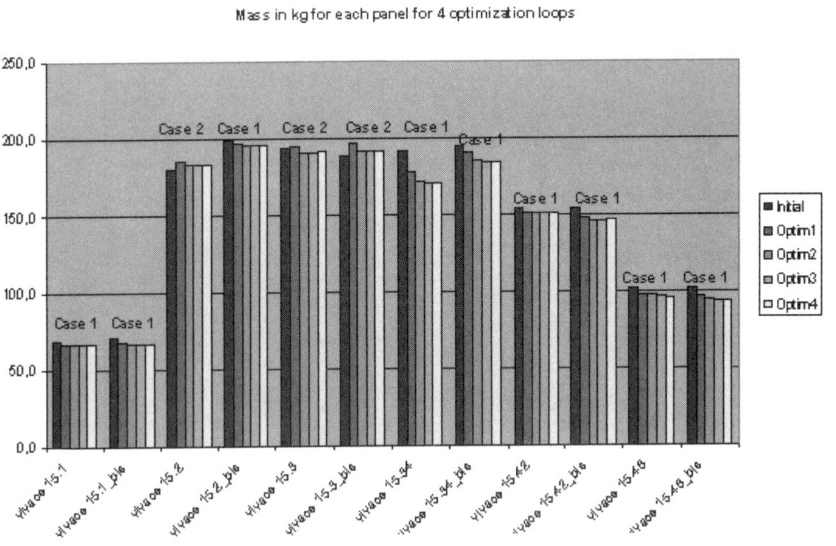

Fig. 31 Convergence history panel per panel. (See also color figure section at the back of the book.)

The mass of panels 15.1, 15.1_bis, 15.46, and 15.46_bis does not decrease a lot. Because these panels correspond to zones that support strong stresses (top and bottom of the fuselage), the optimization does not give an important decreasing of mass.

Six sizing constraints are computed during the optimization. The constraint that contributes most to the sizing of the model is the buckling. When constraints are not active, superstiffener size reaches the minimum, and RF value is not important anymore. See Fig. 32 for a mapping of sizing criteria.

The average superstiffener section is 524 mm^2. As shown in Fig. 33, the superstiffeners with the largest section area are situated in the fuselage bottom, whereas the fuselage middle is characterized by the application of smaller superstiffeners.

The section is larger in the bottom because bottom panels support larger stresses. The high sections found in interframes 38–39 and panels 15.1 and 15.3 are caused by the stress concentrations around cargo door corners in the frame bay 37–38. The high sections found in interframe 45–46 and panels 15.3 and 15.46 are caused by the stress concentrations around center wing-box corners in the frame bay 46–47.

The section is small around the minimum design zone, which supports less stresses than others panels, only a low shear stress (see red cells on the Fig. 33 with section for each superstiffener). The average skin thickness is 2.17 mm. The thickness mapping is fully in agreement with the superstiffener section area mapping as shown on Fig. 34. The same interpretations applies to the section areas.

	Frame 38-39	Frame 39-40	Frame 40-41	Frame 41-42	Frame 42-43	Frame 43-44	Frame 44-45	Frame 45-46	
Stiffener Number	116	117	118	119	120	121	122	123	Panel
2	Buckling	Buckling	Buckling	Buckling	Buckling	Buckling	Buckling	Buckling	15.1
3	Buckling	Buckling	Buckling	Buckling	Buckling	Buckling	Buckling	Buckling	
4	Buckling	Buckling	Buckling	Buckling	Buckling	Buckling	Buckling	Buckling	
5	Buckling	Buckling	Buckling	Buckling	Buckling	Buckling	Buckling	Buckling	
6	Buckling	Buckling	Buckling	Buckling	Buckling	Buckling	Buckling	Buckling	
7	Buckling	Buckling	Buckling	Buckling	Buckling	Buckling	Buckling	Buckling	15.2
8	Buckling	Buckling	Buckling	Buckling	Buckling	Buckling	Buckling	Buckling	
9	Buckling	Buckling	Buckling	Buckling	Buckling	Buckling	Buckling	Buckling	
10	Buckling	Buckling	Buckling	Buckling	Buckling	Buckling	Buckling	Buckling	
11	Min Design	Buckling	Buckling	Buckling	Buckling	Buckling	Buckling	Buckling	
12	Min Design	Min Design	Buckling	Buckling	Buckling	Buckling	Buckling	Buckling	
13	Min Design	Min Design	Buckling	Buckling	Buckling	Buckling	Buckling	Buckling	
14	Min Design	Min Design	Buckling	Buckling	Buckling	Buckling	Buckling	Buckling	
15	Min Design	Min Design	Buckling	Buckling	Buckling	Buckling	Buckling	Buckling	
16	Min Design	Buckling	Buckling	Buckling	Buckling	Buckling	Buckling	Buckling	
17	Buckling	Buckling	Buckling	Buckling	Buckling	Buckling	Buckling	Buckling	
18	Buckling	Buckling	Buckling	Buckling	Buckling	Buckling	Buckling	Buckling	
19	Buckling	Buckling	Buckling	Buckling	Buckling	Buckling	Buckling	Buckling	
20	Buckling	Buckling	Buckling	Buckling	Buckling	Buckling	Buckling	Buckling	
21	Buckling	Buckling	Buckling	Buckling	Buckling	Buckling	Buckling	Buckling	
22	Buckling	Buckling	Buckling	Buckling	Buckling	Buckling	Buckling	Buckling	
23	Buckling	Buckling	Buckling	Buckling	Buckling	Buckling	Buckling	Buckling	15.3
24	Buckling	Buckling	Buckling	Buckling	Buckling	Buckling	Buckling	Buckling	
25	Buckling	Buckling	Buckling	Buckling	Buckling	Buckling	Buckling	Buckling	
26	Buckling	Buckling	Buckling	Buckling	Buckling	Buckling	Buckling	Buckling	
27	Buckling	Buckling	Buckling	Buckling	Buckling	Buckling	Buckling	Buckling	
28	Buckling	Buckling	Buckling	Buckling	Buckling	Buckling	Buckling	Buckling	
29	Buckling	Buckling	Buckling	Buckling	Buckling	Buckling	Buckling	Buckling	
30	Buckling	Buckling	Buckling	Buckling	Buckling	Buckling	Buckling	Buckling	
31	Buckling	Buckling	Buckling	Buckling	Buckling	Buckling	Buckling	Buckling	
32	Buckling	Buckling	Buckling	Buckling	Buckling	Buckling	Buckling	Buckling	
33	Min Design	Min Design	Min Design	Buckling	Buckling	Buckling	Buckling	Buckling	
34	Min Design	Min Design	Min Design	Min Design	Min Design	Min Design	Buckling	Buckling	
35	Min Design	Min Design	Min Design	Min Design	Min Design	Buckling	Min Design	Buckling	
36	Min Design	Min Design	Min Design	Min Design	Min Design	Buckling	Buckling	Buckling	
37	Min Design	Min Design	Min Design	Buckling	Buckling	Buckling	Buckling	Buckling	
38	Min Design	Min Design	Min Design	Buckling	Buckling	Buckling	Buckling	Buckling	
39	Min Design	Min Design	Buckling	Buckling	Buckling	Buckling	Buckling	Buckling	15.34
40	Buckling	Min Design	Buckling	Buckling	Buckling	Buckling	Buckling	Buckling	
41	Buckling	Buckling	Buckling	Buckling	Buckling	Buckling	Buckling	Buckling	
42	Buckling	Buckling	Buckling	Buckling	Buckling	Buckling	Buckling	Buckling	
43	Buckling	Buckling	Buckling	Buckling	Buckling	Buckling	Buckling	Buckling	
44	Buckling	Buckling	Buckling	Buckling	Buckling	Buckling	Buckling	Buckling	
45	Buckling	Buckling	Buckling	Buckling	Buckling	Buckling	Buckling	Buckling	
46	Buckling	Buckling	Buckling	Buckling	Buckling	Buckling	Buckling	Buckling	
47	Buckling	Buckling	Buckling	Buckling	Buckling	Buckling	Buckling	Buckling	
48	Buckling	Buckling	Buckling	Buckling	Buckling	Buckling	Buckling	Buckling	
49	Buckling	Buckling	Buckling	Buckling	Buckling	Buckling	Buckling	Buckling	
50	Buckling	Buckling	Buckling	Buckling	Buckling	Buckling	Buckling	Buckling	
51	Buckling	Buckling	Buckling	Buckling	Buckling	Buckling	Buckling	Buckling	
52	Buckling	Buckling	Buckling	Buckling	Buckling	Buckling	Buckling	Buckling	
53	Buckling	Buckling	Buckling	Buckling	Buckling	Buckling	Buckling	Buckling	15.42
54	Buckling	Buckling	Buckling	Buckling	Buckling	Buckling	Buckling	Buckling	
55	Buckling	Buckling	Buckling	Buckling	Buckling	Buckling	Buckling	Buckling	
56	Buckling	Buckling	Buckling	Buckling	Buckling	Buckling	Buckling	Buckling	
57	Buckling	Buckling	Buckling	Buckling	Buckling	Buckling	Buckling	Buckling	
58	Buckling	Buckling	Buckling	Buckling	Buckling	Buckling	Buckling	Min Design	
59	Buckling	Buckling	Buckling	Buckling	Buckling	Buckling	Buckling	Buckling	
60	Buckling	Buckling	Buckling	Buckling	Buckling	Buckling	Buckling	Buckling	
61	Buckling	Buckling	Buckling	Buckling	Buckling	Buckling	Buckling	Buckling	
62	Buckling	Buckling	Buckling	Buckling	Buckling	Buckling	Buckling	Buckling	
63	Buckling	Buckling	Buckling	Buckling	Buckling	Buckling	Buckling	Buckling	
64	Buckling	Buckling	Buckling	Buckling	Buckling	Buckling	Buckling	Buckling	
65	Buckling	Buckling	Buckling	Buckling	Buckling	Buckling	Buckling	Buckling	
66	Buckling	Buckling	Buckling	Buckling	Buckling	Buckling	Buckling	Buckling	
67	Buckling	Buckling	Buckling	Buckling	Buckling	Buckling	Buckling	Buckling	15.46
68	Buckling	Buckling	Buckling	Buckling	Buckling	Buckling	Buckling	Buckling	
69	Buckling	Buckling	Buckling	Buckling	Buckling	Buckling	Buckling	Buckling	
70	Buckling	Buckling	Buckling	Buckling	Buckling	Buckling	Buckling	Buckling	
71	Buckling	Buckling	Buckling	Buckling	Buckling	Buckling	Buckling	Buckling	
72	Buckling	Buckling	Buckling	Buckling	Buckling	Buckling	Buckling	Buckling	

Fig. 32 Mapping of active sizing criteria for the optimum design.

	Frame 38-39	Frame 39-40	Frame 40-41	Frame 41-42	Frame 42-43	Frame 43-44	Frame 44-45	Frame 45-46	
Stiffener Number	116	117	118	119	120	121	122	123	Panel
2	729.12	705.17	687.93	680.04	676.54	673.92	672.15	671.27	
3	793.05	701.08	680.23	672.48	667.96	665.80	664.73	662.49	
4	801.25	687.69	669.82	660.02	656.36	655.19	655.00	654.52	16.1
5	732.13	670.23	653.31	644.72	642.24	642.55	643.78	645.21	
6	691.38	645.56	630.53	625.61	627.22	627.94	631.61	637.01	
7	667.29	609.03	604.01	603.21	608.58	612.30	617.61	625.16	
8	602.73	572.82	572.56	579.33	587.09	595.61	602.97	609.85	
9	511.31	529.41	539.19	553.21	567.01	578.91	589.61	597.60	
10	449.94	478.86	505.94	527.66	547.77	564.05	577.14	587.11	
11	428.96	441.47	475.97	506.23	531.51	551.34	566.26	578.52	
12	428.96	428.96	467.49	494.07	522.27	543.63	559.54	572.11	
13	438.24	438.24	476.93	516.81	541.41	562.83	578.39	589.88	
14	447.52	469.06	507.23	546.44	570.13	587.08	599.59	608.47	16.2
15	463.43	486.35	525.52	558.27	581.43	594.12	599.92	604.22	
16	476.88	518.57	553.38	578.37	591.87	599.58	600.15	599.29	
17	511.44	553.96	585.27	599.70	604.50	603.87	601.16	595.11	
18	527.65	573.99	596.58	600.88	597.88	592.35	585.12	575.39	
19	537.85	584.47	596.10	592.31	584.11	575.11	564.73	553.73	
20	564.11	608.48	609.73	597.79	584.84	571.35	558.49	546.82	
21	538.57	576.44	568.67	551.65	536.52	522.58	511.03	498.73	
22	533.76	560.67	540.41	518.00	500.88	487.55	477.23	467.81	
23	649.56	619.74	575.05	543.98	521.96	504.67	490.23	479.32	
24	746.99	651.73	583.21	550.31	526.27	505.98	489.52	477.67	
25	825.89	639.40	573.31	542.65	517.63	498.63	483.89	474.97	
26	876.82	580.81	542.22	516.30	494.46	476.41	465.40	459.60	
27	698.51	518.86	494.36	476.97	460.33	447.65	440.40	436.92	
28	484.30	456.68	433.03	421.84	412.13	398.92	395.98	394.47	
29	413.04	389.28	377.37	372.77	370.05	360.96	360.96	360.96	
30	382.88	368.02	366.24	366.24	366.24	366.24	366.24	366.24	16.3
31	374.62	371.52	371.52	371.52	371.52	371.52	371.52	371.52	
32	379.52	379.52	379.52	379.52	379.52	379.52	379.52	379.52	
33	371.52	371.52	371.52	371.52	371.52	371.52	371.52	371.52	
34	355.52	355.52	355.52	355.52	355.52	355.52	355.52	355.52	
35	397.52	397.52	397.52	397.52	405.22	421.64	431.16	397.52	
36	439.44	439.44	439.44	452.92	481.04	510.90	670.79	501.47	
37	439.44	439.44	441.58	473.29	507.44	576.12	802.79	784.91	
38	400.72	400.72	402.86	421.09	463.63	538.39	642.71	860.70	
39	369.92	369.92	369.92	369.92	411.42	445.41	505.16	677.94	
40	389.65	377.92	377.92	377.92	405.39	434.53	500.25	675.28	
41	408.28	388.11	388.18	378.00	407.27	435.30	498.57	714.88	
42	420.55	402.78	389.12	389.21	407.29	431.27	484.15	668.64	
43	397.13	380.05	358.53	369.51	374.76	399.12	432.62	569.09	
44	355.49	346.54	337.70	338.72	341.57	358.07	372.43	467.20	
45	395.00	386.67	379.89	379.71	385.08	388.62	411.42	461.46	16.34
46	466.95	446.98	431.71	426.69	428.75	435.54	461.14	536.70	
47	474.49	449.03	429.78	419.16	414.25	422.08	440.22	533.08	
48	459.57	434.88	414.71	401.43	393.07	395.55	415.10	463.98	
49	462.30	436.40	415.97	399.67	386.48	385.83	397.54	405.67	
50	472.65	447.15	425.07	405.81	388.62	382.70	384.34	377.61	
51	483.93	460.01	435.36	413.52	392.58	380.44	375.62	369.60	
52	444.85	475.55	447.38	422.55	397.92	379.79	371.23	369.60	
53	471.74	494.09	461.83	433.73	405.18	382.40	369.60	369.60	
54	547.88	509.23	477.15	446.72	415.38	385.49	371.75	369.60	
55	551.09	521.77	492.79	469.95	428.89	393.46	371.75	369.60	
56	544.79	532.94	510.46	478.01	444.59	407.55	369.99	369.60	
57	537.16	541.10	528.86	499.87	463.20	425.11	378.44	369.60	
58	533.69	547.09	546.46	525.96	488.43	447.76	399.54	369.60	
59	529.67	551.79	558.51	552.04	520.10	475.05	427.99	374.35	16.42
60	524.59	552.48	573.70	575.00	557.25	512.33	463.02	500.22	
61	513.94	568.36	586.34	597.12	597.15	578.65	512.26	631.72	
62	546.62	578.70	598.85	622.64	642.99	676.47	621.34	650.98	
63	582.06	574.06	609.08	648.14	699.56	730.98	722.28	676.70	
64	559.43	579.00	609.38	656.10	729.77	749.77	771.75	674.32	
65	536.03	576.07	609.57	646.05	729.76	766.34	824.47	692.43	
66	534.36	571.30	608.27	640.22	729.76	763.37	887.17	785.76	
67	549.87	573.73	608.54	655.84	729.76	752.91	883.14	925.12	
68	560.81	570.89	602.35	665.44	729.76	738.73	812.59	1248.22	
69	552.83	562.31	584.32	632.89	711.43	731.40	763.60	1217.5391	
70	544.17	555.22	569.41	593.09	651.57	712.61	713.61	838.52	
71	540.45	551.35	563.48	577.21	604.21	653.85	656.58	655.01	16.46
72	538.57	550.36	561.50	574.77	590.63	606.68	610.94	592.41	
73	538.27	547.67	559.97	573.54	586.95	600.28	607.07	614.93	
74	539.21	546.01	559.65	571.29	591.01	599.44	621.36	633.13	

Fig. 33 Mapping of stringer bay cross-sectional areas for the optimum design.

VI. Loose Coupling [20]

To reach the level of a full fuselage cover optimization, the tight coupling approach can be very greedy in terms of processors. To achieve the same computational times as for a barrel, the hardware need is about 10 times higher, that is, 300 processors. Such a large number of processors is not necessarily available even in a large company. This is the reason why another approach, named loose coupling, was explored.

This approach consists of replacing the call to stress analysis tools (here ASSIST) by a call to a surrogate model. More information on surrogate modelling is provided in the chapter dedicated to optimization. More references can also be found in Simpson et al. [21] and Sellar et al. [22]. Note that surrogate modelling is also referred to as meta-model or response surface.

	Frame 38-39	Frame 39-40	Frame 40-41	Frame 41-42	Frame 42-43	Frame 43-44	Frame 44-45	Frame 45-46	
Stiffener Number	116	117	118	119	120	121	122	123	Panel
2	2.66	2.68	2.64	2.62	2.61	2.60	2.59	2.59	
3	2.90	2.68	2.62	2.59	2.58	2.57	2.57	2.56	
4	2.94	2.63	2.58	2.55	2.54	2.54	2.54	2.54	16.1
5	2.71	2.57	2.53	2.50	2.50	2.50	2.50	2.51	
6	2.64	2.49	2.45	2.44	2.44	2.45	2.46	2.47	
7	2.53	2.37	2.36	2.36	2.37	2.40	2.42	2.43	
8	2.31	2.24	2.24	2.27	2.30	2.34	2.36	2.39	
9	1.99	2.07	2.11	2.17	2.22	2.27	2.31	2.34	
10	1.71	1.85	1.97	2.07	2.15	2.21	2.26	2.30	
11	1.60	1.67	1.84	1.98	2.08	2.16	2.22	2.27	
12	1.60	1.60	1.75	1.92	2.04	2.13	2.19	2.24	
13	1.60	1.60	1.79	1.98	2.07	2.16	2.22	2.26	
14	1.60	1.66	1.89	2.05	2.14	2.21	2.25	2.29	16.2
15	1.63	1.79	1.98	2.10	2.19	2.23	2.25	2.27	
16	1.74	1.95	2.09	2.18	2.23	2.25	2.25	2.25	
17	1.92	2.11	2.21	2.26	2.27	2.27	2.26	2.23	
18	2.05	2.24	2.30	2.31	2.30	2.27	2.25	2.21	
19	2.15	2.33	2.36	2.34	2.30	2.27	2.22	2.18	
20	2.27	2.43	2.42	2.37	2.32	2.26	2.21	2.15	
21	2.31	2.44	2.41	2.33	2.27	2.21	2.16	2.10	
22	2.32	2.41	2.33	2.24	2.17	2.11	2.06	2.02	
23	2.49	2.50	2.36	2.25	2.16	2.08	2.02	1.97	
24	2.72	2.60	2.40	2.28	2.17	2.09	2.02	1.96	
25	3.16	2.96	2.36	2.25	2.15	2.07	2.00	1.95	
26	3.52	2.40	2.26	2.16	2.07	2.00	1.94	1.91	
27	2.88	2.19	2.08	2.00	1.93	1.88	1.84	1.82	
28	2.12	2.00	1.90	1.83	1.79	1.71	1.69	1.68	
29	1.89	1.76	1.71	1.68	1.66	1.60	1.60	1.60	
30	1.68	1.60	1.60	1.60	1.60	1.60	1.60	1.60	16.3
31	1.60	1.60	1.60	1.60	1.60	1.60	1.60	1.60	
32	1.60	1.60	1.60	1.60	1.60	1.60	1.60	1.60	
33	1.60	1.60	1.60	1.60	1.60	1.60	1.60	1.60	
34	1.60	1.60	1.60	1.60	1.60	1.60	1.60	1.60	
35	1.60	1.60	1.60	1.60	1.64	1.72	1.77	1.60	
36	1.60	1.60	1.60	1.67	1.81	1.96	2.43	1.90	
37	1.60	1.60	1.61	1.77	1.94	2.25	2.79	3.00	
38	1.60	1.60	1.61	1.70	1.97	2.32	2.54	3.44	
39	1.60	1.60	1.60	1.60	1.87	2.07	2.31	2.75	
40	1.67	1.60	1.60	1.60	1.77	1.96	2.21	2.63	
41	1.79	1.66	1.66	1.60	1.78	1.96	2.20	2.88	
42	1.87	1.76	1.66	1.67	1.78	1.92	2.14	2.71	
43	1.88	1.76	1.60	1.67	1.69	1.88	2.03	2.45	
44	1.73	1.67	1.60	1.60	1.60	1.73	1.78	2.18	
45	1.78	1.74	1.70	1.70	1.72	1.73	1.82	2.08	16.34
46	2.04	1.94	1.86	1.83	1.84	1.88	2.00	2.33	
47	2.11	2.00	1.90	1.84	1.81	1.85	1.96	2.34	
48	2.08	1.97	1.86	1.79	1.74	1.76	1.86	2.07	
49	2.08	1.97	1.86	1.78	1.70	1.70	1.76	1.81	
50	2.12	2.02	1.90	1.81	1.71	1.68	1.69	1.65	
51	2.17	2.07	1.95	1.85	1.73	1.67	1.64	1.60	
52	1.99	2.11	2.00	1.89	1.76	1.67	1.61	1.60	
53	2.03	2.16	2.06	1.94	1.80	1.68	1.60	1.60	
54	2.29	2.20	2.11	1.99	1.85	1.70	1.61	1.60	
55	2.29	2.23	2.15	2.05	1.92	1.74	1.61	1.60	
56	2.27	2.26	2.20	2.11	1.99	1.81	1.60	1.60	
57	2.27	2.27	2.25	2.17	2.06	1.90	1.65	1.60	
58	2.26	2.29	2.29	2.24	2.14	2.00	1.77	1.60	
59	2.25	2.30	2.32	2.30	2.23	2.10	1.91	1.63	16.42
60	2.24	2.31	2.35	2.35	2.31	2.20	2.05	2.02	
61	2.23	2.35	2.37	2.40	2.39	2.34	2.19	2.44	
62	2.34	2.39	2.39	2.45	2.56	2.75	2.58	2.57	
63	2.42	2.38	2.42	2.55	2.88	3.08	3.03	2.74	
64	2.34	2.38	2.43	2.59	3.08	3.21	3.35	2.72	
65	2.29	2.38	2.43	2.53	3.08	3.31	3.69	2.83	
66	2.30	2.39	2.45	2.50	3.08	3.30	4.10	3.44	
67	2.36	2.40	2.47	2.60	3.08	3.23	4.07	4.35	
68	2.39	2.41	2.47	2.66	3.08	3.14	3.61	6.44	
69	2.37	2.40	2.44	2.96	2.96	3.09	3.29	6.24	
70	2.35	2.38	2.42	2.46	2.67	2.97	2.96	3.78	
71	2.35	2.38	2.41	2.44	2.48	2.69	2.68	2.76	16.48
72	2.35	2.38	2.41	2.44	2.46	2.52	2.54	2.47	
73	2.35	2.38	2.41	2.44	2.47	2.51	2.51	2.52	
74	2.35	2.37	2.41	2.44	2.48	2.51	2.51	2.54	

Fig. 34 Mapping of pocket thicknesses for the optimum design.

A. Surrogate Models for Stress Tools

Initially, the intent was to approximate stress responses, that is to say, to replace stress tools with surrogate models, but two drawbacks had to be considered:

1) The optimization is still to be made. Even if the surrogate models are quicker than ASSIST, there is a large number of optimizations to be done, and it takes time.

2) The number of inputs is quite large to build a surrogate model.

Indeed, with the local optimization problem exposed in Section III, the inputs of the surrogate model are as follows: stringer height, web stringer thickness, free flange width, free flange thickness, attached flanged thickness, skin thickness, F_{XSR}, N_{xy}, pitch, and curvature radius.

Note that the loading has been simplified and that the sizing process (considering that the main sizing criteria for the barrel was buckling) has been reduced to buckling/postbuckling only. The total number of inputs is then 10 as just listed. To cover such a design space, a very large number of samples is needed, which, in turn, requires a very large number of computations. This is the well-known problem of "curse of dimensionality."

Nevertheless, the attempt was made, and a specific piecewise approach was used to consider the nonlinear responses of ASSIST. This piecewise approach consists of mixing various neural networks and is called MoE (mixture of experts). Neural networks are well known to cover highly nonlinear responses with a capability to generalization and extrapolation [23, 24]. The MoE approach gives a pragmatic answer to the curse of dimensionality [25, 26].

However, it took a long time to understand the nonlinearity of the software and to put in place the appropriate MoE. For this reason, this approach was not considered mature enough for industrial application: additional research work is still needed.

B. Design Curves

The alternative approach was to directly approximate the optimization results. The advantage in doing so is that all optimization design variables disappear from the input space. Moreover, the optimization is done a priori and is not to be done inside the global loops: the surrogate models have just to be calculated with appropriate internal loads to get the minimum thickness and minimum section area. If there are multiple load cases, the most conservative values for skin thickness and stringer section area are selected.

The resulting set of inputs to be considered is F_{XSR}, N_{xy}, pitch, and curvature radius. The outputs are restricted to the minimum thickness and the minimum area found with the local optimization. In this case, because of the reduced dimension of the input space, the computational effort becomes much more reasonable, and a surrogate model approach can really be considered for the approximation. Moreover as aforementioned, with this approach, the local optimization is no longer necessary at global level. It is replaced by a simple interrogation of the design curve. It is then much more computationally efficient, as shown in Section VI.C.

This design curve gives for each set (F_{XSR}, N_{XY}, p, R) the minimum skin thickness and the minimum stringer area as resulting from the local optimizations.

On top of that, the detailed design variables can also be derived and the stinger profiles designed. Again some preliminary studies show that NN was the surrogate model most appropriate to the nonlinear responses to be approximated. More information on neural networks is provided in the optimization chapter or in Bishop [23] and Dreyfus et al. [24].

In the first attempt described in paragraph 1 (analysis regression), the dimension of the input space had required the use of filling space DoEs like Latin hypercube. In this new approach, the significant reduction of the input dimension (from 10 to 4) authorizes the use of a full factorial DoE [27, 28].

For the training of the NN, a specific BOSS Quattro session was then used as shown on Fig. 35: it is a double parametric session. The parametric task implements the factorial DoE and drives an optimization task: The optimization is to be run for all quadruplets of the DoE as illustrated on Fig. 36.

Because (pitch, radius curvature) couples are discrete values, a double parametric session is used: parametric 1 for continuous inputs—compression and shear combination; parametric 2 for discrete inputs—{pitch/radius curvature} couples. Upon running of the optimization cases, results are extracted from output files *.log and are implemented in the training set. The training will

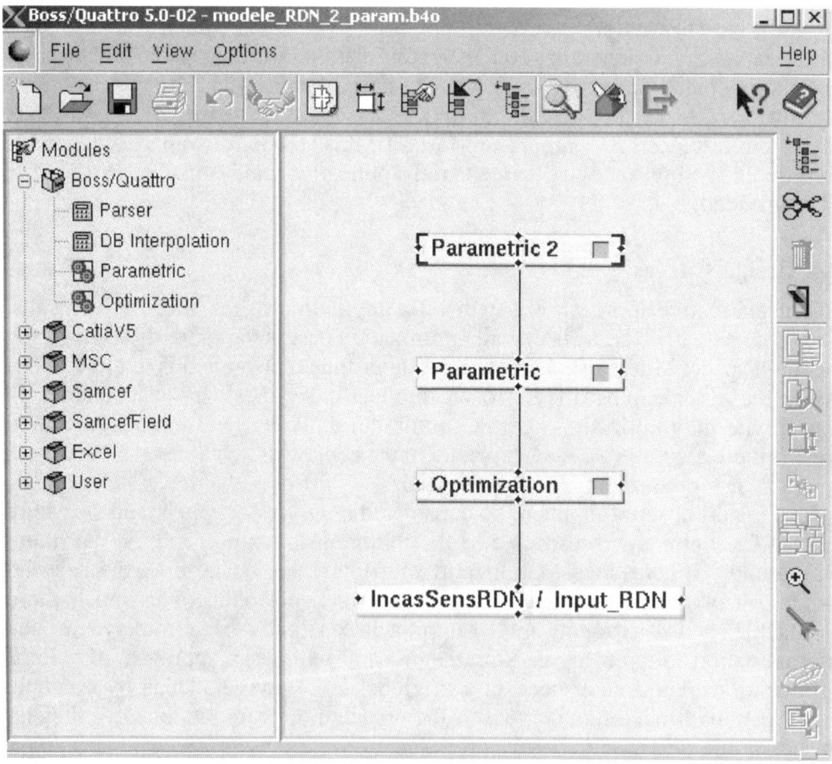

Fig. 35 Main GUI window for neural network.

Compression (Newtons)	10 000	25 000	50 000	75 000	100 000
	125 000	150 000	175 000	200 000	225 000
	250 000	275 000			

Shear (MPa)	5	30	50	70	90
	110	130			

Pitch / radius curvature (mm)	191.1/3300	202.7/3300	154.0/3430	132.4/5220	145.2/5220
	149.2/5220	154.0/5220	158.6/5220	165.2/5220	172.1/5220
	180.4/5220	191.1/5220	197.7/5220		

Fig. 36 Full factorial DoE for the training set.

then be based on a training set with 1092 input cases ($12 \times 7 \times 13$). Seven outputs are extracted from *.log files: web height, web thickness, attached flange width, attached flange thickness, free flange width, free flange thickness, and skin thickness. Then, 7644 output cases (1092×7) will be used to build the neural network. The neural network was built with the help of [23, 24] following specifications: one hidden layer and sigmoidal activation and a linear output layer. *Remark:* Sigmoidal activation refers to the transfer function of each neuron (regularization of a heaviside function). The MATLAB® Neural Network Toolbox was used to build the neural net [29]. Four nets were performed: with 40 neurons, with 50 neurons, with 60 neurons, and with 70 neurons. The convergence of the training process is illustrated in Fig. 37. The quality of the neural network can be checked using three methods: MSE value, comparison between training set output and neural net output, and comparison between neural net output and BOSS QUATTRO session using the same inputs (test base). More information on the quality of surrogate model fits can be found in the optimization chapter.

Comparison is made on the superstiffener section area that is computed using each output as shown on Table 4. The accuracy target that was pursued was a min/max error of less than 5%. However, it was not possible to directly drive this error measure and find automatically the optimum number of neurons. This is the reason why the several NNs were built for 40, 50, 60, and 70 neurons.

Figure 38 shows that a 70-neurons neural net seems to give a satisfactory accuracy (min and max value under 4% with test-base comparison). Figure 39 show response surfaces of superstiffener section area for target (left) and 70-neurons neural net (right).

C. Demonstration

A demonstration was then made on the full-barrel test case. For that a new version of STIFFOPT was developed based on this loose coupling approach. It just consists of replacing calls to BOSS Quattro by a call to the design curves. The sizing is then performed internally to PATRAN based on a PCL. The optimization speed is then becoming very fast.

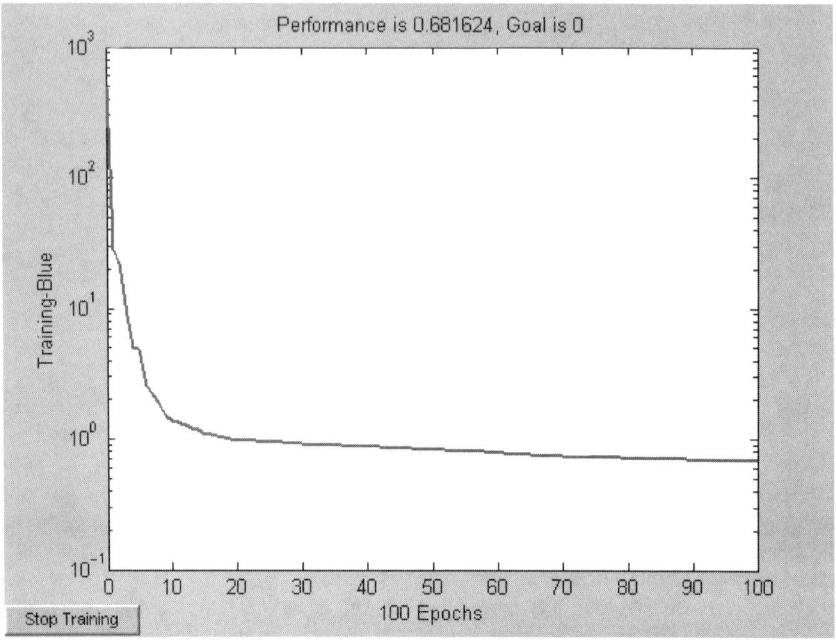

Fig. 37 MSE for 40 neurons.

A new window (Fig. 40) has been added in GUI STIFFOPT. The user can choose neural-net method when he defines optimization method.

This new approach is, of course, very competitive with respect to the tight coupling one. However, it is worth mentioning that the accuracy will always be questionable. This is the reason why this approach is not recommended for detailed design when the utmost weight savings are sought for the structure.

Table 4 Training results

	40 neurons	50 neurons	60 neurons	70 neurons
MSE	0.68	0.43	0.36	0.35
Difference training set/neural-net output				
Min/max	−23/16.7	−12.4/14.6	−14.1/9.1	−12.2/16.1
Relative mean error	−0.02	−0.05	−0.038	−0.042
Absolute mean error	2.44	2.09	1.67	1.72
Difference neural-net output/test base				
Min/max	−6.5/6.9	−4.4/5.9	−3.7/8.0	−2.4/4.0
Relative mean error	−0.16	−0.32	0.028	0.028
Absolute mean error	2.15	1.94	1.74	1.26

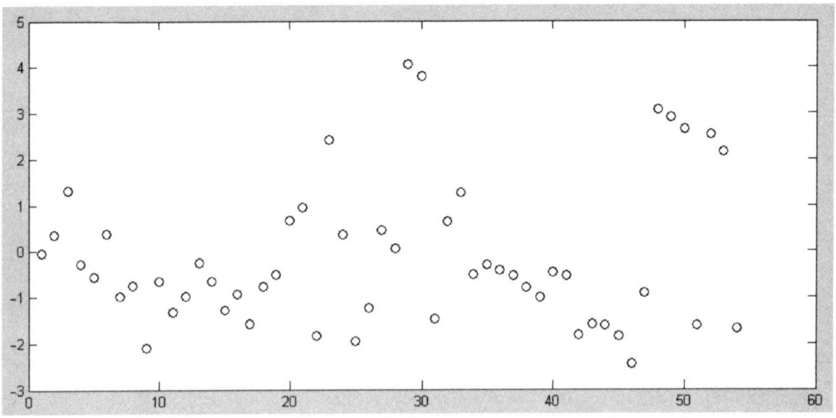

Fig. 38 Training set target and test-base error with 70 neurons.

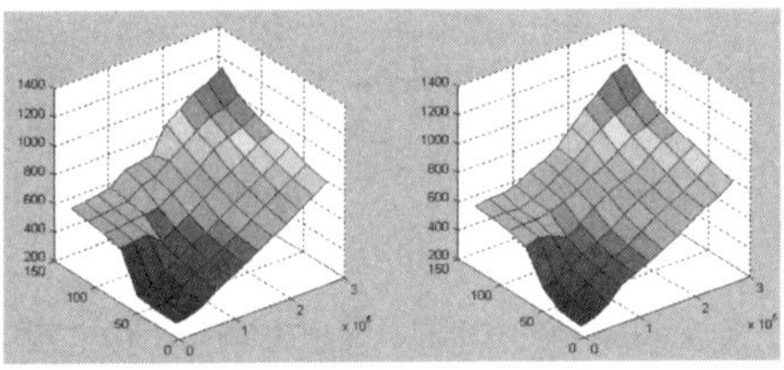

Superstiffener section area 132/5520

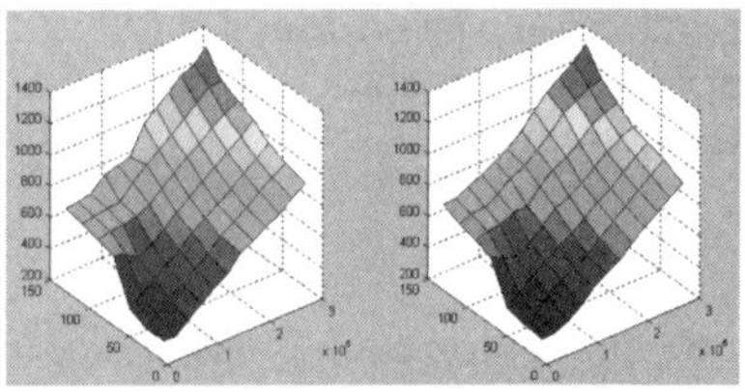

Superstiffener section area 154/3430

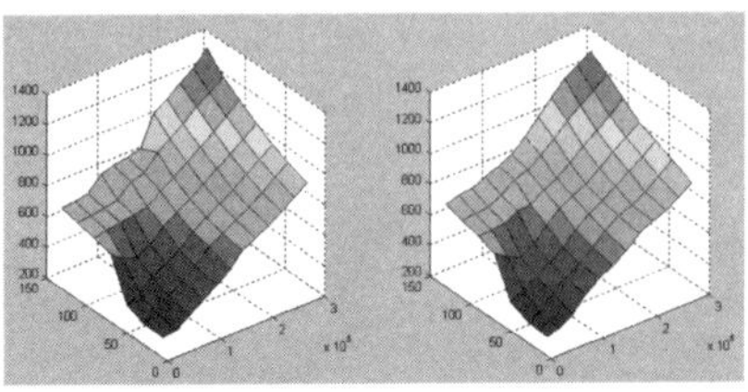

Superstiffener section area 158/5220

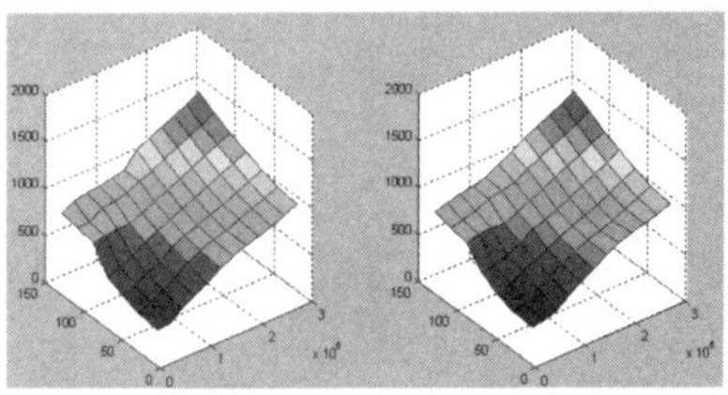

Superstiffener section area 180/5220

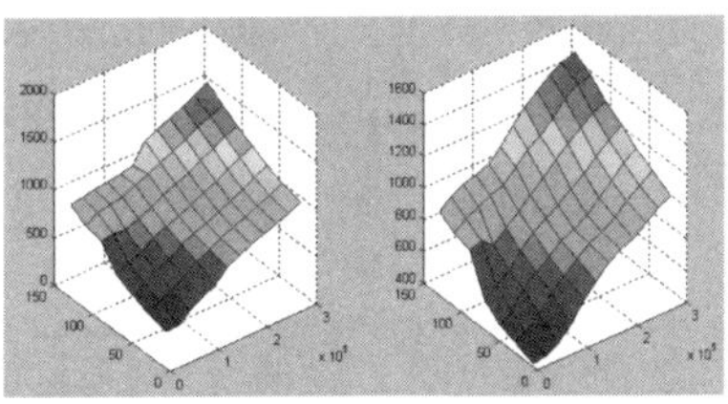

Superstiffener section area 191/3300

Fig. 39 Response surface for section area found for various (pitch, curvature radius) couples. (See also color figure section at the back of the book.)

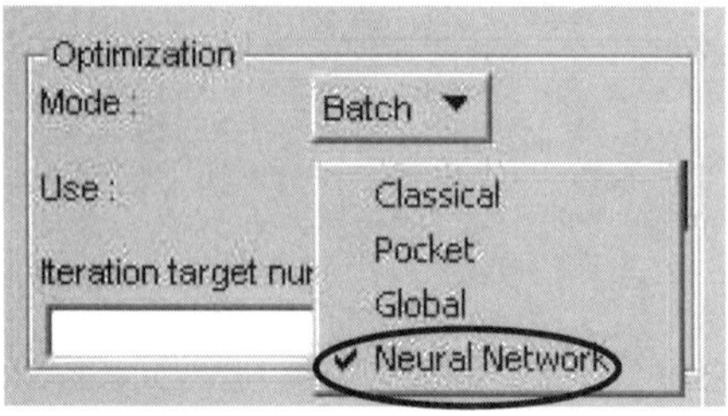

Fig. 40 STIFFOPT optimization window.

1. *Tight Coupling Convergence*

Figure 41 depicts the results of the three optimization loops that were performed with tight coupling. The first optimization is made with initial values identical to neural-net session initial values. Optimization 2 and 3 are computed with optimized results values (respectively, optimization 1 results and optimization 2 results).

Mass decreases by 8.1% after the first optimization, and then decreases again by 0.5%. Convergence is reached after the third optimization loop. This first optimization is performed with high initial values, which do not lead to an optimum. The second optimization is made with values already optimized, and the optimum is consequently lower.

2. *Convergence with Neural-Net Optimization*

Seven optimization loops are performed using multi-init neural net. As it can be seen from Fig. 42, convergence is reached as soon as the second optimization. Mass decreases or increases slightly within 0.1% after the second optimization.

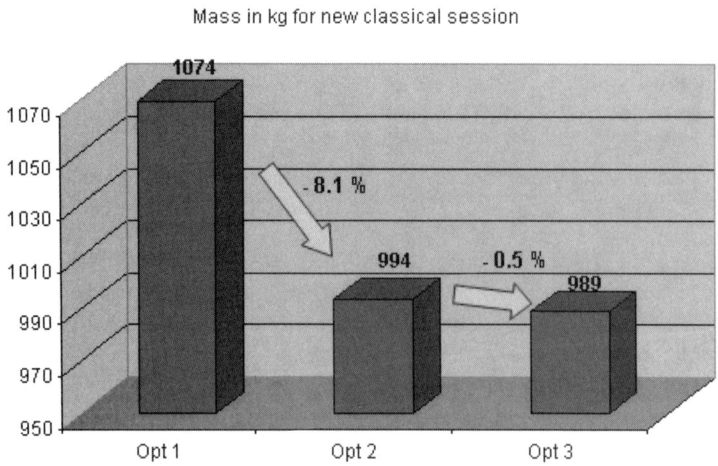

Fig. 41 Mass decrease for three optimization loops.

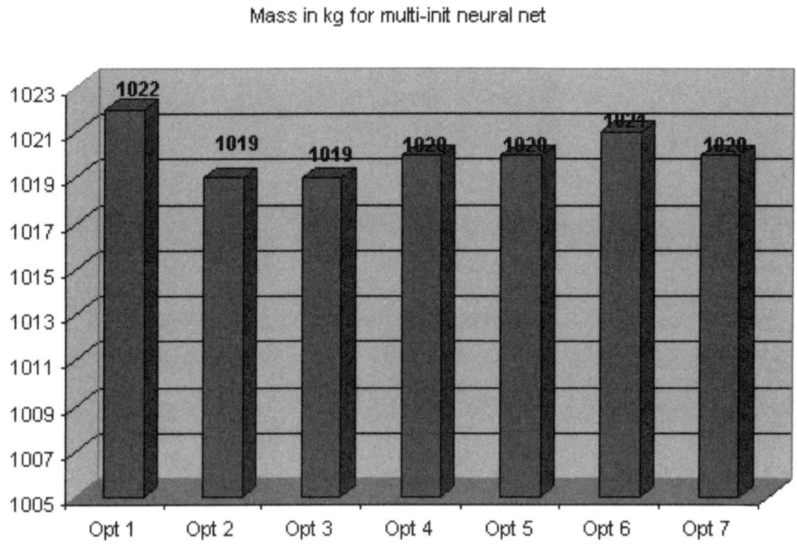

Fig. 42 Mass decrease for seven optimization loops with neural-net multi-init.

Even if global mass is stabilized after the second optimization loop, Fig. 43 shows that there is a slight redistribution of internal loads and superstiffener section areas along subsequent iterations.

3. *Comparison Classical and Neural-Net Optimization*

After three optimization loops, a difference of about 30 kg still exists between new classical results and neural-net results (see Table 5). Despite multi-init use allowing the avoidance of local minima, an optimization with a neural net is limited by initial values: the convergence of local optimizations is probably better because the same optimization is reconsidered several times along upper iterations with close starting points. This enforces the convergence. Another reason is also an insufficient accuracy of the neural net, which was found to be around 4% (see Table 4).

VII. Additional Considerations

Other developments have been made to refine and complete the STIFFOPT optimization process. A specific option has been implemented to focus the optimization on skin pockets each time the pockets are sized by a buckling on-set criterion. This is complemented by a stringer-only optimization, which can be run subsequently to complete the detailed sizing.

Another option has also been added to smooth the design after optimization. Because the optimization is local, it is necessary at the end to harmonize the stringer dimensions, for example, in order to have a constant stringer height.

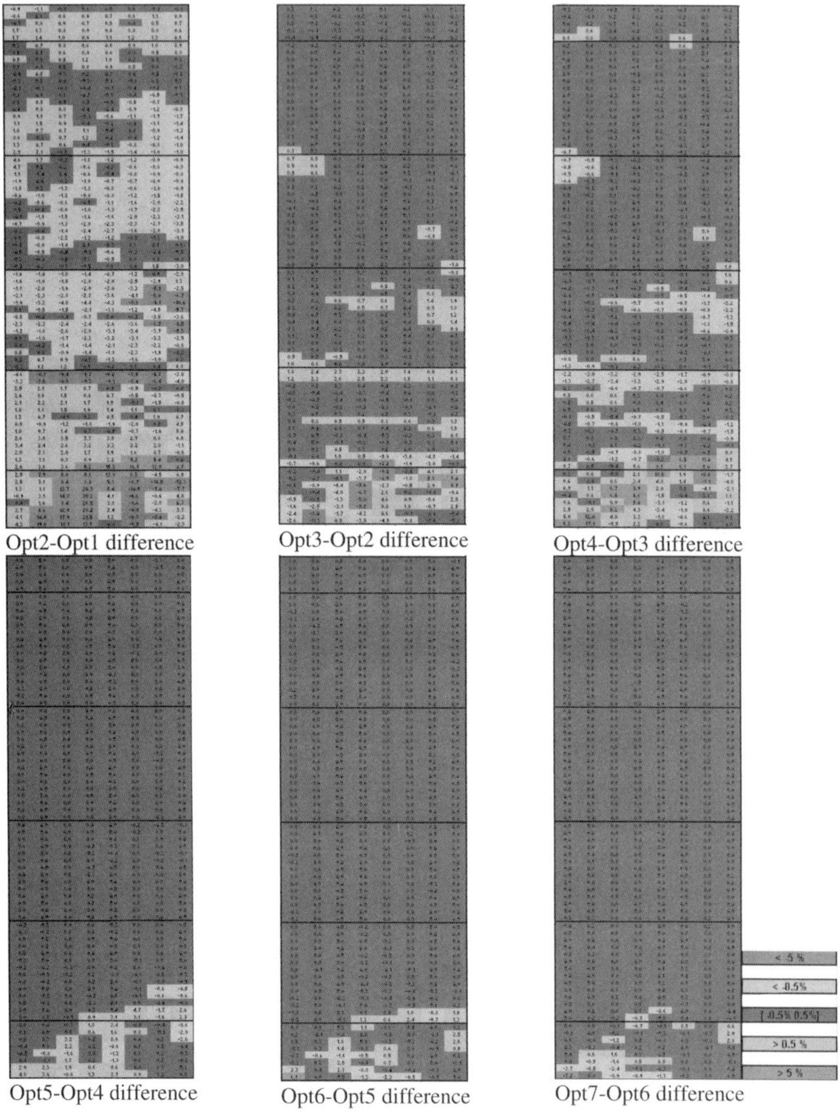

Fig. 43 Section difference for each superstiffener between two optimization loops.

Table 5 Comparison between tight and loose coupling

Mass in kg	New classical session	Multi-init neural net	Delta, %
Optimization 1	1074	1022	−4.2
Optimization 2	993.6	1019	2.4
Optimization 3	988.7	1019	2.9

Finally, a comparison was made between the STIFFOPT optimization process and a more complete one with an optimum control of the internal load redistribution. This was implemented by including the neural-net equations as constraints inside the optimization module of NASTRAN: SOL200. The comparisons made show that both sizing optimization approaches give similar results. However, caution needs to be exercised: this result does not mean that the optimization processes are equivalent, but only that in case of the considered test case, they are comparable. It is well known that there is no equivalence, particularly when some stiffness criteria drive the design.

VIII. Refinement of Local Optimization [10, 30]

As mentioned before, the tight coupling optimization is particularly recommended for detailed design, whereas the surrogate model approach is probably the most suited for rapid sizing optimization both for reason of computational times and accuracy. However, considering that detailed design is usually trying to find the lightest structure possible within a giving margin policy, the question of whether to use higher-fidelity methods rather than analytical engineer-based ones can be asked.

Indeed, numerical simulation of stability is becoming more and more common practice, and the trend is now to replace tests with such numerical simulation approaches (virtual testing). The next step will probably be to perform optimization based on virtual testing. There are few references in the literature to address these kinds of topics, and the complement of tests is often necessary to the optimization process [31].

This is the goal of this section to present such an approach developed inside a COTS tool [31] and demonstrated at panel level [32]. The idea behind is simply to replace the ASSIST analysis by a COTS tool [31] finite element based simulation, in order to have the STIFFOPT framework based on virtual testing approaches.

A. Panel Use Case

It is worth mentioning first the targeted application. The fuselage barrel is considered to be split into local finite element models to analyze the stability. The buckling/postbuckling behavior is considered local, which means that an instability can only occur within a frame bay. In fact the local finite element model is a kind of numerically based ASSIST.

Because of the current trends followed by the civil aeronautical industry, including this project, it was also decided to focus on composite materials.

The panel use-case used here is a panel representative of the behavior of a carbon-fiber-reinforced plastic (CFRP) fuselage with seven omega stringers (Fig. 44). Boundary conditions are simplified for demonstration to single supported sides. The corresponding finite element model was built with a COTS tool [31]. Note that this model has 17,326 nodes; 16,000 cells; and 109,777 degrees of freedom.

The definition of the optimization problem was intentionally simplified to focus the demonstration on the integration of nonlinear analysis in a gradient-based optimization tool. The angles are considered as fixed to standard

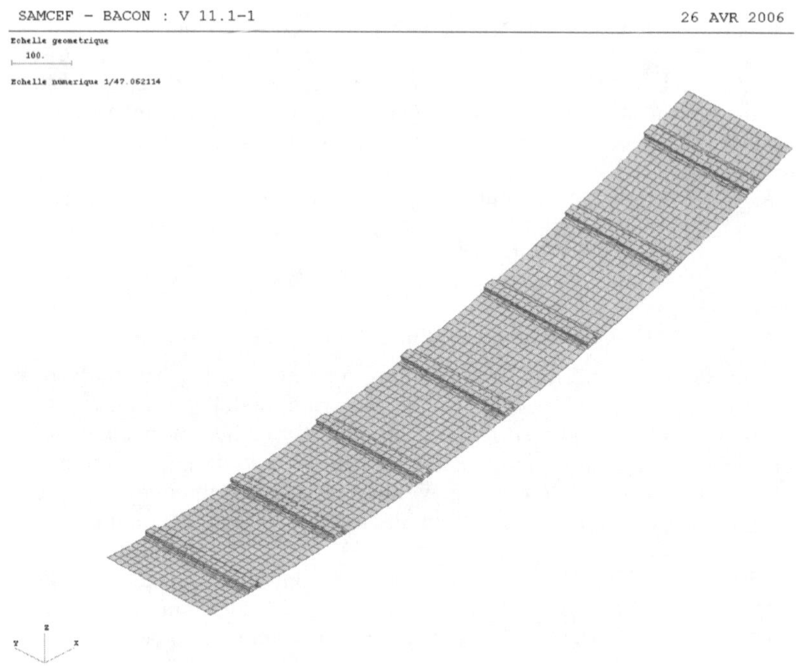

Fig. 44 Panel use case.

orientations: 0, −45, 45, and 90 deg. The difficulty to address angle variables directly has been demonstrated by Bruyneel [33].

The stacking sequence itself is not considered as designable, even if we have developed at AIRBUS a simplified optimization process linking the stacking sequence to the thickness and producing manufacturable covers [34, 35]. We finally have based there here on a homogenized representation of the composite layup using only angle thicknesses as design variables.

Design variables are ply thicknesses for each ply orientation (0, 45, and 90 deg), for each one of the seven superstringers. A distinction is made between thicknesses for skin panels and for stringers. This amounts to considering $3 \times 7 \times 2 = 42$ design variables. In the sequel, we denote them as follows:

1) For the skin panels, $t_{angle}^{SKIN,i}$, with $i \in \{1, \ldots, 7\}$ and $angle = 0, 45, \text{or } 90$ deg.

2) For the stringers, $t_{angle}^{STRINGER,i}$, with $i \in \{1, \ldots, 7\}$ and $angle = 0$, 45, or 90 deg.

Lower and upper bounds on these variables are set to 0.4 and 2 mm, respectively. The objective function is the weight—to be minimized. Constraints are expressed as follows:

1) *Buckling reserve factor:*

$$\mathrm{RF}_{\mathrm{buckling}} \geq 0.76$$

2) *Collapse reserve factor:*

$$\mathrm{RF}_{\mathrm{collapse}} \geq 1$$

The objective function, the buckling RF, and their sensitivities are computed by a linear finite element analysis while the collapse reserve factor and its sensitivity are provided by a nonlinear analysis.

B. Development of Sensitivities for Nonlinear Finite Elements Analysis

The nonlinear analysis is performed by a COTS tool [31], a finite element software package that solves nonlinear structural and mechanical problems. The general framework of the research work and results we describe here is solving the optimization problem defined in Section II, where one of the constraints takes the form:

$$\mathrm{RF}_{\mathrm{collapse}} \geq 1$$

Our objective was twofold:

1) Find a suitable way to compute this result on the basis of results provided by the nonlinear analysis.

2) Ensure that the sensitivities of this result (with respect to all design variables) can be computed.

An obvious choice for the collapse RF is the load factor, which we denote by λ_{f} in the sequel. As an illustration, let us consider a simple superstiffener subject to both compression and shear loads. Assume that loads are applied progressively, the full loads (100%) corresponding to time 1. The nonlinear analysis terminates at time $t \approx 0.566$ (that is, largely before the full loads are applied) because time steps become too small. This means that for the given value of the design variables only 56.6% of the loads can be applied before collapse occurs. In this case we could thus have taken

$$\mathrm{RF}_{\mathrm{collapse}} = \lambda_{\mathrm{f}} = t \approx 0.566$$

Figure 45 shows the displacement of a node (belonging to the skin panel of the superstiffener) along the z axis.

However, this way to compute the collapse RF is not fully satisfactory because the sensitivity of λ_{f} is not directly available from a nonlinear analysis. This is why we have chosen to derive such a sensitivity using another method for the nonlinear analysis, namely, Riks continuation method [36]. Although classical Newton methods could have problems when passing a limit-point (because the generalized load displacement curve can have a decreasing time along the curve), continuation methods (also called *arc-length* or *Riks* methods) involve an additional parameter, namely, the arc length (denoted by s in the sequel), which is controlled in place of the time.

Because of this additional variable being introduced, an additional equation is added to the system of equations to describe the relation between the generalized

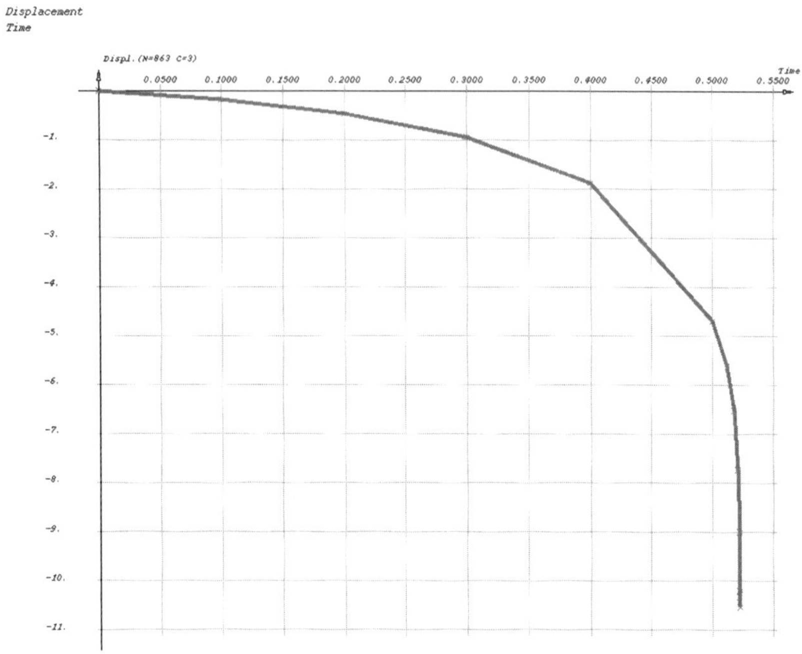

Fig. 45 Load vs displacement diagram for the nonlinear analysis.

displacements q and load λ_f on the one hand and the arc length s on the other. The simplest form of this constraint equation, corresponding to a hyperplane perpendicular to the predictor, was first introduced by Riks.

In a similar way, we added one equation allowing computing the sensitivity:

$$\frac{\partial \lambda_f}{\partial a}$$

We also constrained the unknown vector to be orthogonal to the load-displacement curve rather than a simple measure based on the vertical gap $\Delta\lambda_f$, which allows a better accuracy of the sensitivity measure as shown on Fig. 46.

Altogether, this methodology allowed us to derive a suitable algorithmic process for computing the value of the reserve factor and its sensitivity. This

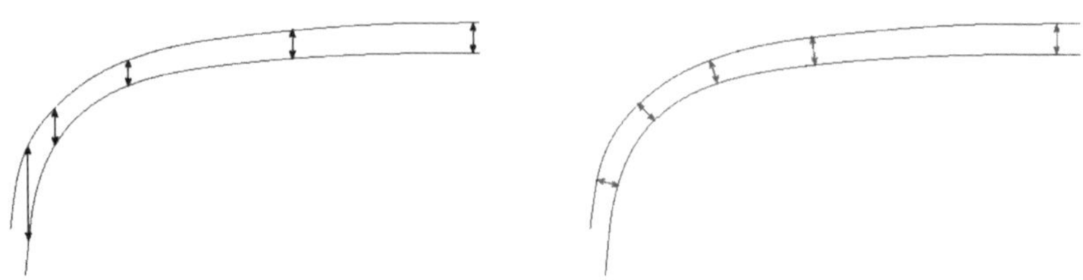

Fig. 46 Sensitivity analysis with Riks method.

process was successfully implemented within a COTS tool [31] and tested on a variety of examples.

C. Panel Optimization

In this paragraph we show how the industrial application described in Section VIII.A can be solved in practice. We first briefly describe the problem formulation in the framework of an application manager before showing the results we obtained. To have a step-by-step approach, buckling constraints only were considered in a first time.

1. Optimization Session

A complete computational process has been created, involving as many external tasks as the number of analyses, the latter being connected to the optimization task, as illustrated on the screenshot in Fig. 47.

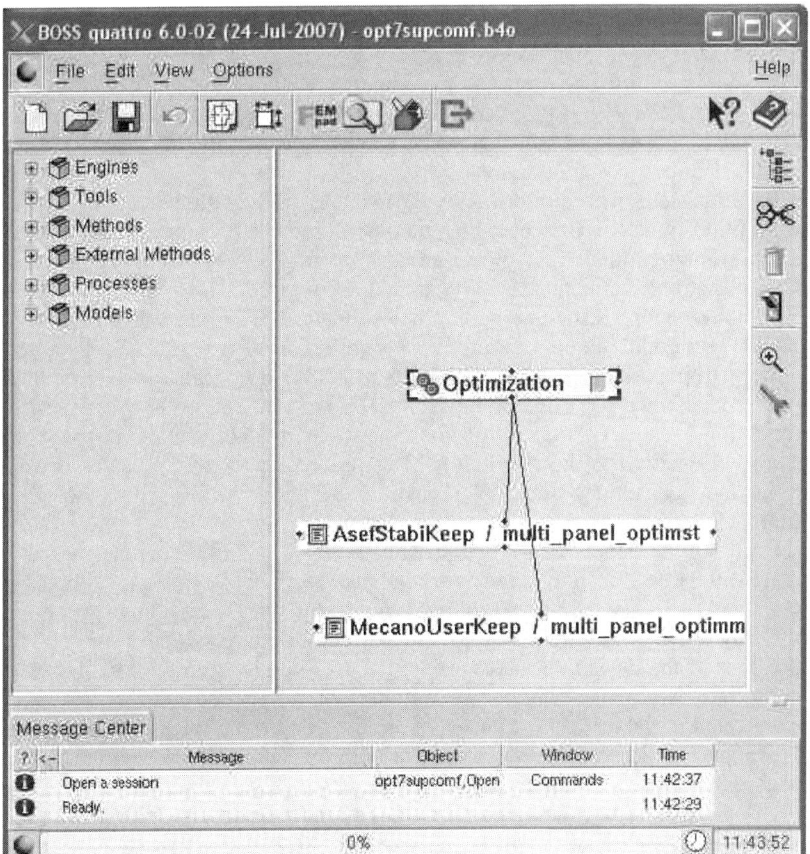

Fig. 47 Main window with task tree of the optimization process.

Two different kinds of analyses are used: 1) linear buckling analysis based on a COTS tool [31]–Asef stabi to capture the buckling on-set for skin and 2) nonlinear finite element analysis to capture the panel collapse. Of course, any branch of the task tree can be removed in order to concentrate the optimization on one criterion.

2. *Linear Buckling*

In this section, we first focus on the solution of the linear buckling optimization problem, removing the collapse constraint from the preceding formulation. This shows a possible cause of erratic convergence, which is shown to come from an incomplete formulation of the optimization problem. From a practical point of view, the buckling reserve factor is computed with a COTS tool [31], which is dedicated to the computation of buckling modes and related numerical results.

By definition, the first buckling load is of interest when designing a structure to withstand instability, and this single value corresponds to the $RF_{buckling}$ constraint. However, because of mode switching, there is no guarantee that the first buckling load always corresponds to the same buckling mode, and as a consequence the related sensitivities are not necessarily relevant to the subsequent steps and can cause erratic convergence. This is the reason why, instead of using mode-tracking techniques, a small set of say n buckling loads is often actually computed, the $RF_{buckling}$ constraint being then a vector-valued result. Because all of these n constraints must now be satisfied, mode switching inside those n values is not an issue anymore.

Our initial tests were performed with $n = 12$, and they allowed us to see that, for some designs, the first buckling modes might not be representative of the overall structure. Indeed, it turns out that, at a given iteration, the buckling modes might only influence a small part of the structure, which will be designed, while the remaining structural parts are not sensitive. The panel thickness in the sensitive part could increase to satisfy the stability criteria, while the thickness in the insensitive part will certainly reach its lower bound because it is to be minimized. At the next iteration, the low-thickness part is likely to become sensitive to buckling because of a small local stiffness, whereas the remaining part could become insensitive to the restrictions. If repeated, this scenario leads to oscillations and deteriorates the convergence of the optimization process, as illustrated in Fig. 48.

When simple panels of limited size including few stiffeners are studied, or when the thickness design variables are defined over wide regions, local buckling modes are less likely to appear. Should this occur, they would anyway be supported by a design variable that covers a wide structural part. It turns out that the values of the design variables are not only driven by a target on a minimum weight, but also by buckling considerations. When larger structures are studied, some design variables can become blind to the first (local) buckling modes used in the optimization problem. In such a case, the convergence difficulty just discussed is likely to happen.

Following this, we increased the value of n and set $n = 100$. The results obtained in this case were much better in the sense that only six iterations of the optimization process were required for convergence, as shown in Fig. 49.

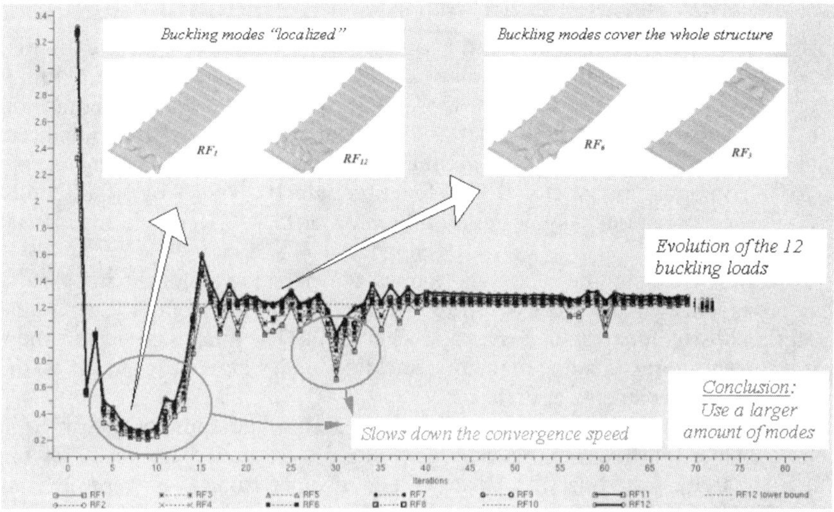

Fig. 48 Optimization convergence history for buckling reserve factors (initial).

This shows that a wide range of the first buckling loads must be used in the practical formulation of a linear buckling optimization problem. As was further illustrated by [37], using those larger sets not only makes the whole structure sensitive to buckling, but it also allows avoiding oscillations and/or slow convergence of the optimization process.

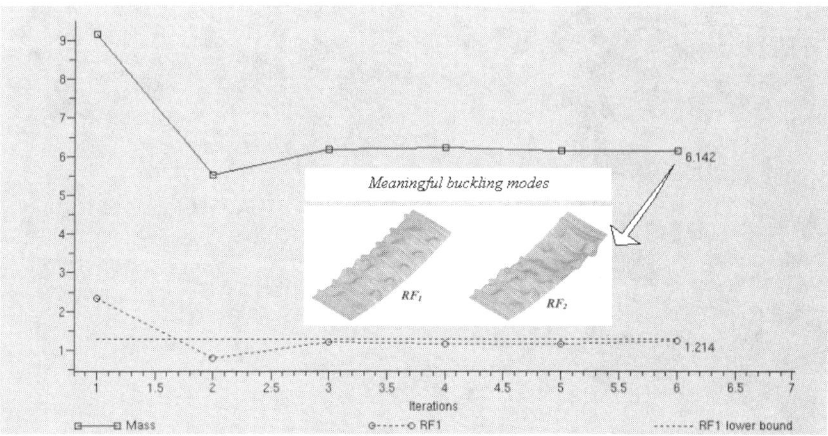

Fig. 49 Optimization convergence history for buckling reserve factors (improved).

3. Nonlinear Finite Element Analysis

The optimization run converged properly in 17 iterations. The curves displayed in Fig. 49 show the evolution of all three functions defining the optimization process; values displayed are those at optimum. Note that a tolerance on constraint violation (set to 2.5%) was used in accordance with AIRBUS practices. All constraints are thus satisfied at optimum. Also recall that the buckling reserve factor is computed as a vector-valued function with 100 values by a COTS tool [31] (hence the bars displayed in Fig. 50 represent these successive sets of 100 values).

Both runs yield significant weight savings while reserve factors do not perturb the convergence of the process. The sequences of images in Fig. 51 show the evolution of structure response over a few iterations. The images on the left show displacements corresponding to the first buckling load, whereas the images on the right show displacements at collapse.

Figure 52 shows the final values of the 42 design variables for both margin policies. Note that the variable names in Fig. 52 are built with the following rules: 1) variables starting with R*i* are related to superstringer number *i*, 2) variables R*_HOM* are ply thicknesses for the *skin*, 3) variables R*_HOMSTY* are ply thicknesses for the *stringers*, and 4) the last digit is related to ply orientation (1 for 0 deg, 2 for 45 deg, and 4 for 90 deg). Furthermore, because a symmetry is assumed between 45- and −45-deg plies, the values R*_HOM_1_3_2 and R*_HOMSTY_3_2 have to be multiplied by two when considering total thickness.

Table 6 shows the same results, but aggregated at the level of skin panels and stringers.

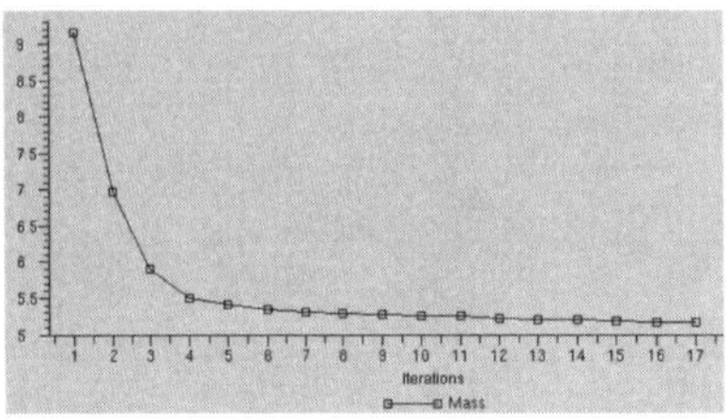

Weight = 5.1513

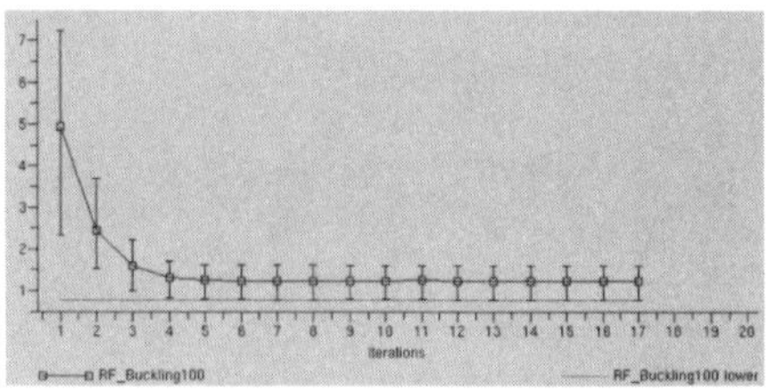

$RF_{buckling} = 0.7531$

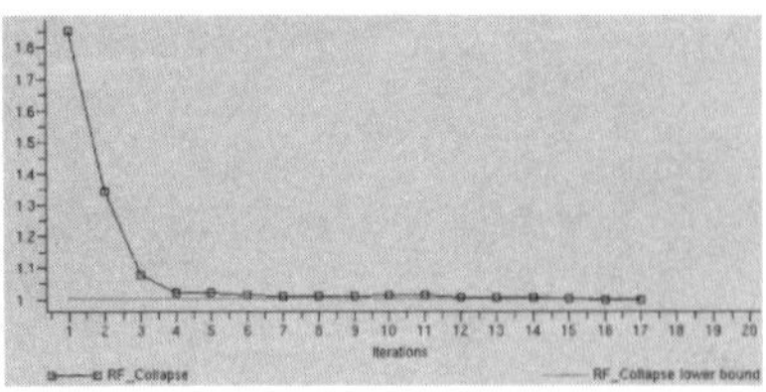

$RF_{collapse} = 0.9951$

Fig. 50 Convergence history for the various responses.

Second run (margin policy: 0.76)

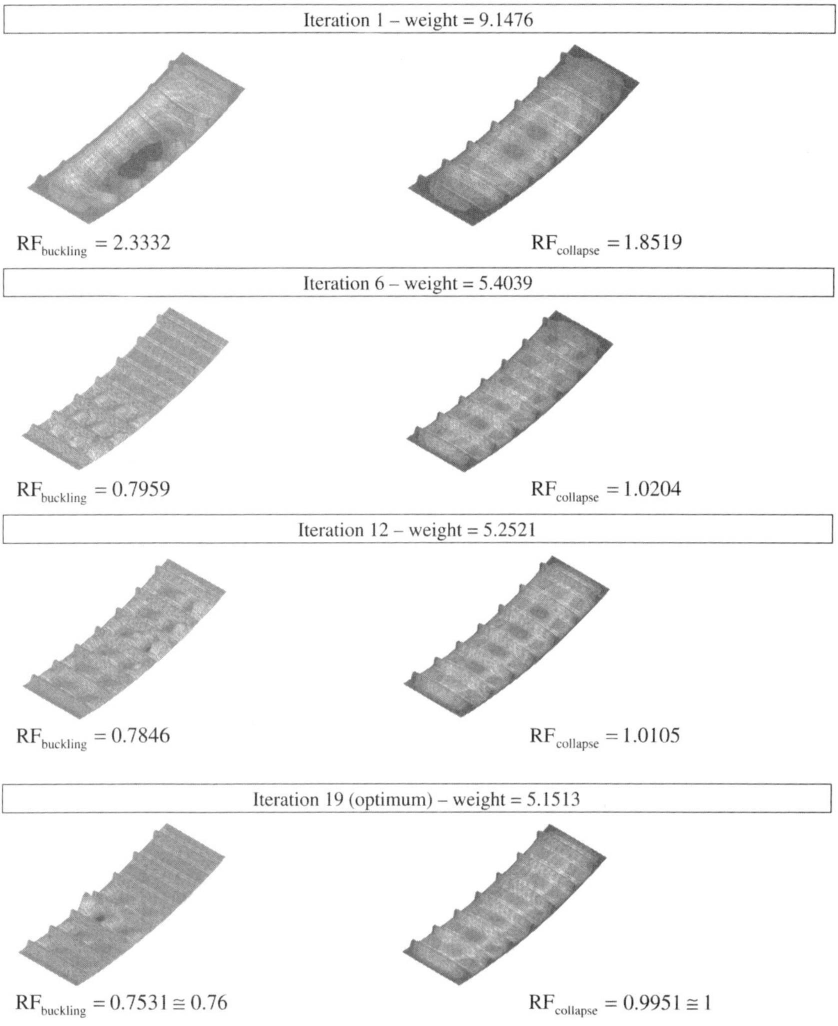

Fig. 51 Buckling on-set and collapse modes for various iterations.

IX. Summary and Conclusions

Presented in this chapter is a flexible and robust approach to structural sizing optimization at different stages of product design maturity. For early sizing stages a rapid sizing approach has been proposed, based on efficient neural-network regression of local optimization results. These neural networks can be used either in the STIFFOPT tool for quick automated sizing of a full structure cover or can be interrogated to give optimum design results at structural element level. Improvements have been applied to the neural-net construction

Number	Variable name	Value for margin 0.76
0	_R1_HOMSTY_3_1	0,4
1	_R1_HOMSTY_3_2	0,4
2	_R1_HOMSTY_3_4	0,4
3	_R1_HOM1_3_1	0,4
4	_R1_HOM1_3_2	1,031272354
5	_R1_HOM1_3_4	0,4
6	_R2_HOMSTY_3_1	0,408197074
7	_R2_HOMSTY_3_2	0,4
8	_R2_HOMSTY_3_4	0,4
9	_R2_HOM1_3_1	0,4
10	_R2_HOM1_3_2	1,096481502
11	_R2_HOM1_3_4	0,4
12	_R3_HOMSTY_3_1	0,738890995
13	_R3_HOMSTY_3_2	0,4
14	_R3_HOMSTY_3_4	0,4
15	_R3_HOM1_3_1	0,4
16	_R3_HOM1_3_2	0,968944711
17	_R3_HOM1_3_4	0,470371556
18	_R4_HOMSTY_3_1	0,609791644
19	_R4_HOMSTY_3_2	0,4
20	_R4_HOMSTY_3_4	0,4
21	_R4_HOM1_3_1	0,4
22	_R4_HOM1_3_2	0,998656614
23	_R4_HOM1_3_4	0,465830822
24	_R5_HOMSTY_3_1	0,743507517
25	_R5_HOMSTY_3_2	0,406364241
26	_R5_HOMSTY_3_4	0,4
27	_R5_HOM1_3_1	0,4
28	_R5_HOM1_3_2	0,97919137
29	_R5_HOM1_3_4	0,496159837
30	_R6_HOMSTY_3_1	0,421884952
31	_R6_HOMSTY_3_2	0,4
32	_R6_HOMSTY_3_4	0,4
33	_R6_HOM1_3_1	0,4
34	_R6_HOM1_3_2	1,095987273
35	_R6_HOM1_3_4	0,4
36	_R7_HOMSTY_3_1	0,4
37	_R7_HOMSTY_3_2	0,4
38	_R7_HOMSTY_3_4	0,4
39	_R7_HOM1_3_1	0,4
40	_R7_HOM1_3_2	1,033261722
41	_R7_HOM1_3_4	0,4

Fig. 52 Values of design variables at convergence.

and the design of experiments in order to obtain more robust results and to consider higher dimensions of the input space.

For preliminary sizing stages, all local optimizations are performed in the frame of the sizing process, evoking specialized tools in a two-level process with two systems of iterations. A PC cluster has been used for computational efficiency. The proposed approach is particularly suited to parallelization because all optimizations are independent from each other. For detailed sizing stages a new

Table 6 Values of design variables at convergence

	Total thicknesses
Stringer 1	1,6
Skin panel 1	2,862544708
Stringer 2	1,608197074
Skin panel 2	2,992963004
Stringer 3	1,938890995
Skin panel 3	2,808260978
Stringer 4	1,809791644
Skin panel 4	2,86314405
Stringer 5	1,956236
Skin panel 5	2,854542576
Stringer 6	1,621884952
Skin panel 6	2,991974547

optimization process was put in place at local level, considering nonlinear finite element analysis to evaluate the postbuckling behavior of each panel. An original hybrid formulation was adopted to formulate the optimization problem, combining together linear buckling for skin buckling on-set and nonlinear analysis for panel collapse. To solve the optimization problem, semi-analytical sensitivities was developed and embedded in a gradient-based optimization process. The efficiency and robustness of the full optimization process were demonstrated. This work is one of the few examples showing an accurate gradient-based optimization processes applied to nonlinear finite element analysis. In this third sizing stage, the need for HPC is even greater, considering the long time necessary for each optimization (20 h).

The implementation of the three sizing stages is to be completed in the near future. The same framework will be used for integration. A full multilevel optimization process will need to be developed controlling the overall optimality, that is, considering optimal load redistribution, design continuity, and more general global constraints (such as stiffness constraints). The optimization process will have to be properly decomposed and coordinated and implemented based on HPC architectures (massive parallelism). The extended enterprise context is also to be considered with a structural design distributed around the world to different partners. All of these different areas of investigation are being pursued in a recently started European Project named MAAXIMUS and focused on virtual design and testing of a composite fuselage.

References

[1] Schmit, L. A., and Mallet, R. H., "Structural Synthesis and Design Parameter Hierarchy," *Proceedings of the 3rd ASCE Conference on Electronic Computation*, American Society of Civil Engineers, Structural Division, Reston, VA, 1963, pp. 269–300.

[2] "BOSS Quattro User's Handbook – Release 6.0-02," SAMTECH, Liège (Angleur), Belgium, 2008.

[3] Remouchamps, A., and Radovcic, Y., "BOSS Quattro: an Open System for Parametric Design," *Structural and Multidisciplinary Optimization*, Vol. 23, No. 2, March 2002, pp. 140–152.

[4] Remouchamps, A., Grihon, S., Raick, C., Colson, B., and Bruyneel, M., "BOSS Quattro: an Open System for Parametric Design," *Structural and Multidisciplinary Optimization*, Vol. 23, No. 2, March 2002, pp. 140–152.

[5] Schmit, L. A., Jr., and Ramanathan, R. K., "Multilevel Approach to Minimum Weight Design Including Buckling Constraints," *AIAA Journal*, Vol. 16, No. 2, Feb. 1978, pp. 97–104.

[6] Balling, R. J., and Sobieszczanski-Sobieski, J., "An Algorithm for Solving the System-Level Problem in Multi-Level Optimization," NASA Contractor Report 195015, Dec. 1994.

[7] Timoshenko, S., and Gere E., *Theory of Elastic Stability*, McGraw-Hill, New York, 1936.

[8] Merval, A., Samuelidès, M., and Grihon, S., "Multi-Level Optimization with Local Mass Minimization," *Second European Conference for Aerospace Sciences*, Paper number 1.02.04, Published as CD-ROM by von Karman Institute, Brussels, Belgium in cooperation with ULB, Université Libre de Bruxelles, Belgium, 2007.

[9] Merval, A., Samuelidès, M., and Grihon, S., "Lagrange Kuhn-Tucker Coordination for Multilevel Sizing of Aeronautical Structures," AIAA Paper 2008-2223, April 2008.

[10] Remouchamps, A., Grihon, S., Raick, C., Colson, B., and Bruyneel, M., "Numerical Optimization: A Design Space Odyssey," *International Workshop, 2007: Advancements in Design Optimization of Materials, Structures and Mechanical Systems*, Northwestern Polytechnical University (NPU), Xi'an, China, 2007.

[11] *NASTRAN Linear Static Analysis User's Guide*, MSC.Software Corp., Santa Ana, CA, 2000.

[12] "MTS004: Manuel de Calcul Statique pour Matériaux Métalliques," AIRBUS France, Toulouse, France, 2000.

[13] "The NAG FORTRAN Library," Numerical Algorithms Group, Oxford, U.K., http://www.nag.co.uk/numeric/fl/fldescription.asp [accessed July 2009].

[14] Fleury, C., "Dual Methods for Convex Separable Problems," *Optimization of Large Structural Systems*, Vol. I, edited by G. I. N. Rozvany, Kluwer Academic, Amsterdam, 1993, pp. 509–530.

[15] Bruyneel, M., Duysinx, P., and Fleury, C., "A Family of MMA Approximations for Structural Optimization," *Structural and Multidisciplinary Optimization*, Vol. 24, No. 4, Oct. 2002, pp. 263–276.

[16] *PATRAN Programming Command Language Users Guide*, MSC.Software Corp., Santa Ana, CA, 2000.

[17] "MAAXIMUS," Integrated Project, 7th Framework, 2008–2013, http://www.maaximus.eu/ [retrieved July 2009].

[18] Vaswani, V., *PHP Programming Solutions*, McGraw-Hill, New York, 2007.

[19] "PHP—Hypertext Processor," http://nl.php.net/tut.php.

[20] Merval, A., Samuelides, M., and Grihon, S., "Application of Response Surface Methodology to Stiffened Panel Optimization," AIAA Paper 2006-1815, May 2006.

[21] Simpson, T. W., Mauery, T. M., Korte, J. J., and Mistree, F., "Comparison of Response Surfaces and Kriging Models for Multidisciplinary Design Optimization," AIAA Paper 1998-4755, Sept. 1998.

[22] Sellar, R. S., Batill, S. M., and Renaud, J. E., "Response Surface Based, Concurrent Subspace Optimization for Multidisciplinary System Design," AIAA Paper 1996-0714, Jan. 1996.

[23] Bishop, C. M., *Neural Networks for Pattern Recognition*, Clarendon Press, Oxford, U.K., 1997.

[24] Dreyfus, G., Samuelides, M., and Gordon, M., *Réseaux de Neurones: Méthodologie et Applications*, edited by Eyrolles, Paris, France, 2002.

[25] Jordan, M. I., Jacobs, R. A., Nowlan, S. J., and Hinton, G. E., "Adaptive Mixture of Local Experts," *Neural Computation*, Vol. 3, No. 1, Spring 1991, pp. 79–87.

[26] Jordan, M. I., and Jacobs, R. A., "Hierarchical Mixtures of Experts and the EM Algorithm," *Neural Computation*, Vol. 6, No. 2, March 1994, pp. 181–214.

[27] Wang, G. G., "Adaptive Response Surface Method Using Inherited Latin Hypercube Design Points," *Journal of Mechanical Design*, Vol. 125, June 2003, pp. 210–220.

[28] Cohn, D. A., "Neural Network Exploration Using Optimal Experimental Design," MIT CBCL Paper 99, June 1994.

[29] "MATLAB User's Guide–Release R2008b–Neural Network Toolbox," The Mathworks, Natick, MA, 2008.

[30] Colson, B., Bruyneel, M., Grihon, S., Jetteur, P., and Remouchamps, A., "Composite Panel Optimization with Nonlinear Finite Element Analysis and Semi-Analytical Sensitivities," *NAFEMS Seminar–Simulating Composite Materials and Structures*, Nov. 2007, http://www.nafems.org/events/nafems/2007/compositeDACH/ (in German).

[31] "A COTS tool: Software Package for Finite Element Analysis of Structures," SAMTECH, S.A., Liège Belgium, http://ftp.samtech.com/en/pss.php?ID=5&W=products [retrieved July 2009].

[32] Lanzi, L., and Giavotto, V., "Post-Buckling Optimization of Composite Stiffened Panels: Computations and Experiments," *Composite Structures*, Vol. 73, No. 2, May 2006, pp. 208–220.

[33] Bruyneel, M., "A General and Effective Approach for the Optimal Design of Fiber Reinforced Composite Structures," *Composites Science and Technology*, Vol. 66, No. 10, 2006, pp. 1303–1314.

[34] Carpentier, A., Barrau, J.-J., Michel, L., and Grihon, S., "Buckling Optimization of Composite Panels via Lay-up Tables," *III European Conference on Computational Mechanics: Solids, Structures and Coupled Problems in Engineering: Book of Abstracts*, edited by C. A. Mota Soares, et al., Springer, Netherlands, 2006, pp. 226.

[35] Carpentier, A., Barrau, J.-J., Michel, L., and Grihon, S., "Optimization Methodology of Composite Panels," *12th European Conference on Computational Mechanics, Solids, Structures and Coupled Problems in Engineering*, Paper number 142, 29 Aug.–1 Sept. 2006, Biarritz, France.

[36] Riks, E., Rankin, C., and Brogan, F., "On the Solution of Mode Jumping Phenomena in Thin Walled Shell Structures," *Computer Methods in Applied Mechanics and Engineering*, Vol. 136, No. 1, 1996, pp. 59–92.

[37] Bruyneel, M., Colson, B., and Remouchamps, A., "Discussion on Some Convergence Problems in Buckling Optimization," *Structural and Multidisciplinary Optimization*, Vol. 35, No. 2, Feb. 2008, pp. 181–186.

Chapter 8

Pylon Multidisciplinary Optimization

Stéphane Grihon,* Mathieu Meaux,† Albert Lucchetti,‡ and Patrick Sarouille§

AIRBUS, Toulouse, France

Julien Laurenceau¶

CERFACS, Toulouse, France

and

Gérald Carrier** and Sylvain Mouton**

ONERA, Paris, France

Abbreviations

CFD = computational fluid dynamics
COTS = commercial-off-the-shelf
DOC = direct operating cost
DoE = design of experiments
FBO = fan-blade-off
GFEM = global finite element model
MDO = multidisciplinary optimization
MMFD = modified method of feasible directions
RSM = response surface mode

*Expert in Structure and Multidisciplinary Optimization.
†Doctor Engineer in Aerodynamic Optimization.
‡Loads Engineer.
§Aeroelastics and Loads Engineer.
¶Ph.D. Student.
**Doctor Engineer in Aerodynamics.

Nomenclature

A = matrix of aerodynamic forces
$[AGFM(\omega)][AGFC(\omega)][FGT(\omega)]$ = aerodynamic generalized forces of motion, of control surfaces, and of turbulence
$\{a\}$ = acceleration vector
a,b = trade coefficients for stress-aerodynamic optimization
$[B]$ = damping matrix
$C_D(\alpha)$ = drag coefficient
$C_L(\alpha)$ = lift coefficient
D = drag
$D(\alpha)$ = drag
$\{F\}$ = other input forces written in modal base fixed internal loads
$\{g\}$ = field of accelerations due to the weight
$g(\lambda)$ = geometrical constraints
$J(\alpha)$ = aerodynamic objective function
$[K]$ = stiffness matrix
$L(\alpha)$ = lift
$[M]$ = mass matrix
$\{N\}$ = nonlinear forces
$\{P\}$ = applied forces
$\bar{q}$ = dynamic pressure
$\{q\}\{\dot{q}\}\{\ddot{q}\}$ = generalized coordinates (position, speed, acceleration)
U/V = gust profile
$\{u\}$ = displacement vector
$\{v\}$ = velocity vector
W = pylon width
$\tilde{W}$ = GFEM weight
$\tilde{W}(\lambda)$ = weight
X,Z = horizontal and vertical position of the engine
α = vector of optimization shape variables with lower bounds $\alpha_{\min}$ and upper bounds $\alpha_{\max}$
α, $\alpha_{\min}$, $\alpha_{\max}$ = vector of aerodynamic shape variables and bounds
α_{aero} = aerodynamic variables
α_{MD} = multidisciplinary variables
α_{struct} = structure variables
$\{\delta\}$ = control surface motion (input)
λ, $\lambda_{\min}$, $\lambda_{\max}$ = vector of structural design variables and bounds
Φ = modal basis/matrix
$\sigma(\lambda)$, $\bar{\sigma}$ = stresses and stress allowables
∇ = gradient
$[\mu][\beta][\gamma]$ = mass, damping, and stiffness matrix in a modal basis

I. Introduction

A. Context

THE aeronautical industry is a business governed by increasing demands of the customers for a better product performance and at the same time cost and time-to-market reduction. For the European industry values for these objectives have been sent in the vision 2020 [1]. Numerical optimization is one of the key technologies contributing to the achievement of these goals. It incorporates mathematical theories and associated algorithms that guarantee an optimum performance, and software implementation that supposes integration and automation of design stages and ensures reduction of lead times. However, the demand for a higher product performance encourages each discipline to concentrate on itself, while the growing complexity of the product requires multidisciplinary design.

Hence, given that the single discipline optimization tools are sufficiently mature and intensively used by aircraft projects, the purpose now is to improve their respective performance and connectivity and to build integrated environments for the global aircraft design at each stage of the development process.

The design of the engine pylon (Fig. 1), as part of the powerplant installation definition, is an inherently multidisciplinary problem and a reasonable first target for multidisciplinary optimization development in the aircraft industry.

B. Design Process

Currently, the design process of an engine pylon is mostly driven by structures and systems. After the selection of an engine for the aircraft, the engine is rapidly positioned along the wing span by the future project office (department in charge of aircraft feasibility phase at AIRBUS) considering various criteria such as engine failure, size of the empennage, impact on loads, and wing weight. Then, a first assumption is made regarding the horizontal and vertical position, based on basic aerodynamic design rules. Various pylon structures are then proposed depending on the selected configuration for pylon-to-engine and pylon-to-wing attachments: these attachments are closely linked to the requirements in terms of overall flexibility for the pylon/engine ensemble. System installation is also investigated inside and around the pylon structure completing the

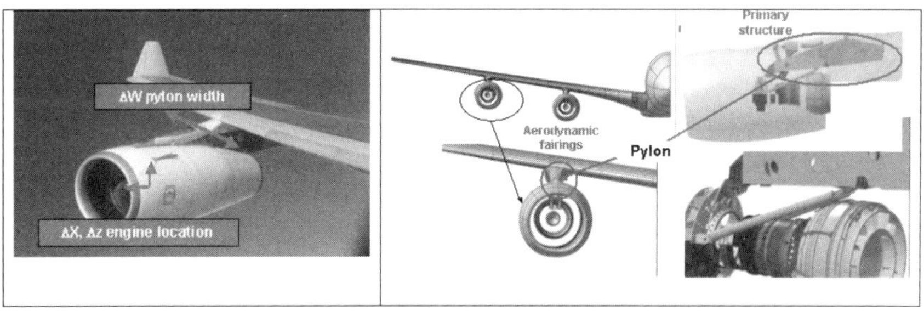

Fig. 1 Pylon design case study.

architecture of the component. Finally, the aerodynamics is considered in more details: pylon fairings are designed and optimized essentially to limit shock effects (wave drag). Horizontal and vertical engine positioning can be slightly reconsidered. Direct stiffness effects through sizing on loads are not necessarily studied.

Of course, this design process is not as sequential as just presented, and iterations do take place between disciplines across multidisciplinary design reviews including all parties (see [2, 3] for more practical details about the industrial design process of powerplants). However, the level of the integration can be much improved. The intent of this research is to advocate for a more integrated process fully supported by numerical simulation and optimization techniques.

Because it is still prohibitively difficult to cope directly with all design variables and disciplines, the following assumptions and simplifications are made:

1) The engine wing-span positioning was considered as already frozen.

2) The configuration design variables were limited to horizontal and vertical position of the engine and the pylon width. These are the main interaction variables between aerodynamic and structure.

3) The multidisciplinary optimization process was investigated through bidisciplinary scenarios.

The rest of the chapter is organized as follows. The selected case study is detailed in Section II. The monodisciplinary optimization processes are presented (structure and aerodynamic optimization) in Section III while the integration of stress and aerodynamic optimization is described in Section IV. The integration of stress and loads optimization is addressed in Section V.

II. Presentation of the Test Case

The test case consists of an outer powerplant of a very large civil transport aircraft. The optimization is mostly concentrated on the pylon. In all multidisciplinary tasks the pylon sizing is considered as a basis for a weight indicator. The impact of shape variations on the pylon secondary structures will not be considered. The engine is considered as frozen. Please refer to Fig. 1 for an overview of the wing-pylon configuration.

Regarding the aerodynamic aspects, only the shape of the pylon fairings is considered here. The specific aerodynamic design variables are assumed to be mainly independent and are further detailed later (Section III.B.1; Fig. 5).

Regarding coupled multidisciplinary variables, the horizontal position X and the vertical position Z of the engine are considered, as well as the pylon width W. All have impact on both stress and aerodynamic behavior. However, the variations will be assumed small enough so that their effect on the aircraft loads can be neglected.

III. Single Discipline Optimizations

A. Structure Optimization

1. Definition

The selected structure optimization process has been intentionally simplified considering that stress criteria are represented by stress allowables. This is an approach commonly used for pylon rapid sizing [4]. These stress allowables are dependent on the material, the load case (fatigue or static case), and on the temperature.

The optimization problem to be solved can be written as follows:

$$\begin{cases} \text{Min } \tilde{W}(\lambda) \\ \lambda_{\min} \leq \lambda \leq \lambda_{\max} \\ \sigma(\lambda) \leq \overline{\sigma} \\ g(\lambda) \leq 0 \end{cases}$$

For λ, the vector of optimization design variables with lower bounds $\lambda_{\min}$ and upper bounds $\lambda_{\max}$, these design variables are thicknesses of given regions whose thickness is considered as uniform. They can also be scaling factors over all element sizing properties of a region. In this second case, sizing properties do not need to be uniform. The optimization constraints, $\sigma(\lambda)$, are inequality constraints giving upper or lower bounds to specific stress responses. This simplified formulation allows the use of the optimization module of NASTRAN: SOL200 [5] and the performing of an automated sizing of the global finite element model (GFEM).

The refinement of the sizing process in further stages requires taking into account stability (buckling/postbuckling), but also additional geometrical constraints.

Objective function. The objective function is the GFEM weight $\tilde{W}(\lambda)$ and is considered as a representative weight indicator. However, it is important to stress that it is far from the real weight because not all elements can be modelled (example giving, some joints are not modelled), the geometry is approximated, and additional weight needs to be considered (with regard to structure fasteners, treatment, secondary structures, and systems). We have however assumed that the delta weight is well represented by the delta weight given by the sized GFEM.

Design variables. Twenty-seven thicknesses in various regions were considered with specific lower and upper bounds. Regions are illustrated per subcomponent in Fig. 2.

Constraints. The constraints are stress allowables defined according to the load case:

1) Fan-blade-off (FBO) cases are failure cases linked to the break of a fan blade. Material failure allowables are considered for these static load cases. Maneuver load cases are also considered as ultimate load cases, but they are not really critical with respect to FBO cases.

2) Flight cases are essentially considered as limit load cases for fatigue allowables (reduced allowables based on equivalent stresses). Fatigue allowables are of course more demanding (stringent) than static allowables. Thermal effects are also taken into account on both kinds of stress allowables. Major, minor principle, and shear stresses are controlled in panel and spar webs. Longitudinal stresses (traction/compression) are controlled in flanges. Specific gauge elements have been implemented to control the stress in flange edges.

There are also geometrical constraints linked to manufacturability: spar flange thickness > spar web thickness and panel side thickness > panel web thickness.

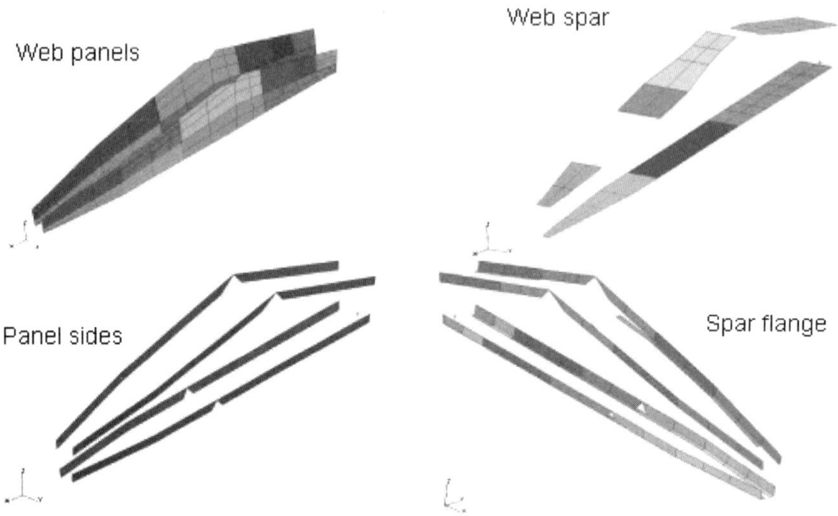

Fig. 2 Definition of pylon structural design variables.

2. Implementation

The optimization problem is defined in a NASTRAN SOL200 bulk data file. In fact, the generation of such a bulk data file has been linked to the GFEM parameterization and automated. A specific development has been made in the past at AIRBUS using the PATRAN programming command language to parameterize the GFEM with respect to external shape and topology and then chain automatically with NASTRAN SOL200 [6]. This tool is illustrated in Fig. 3.

3. Demonstration

Described next are the typical results found with the sizing optimization process. Figure 4a shows the thickness distribution found after optimization. Thickness varies between 2.5 mm on spar webs up to 22 mm in spar flanges and panel sides. Figure 4b shows the stress distribution after optimization.

The fatigue flight load cases are the dimensioning load cases for flanges with respect to longitudinal stresses as measured in gauges. The FBO static load cases are the dimensioning load cases in panel webs with respect to the major principal stress allowable and in the spar webs with respect to the shear stress allowable.

B. Aerodynamic Optimization

1. Definition

The aerodynamic optimization process aims at changing the shape of pylon fairings in order to minimize the drag. These fairings are supposed to envelop the pylon shape without any interference with the internal structure and

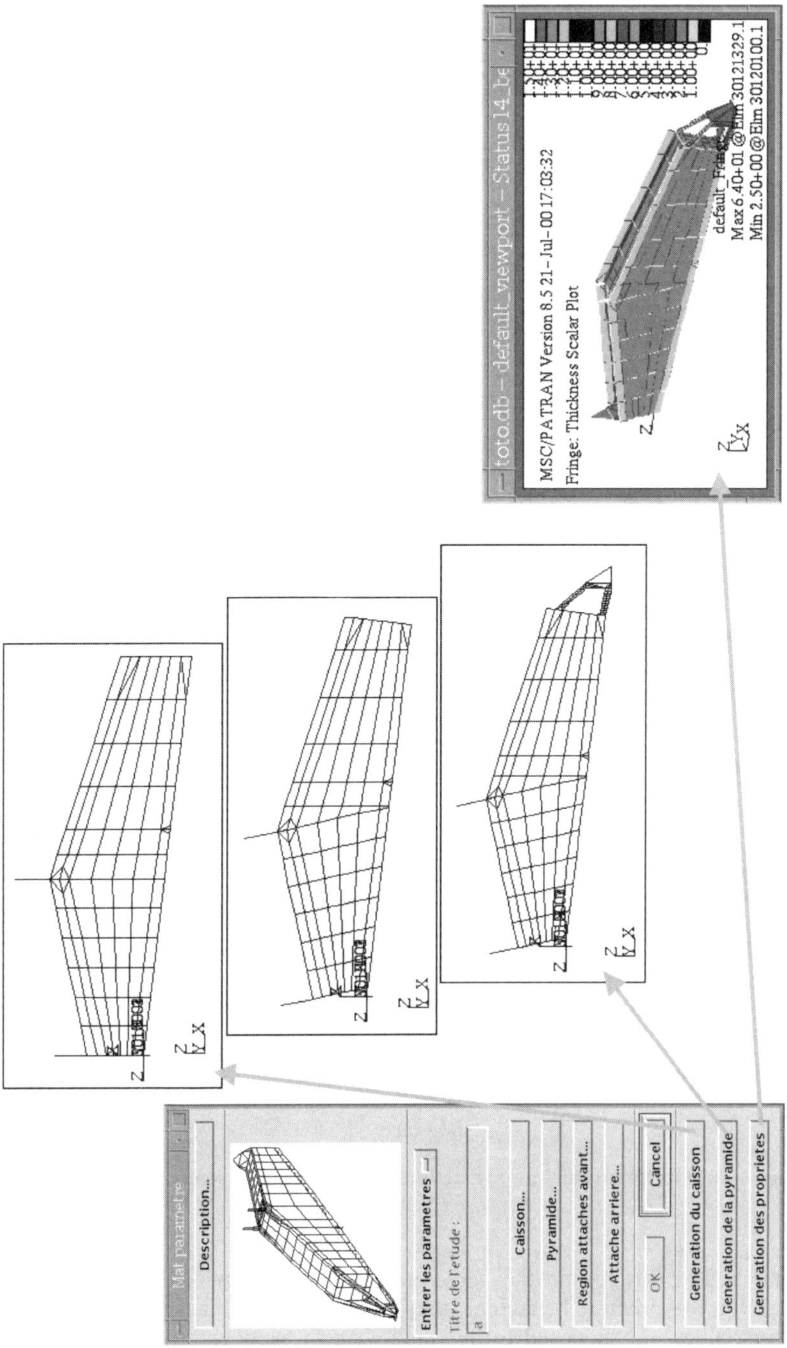

Fig. 3 Parameterized pylon model generation and rapid sizing.

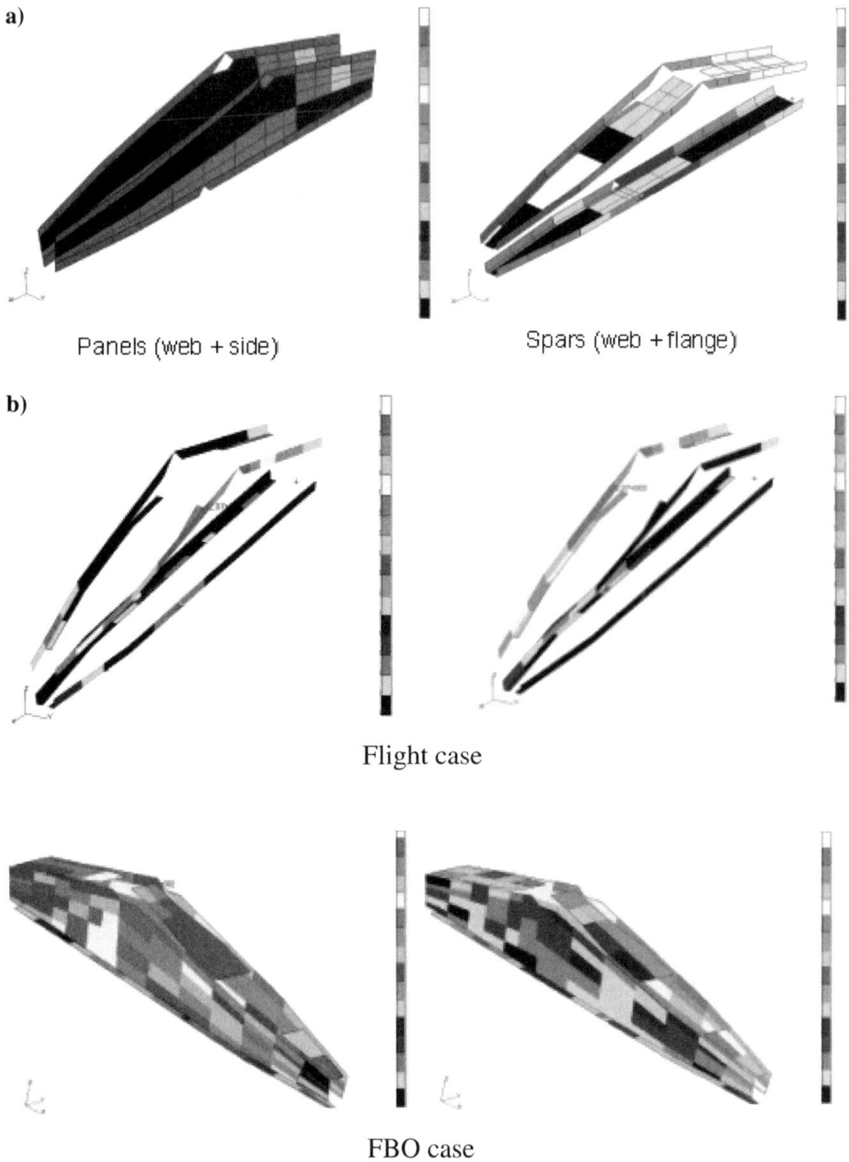

Fig. 4 a) Optimum thickness distribution. b) Optimum stress distribution.

equipment. For this reason the aerodynamic design variables have no direct influence on the structure behavior if they are bounded within limited variations. These variables are supposed to be small shape variations around the overall pylon envelope driven by the top-level variables *X*,*Z*,*W*.

The optimization problem to be solved can be written as follows:

$$\begin{cases} \text{Min } D(\alpha) \\ \alpha_{\min} \leq \alpha \leq \alpha_{\max} \end{cases}$$

In the general case it would be necessary to add wing aerodynamic design variables. The optimization of the pylon cannot be done independently from the wing because of the aerodynamic interference between both components. This is to be considered in future work.

Objective function. In this research work, taking into account the small expected lift changes from one aerodynamic shape to another, the angle of attack has been kept constant, and the explicit lift constraint in the optimization problem has been replaced by a penalization term added to the objective function D. This penalty accounts for the residual induced drag variation resulting from the weak lift variation induced by the pylon shape modification. The objective function of the actual optimization problem reads

$$J = C_D - \frac{\mathrm{d}C_D}{\mathrm{d}C_L} C_L$$

During postprocessing, aerodynamic forces can be computed by integrating elementary forces on the skin of the aircraft (most obvious method called near-field approach) or alternatively by balancing forces in the whole computational volume, which allows the separation of physical contributions (far-field approach method) [7]. The near-field approach has been selected because we do not need to separate the various drag contributions: the overall drag is considered in the optimization.

It is considered that changes in friction drag are insignificant. Predicted changes in friction drag should therefore be ignored to avoid unnecessary numerical noise in the optimizer. Lift caused by friction is always negligible, and therefore only pressure forces are to be considered. Finally, $\mathrm{d}C_D/\mathrm{d}C_L$ is regarded as constant during the optimization and assessed around baseline shape. The objective function can then be rewritten as

$$J(\alpha) = C_{\mathrm{Dp}}(\alpha) - \left(\frac{\mathrm{d}C_{\mathrm{Dp}}}{\mathrm{d}C_{\mathrm{Lp}}}\right)_{\text{baseline}} C_{\mathrm{Lp}}(\alpha)$$

and its gradient with respect to shape parameters α:

$$\nabla_\alpha J(\alpha) = \nabla_\alpha C_{\mathrm{Dp}}(\alpha) - \left(\frac{\mathrm{d}C_{\mathrm{Dp}}}{\mathrm{d}C_{\mathrm{Lp}}}\right)_{\text{baseline}} \nabla_\alpha C_{\mathrm{Lp}}(\alpha)$$

Drag improvements are expected from first order—viscous pressure, induced drag (caused by change in load distribution on wing); second order—wave drag; and insignificant—friction drag.

Design variables. We selected a solution with 19 parameters controlling 17 Hicks–Henne bumps [8] of various degrees spread on the pylon in areas numbered from 1 to 6 in Fig. 5.

The "beret Basque" French nickname stands for a specific pylon side deformation. It is useful mainly for low-speed performance to enhance slat air tightness. The beret Basque increases the cross-sectional area of the front of the pylon; the area has to decrease downstream without causing the flow to separate and without creating speed peaks.

Bounds (α_{min}, α_{max}) are given to all of these design variables. Lower bounds are used to avoid interferences between pylon fairings and systems or pylon primary structure. Upper bounds are set large enough so as not to be active. Extreme shapes for the modification of the wing-pylon intersection region are shown on Fig. 6.

Constraints. Considering the definition of the objective function J, which adds a correction term for the residual lift variations to the actual drag, no explicit constraint is imposed during the optimization, except that the design variables are bounded. The constraint of constant lift is explicitly taken into account in the objective function as described before.

2. *Implementation*

The optimization is performed with the optimization library DOT from G. Vanderplaats [9]. The optimization algorithm used is the modified method of feasible directions (MMFD) [10]. The overall aerodynamic optimization process has been implemented inside OptaliA, the AIRBUS aerodynamic optimization framework [11]. This framework enables parameterization, mesh deformation, aerodynamic analyses [computational fluid dynamics (CFD)], postprocessing, and also includes the optimization algorithm. The use of a specific grammar and XML files allows one to easily define design variables

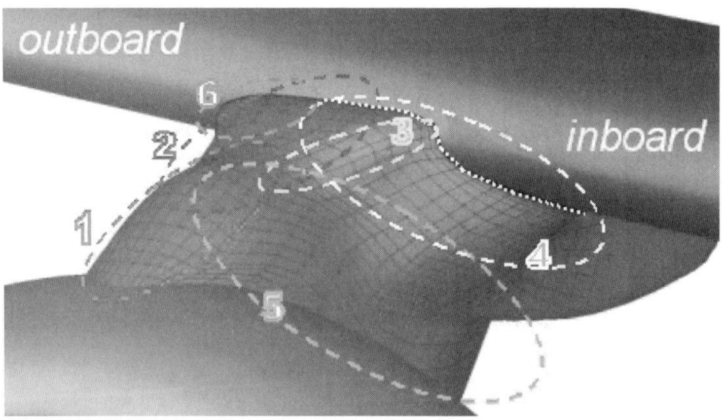

Fig. 5 Illustration of pylon design variables: 1, lower leading edge; 2, upper leading edge; 3, "béret basque"; 4, wing-pylon inboard intersection; 5, pylon inboard flange; and 6, wing-pylon outboard intersection.

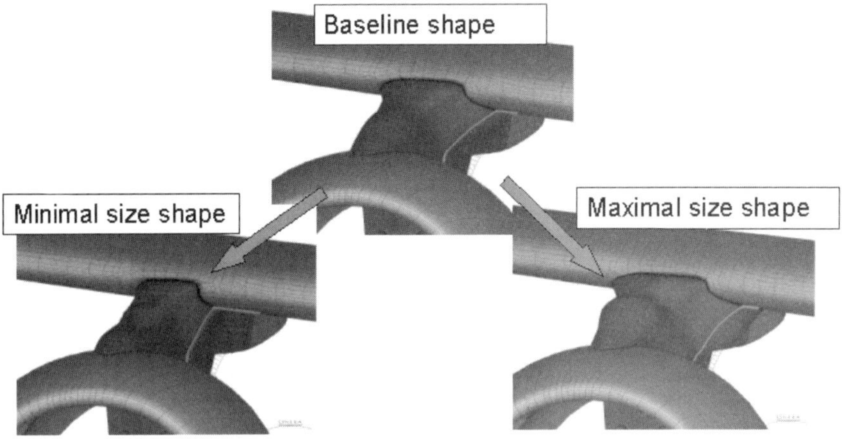

Fig. 6 Illustration of shape variations.

and to share these across the framework. It also allows one to generate automatically the interface for creating these design variables.

CFD analysis. Complex transonic flow and a large contribution expected from viscous pressure drag advocate for a Reynolds-averaged Navier–Stokes physical modelling. A Spalart–Allmaras turbulence model and a second-order Roe scheme are used for the CFD computations, which were carried out with the *elsA* software, a multi-application flow solver jointly developed by ONERA, CERFACS, and AIRBUS [12].

To maintain the computational time within reasonable limits, considering the numerous flow calculations required during the optimization, the boundary layer is modelled only on the wing and on the outboard pylon (see Fig. 7). For the same reason, the original mesh was coarsened by one point out of two in all directions. After several attempts, final mesh size reached 1.5 millions nodes in 134 structured blocks. It was demonstrated that mesh smoothing dramatically improves convergence history and spurious drag level, allowing the reaching of a converged solution (within 0.1 drag count) in 500 multigrid cycles, as presented in Fig. 8.

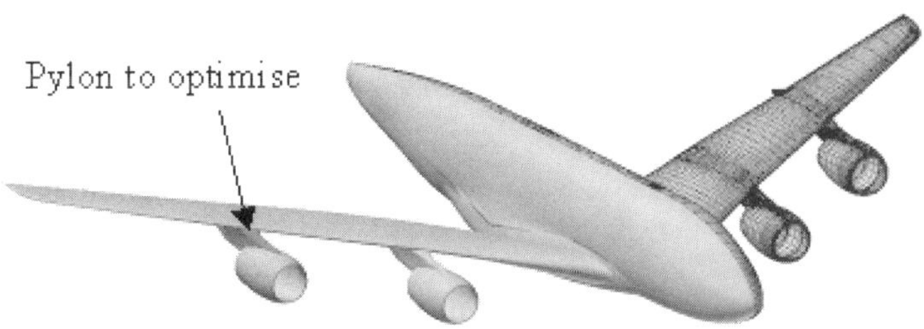

Fig. 7 Overview of aircraft geometry and mesh.

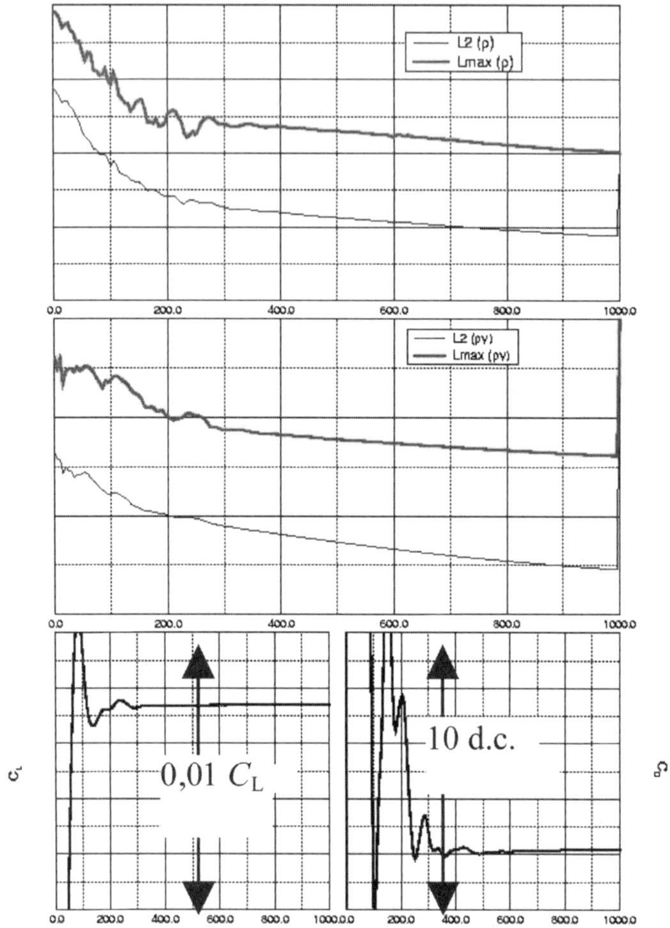

Fig. 8 Convergence history of computation on baseline configuration. (See also color figure section at the back of the book.)

Because of the constraint on computational time for optimization purposes, the mesh is kept very coarse according to the case complexity, which reduces absolute precision, but nevertheless, is expected to enable relative improvement from a shape to another one.

Sensitivity analysis. The sensitivity analysis is based on a formal differentiation method called the adjoint method [12, 13]. Usually the calculation of response sensitivities for N design variables comes at approximately the cost of N analyses. Instead in the adjoint method the sensitivity equations are transposed. Thus, if M is the number of responses, the sensitivity analysis now takes only the time of M analyses. Hence, the adjoint method is very appealing for aerodynamics because the number of responses to be differentiated is very small: the overall drag and potentially the overall lift of the aircraft. In contrast, the adjoint approach is somewhat less applicable to structural analysis where the number of responses is larger than the number of design variables.

The adjoint method for sensitivity analysis is a feature of *elsA* software [11] and is used by the Airbus proprietary aerodynamic optimization framework OptAlia [14].

In addition to the effort on multidisciplinary optimization, significant work was done to make the sensitivity analysis performed with the adjoint approach more robust. Specific numerical schemes were applied with artificial dissipation terms used in the iterative solution of the linear system coming from the adjoint method approach.

The work presented in this paper is a rare example of an aerodynamic optimization problem based on adjoint method sensitivity analysis and applied to a complex three-dimensional Navier–Stokes.

3. Demonstration

The optimization required approximately 20 iterations and 70 CFD analyses (including additional adjoint resolutions for sensitivity analysis). This is illustrated in Fig. 9. Each iteration requires several direct analyses because of the line search approach, which is necessary to guarantee the convergence. Anyhow, other methods exist, such as trust regions, which could be less demanding in terms of the number of CFD analyses. Also, the number of CFD analyses, which would be required with a finite difference approach, could reach about

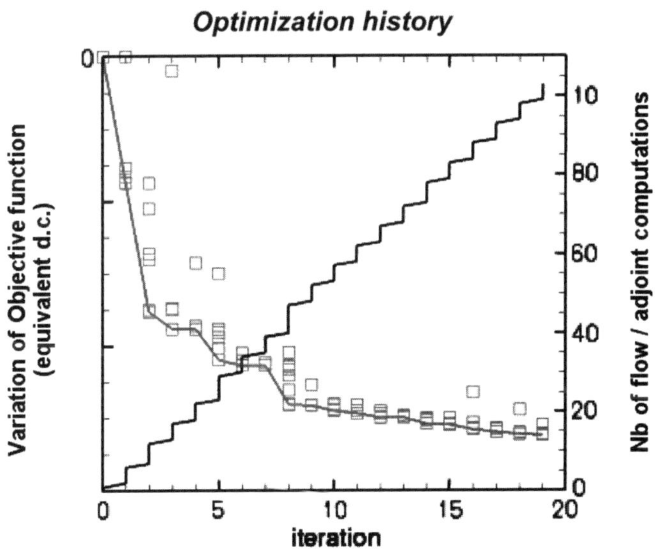

Fig. 9 Optimization history.

200. The new design obtained is illustrated in Fig. 10. The following comments are made:

1) A new shape is proposed for the inboard wing/pylon intersection.
2) The radius of curvature for the outboard intersection is increased.
3) The pylon leading edge is deflated.

These geometry changes produce a reduction of 0.6 drag count for the overall aircraft. The reduction comes essentially from the pressure drag contribution.

IV. Integration of Stress and Aerodynamic Optimization [15, 16]

A. Purpose

Proposed in this section is an optimization process aiming to optimize the powerplant installation with respect to both disciplines: structure and aerodynamics. There is a compromise between structural and aerodynamic performance. The pylon width is a good example of the compromise to be reached: the wider the pylon is, the lower the weight because of a stiffer pylon structure. However, a larger pylon width has a negative effect on aerodynamic performance because of larger velocity on the pylon's sides, which increases the magnitude of interference with wing.

The multidisciplinary optimization problem to be solved can then be defined as follows.

1. Design Variables

The test case was reduced to three parameters in order to demonstrate the methodology, including two engine position parameters, *X*-position and *Z*-position, and the pylon width. The *Y* position of the engine is not considered because it has a large effect on the overall aircraft balance and on the external loads: other disciplines should have been considered to address this new variable.

2. Objective Function

To ensure the best compromised pylon design, a multidisciplinary optimization criterion mixing weight and drag objectives has to be derived knowing

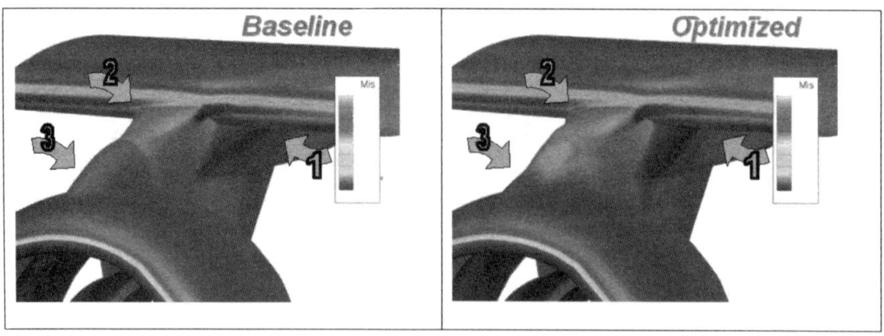

Fig. 10 Optimization results—comparison of the baseline and the optimum shape.

the relative impact of structure and aerodynamics on the aircraft performance. This impact can be expressed with tradeoff coefficients supplied by the future project office using their overall aircraft assessment tools. The tradeoff coefficients, a and b, enable the definition of a single multidisciplinary objective function as an indicator of direct operating cost (DOC):

$$\mathrm{DOC} = aD + b\tilde{W}$$

3. *Constraints*

No constraint is imposed except lower and upper bounds on the design variables, consistent with the design of experiments defined next.

4. *Loose Coupling Approach*

The purpose is to find the best engine position together with the best pylon width with respect to a combined weight and drag criterion. The design variables are illustrated in Fig. 11. Of course, weight and drag can also be optimized as explained in Sections II and III with respect to structural sizing design variables and refined aerodynamic shape design variables. Moreover, it is very clear that the sizing design variables have no significant effect on drag. In the same way, if changed in a reasonable extent, refined aerodynamic shape variables have a negligible effect on the structural behavior (stiffness).

Hence, the multidisciplinary optimization problem can be stated as

$$\min_{\alpha_{\mathrm{MD}}, \alpha_{\mathrm{struct}}, \alpha_{\mathrm{aero}}} \quad a\, D(\alpha_{\mathrm{MD}}, \alpha_{\mathrm{aero}}) + b\tilde{W}(\alpha_{\mathrm{MD}}, \alpha_{\mathrm{struct}})$$

This is mathematically equivalent to the following problem:

$$\min_{\alpha_{\mathrm{MD}}} a\, D^{*}(\alpha_{\mathrm{MD}}) + b\tilde{W}^{*}(\alpha_{\mathrm{MD}})$$

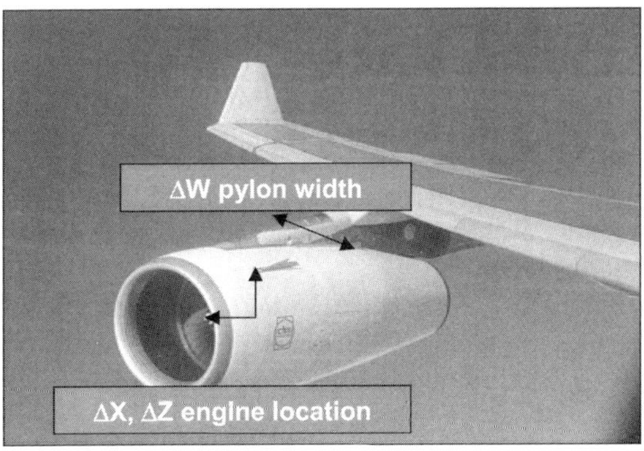

Fig. 11 Illustration of multidisciplinary design variables for the multidisciplinary optimization of powerplant.

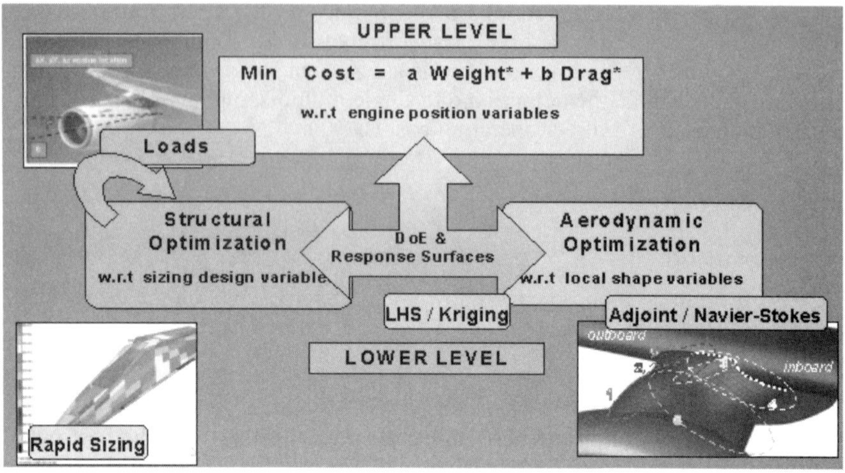

Fig. 12 Illustration of the multilevel and multidisciplinary optimization process. (See also color figure section at the back of the book.)

with

$$\begin{cases} D^{*}(\alpha_{\text{MD}}) = \underset{\alpha_{\text{aero}}}{\text{Min}}\; D(\alpha_{\text{MD}}, \alpha_{\text{aero}}) \\ \tilde{W}^{*}(\alpha_{\text{MD}}) = \underset{\alpha_{\text{struct}}}{\text{Min}}\; \tilde{W}(\alpha_{\text{MD}}, \alpha_{\text{struct}}) \end{cases}$$

There are two embedded optimization levels (see also Fig. 12): the upper level is with respect to the multidisciplinary design variables α_{MD}, and the lower level is with respect to the monodisciplinary design variables α_{aero}, α_{struct}. The loose coupling approach consists of building explicitly the function D^* and $\tilde{W}^*$ to decouple both optimizations. Of course, this is only possible by using an approximation. To do that, a surrogate model approach is used with 1) the definition of a design of experiments (DoE) related to the multidisciplinary design variables α_{MD}, that is to say, a table of various configurations of engine position and pylon width, and 2) the construction of an appropriate surrogate model for each response (drag and weight). It is well known that DoE and surrogate modelling can be combined, especially when polynomial surrogate models (also known as response surfaces) are used.

The surrogate models used for the minimum drag and the minimum weight response surfaces as presented next. For an in-depth discussion on response surfaces, please refer to the optimization chapter.

The loose coupling approach is summarized in the following algorithm depicted in Fig. 12:

1) Definition of a design of experiments for each discipline with respect to upper-level design variables

2) Optimization of each experiment with respect to lower-level design variables

3) Construction of surrogate models for lower-level optimization results with respect to upper-level design variables

4) Combination of the surrogate models

5) Optimization with respect to upper-level design variables

Such an approach can be viewed as a simplified approach of the BLISS scheme [17] and was already explored with CFD analysis (and not optimization) at the lower level in the frame of a past European project (MDO Brite-Euram project 96–97) [18].

B. Design Space

The design space is based on midrange variations of the three design variables considered. These variations are about 20% and are considered for demonstration. The impacts on the pylon shape are summarized in the Fig. 13.

C. Structure Surrogate Model

1. DoE and Surrogate Model

The surrogate model considered for the minimum weight results is a simple polynomial response surface. This choice is dictated by the fact that the weight is considered as quite smooth in the selected range (about $+/-20\%$ variation). A second-order polynomial (quadratic function) was selected to approximate the weight.

A specific DoE, the Taylor design of experiment, was defined to exactly interpolate a second-order polynomial. It was called "Taylor" because the coefficients of the polynomial can be considered as finite difference approximations of partial derivatives of the function. This method can be generalized to build any polynomial with multiple variables. The principle is illustrated in Fig. 14 in the case of three design variables.

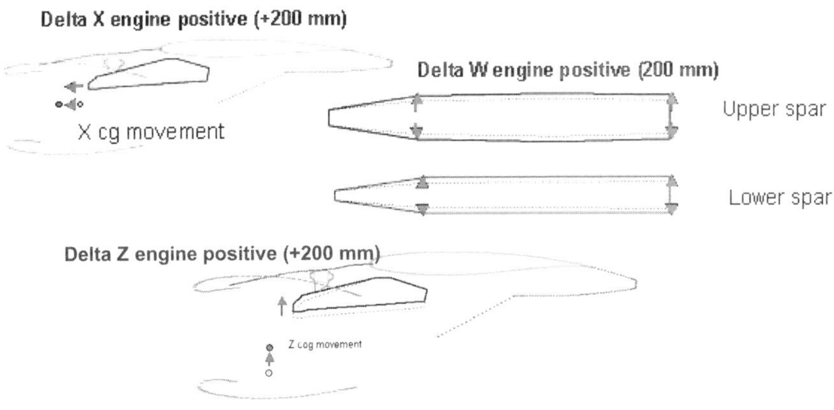

Fig. 13 Illustration of pylon shape variations.

$$
\begin{bmatrix}
0 & 0 & 0 \\
+ & 0 & 0 \\
0 & + & 0 \\
0 & 0 & + \\
- & 0 & 0 \\
0 & - & 0 \\
0 & 0 & - \\
+ & - & 0 \\
+ & 0 & - \\
0 & + & -
\end{bmatrix}
$$

constant term	a_0
first order terms	a_i
squared terms	a_{ii}
rectangular terms	$a_{ij}\,(i \neq j)$

$$\widetilde{W}(\lambda) = a_0 + \sum_{i=1,3} a_{ii}.\lambda_i + \sum_{i=1,3} a_{ii}.\lambda_i^2 + \sum_{\substack{i=1,3 \\ j=1,3 \\ i<j}} a_{ij}.\lambda_i\lambda_j$$

Fig. 14 Illustration of the design of experiment and response surface model.

The DoE is suited to a second-order polynomial response surface mode (RSM) (quadratic form) as follows:

The baseline experiment (center point) gives a_0, N (=6) positive direct effects give a_i (linear contribution), N negative direct effects give a_{ii} (squared terms), $N(N-1)/2$ interactions give a_{ij} with $i \neq j$ (rectangular terms). A total of $N(N+1)/2$ calculations are required to get a second-order RSM.

2. *Demonstration*

The lower-level design variables are defined as in Section II, and the optimization for each experiment is performed as described in Section III. The result found for the second-order interpolation is shown in Fig. 15. X_1 designates the X engine position, X_2 the Z engine position, and X_3 stands for the pylon width W. This regular linear system is solved for the unknowns a_i that give the

X1	X2	X3	(X1)²	(X2)²	(X3)²	(X1*X2)	(X1*X3)	(X2*X3)	cte							
0	0	0	0	0	0	0	0	0	1		a1		W0		902,15	
+	0	0	+	0	0	0	0	0	1		a2		W1		938,36	
0	+	0	0	+	0	0	0	0	1		a3		W2		889,99	**Weight Wi**
0	0	+	0	0	+	0	0	0	1		a4		W3		851,42	**in kg**
−	0	0	+	0	0	0	0	0	1	*	a5	=	W4	=	845,17	
0	−	0	0	+	0	0	0	0	1		a6		W5		915,77	
0	0	−	0	0	+	0	0	0	1		a7		W6		975,05	
+	−	0	+	+	0	−	0	0	1		a8		W7		963,39	
+	0	−	+	0	+	0	−	0	1		a9		W8		1036,64	
0	+	−	0	+	+	0	0	−	1		a0		W9		966,97	

0 : initial value of Xi   + : positive perturbation of Xi   −: negative perturbation of Xi

Fig. 15 Second-order polynomial interpolation of weight.

second-order polynomial. The right-hand side of the equation in Fig. 15 is obtained from NASTRAN SOL200 sizing optimizations.

The consistency of weight results is clear and can be interpreted as follows: the X position has a weight increase effect caused by a larger lever arm, the Z position has a weight decrease effect caused by a smaller lever arm, and the pylon width has a weight decrease effect caused by lower inertia. Finally, the best configuration is obtained for a minimum engine X position, whereas the worst one is found for the most forward engine and the minimum width (the most slender pylon geometry). This configuration is found when performing weight minimization based on the response surface showing also that the weight is quite a monotonous function.

The accuracy of the interpolation was then verified to be within 1% based on a random sampling of the design space. Thus, the choice of a second-order polynomial is well confirmed by this validation, and it is therefore not necessary to introduce a more complex surrogate model.

D. Aerodynamic Surrogate Model

1. DoE and Surrogate Model

The behavior of the drag surrogate model is expected to be much more nonlinear compared to the one for the weight. It was therefore decided to use a more advanced surrogate model. The choice of the aerodynamicists favored the kriging models (Fig. 16) because these are proven to reduce the level of error. More information on kriging models can be found in the optimization Chapter 6 and in [19].

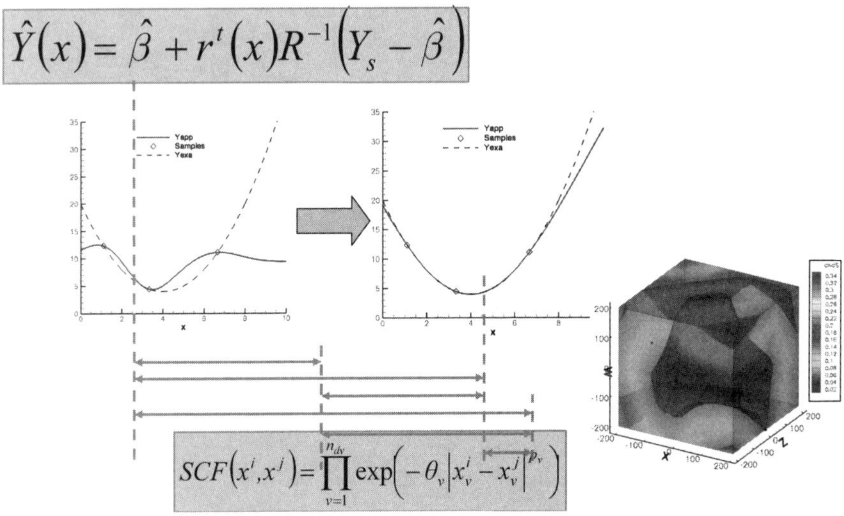

Fig. 16 Illustration of Kriging in two (educational case) and three-dimensions (our test case).

The kriging surrogate model is based on a statistical estimate of correlation between samples (matrix R in the top equation of Fig. 16) assuming a Gaussian probability distribution (function SCF in the bottom equation of Fig. 16). Depending on the parameters of this Gaussian distribution θ_ν, the smoothness of the response can be favored with respect to the accuracy of the fitting. Highly nonlinear responses can be approximated thanks to kriging.

2. *Demonstration*

A response surface obtained from kriging surrogate models is illustrated on the left side of Fig. 17. This is a dummy test, based on the variations of the wave drag of an airfoil with respect to two "bump" design variables. The figure shows that the kriging model is able to capture high nonlinearities.

A response surface for our test case is presented on the right side of Fig. 17. Because there are three upper-level variables, the values are represented in three dimensions based on a color scale.

E. MDO Demonstration

Finally, the MDO process was demonstrated by combining the polynomial surrogate model for structural weight with the kriging surrogate model for aerodynamic drag (Fig. 18). The multidisciplinary optimum is found for

$$\begin{cases} \Delta X = +90\,\text{mm} \\ \Delta Z = +200\,\text{mm} \\ \Delta W = +200\,\text{mm} \end{cases}$$

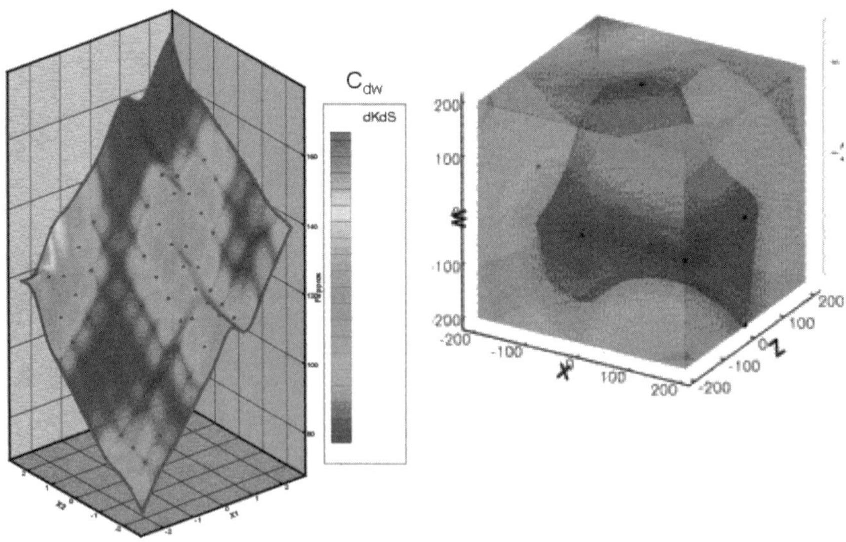

Fig. 17 Illustrations of kriging surrogate models.

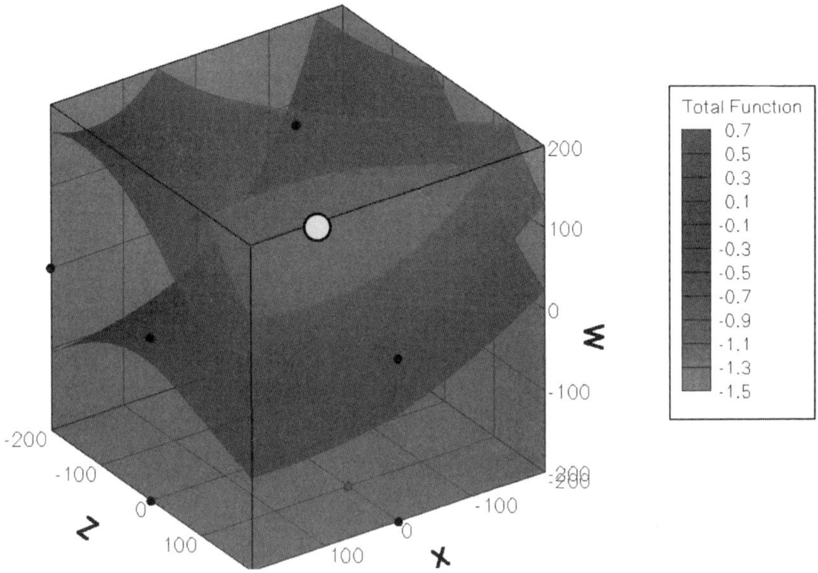

Fig. 18 MDO demonstration. (See also color figure section at the back of the book.)

This is the same optimum as the one obtained for drag in a single-objective optimization. This means that drag is the main driver of the design. It is also observed that this is a surprising configuration where an increase of pylon width benefits drag. This could be explained by a local increase of the air speed as a result of a thinner pylon.

However, it is worth mentioning that this increase in width is probably detrimental to the buffet margin at high Mach numbers, which was not examined in this framework. The overall aircraft performance improvement is given by delta weight of -77 kg and delta drag of -0.9 dc. The structure optimum was found for

$$\begin{cases} \Delta X = -200\,\text{mm} \\ \Delta Z = +200\,\text{mm} \\ \Delta W = +200\,\text{mm} \end{cases}$$

The weight saving was about -83 kg. As the multidisciplinary optimum is found by increasing W and Z, there is not only a drag reduction but also a weight reduction.

V. Integration of Stress and Loads Optimization

A. Purpose

The purpose of this section is to demonstrate a more complete structural optimization process taking benefit of the effect of stiffness on loads. The objective is to perform multidisciplinary optimization where stress and loads disciplines are considered. In this case the variables are the same and are structural design variables, but with two different effects: on structural responses (stress and internal loads) and on loads through the overall aircraft stiffness.

Various scenarios have been considered and differentiated considering the optimization of the flexibility effect. Essentially two scenarios are compared next. The first scenario is the one generally experimented across an aircraft development and is based on passive load loops, updating the sizing with respect to loads and then the loads with respect to stiffness up to convergence. The second scenario is to drive this iterative process based on a weight minimization principle.

Two kinds of loads are considered: maneuver loads that are computationally cheap and FBO loads based on nonlinear transient analysis. In the second case, as the computational process is very heavy, it is better to develop a loose coupling approach than the tight coupling approach developed for the maneuver loads. In the first case the sensitivity analysis is performed based on a semi-analytical approach; in the second case a response surface is built and is used to compute the derivatives. Both scenarios are compared for each case. The superiority of the second scenario is found to be more weight efficient as demonstrated in Section V.B.

B. Maneuver Loads [20]

The purpose was to demonstrate the stress and load interaction on a simple and easy-to-understand stress and load effect resulting from basic static aeroelastic phenomena. For this test case, it was necessary to take into account a more comprehensive structure than the pylon itself because pylon flexibility does not have a

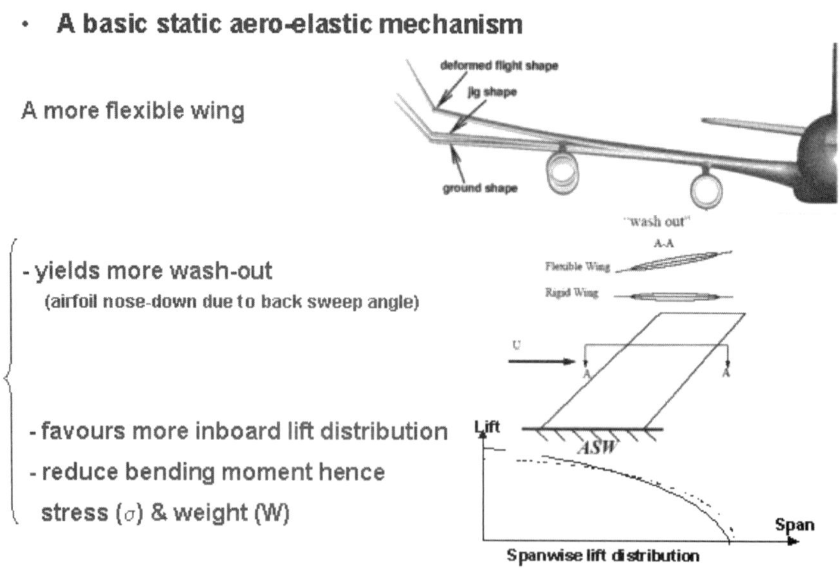

Fig. 19 Basic phenomenon for static aeroelastic load alleviation.

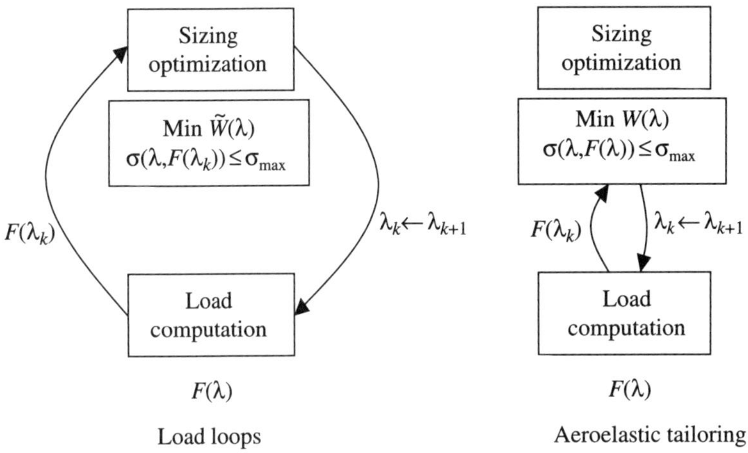

Fig. 20 Illustration of both scenarios.

significant effect on a steady maneuver load case. For this reason we extended the test case to include a wing and pylons. Considered is a maximum bending case (2.5-g pull-up maneuver), which is generally close to the most critical case for a wing. This basic phenomenon is illustrated in Fig. 19.

Given this phenomenon, we can expect from a stress and loads optimization that the trend will be to move the thickness distribution forward and towards the root in order to increase the bending and nose-down torsion flexibility, respectively. This is, in fact, what will be confirmed later, but to a different extent, depending on the scenario considered with regard to coupling stress and loads in the optimization process.

1. *Two Tight Coupling Scenarios*

The two scenarios already mentioned are described in Fig. 20. Notations are the same as for the structural optimization problem in Section III, except that here the external loads are function of sizing design variables $F(\lambda)$; k stands for an iteration index.

The "load loop" is the scenario run in an aircraft development and is performed at the later design stages (i.e., after the baseline has been defined) because the level of detail does not allow for keeping a full control of the overall aircraft optimum. The so-called "aeroelastic tailoring" scenario is the proper way of addressing the structure optimization process in the presence of loads depending on sizing. It is believed that this approach has to be promoted in early design stages to ensure the right model stiffness and loads. We can then expect that the following design stages are only changing design details, which do not affect this optimum setting and the overall aircraft optimum too much.

This dependency of loads on sizing is to be taken into account in the stress responses. Furthermore, we will see that it is necessary to add a sensitivity contribution term for structure sizing in order to obtain a proper update.

2. *Load Model and Sensitivity Analysis*

The basic equations to compute airframe external loads are the full aircraft equilibrium equations know as Lagrange equations:

$$[\mu]\cdot\{\ddot{q}\}+[\beta]\cdot\{\dot{q}\}+[\gamma]\{q\}=\bar{q}\cdot[AGFM(\omega)]\cdot\{q\}+\bar{q}\cdot[AGFC(\omega)]\cdot\{\delta\}$$
$$+\bar{q}\cdot[FGT(\omega)]\cdot\frac{U}{V}+\{F\}$$

This is the most complete representation of Lagrange equations. In our simpler case of a pull-up steady maneuver, we can reduce these equations by removing the gust term, considering the control surface effect as part of the other input forces as well as the inertia force induced by a 2.5-g acceleration; we then have

$$\gamma q=\Phi^T A\Phi q+\Phi^T f$$

with

$$\Phi^T A\Phi=\bar{q}[AGFM(\omega)] \qquad \text{and} \qquad \Phi^T f=\bar{q}[AGFC(\omega)]\,\{\delta\}-2.5[M]\{g\}+F$$

Past work at AIRBUS has shown that for sensitivity analysis, considering the modal basis as a fixed displacement basis can give satisfactory results [21, 22]. Other works have shown that it is possible to use a modal basis for static analysis and optimization [23]. Then the derivation with respect to structural sizing variables λ is quite straightforward considering that only the generalized stiffness matrix is depending on λ through the stiffness matrix K:

$$\gamma(\lambda)=\Phi^T K(\lambda)\Phi \qquad \text{gives} \qquad \frac{\partial\gamma(\lambda)}{\partial\lambda_i}=\Phi^T\frac{\partial K(\lambda)}{\partial\lambda_i}\Phi$$

And finally considering the precedings equation, we obtain

$$\frac{\partial q}{\partial\lambda}=-\left(\gamma-\Phi^T A\Phi\right)^{-1}\Phi^T\frac{\partial K}{\partial\lambda}\Phi q$$

The derivation of the loads is then computed as

$$\frac{\partial F}{\partial\lambda}=A\Phi\frac{\partial q}{\partial\lambda}$$

It is important now to understand how these sensitivities contribute to the structural optimization process. If we consider, as we did up to now, that the sizing optimization constraints can be formulated as stress allowables, the purpose is to correctly represent the double dependency of stresses with respect to sizing:

$$\sigma = \sigma[\lambda, F(\lambda)]$$

To represent this double dependency as all responses are approximated in the optimization process by their values and their first-order derivatives, it is sufficient to provide to the optimization algorithm the values of stresses (with the proper external loads update) and their proper derivatives:

$$\frac{\mathrm{d}\sigma}{\mathrm{d}\lambda} = \frac{\partial\sigma}{\partial\lambda} + \frac{\partial\sigma}{\partial F}\frac{\partial F}{\partial\lambda}$$

To perform a proper aeroelastic tailoring, we just then need to add to the standard sensitivities the complementary term $\partial\sigma/\partial F\,\partial F/\partial\lambda$, which is the "flexibility effect" on stress responses.

It is this term, which has to be used to reduce the stress levels and then the weight.

Moreover, this term is quite easy to compute if we remark that $\partial\sigma/\partial F$ is nothing else than the linear operator that allows computing stresses from external loads. The complementary sensitivity term is in fact made of additional static linear analysis based on the pseudoload $\partial F_k/\partial\lambda_i$ (as many as load cases time design variables).

Thus, the full sensitivities can be computed from a SOL200 and a SOL101 with multiple load cases.

3. *Implementation*

The full computational process for the aeroelastic tailoring solution is illustrated in Fig. 21. As can be seen in Fig. 21, various software solutions are used:

1) MATLAB® is used to compute loads and their sensitivities from structural (modal basis and generalized stiffness matrix) and aerodynamic information (aerodynamic influence coefficients).

2) NASTRAN is used to compute stresses and their sensitivities as well as the data for loads. Proprietary FORTRAN code is used to finalize the stress sensitivity analysis.

For this reason, an optimization software framework was used to integrate the full process and to connect it to an optimization algorithm based on first-order information (values and their sensitivities). The BOSS Quattro software [24, 25] was found adequate for this task, particularly because of its convex linearization approaches, which requires only a first-order information for the objective and constraints functions.

For the sake of simplicity, the aeroelastic analysis was implemented in a single script, which is directly connected to the optimization algorithm (see central window in Fig. 22). Nevertheless it is possible to make it appear as a decomposed task tree showing all steps of the process. Windows shown on the left and right of Fig. 22 show the capture of results and the formulation of the optimization problem (objective function and constraints).

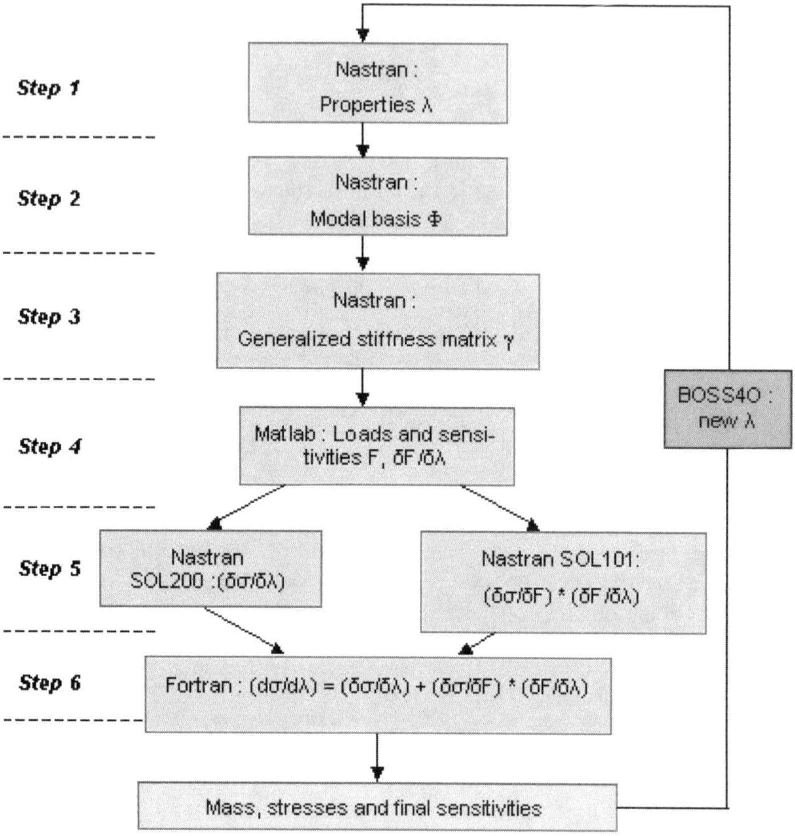

Fig. 21 Illustration of the aeroelastic tailoring process.

It is very simple to simulate the load-loop approach based on the aeroelastic tailoring implementation: one just has to cancel the load sensitivities, and everything works as if there were just a sequential structure optimization process with load updates at each iteration. However, to strictly implement the load-loop scenario each structure optimization should be completed before load updating, and the loads should not be updated at each iteration. Meanwhile some experiments performed in parallel of this study have shown that the results are worse with such an approach and that a more premature convergence is observed. It is easy to understand that it is better to perform the load update at each iteration in order to obtain a smoother variation of loads.

4. Demonstration

As already stated, the demonstration was made on a wing with pylon. The principle of variable definition already presented in Section III was extended to

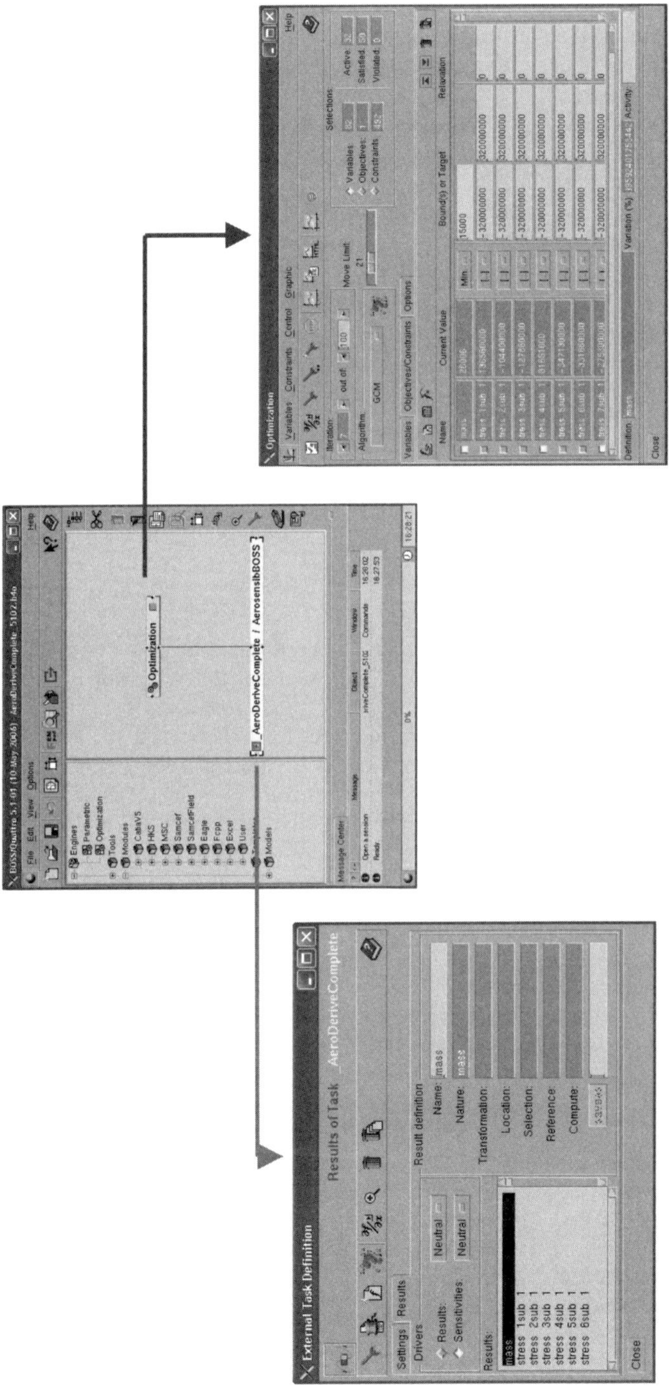

Fig. 22 Aeroelastic tailoring process: BOSS Quattro implementation.

the wing. The optimization problem definition is similar to the one formulated for the pylon optimization. An illustration of the 82 regions of the wing defined to support the optimization design variables is presented in Fig. 23 resulting in a 82-variable design space.

A simplified optimization problem was formulated with a longitudinal stress allowable of approximately 320 MPa on lower skin and −320 MPa on upper skin. A representative element was selected in each region in order to simplify the optimization to an X variable design space. As a result, the full structure is not guaranteed to be feasible at the end of the optimization process: some elements can slightly violate their respective constraints. However, this simplification was found acceptable in terms of representative sizing and weight.

Six load cases are considered: $+2.5g$, Mach 0.5; $+2.5g$, Mach 0.85; $+1.67g$ + balanced sideslip, Mach 0.5; $+1.67g$ + balanced sideslip, Mach 0.85; $-1g$, Mach 0.5; and $-1g$, Mach 0.85. Three scenarios have been run for comparison: optimization with fixed loads, optimization with flexible loads following the load-loops scenario, and optimization with flexible loads following the aeroelastic tailoring scenario.

The results obtained are presented in Figs. 24–26. Note that the colors are related to a scaling factor with respect to the initial sizing in each region. This scaling factor was bounded between 0.7 and 1.5. The white regions in the following figures then refer to a part of the wing 30% more flexible than the original one.

The effect of the load update on the structural optimization process is quite clear. As anticipated in the introduction of this section, the wing tip is made more flexible with less material on the rear part of the wing.

As explained in Fig. 19, this is a classical effect for a backward swept wing. Because of the sweep angle, a larger vertical deflection is seen in the flight direction as a larger nose-down twist of each airfoil. This larger nose-down twist involves a lift reduction and a bending moment reduction. Moreover, the fact that the thickness is reduced at the rear part of the wing box (chord-wise) leads to an increase of the nose-down effect (because of differential bending). The weight saving as a result of load updates is about 1.4 tons.

When applying the aeroelastic tailoring process including load sensitivities, the same phenomenon appears for sizing redistribution, but to a greater extent.

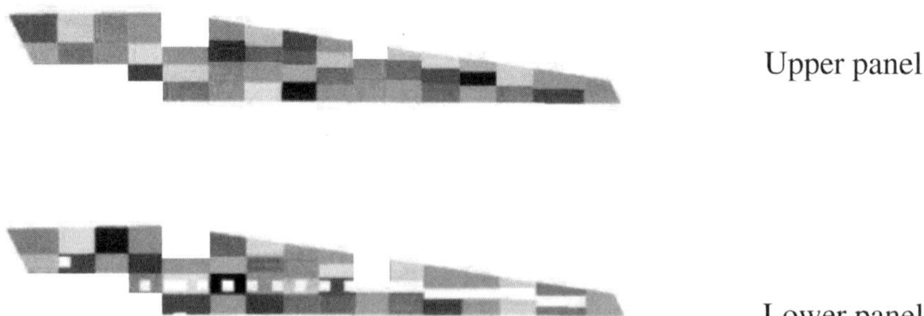

Fig. 23 Illustration of regions for design variables.

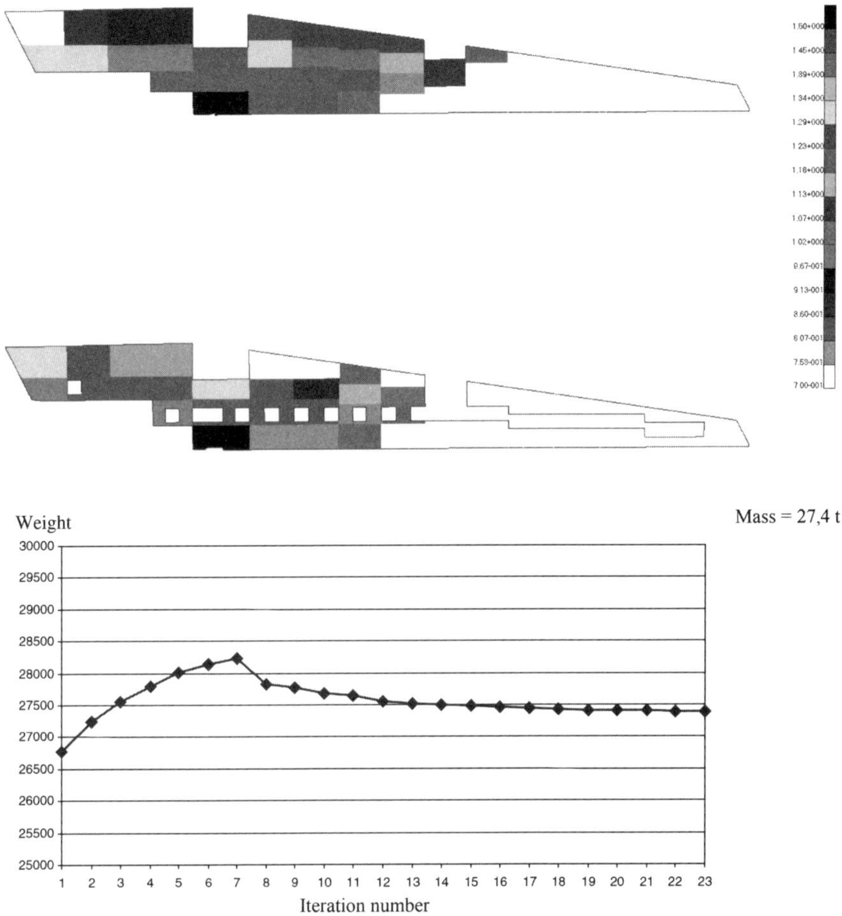

Fig. 24 Optimization with fixed loads.

This results in an additional weight saving of 200 kg. This demonstrates the benefit of a proper stress and loads optimization including load sensitivities.

C. Fan-Blade-Off Loads [26]

The optimization process demonstrated in the preceding section is much more difficult to address when dealing with complex dynamic loads. An attempt was made to integrate such dynamic loads in an optimization process comparable with the one illustrated in the preceding section in case of steady maneuver loads. Again the interest is to investigate a proper treatment of the flexibility effect on loads, via a comparison of a load-loop approach with an aeroelastic tailoring approach.

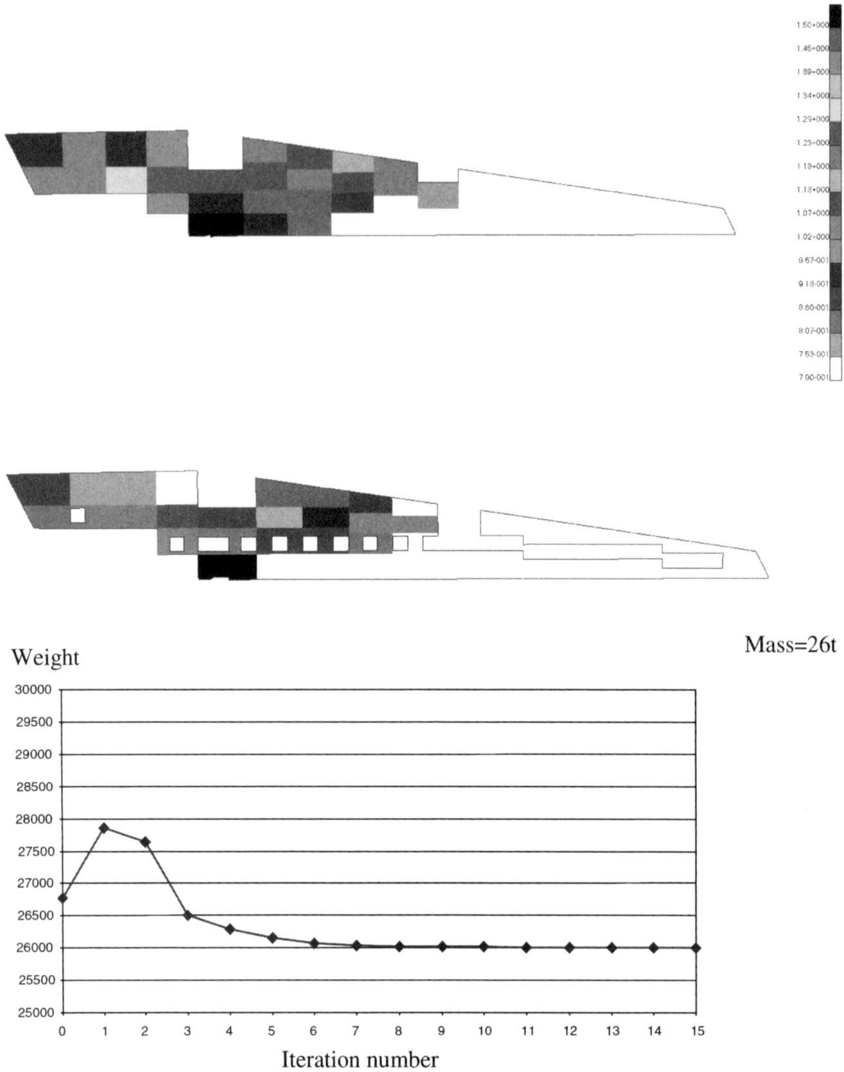

Fig. 25 Optimization with load loops.

1. Tight or Loose Coupling for Both Scenarios

For this purpose, it was decided to base the investigation on probably one of the most complex dynamic load simulation processes: FBO loads. These loads result from the break of a fan blade and the resulting highly nonlinear dynamical behavior of the powerplant. Two aircraft mass configurations and four blade release angle scenarios (0, 90, 180, 270 deg) are considered. Therefore, FBO simulations require for their computation nonlinear transient dynamic analyses (based on NASTRAN SOL129) with hours of computation on a 16-processors machine.

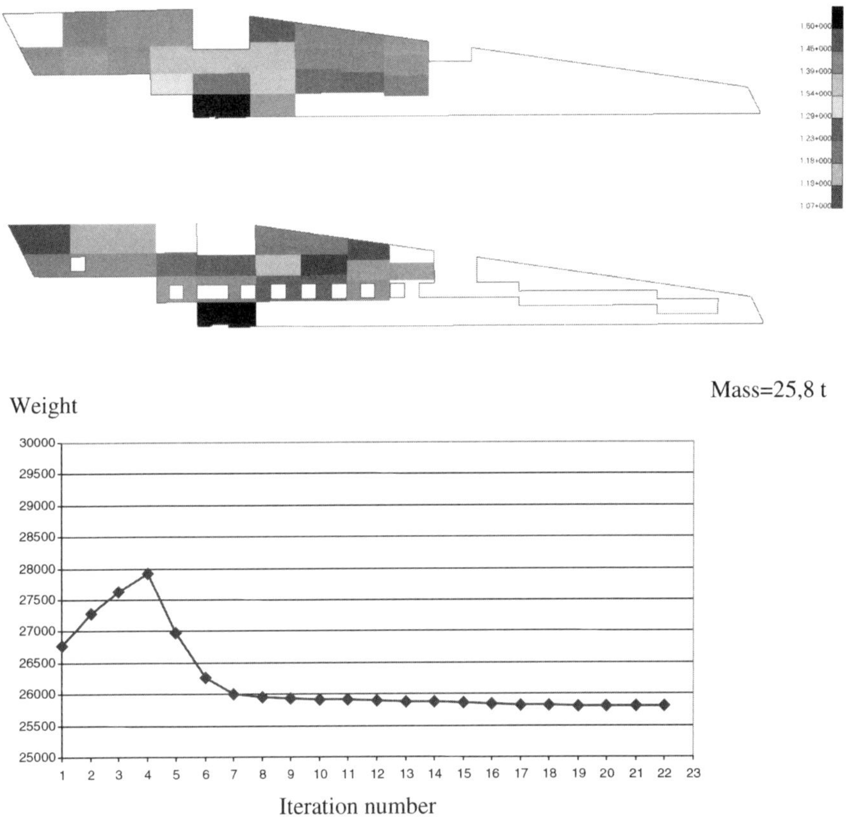

Fig. 26 Optimization with load sensitivities (aeroelastic tailoring).

Moreover, because of the complexity of equations and their nonlinearity (including shocks), it is very difficult to compute analytical sensitivities. Addressing these sensitivities with finite differences inside an optimization process for which we do not know the number of iterations required as well as its robustness was not found to be very realistic. This is the reason why the work was directed toward a loose coupling approach, that is to say, a surrogate model approach. With such an approach, the number of calculations to be made is strictly under control and is based on a predefined table of configurations (DoE).

2. *Load Model and Surrogate Model*

Load model. Again Lagrange equations for the full free-free aircraft constitute the basis for load calculations:

$$[M]\{a\} + [B]\{v\} + [K]\{u\} = \{P\} + \{N\}$$

Here the Lagrange equations are considered in the time domain, with structure nonlinearities linked to shocks, contacts, and gaps. The simulation is performed with NASTRAN SOL129 (no semi-analytical sensitivities available as for most of iterative solutions of the commercial-off-the-shelf (COTS) software solver used) based on a finite element model of the full powerplant (see Fig. 27).

In this case, the effect of the sizing design variables is not through an aeroelastic phenomenon but through a dynamic effect via the stiffness matrix. However, this effect is known to be significant, and a more flexible pylon usually gives reduced FBO loads. This is what is looked for in this study.

This time, however, it is not possible to use a modal basis reduction because of the nonlinearity of the behavior. The only model reduction possible is a modal synthesis with a Guyan reduction of the rest of the aircraft.

The FBO simulation process is then made of two stages:

1) The first stage is a model reduction of the engine and the pylon with SOL 103. The external engine and the wing are already reduced (Guyan reduction).

2) Stage two is a nonlinear transient solution (SOL 129) [27]. This latter calculation is divided into two "subcases":

a) The first stage represents the engine evolution to the maximal thrust. There is no failure at this step. This subcase establishes the initial condition of the next subcase.

b) The second subcase is the FBO properly. It corresponds to the engine failure.

The computational process to access to FBO loads is then illustrated in Fig. 28.

Different results are requested from the MSC Nastran solver at the end of the nonlinear transient solution. Two types of FBO results are considered to define critical loads because all pylon attachments have to be taken into account: pylon-to-wing and pylon-to-engine.

For each attachment, forces and moments are measured at each time step of the nonlinear transient solution. They are transferred to the center of gravity of the engine, which is the reference point for the pylon structure loading. Then a load envelope is computed. A load is selected when one force or moment component is extremum for any of the four blade release angles.

Surrogate model. To build a surrogate model, it was decided to use a reduced parameterization of the pylon in order to focus only on stiffness

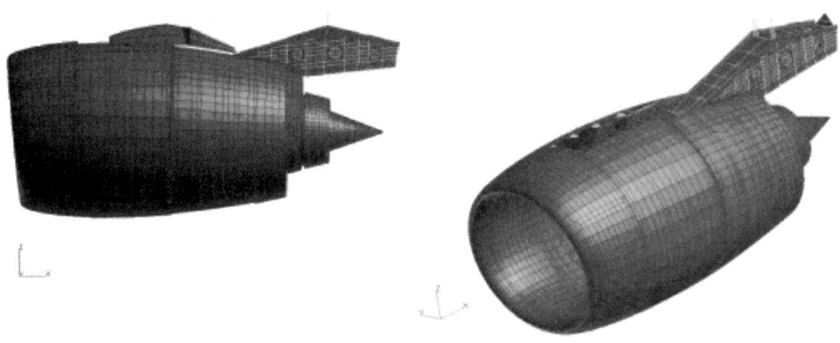

Fig. 27 Powerplant finite element model for FBO simulation.

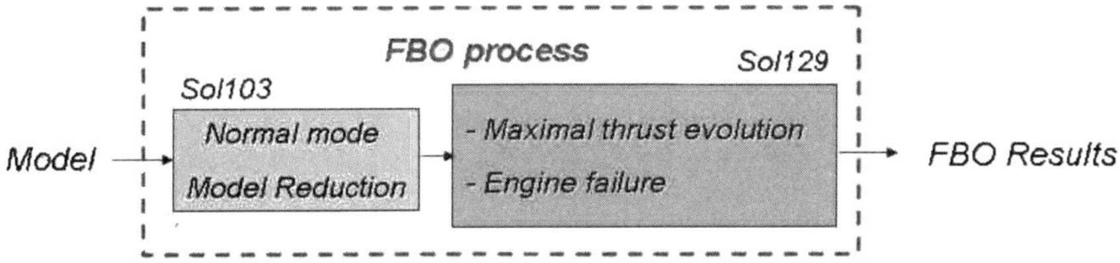

Fig. 28 FBO simulation process.

drivers. In reality, this reduced parameterization should be mixed with a finer one, focused on stress control, but for sake of simplicity in this first attempt, it was decided to take the same parameterization for stiffness and stress control. The parameterization is reduced to the six design variables illustrated in Fig. 29.

It was decided to use the same kind of DoE and second-order polynomial as used to represent the weight in the stress and aerodynamic optimization (see Section III). As shape variations are reasonable (+/−20%), it was expected that a second-order Taylor approximation could be acceptable in this range. The setting of the "Taylor" DoE is shown in Fig. 30. The same DoE is used for the 12 components of the FBO loads (at engine and wing interfaces).

The second-order polynomial surrogate model was then checked by plotting two variable cuts and comparing with engineering judgement (Fig. 31) and comparing with exact results for one configuration. The approximation error was considered as acceptable for the demonstration. Normally the second check should have been done for a random sampling with a "sufficient"

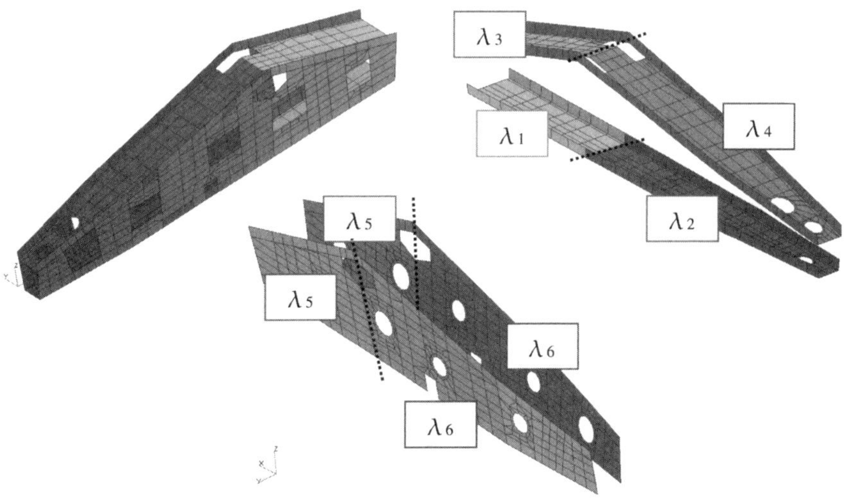

Fig. 29 Illustration of sizing design variables.

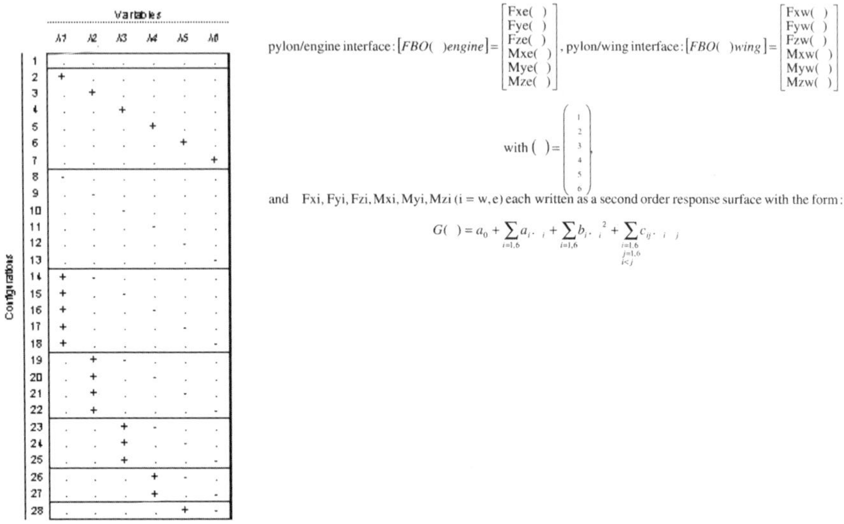

Fig. 30 Illustration of the design of experiment and response surface model.

number of individuals. But time constraints have precluded the accomplishment of this comparison.

3. *Implementation*

As for maneuver loads, the optimization process was integrated in COTS optimization software BOSS Quattro (Fig. 32).

4. *Demonstration*

The comparison of the two approaches (load loops and aeroelastic tailoring) indicates that finally they produce the same results. This can be explained by the fact that the FBO variations with respect to stiffness variables are quite

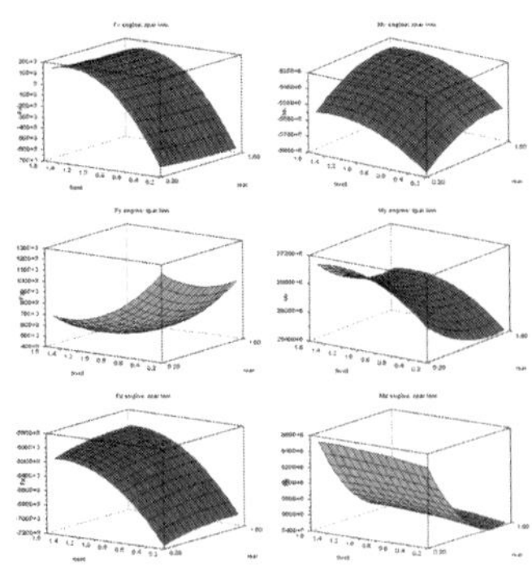

	Numerical calculation	Response Surface	Variation
Fx (N)	4,95E+03	4,54E+03	8%
Fy (N)	3,50E+04	3,56E+04	-2%
Fz (N)	-6,13E+05	-6,10E+05	0%
Mx (N.mm)	-5,48E+08	-5,49E+08	0%
My (N.mm)	2,70E+09	2,71E+09	-1%
Mz (N.mm)	5,89E+08	6,05E+08	-3%

Comparison for $\lambda = (0.8\ 1.2\ 0.8\ 1.2\ 0.8\ 1.2)^T$

Fig. 31 Surrogate model checking.

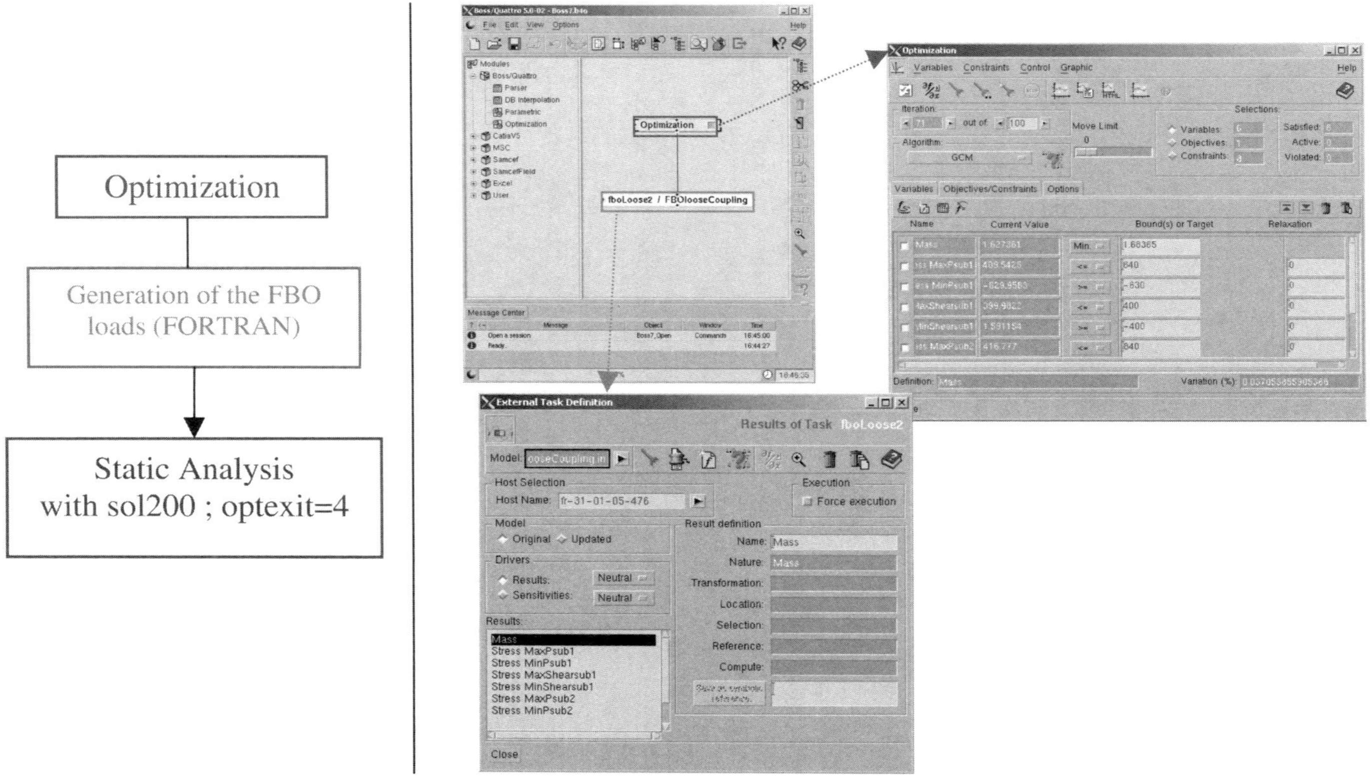

Fig. 32 Integration of the FBO optimization process inside COTS software.

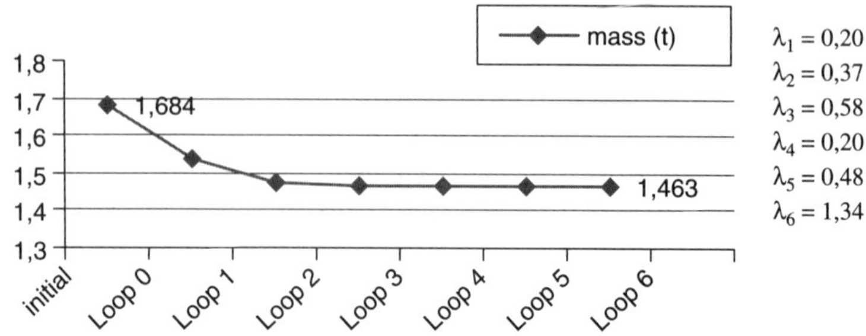

Fig. 33 Convergence of the FBO loads and stress optimization process.

monotone. The convergence is very fast in three to four iterations as shown in Fig. 33.

VI. Summary and Conclusions

In this research work, two main bidisciplinary optimization processes have been considered: stress and aerodynamic optimization and stress and loads optimization. The main reason why only two disciplines have been considered is because otherwise it would have been too ambitious for practical purposes to involve more disciplines in a first attempt.

In each case, appropriate methods have been developed, linked to the level of coupling between disciplines and to the complexity of the involved computations. The main findings are as follows:

1) *Stress and aerodynamic optimization* can be easily split in a bilevel optimization process as a result of the weak interaction between structure sizing variables and refined aerodynamic shape variables. Single discipline optimizers have been used to derive optimum results with respect to these variables. Then to solve the MDO problem for coupled variables, an approach based on design of experiments and response surface models has been successfully used. These DoE/RSMs have been adapted to the complexity of the responses using Taylor DoEs and second-order polynomial for structural weight (low nonlinearity) and Latin hypercube and kriging models for the (highly nonlinear) aerodynamic drag. The method was then successfully demonstrated on an engine powerplant. The lesson learned is that this approach is very simple, keeps discipline autonomy, and benefits from fit for purpose single discipline optimization capabilities. Thus, it is very suited to a rapid industrial implementation.

2) *Stress and loads optimization* has been studied for two kinds of loads with different levels of complexity, that is, steady maneuver loads and FBO loads. In the first case it was possible to utilize a full analytical approach based on gradients where loads have been differentiated and sensitivities have been combined with structure sensitivities. The combined sensitivities were then used in a gradient-based optimization process embedding the full analysis chain. In the second case, FBO loads were obtained from a very complex nonlinear simulation, which was possible, but difficult to differentiate. Considering also the computational

times, it was found to be more efficient to build a response surface model for a small number of parameters.

In both cases, a comparison was made between a standard load-loop scenario and an approach driven by load sensitivities. The second approach was found to be better, especially in the case with maneuver loads, resulting in a significant weight benefit. This is an important result strengthening the case for an industrial implementation of this approach and leading to an improvement of the aerostructure optimization process.

Significant progress has been made in this research work regarding the understanding of interactions between disciplines. Dedicated approaches for DoE and RSM have been developed and successfully realized. Complex gradient-based optimization processes with analytical differentiation have been developed and demonstrated. Significant progress has also been made in the frame of monodisciplinary optimization, which is now at an automation level, particularly in the case of aerodynamic optimization, where many improvements have been made to the adjoint method for sensitivity analysis. These improvements were necessary to address the three-dimensional Navier–Stokes simulations of powerplant. All of these achievements contribute to building industrial capability for a numerical MDO process.

Regarding perspectives for future research and development, although both bidisciplinary processes described in this chapter are meaningful in terms of aircraft application, it is clear that an added value would be to couple stress, aerodynamics, and loads in order to achieve a full aerostructure numerical optimization process. This is particularly relevant to wing optimization for which the stiffness can also have an impact on drag performance (multipoint optimization) [28]. For a powerplant the coupling between structure and aerodynamics through stiffness is negligible.

Consideration of active load-alleviation mechanisms (through control surfaces) is another important topic targeting a full aeroservoelastic numerical optimization capability [29].

References

[1] "European Aeronautics: A Vision for 2020," ACARE (Advisory Council for Aeronautic Research in Europe), Office for Official Publications of the European Communities, C-2985, Luxembourg, Jan. 2001.

[2] Bècle, J. P., and Mogilka, P., "Powerplant Integration for Transport Aircraft at Cruise Conditions—French Experience," ONERA French Aerospace Lab, Technical Paper 1996-8, May 1995.

[3] Garcia, A., "Intégration des Systèmes Propulsifs, Choix et Compromis," *colloque AAAF Aéropropulsion 1990*, Paris, France, 20–21 March 1990.

[4] Grihon, S., "Application of Structural Optimisation to AIRBUS Engine & Pylon Designs," *AERO 2002 MSC Software Worldwide Aerospace Conference & Technology Showcase*, Paper 122, Toulouse, France, 8–10 April 2002.

[5] Moore, G. J., "MSC Nastran User's Guide for v68: Design Sensitivity and Optimization," MSC.Software Corp., Santa Ana, CA, April 1994.

[6] Finette, S., "Presentation of Parametric Model of A380 Engine Power Plant," *AERO 2002 MSC Software Worldwide Aerospace Conference & Technology Showcase*, Paper 104, Toulouse, France, 8–10 April 2002.
[7] Van der Vooren, J., and Destarac, D., "Drag/Thrust Analysis of Jet-Propelled Transonic Transport Aircraft: Definitions of Physical Drag Components," *Aerospace Science and Technology*, Vol. 8, No. 7, Oct. 2004, p. 671.
[8] Hicks, R. M., and Henne, P. A., "Wing Design by Numerical Optimisation," *Journal of Aircraft*, Vol. 15, No. 7, 1978, pp. 407–412.
[9] *D.O.T User's Manual*, Vanderplaats Research & Development, Colorado Springs, CO, 1995.
[10] Zoutendjik, G., *Method of Feasible Directions*, Elsevier, Amsterdam, The Netherlands, 1960.
[11] Lazareff, M., Vuillot, A-M., and Cambier, L., *Manuel Théorique ElsA (ElsA Theoretical Manual)*, ONERA French Aerospace Lab, Chatillon Cedex, France, 2002.
[12] Jameson, A., "Aerodynamic Design via Control Theory," *Journal of Scientific Computing*, Vol. 3, No. 3, Sept. 1988, pp. 233–260.
[13] Peter, J., "Discrete Adjoint Techniques in elsA (Part I): Method/Theory," *Proceedings of 7th ONERA/DLR Aerospace Symposium*, ODAS 2006, Toulouse, France, 4–6 Oct. 2006.
[14] Meaux, M., and Cormery, M., "Viscous Aerodynamic Shape Optimization Based on the Discrete Adjoint State for 3D Industrial Configurations," European Congress on Computational Methods in Applied Sciences and Engineering (ECCOMAS) 2004, July 2004, http://www.mit.jyu.fi/eccomas2004/proceedings/pdf/897.pdf [retrieved July 2009].
[15] Mouton, S., Carrier, G., Meaux, M., and Laurenceau, J., "Multidisciplinary Optimizaton of Powerplant," VIVACE, Forum II, World Forum Convention Centre, Den Haag, The Netherlands, 24–26 Oct. 2006.
[16] Mouton, S., Laurenceau, J., and Carrier, G., "Aerodynamic and Structural Optimization of Power Plant Integration on the Wing of a Transonic Transport Aircraft," *42ème colloque d'Aérodynamique Appliquée AAAF*, Sophia-Antipolis, France, 19–21 March 2007.
[17] Sobieszcanski-Sobieski, J., Agte, J. S., and Sandusky, R., Jr., "Bi-Level Integrated System Synthesis (BLISS)," AIAA Paper 1998-4916, Sept. 1998.
[18] Grihon, S., and Cormery, M., "Multidisciplinary Optimization Applied to a 500 Seat A/C," *Proceedings of the 4th ECCOMAS Conference*, European Congress on Computational Methods in Applied Sciences and Engineering (ECCOMAS), edited by K. D. Papailiou et al., Wiley, 1998.
[19] Jones, D. R., Schlonlau, M., and Welch, W. T., "Efficient Global Optimization of Expensive Black-Box Functions," *Journal of Global Optimization*, Vol. 13, No. 4, Dec. 1992, pp. 455–492.
[20] Grihon, S., "Stress & Loads Optimisation of a Wing & Pylon with Manœuvre Loads," *VIVACE Forum 3*, Oct. 2007, http://www.vivaceproject.com/content/forum3/6.3%20Grihon.pdf [retrieved July 2009].
[21] Grihon, S., and Esquerré, J. P., "A Reduced Basis Model for Aeroelastic Optimization," *CEAS/AIAA/ICASE/NASA Langley International Forum on Aeroelasticity and Structural Dynamics*, June 1999, Pt.1, pp. 325–332.
[22] Grihon, S., Mahé, M., and Seigneuret, F., "Une Méthode de Réduction pour la Réanalyse Modale et Aéroélastique de Structures Avion," *Actes du Colloque*

National en Calcul des Structures de Giens, CSMA-AFM, Giens, France, 1999, pp. 173–178.

[23] Karpel, M., "Stress Consideration in Reduced-Size Aeroelastic Optimization," AIAA Paper 94-1480-CP, 1994.

[24] *BOSS Quattro User's Handbook, Release 6.0-02*, SAMTECH, Liège (Angleur), Belgium, 2008.

[25] "BOSS Quattro: An Open System for Parametric Design," *Structural and Multidisciplinary Optimization*, Vol. 23, No. 2, March 2002, pp. 140–152.

[26] Grihon, S., "Pylon Design Optimisation," *VIVACE Forum 1*, Sept. 2005, http://www.vivaceproject.com/content/forum1/4-5.pdf [retrieved July 2009].

[27] "MSC.Nastran SOL129 Explicit Nonlinear Users Guide," MSC.Software Corp., Santa Ana, CA, 1994.

[28] Martins, J. R. R. A., Alonso, J. J., and Reuther, J., "Aero-Structural Wing Design Optimization Using High-Fidelity Sensitivity Analysis," Confederation of European Aerospace Societies (CEAS), Cologne, Germany, June 2001, pp. 211–217.

[29] Krammer, J., Höhnlinger, H. G., and Stettner, M., "MDO Technology Needs in Aeroelastic Structural Design," AIAA Paper 1998-4731, Sept. 1998.

Chapter 9

Engine MDO Deployed on a Two-Stage Turbine

Roland Parchem* and Bernd Meissner*

Rolls-Royce Deutschland, Dahlewitz, Germany

Marc Nagel†

MTU, Munich, Germany

and

Ernst Kesseler‡

Katern 09, Amsterdam, The Netherlands

Abbreviations

ASA = adaptive simulated annealing
B2B = business-to-business connection
CAD = computer-aided design
CAE = computer-aided engineering
CFD = computational fluid dynamics
COTS = commercial-off-the-shelf
DoE = design of experiments
FE = finite element
GSP = gas turbine simulation program
GUI = graphical user interface
HPT = high-pressure turbine
IGES = initial graphics exchange specification
IPR = intellectual property rights
IT = information technology

*Mechanical Methods.
†Aerodynamic Methods.
‡Director.

MDO = multidisciplinary design optimization
MTO = maximum takeoff
NSGA = nondominated sorting genetic algorithm
PLM = product life-cycle management
POF = probability of failure
RD = robust design
RMSE = root-mean-square error
SaS = share a space, EUROSTEP PLM server in Stockholm
SFC = specific fuel consumption
SQP = sequential quadratic programming

Nomenclature

$c(T)$ = stress coefficient
F = robust function
G = robust constraint function
gk = kth inequality constraint
$m(T)$ = stress exponent
N_f = number of life cycles
T = temperature
Xi = design variable i
$\sigma(T)$ = temperature-dependent stress

I. Introduction

A MULTILEVEL automated distributed design process in a collaborative engineering environment considering multidisciplinary design optimization (MDO) and robust design (RD) is a major step forward with respect to the current state-of-the-art but also a big challenge. The intention within a European virtual enterprise was to develop and demonstrate distributed collaborative design studies embedded in the context of an integrator and a number of connected suppliers. In this context, individual design frameworks operate at each participating partner of the virtual enterprise with the necessary optimization and robust design tools and processes based on specific needs and existing expertise. Linking these individual design frameworks together, in order to create a combined integrated design framework, offers the possibility to share the optimization and robust design tasks, allowing many design iterations within one closed loop. In contrast, the current way of collaborative engineering involving multiple partners is much more fragmented and manually driven (exchange of design request via emails and paper documents), which results in a limited number of design iterations within the given time schedules and consequently limited design space exploration.

This chapter describes the aeroengine MDO and RD deployed on a virtual two-stage turbine in the virtual enterprise as introduced in Chapter 1. The MDO objectives have been aligned to the aeroengine customer requirements, such as better performance, lower weight, longer life cycles, and lower costs. Such requirements are the translation of the vision 2020 [1] objectives. Aerodynamic and thermomechanical analyses are required in order to assess these multiple objectives. The investigated two-stage turbine assembly consists mainly of discs and turbine blades. The analysis, including optimization and robustness studies of these components, was distributed within the virtual enterprise. The optimization and robust design tasks were performed individually as well as coupled between the partners using flexible interfaces controlled by the integrator. The individual design frameworks for optimization and robust design at each partner (locally) were connected to the main workflow, which was controlled and activated by the integrator (globally) as described in Chapter 10 while respecting all security aspects and proprietary rights. The common data access (interface data) between the partners was organized using a virtual enterprise collaboration hub architecture concept. The analyses at each contributing partner were performed using computer-aided engineering (CAE) tools and also surrogate modeling techniques, such as, for example, the response surface method. The latter enabled the speeding up of the optimization and robust design processes for the parameterized two-stage turbine assembly. The design process was demonstrated for a coupled preliminary design and detailed design phase in the first step and for a design concept study in the second step. Thus an aeroengine business-case-related scenario could be demonstrated online in a collaborative environment. Particularly a solution was realized for a global integrating framework, including MDO and RD capabilities, based on individual design framework solutions from different partners.

The following sections will describe the requirements, the overall workflows, the individual partner's optimization and robust design tasks, and the experience gained from the engine MDO and RD in a virtual enterprise.

II. Requirements for Engine MDO and Robust Design in the Virtual Enterprise

One major requirement for a cross-partner analysis is the availability of (local) generic analysis processes, respectively design frameworks, suitable for optimization and robust design purposes. Historically the contributing partners have developed individual design frameworks for MDO and RD. Each partner has implemented its own company-specific tools, such as computational fluid dynamics (CFD), computer-aided design (CAD), finite element (FE), data evaluation tools, optimization, and robust design algorithm, into an automated design analysis process. Different process integration software such as commercial-off-the-shelf (COTS) software (data available online at http://www.engineous.com), proprietary solutions (data available online at http://www.spineware.nl/frm-pr.html) and UNIX scripting were used by the partners. MDO and RD expertise was first demonstrated on the component level and then extended to the assembly level. The partners have gained experience in MDO and RD strategies and surrogate modeling techniques and were faced with all related problems, such as stable CAD modeling, stable parameterization, stable FE meshing, data evaluation, and dealing with a large number of parameters within an adequate analysis time. Figure 1 shows an example of a local design processes, which includes a parameterized model of a component or assembly (CAD); thermomechanical analysis of the component or assembly (FE); and data evaluation—life cycle and mass calculation of the component or assembly to be used as objective functions.

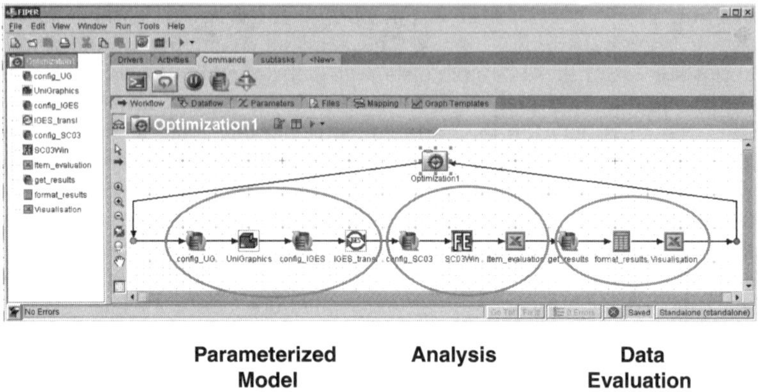

Fig. 1 Example of an automated local design analysis process.

A. MDO and Robust Design Objectives

A virtual two-stage turbine was the subject of investigation as shown in Fig. 2. The two-stage high-pressure turbine consists of two discs with blades and two rows of nozzle guide vanes. High temperature and pressure distributions in addition to centrifugal forces are the main loadings on the turbine, which is why such technically demanding assembly has been selected for this case study.

The major design objectives for the turbine are the increased performance of the turbine (i.e., reduction of the loss factor), reduced weight, and increased life cycles of the components (discs, blades, and vanes).

The major constraints are expressed as geometry dimensions limitations, material strength and temperatures limitations, required robustness and reliability [expressed as probability of failures (POF)], and costs (selection of material, manufacturing capabilities).

The required analyses include the aerodynamic analysis of the blades and vanes and the thermomechanical analysis of the blades, vanes, and discs.

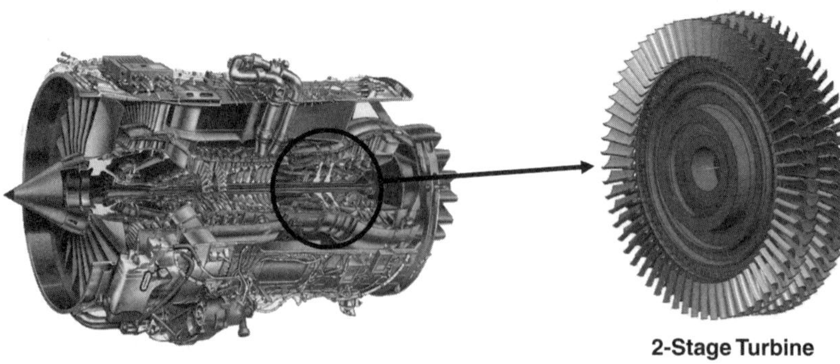

Fig. 2 Two-stage turbine section of an aeroengine.

Robust design and optimization methods are included in the automated design and analysis process.

The results of the analyses, such as the stress value distribution in the disc, are used to calculate the design objectives, for example, the life cycles of the disc, which in turn requires life-cycle assessment methods and data (e.g., lifing methods, material data, etc.). These are usually the intellectual property of each company.

The design is influenced by uncertainties and variability with regard to the manufacturing tolerances, material properties, the loading conditions, environmental changes, any noise factors, and so forth. These uncertainties have to be considered for the robustness and the reliability of the design solution. The probability density functions of the variability and uncertainty are used to determine the variation on the objectives respectively to assess to what extent the customer requirements will be met.

Finally, the multiobjective optimization provides Pareto front results, for example, mass (weight) vs nominal life (life cycles) as illustrated in Fig. 3. Such tradeoff points can be assessed against the customer requirements in order to define a design solution. The uncertainties and variability of the design influencing parameters *Xi* as shown in Fig. 3 (manufacturing tolerances, materials property distributions, etc.) have to be taken into account in order to determine the probability density functions for the objectives *Yi* (life cycles, weight). This allows one to assess to which percentage the customer requirements will be met.

B. Modeling and Parameterization

A successful optimization process requires adequately parameterized models of the components or the assembly. In this view, the analysis models need not fully represent the manufacturing models with all design features in detail. Fewer features are allowed if the model represents the relevant thermomechanical and aerodynamic behavior. The topology of the model must be ensured to be stable during the design iterations within the optimization loop. More details are given in Section V.

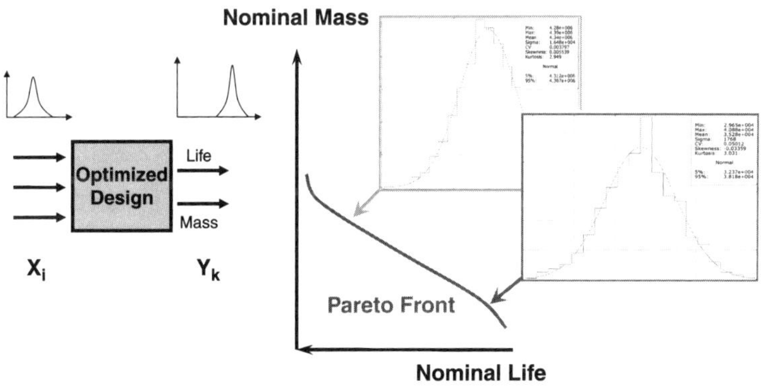

Fig. 3 Example of Pareto front with robustness assessment.

C. Analysis Aspects

A successful automated setting of the boundary conditions, as for instance pressure or temperature distributions onto named selections of component surfaces, is required for a stable analysis run. Access to result data and automated data evaluation focusing on the design objectives and data storage for further assessment criteria must be considered additionally.

D. Collaborative Design Process MDO and Robust Design

Efficacious locally running design frameworks at each of the partners are a prerequisite for the successful connection of these frameworks into an integrated main process flow. The basic requirements are listed here, but more details regarding the information-technology (IT) architecture and the overall process integration are found in Chapter 10:

1) Access to a common data storage, respectively a collaboration hub in order to exchange data with the partners (VEC-hub concept)
2) Access to the main design process flow (Fig. 4) in any way (Web services)
3) Automated start of local design processes activated by the main flow
4) Partners agreement to common data formats and interface definitions
5) Consideration of proprietary rights and each partner's security policies

In addition, a concept of how to share the design tasks between the partners, and in which order, must be agreed within the engineering virtual enterprise.

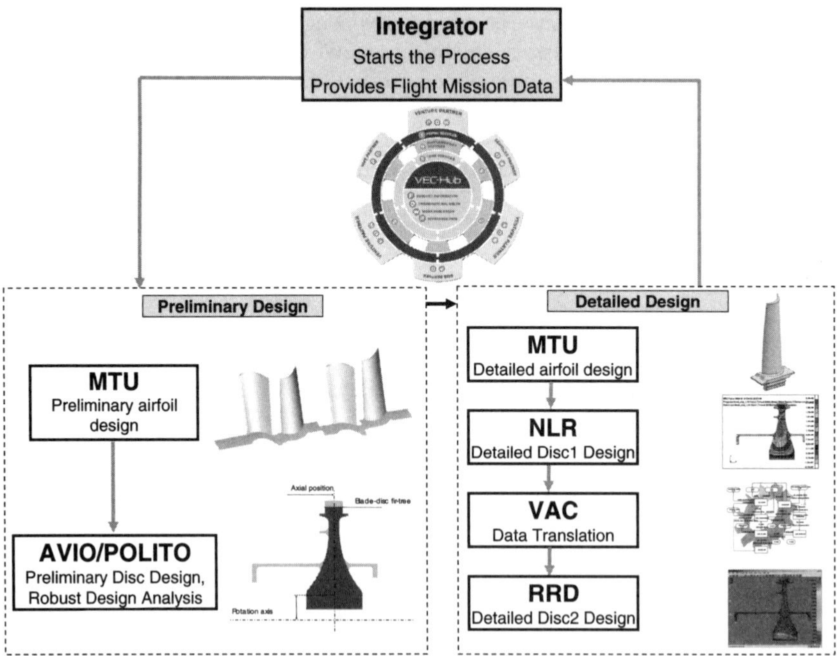

Fig. 4 Preliminary and detailed design flow embedded into main flow.

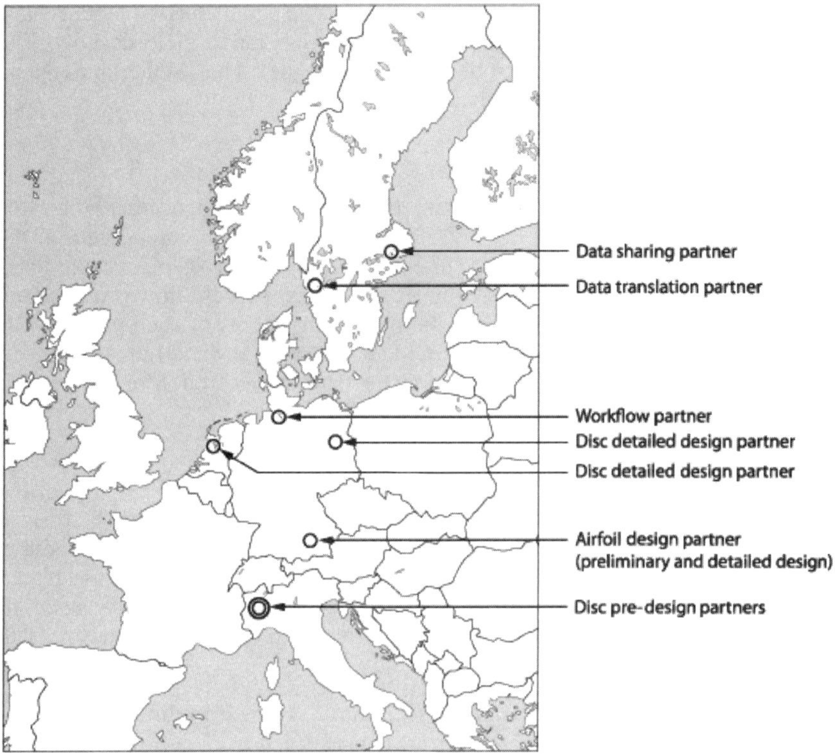

Fig. 5 Collaboration team distributed over Europe.

III. Realization of the Engine MDO and RD in the Virtual Enterprise

In accordance to an engine development program, the distributed design tasks within the engineering consortium were demonstrated for two cases: first, the coupled preliminary design and detailed design and second, a design concept study based on a business scenario.

Figure 4 shows the concept of the principal design flow controlled by an integrator and divided into the preliminary design phase and the detailed design phase considering a collaboration hub accessible by the partners. A permanent access to the main process flow and to a common data repository is required. This was achieved by using a shared data server (data available online at http://www.eurostep.com) in Stockholm Sweden under consideration of the VEC-hub concept (more details in Chapter 10). The overall multicompany process integration (connection to the main flow) and to the VEC hub was performed with the COTS software (data avaliable online at http://www.engineous.com).

A. Preliminary and Detailed Design Process

The analysis steps with regard to the preliminary design and detailed design are shown in Fig. 4. The partner's geographic locations are shown in Fig. 5, whereas the IT support is detailed in Chapter 10. The parameterized models have been prepared by the partners, ready for the collaborative analysis flow. The process was started and controlled at The Hague, The Netherlands (data available online at http://www.vivaceproject.com).

1. Integrator

The integrator provides all relevant flight mission data necessary for the multidisciplinary analysis of the two-stage turbine. The data are stored on the shared data server (data available online at http://www.eurostep.com) in Stockholm (see Fig. 5). The integrator controls the design process flow (online from The Hague) without infringing upon the proprietary rights of each partner. The integrator is not able to control the individual processes at the different partners' locations unless given the right to do so by the partner involved. More details can be found in Chapter 10.

2. Preliminary Design Process

Within the preliminary design process it was the intention to perform a first optimization of the turbine blades airfoils, which results in preliminary analysis input data for the disc design because it is the discs where the blades are connected. In addition, robust design data (distribution density functions of the uncertainties) were also provided in order to assess the robustness of the discs.

Airfoil design partner at Munich, Germany. Here a preliminary aerodynamic optimization of the blades (airfoil) is performed using a proprietary optimization process of the airfoil design partner. The relevant analysis data are provided via the main process. The resulting data (weight of the blades, outer diameter of the disc, rotation speed) are stored back onto the shared repository ready to use for the next partner in the design workflow. More details are given in the MDO airfoil design section of this chapter.

Disc design partner at Turin, Italy. This partner performs preliminary disc analysis and first robustness assessment based on the results data from the airfoil design partner in Munich. The local design process at Turin was activated by the integrator, and the result data are stored back onto the shared data server.

3. Detailed Design Process

Airfoil design partner at Munich, Germany. This partner performs detailed blade (airfoil) design. The result data are stored back onto the shared data server via the main process ready to use for the next partner in the design workflow.

Disc design partner at Amsterdam, The Netherlands. This partner performs detailed design optimization of the first disc using the internal analysis and design optimization process invoked by the integrator. The result data (weight of the disc, life cycles) are uploaded onto the shared data server.

Partner at Trollhättan, Sweden. This partner performs translation of CAD data between the different partners' software packages. More information on data representation and translation is provided in Chapter 10.

Disc design partner at Berlin, Germany. This partner performs detailed design optimization of the second disc using the internal analysis and design optimization process invoked by the integrator. Results data (weight of the disc, life cycles) are uploaded onto the shared data server.

4. Preliminary and Detailed Design Process Comments

At the end of the full process, several design iterations for airfoils (blades) and discs, including some robustness analysis for the required flight mission, were completed for each turbine stage. The data structure used in the process was organized suitable for use in a product life-cycle management (PLM) system (for more details see Chapter 12).

Surrogate modeling techniques as well as online FE procedures were used for this design process. The number of parameters for the individual components (blades, discs) was reduced in order to speed up the design optimization process for the demonstration of this case study. The ability to analyze these components with a higher number of parameters and in a much more detailed manner depends mainly on the acceptable response time for the collaborative workflow. The objective of this particular distributed design process was to demonstrate the MDO and RD capability in a collaborative engineering environment rather than performing a comprehensive optimization of the turbine components. The optimization of the components was performed individually at each partner's framework based on the interface data obtained from the previous optimization of the corresponding connected component made at another company. In the case of the airfoil design partner at Munich, the preliminary airfoil data have been used as a starting point for the detailed design optimization. The two discs analyzed at Amsterdam and at Berlin were coupled via a fixed flange position. The output data from the airfoil design partners were used for both discs. Therefore, a coupling exists to the blade designer. However, no real coupled optimization between the two disc design partners was demonstrated at this stage.

B. Design Concept Study

The second realization of the MDO and RD in the virtual enterprise was based on a business case scenario and online demonstrated in Toulouse, France (data available online at http://www.vivaceproject.com). The collaboration team was again distributed over Europe as shown in Fig. 5. The shared data server at Stockholm, Sweden, was used again to store the relevant data for the analysis and achieved from the results.

Again, the aerodynamic and the mechanical design of a two-stage high-pressure turbine was the subject of interest (Fig. 2). A major objective was to provide a design concept choice on the basis of a multidisciplinary assessment performed by all collaborating partners. Two business case scenarios as explained in Fig. 6 were investigated: concept 1—increased aeroefficiency concept and concept 2—cost-reduction concept.

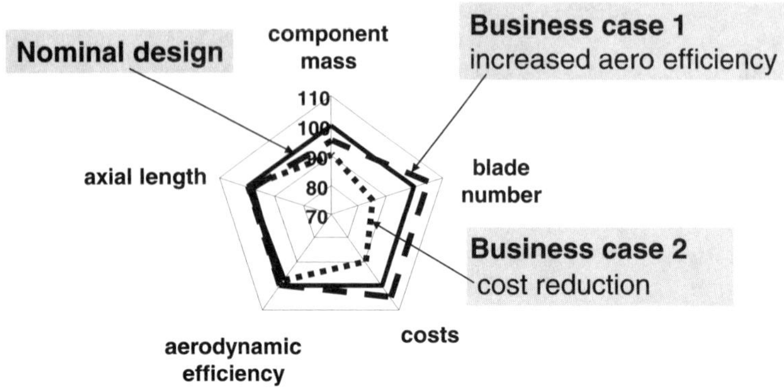

Fig. 6 Spider diagram related to the business case scenarios (differences overdrawn).

In comparison to the nominal design solution as achieved from the preliminary and detailed design process as mentioned in the subsection before, scenario 1 is expected to lead to a better performance [specific fuel consumption (SFC)], but also to an increased component mass, higher number of blades, and therefore increased part costs. Scenario 2 is expected to lead to a cost reduction because of a smaller number of blades and lower component mass; however, this benefit will be achieved at the expense of a lower performance.

The design work flow of this concept study is similar to the one as shown in Fig. 4. The results obtained from the concept study are summarized in Fig. 7. These results offered a basis for a design decision although the intention was mainly to demonstrate to ability of such a distributed, collaborative concept study to provide a basis for a design assessment. Concept 1 seems suitable for long-distance missions and concept 2 for short-distance missions.

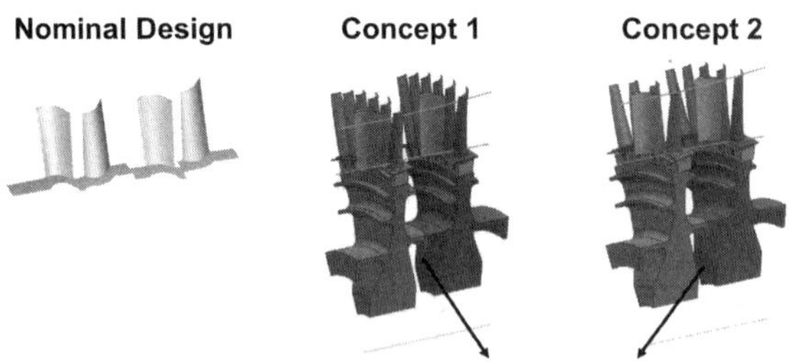

Efficiency	+ 0.1%	- 0.3%
Assembly Weight	+ 0.0%	- 0.0%
Robustness	- 0.3%	+ 3.8%
Part Costs	+ 1.0 %	- 3.0 %

Fig. 7 Results from the detailed concept study.

In summary, the second realization of the MDO and RD in the virtual enterprise was performed closer related to a business case scenario. The integration process was fully automated; the access to the individual partner's design frameworks and vice versa was further improved, also taking into account all security aspects. None of the partner's proprietary rights were infringed, and none of the locally used design methods was exposed to the other partners'. Only agreed design data and results were exchanged. Details of the setup of this integrate work flow are described in Chapter 10.

In the following sections, the turbine blade and disc design is described including information of the individual design frameworks used for this distributed design processes.

IV. MDO at Airfoil Design Partner in Munich, Germany

A. Context

This section focuses on the aerodynamic design and in particular on the configuration of the optimization system and the aerodynamic optimization of the rotor blades.

The aerodynamic optimization of the two-stage turbine is running locally to enable the partner to operate with high-fidelity confidential data and state-of-the-art confidential tools in the optimization processes. The IT aspects of the integration of this local optimization system with the main loop of the process integrator are described in Chapter 10. The integrator provides the flight mission data (rotational speed, mass flow, total inlet temperature, etc.). The process returns also the description of the two rotor blades (including CAD model, analysis data such as mass, center of gravity, etc.), which is used by the partners who concentrate on the rotor disc design within the scope of the virtual enterprise. Subsequently, two of the partners design the pertinent rotor discs while another partner concentrates on the robust design aspects as depicted in Fig. 4.

The coupling of aerodynamic and structural design was chosen to prove the feasibility of a multidisciplinary joint effort. The goals of this joint approach are to reduce iterations between the disciplines and also to reduce design time and design costs. Figure 8 shows the relationship between normal department-orientated multidisciplinary design optimization vs the integrated and interdisciplinary design optimization. An interdisciplinary design implies that every design step is evaluated by all disciplines involved. Often a coupling of detail level processes (A) from one discipline with normal processes (B) from other disciplines is quite efficient, depending on the design task.

The interaction of aerodynamics and structure mechanics leads to a direct consideration of the blade mass and rotational speed of the aerodynamically optimized blade shape during each loop of the disc design. An example for the coupling of aerodynamics and structure mechanics within one partner could be an aerodynamic design already including the constraints of the Campbell diagram or maximal stresses to ensure a resonance-free blade.

This approach requires not only an intensive data exchange between departments, but also the coordination of the tools and processes to be used under the joint design effort.

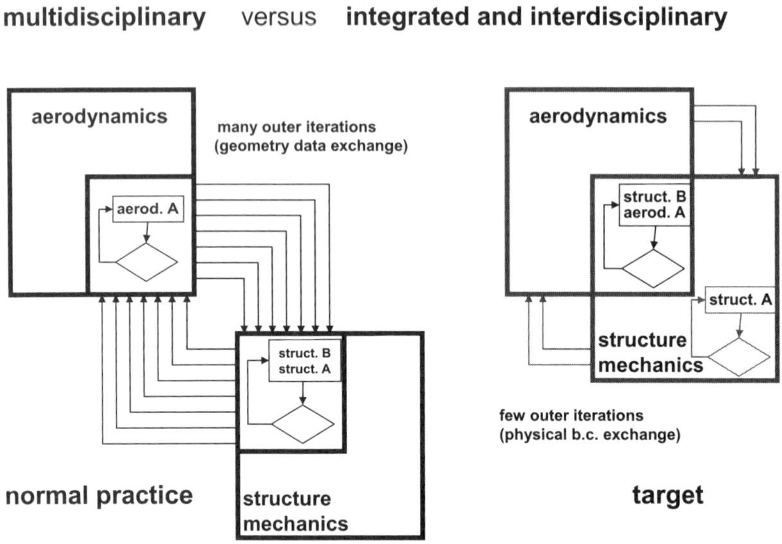

Fig. 8 Multidisciplinary vs integrated and interdisciplinary.

Traditionally only data are exchanged between the design departments. To operate an interdisciplinary design, it is necessary to have a direct coupling of all tools and methods (see Fig. 9).

The main focus is on building a modular optimization framework that allows direct multidisciplinary process chains. The universal applicability of

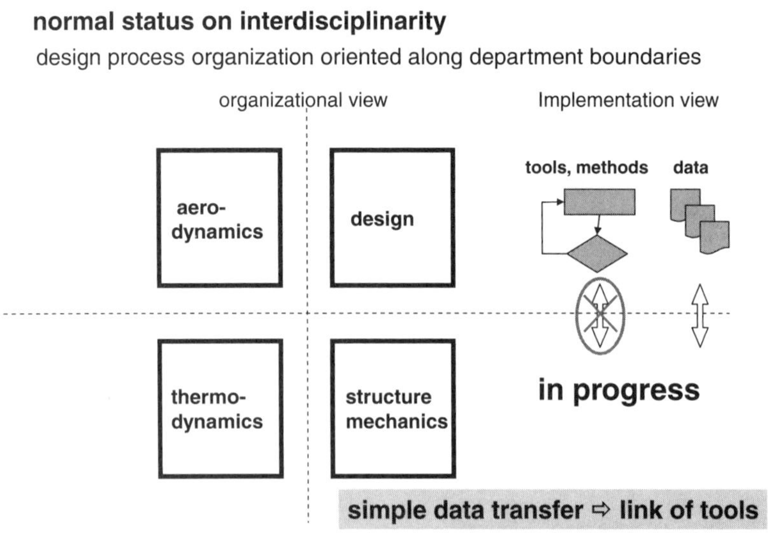

Fig. 9 Overview of current industrial interdisciplinary exchange status.

this framework to various optimization problems across different design levels with different process chains must be ensured. An additional target is the possibility to cooperate with external partners by defining externally accessible interfaces to the in-house optimization framework. The aerodynamic optimization introduced next is an example for the last step in the detailed aerodynamic airfoil design effort.

B. Turbine Optimization Within the Collaborative Environment

The focus of the investigation is on the design of the two rotor blades. The blade geometry of the basic turbine is shown in Fig. 10.

The optimization of the rotor blades was carried out as two three-dimensional single airfoil optimizations using a Navier–Stokes solver. The process chain activated in the optimization framework covers the following process steps (Fig. 11):

1) Step one is three-dimensional parameterization of the airfoil geometry. The blade geometry is defined on two-dimensional blade sections along streamlines. Each blade section is defined by four splines (leading edge, trailing edge, suction side, pressure side), controlled by nondimensionalized parameters. These parameters are radially linked by an additional radial parameterization based on polynoms to ensure a radially smooth three-dimensional blade. The three-dimensional airfoil shape is controlled by the radial parameterization during the optimization loops.

2) Next is processing of the geometry using CAD tools (parasolid, Unigraphics).

3) Step 3 is three-dimensional multiblock low-Reynolds-number mesh generation.

4) Three-dimensional Navier–Stokes flow solution with the Navier–Stokes code TRACE, using a transition criterion, is the next step. The calculation is parallelized on a Linux cluster.

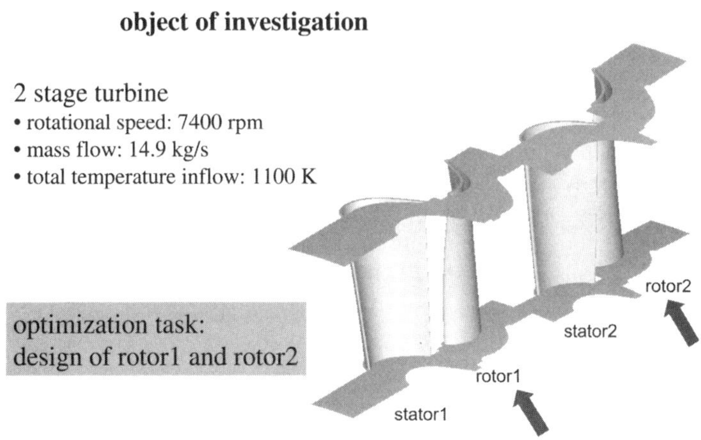

Fig. 10 Basic geometry of the two-stage turbine blades (rotor) and vanes (stator).

5) Postprocessing of integral flow criteria is the fifth step.

6) Last is postprocessing of local flow phenomena, for example, vortices, flow angle distributions, etc.

The targets of the optimizations' objectives were a reduction of losses and a reduction of the three-dimensional flow phenomena, for example, secondary flows.

The optimizations were carried out using the following settings: 29 parameters (geometry parameters of the blade) using 34 constraints. The target value (objective function) consisted of two parts. A gradient-based optimization algorithm sequential quadratic programming (SQP) was used for the optimization runs. The selected procedure allowed a clear improvement of the target function, with all constraints fulfilled.

At the end of the optimization, the final airfoils are combined with blade platform, fillet radius, and fir-tree root. The three-dimensional model of the blade is then analyzed to obtain result values, for example, blade mass (Fig. 12).

The entire process with the optimization framework was integrated in a Web service, which is described with the central workflow (see Chapter 10). Currently the process operates at prototyping phase. It is planned to transfer the framework to a production prototype after additional testing and improvements, for example, for handling aspects.

The integration of the optimization framework in the actual data management (or PLM) and design environment requires data-handling aspects to be taken into account. There are plans to run the optimization loops in a separate operational area. The required data shall be provided automatically after the run in the "prepare" mode via an export process. Following the optimization loop, the optimized configuration shall be returned by importing the files to the data management system.

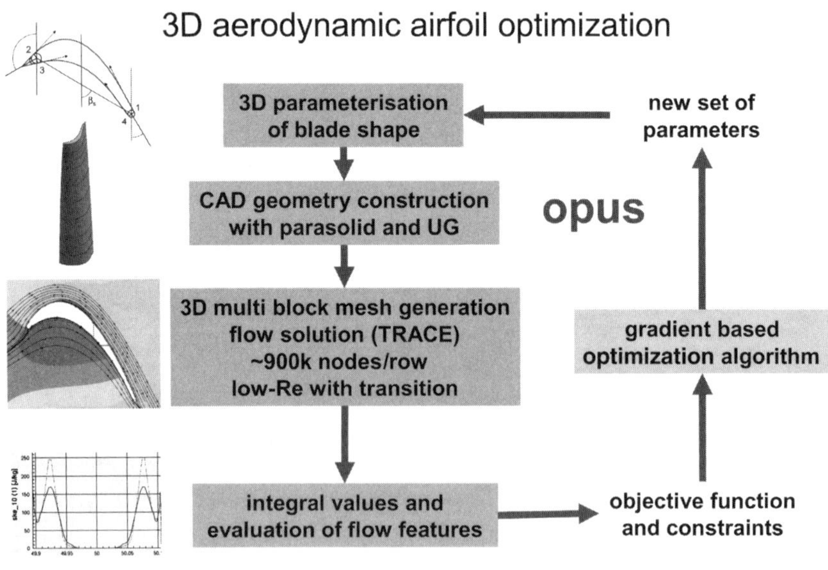

Fig. 11 Process chart for airfoil optimization.

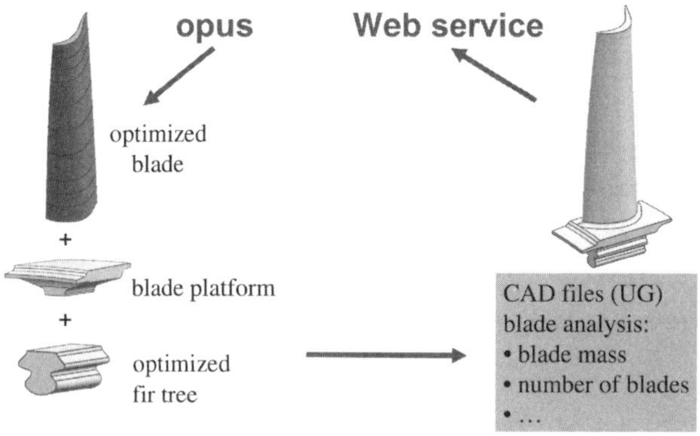

Fig. 12 Combination of airfoil, blade platform, and fir tree.

C. Process Requirements for Optimization Framework

The complexity of the design task requires the development of a generic framework. Because various tools are involved in the intended multidisciplinary design system, a number of problems must be addressed when integrating the process within the scope of an industrial application.

The process chain that is running during the optimization must be easily definable by the design experts involved. The user should be able to select from several alternative tools to define the process chain (considering level of detail, potential alternatives for part processes, e.g., solver, mesh generator, etc.). Also the definition of the variables' ranges, the targets, and constraints must be easily selectable. It should be possible to establish defaults for the user settings and to save the design experience acquired. In addition, full automation of the optimization framework should be possible. A high operational availability of the optimization framework is required. Because of the high computational loads, all parallelization potential should be exploited. This is essential for time-consuming detail design calculations in particular. The process should be adapted to the in-corporate IT infrastructure for maximum utilization of computational resources (work stations, Linux cluster, batch system, etc.). Implementation and maintenance of the optimization framework must be possible within a reasonable period of time. The competence for the single-discipline subprocesses must remain in the respective department. No exclusive bonding to a commercial tool is desired because of issues such as license costs, vendor lock-in, and so forth.

D. Process Integration of Optimization Framework

The framework is built up in a strictly modular way to ensure high flexibility. The main components are a central framework tool, process modules, optimization algorithms, a graphical user interface, and interfaces to commercial optimization tools. Figure 13 illustrates the current framework.

To meet all demands, the structure of the optimization framework is as follows: The core of the system is implemented in a central tool. This tool hosts the structure of the process chain, that is, calls the modules of the process chain.

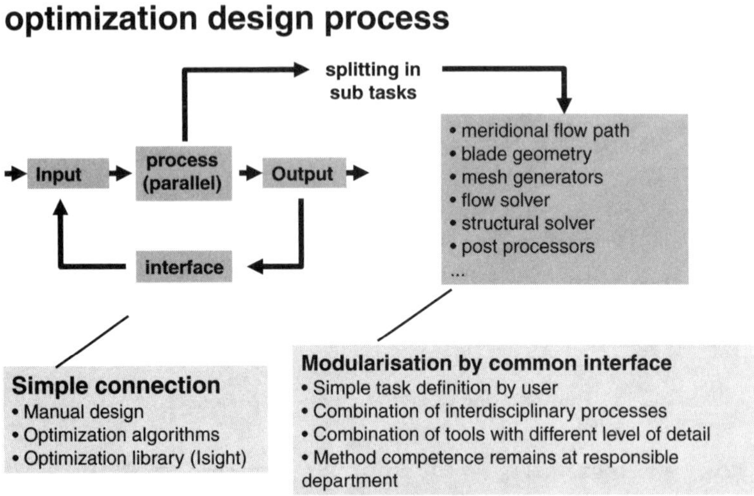

Fig. 13 Current MDO process framework implementation.

The process chain is split in exchangeable modules for every task, for example, geometry parameterization, mesh generator, aerodynamic flow solver, structural solver, etc. All calculations are parallelized, including multilevel nesting. The calculation of process chains, for example, the parameter vectors during gradient search, as well as modules within the process chain, for example, aerodynamic and structural evaluation, or evaluation of two-dimensional aerodynamic calculations are parallelized (nested forks). The nested parallelization allows the running of external programs as well as internal subroutines in parallel as different instances. Basic variables can be accessed directly from all (parallel) running subroutines. This allows for a reduction of redundant information. All parallel running processes can be stopped by means of a universal signal handling through all process levels. The information exchanged between the central tool and the process modules, for example, variables or result values, is written in files with fixed data format. In addition, library functions are provided for all process modules to harmonize and simplify auxiliary tasks such as file handling, data logging, etc. The central tool provides all framework functions. These functions can be operated by the graphical user interface or can be fully automated in batch mode. The handling of the optimization variables (e.g., adjusting range), the result values of all process components, and the calculation of the objective function are performed centrally. Also, the management of the directories for all running process chains (e.g., geometry) and the check of the required data are done by the central tool. All necessary files are copied or generated in these directories while running through the process chain. The amount of data is controlled by assessable clean-up routines, which remove files/directories automatically dependent on the quality of the optimization step. The interface between the central tool and the graphical user interface is an xml-file. It includes the process tree, all necessary possible menu functions, etc., in a generic way and is vendor independent. Changes in the process chain, additional menu functions, or process steps can be implemented by easy modifications in the xml-file without program changes of the graphical user interface. An easy-to-manage interface to the optimization techniques or commercial tools via ASCII-files is provided. In this way the framework allows an easy connection to almost every other optimization technique or tool. The framework processes the calculation task of the selected optimization technique in parallel and calculates the objective function. All relevant information is logged in a database. The framework provides easy-to-use interfaces to the corporate IT infrastructure.

The process modules (e.g., geometry generation processes, mesh generators, solvers, postprocessing, etc.) map the program run logic of the respective subprocess and start the corresponding tools. All process modules use the library functions of the framework, for example, all file operations concerning the variable and result value transfer. Building of new process modules is easily possible. All processes are activated by a switch, controlled by the central tool via the graphical user interface (GUI). Additional control values for each process can be provided as well, if necessary. The processes provide their respective variables and results to the central process by using library functions to ensure compliance with the fixed format. Thus integration of process modules in the central tool, process tree, and parallelization is easy to implement.

The GUI represents the (only) interface to the user. It is defined fully generic. The process tree, the function menu, all buttons, etc., are defined by an xml-file created by the central tool. Type and configuration of the desired design tools are selectable by the user. The GUI shows the process tree with all appropriate controls. The user is guided through the course of action by buttons for the different run modes. Different views are provided for the distributions of all variables and result values running during the optimization. Views containing data provided by the process modules (e.g., profile pressure distribution) with the possibility of comparison for the different optimization steps are provided as well. The GUI permits the editing and adopting of presettings for easy setup of optimizations.

Regarding optimization techniques, several optimization algorithms have been integrated in the framework through their adapted interfaces. The following techniques have been adopted: a gradient-based algorithm sequential quadratic programming (SQP), a parallelized genetic algorithm, an adaptive simulated annealing algorithm (ASA), and a partly parallelized interface to the commercial optimization software (data available online at http://www.vivaceproject.com). These commercial algorithms are used in batch mode as a library of optimization algorithms and not for process integration (data available online at http://www.vivaceproject.com).

The framework is implemented in-house. This approach saves license costs and is flexible and customizable for various (multidisciplinary) design tasks.

Disadvantages of this approach are the implementation costs, including the costs for implementing the graphical user interfaces and the lead time.

The central tool and all program modules are implemented in Perl. This allows platform-independent running of the framework. The GUI is implemented in Qt-4 (data available online at http://trolltech.com).

Time-consuming programs are sent in parallelized condition to a Linux cluster via a batch system (LSF; data available online at http://www.science-computing.de). To swiftly run parallel calculations with a short response time,

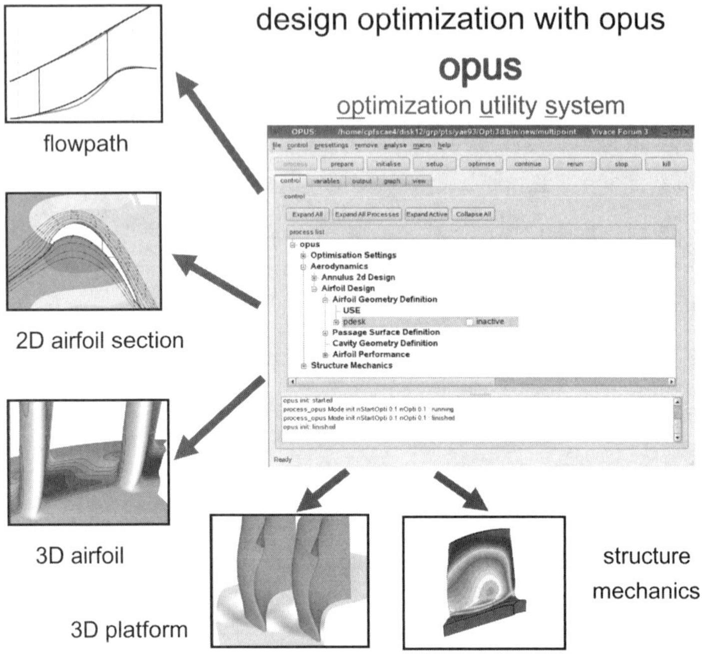

Fig. 14 Opus GUI and case studies.

a number of processors are allocated exclusively in the batch system to run the framework job.

The main advantage of the implementation is the high stability and (nested) parallelization of the framework and its subprocesses. Coupling of multidisciplinary processes is possible if appropriate process modules are available.

The name of the framework is opus (optimization utility system). Figure 14 shows the GUI of opus.

E. Workflow of Optimization Framework

A specific procedure was defined for the operation of the optimization framework. This procedure is the same for all desired calculation tasks. The user has to start a fixed sequence of run modes, as shown in Fig. 15.

The procedure to drive an optimization includes the following steps. First, the user has to define the design task by selecting the desired process modules/design tools (parameterization, mesh generators, solvers, postprocessing, etc.), which need to run in the process chain. Then the corresponding control values (e.g., version numbers, profile section numbers) must be set for the respective processes and also central control factors such as parallelization, data clean up, and so forth. The optimization strategies must be selected according to the calculation process, the design problem, and the time target. Standard gradient-based algorithms seem to be adequate for time-consuming aerodynamic and structural three-dimensional detail optimizations. Less time-consuming optimization tasks, for example, two-dimensional aerodynamic evaluations, also allow the use of stochastic search techniques.

Running the framework in the "prepare" mode directs all processes to report their pre-execution requirements (e.g., files). The user must provide all files, for example, by choosing and checking out the files from the productive data management and design system. The required version management is part of the PLM.

Running the framework in the "initialize" mode starts the parameterization (and if necessary approximation) of the geometry, the calculation of the initial (reference) solution based on the start files and also provides all variables and result values to the user.

The user has to specify the optimization problem by selecting all active design variables and fix their adjustment range, or choose a presetting, if available. Also the user has to select all boundary conditions and the weights on result values, or choose a presetting, if available for the target function.

Running the framework in the "setup" mode checks and freezes the definition of the optimization problem.

Running the framework in the "optimize" mode activates the chosen optimization technique by the framework. The optimization process can read the setup information, write the desired parameter vector(s), and start the framework in the "run" mode (also possible to be run manually by the user) iteratively. The framework reads the appropriate optimization jobs, creates the directory(ies),

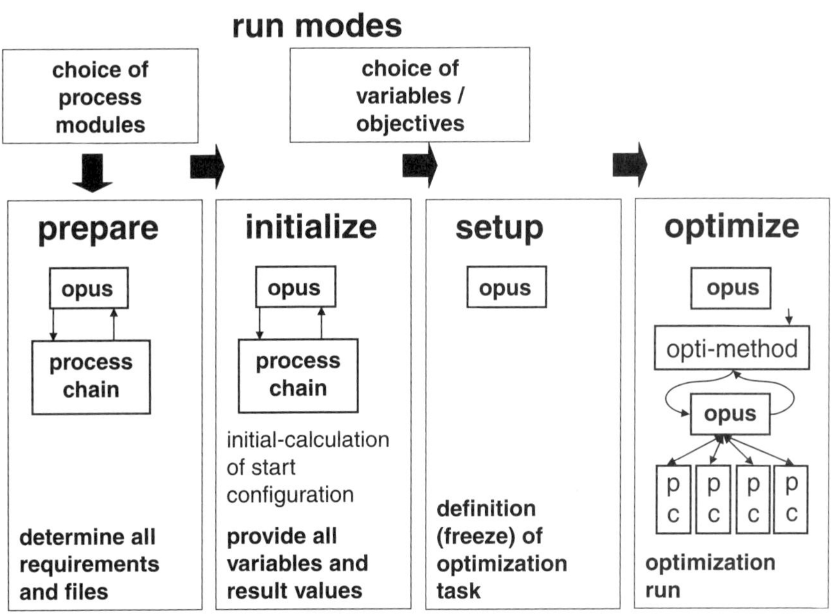

Fig. 15 Fixed sequence of steps to use the optimization framework.

runs the parameter vector or multiple vectors in parallel, calculates and writes the objective functions, and logs the data.

Finally, the user has to analyze the running data, which are automatically updated by the GUI, and perform comparisons of the iteration steps.

F. Major Tools Integrated in Optimization Framework

The process modules of the optimization framework use various tools. All programs run automatically in batch mode. The process modules use central library functions and some fixed conventions to cooperate with each other.

Important proprietary components include geometry parameterization tools for flowpath, blade rows, and platforms; geometry processing within CAD tools, for example, NX (data available online at http://www.science-computing.de) and Parasolid (data available online at http://www.plm.automation.siemens.com/en_us/products/nx/index.shtml); aerodynamic multiblock mesh generation for several blade rows; aerodynamic flow solver (TRACE); aerodynamic postprocessing; structural geometry assembly, for example, turning knobs for compressor blades; structural mesh generation; structural solver Calculix (data available online at http://www.plm.automation.siemens.com/en_us/products/open/); structural mode tracking; and structural postprocessing.

The structure of the optimization framework allows a very flexible composition of the process chain taking the demands of the respective design task into account. Each tool must be integrated only once. They can participate in every optimization task performed by the framework. Optimizations on different application ranges have been tested with the framework.

V. High-Pressure Turbine Disc Assembly Design Optimization

A. Context

This section describes the basics of the two-stage high-pressure turbine (HPT) disc assembly design process considering all aspects of parametric modeling, meshing, the analysis disciplines, and also optimization and robust design aspects. The design process is shared between two collaborative partners (disc design partner at Berlin, disc design partner at Amsterdam; see Fig. 5) using their individual design frameworks, which are connected to the overall main flow as described in Chapter 10. Therefore, this section combines the experience gained by the two disc design partners. A description of the individual design frameworks is given at the end of this section.

Several constraints have to be considered for the disc assembly design. The first set of constraints for the disc assembly results from the upper-level two-stage engine assembly predesign. The second set of constraints results from the detailed blade and vane design as described in the airfoil design section of this chapter. The third set of constraints results from the interface connection between the two discs via the connecting flange position. The result is a design space defined with many constraints. Geometrical constraints include disc width, stage outer diameter, and shaft diameter. Engine characteristics include the disc's temperature distribution and mechanical blade loads. Within these constraints a lower level of two single-disc design optimizations is performed and demonstrated in two ways: individually and combined. The result is provided back to the upper-level design, resulting in a multilevel design capability. Such multilevel design is typical for complex designs. It puts additional requirements on the computational efficiency of the lower-level design tool chains.

B. Modeling and Parameterization

The optimization process requires parameterized CAD models of the assembly respectively the components. A suitable parameterization addressing the analysis objectives is essential for a successful optimization process. Those parameterized models, suitable for analysis purposes, can differ to those CAD models that will be used for manufacturing. Therefore these—in some cases less featured—models are called *analysis models* or *working models*. The topology of these models must be stable at each design iteration step when the driving parameters for the geometry are changed.

Parametrical three-dimensional disc designs are provided and available in a COTS CAD package [CATIA or Unigraphics (UG)] (data available online at http://www.plm.automation.siemens.com/en_us/products/nx/index.shtml). With regard to the creation of an analysis model and also because of relatively early design stage as investigated in this collaboration study, the required accuracy of the simulated results allows one to make some simplifying assumptions that will significantly reduce the computational requirements of the analysis and also reduce the modeling effort. Simplification examples are the omission of blade-disc analysis and the use of a two-dimensional axially symmetrical description of the disc. In an actual design, the full three-dimensional disc geometry will be stored in a PLM system. More information on such PLM systems is provided in Chapter 12. From this three-dimensional description the two-dimensional description needs to be extracted. Figure 16 shows how the axially symmetric two-dimensional geometry description relates to the three-dimesional disc geometry. To simplify this description, it is assumed that the data extraction is performed stand alone, without the support of a PLM system. For the exchange of data, a standard data format is used in line with the PLM approach. In this case initial graphics exchange specification (IGES) format is used.

Figure 17 shows the parameterization of such a two-dimensional (axis symmetric) turbine disc as used for the optimization and robust design studies.

Achieving the best parameterization setup for optimization purposes requires manual intervention in an iterative process. The authors could not identify common rules such as, for example, the highest and the lowest dependency between the disc design parameters. Each individual component has its own parameterization rules depending on the objectives of the optimization. Figures 18a–18c show an example of an distorted geometry, which can be avoided by setting rules for the parameters (P1, P2, P3).

C. Meshing

Stable FE meshing is essential for optimization and robust design purposes because, due to the change of the geometry, the topology of the mesh can change and therefore, depending on the mesh quality, the results can vary unrelated to the

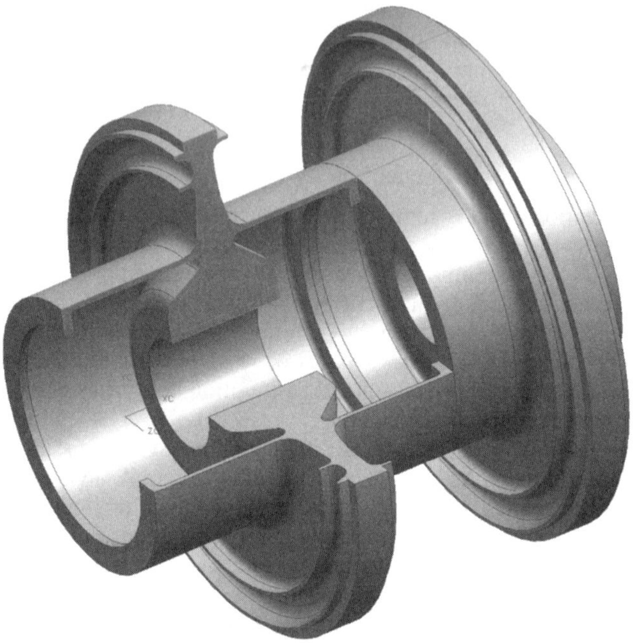

Fig. 16 Position of axially symmetric two-dimensional geometry within three-dimensional HPT disc.

physics of the disc design. This can be avoided by considering mesh errors respectively to ensure that the mesh quality is sufficient. Otherwise, the optimization algorithm will not be stable or will lead to wrong results. Figure 19 illustrates this effect where a coarse mesh does not give a continuous response on a continuous change of the parameter. This negative effect can be significantly increased if the objective function is already very sensitive to the analysis result (stress values). This happens for example for the life-cycle number, which does depend exponentially on the stress result. Therefore, small changes in the stress values can result in a large change for the predicted life cycles. Using such results for surrogate modeling techniques, for example, the response surface method, will represent the wrong physical behavior.

Although the data transfer between different software tools is simplified using standard data formats, in practice the transfer is not always straightforward. Especially, if the geometry contains multiple curves with a relatively small radius, as in the case of the disc, then meshing and load application based on curve identification is difficult. Small changes to the disc design can lead to different curve definitions and numbering. Load application and mesh size definitions based on the curve definitions might therefore not be valid for all generated disc designs and might remain unnoticed in an automated design process, thus leading to erroneous results. An example is shown in Figs. 20a and 20b. Modifying the design variable X_2 (see Fig. 21) leads to a geometry where the used COTS software in this case has difficulties in defining a proper curve and a small curve

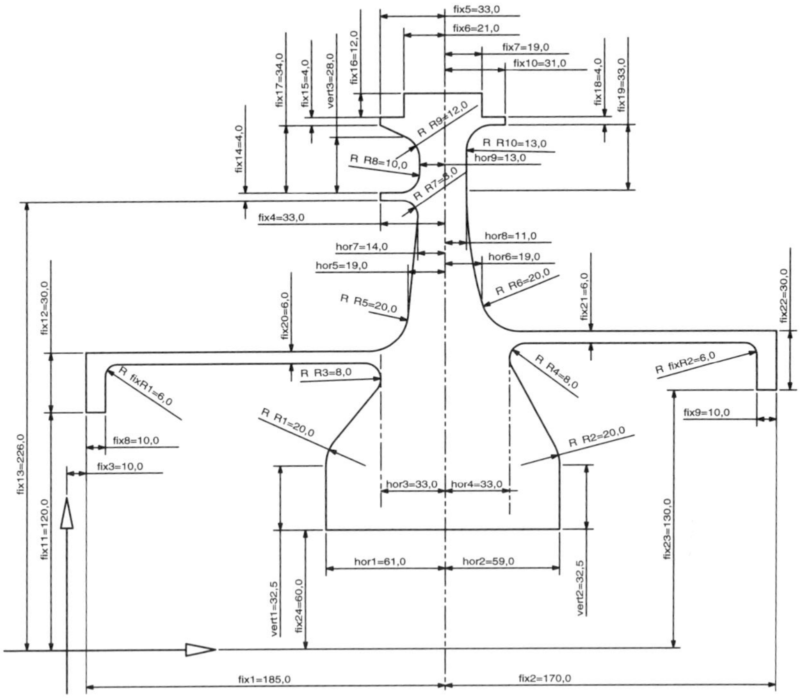

Fig. 17 Parameterized two-dimensional turbine disc model.

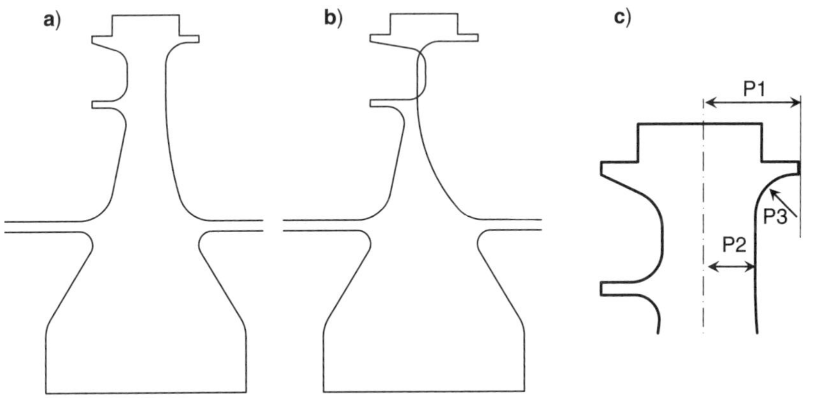

Fig. 18 a) Topology. b) Distorted geometry. c) Parameterization rules P2 + P3 < P1.

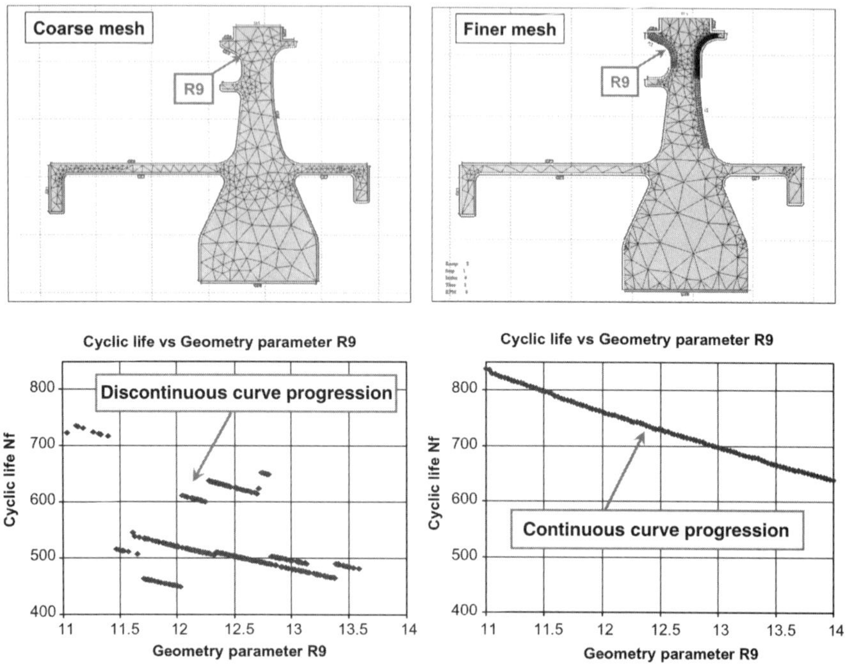

Fig. 19 Mesh influence on results evaluation (life cycles).

where the large radius is placed. This changes the internal numbering of the curves. If in this case a load is applied to a certain curve number, the loading is placed incorrectly (Fig. 20b) yielding erroneous results.

It is therefore necessary to take great care when using the finite elements software to automatically mesh the two-dimensional geometry. The disc design partners (Amsterdam, Berlin) have used different CAD systems and different finite elements software.

D. Thermomechanical Analysis

The major design objectives for the turbine disc are the life (fatigue life expressed in the life-cycle number, also called lifing) and the weight (mass) of the disc.

The major loads during operating conditions (idle, maximum takeoff, cruise) are the thermal loads (derived from the temperature distributions) and the mechanical loads (derived from the centrifugal loads caused by rotation).

Therefore, a thermomechanical analysis is required in order to assess the design objectives. The results of the thermomechanical analyses are temperature distributions, stresses, and displacements at the disc. The stress results can then be calculated into life-cycle numbers by using the so-called "lifing methods," which are implemented into the analysis process (data evaluation). The lifing methods are usually proprietary of each partner and developed for specific design

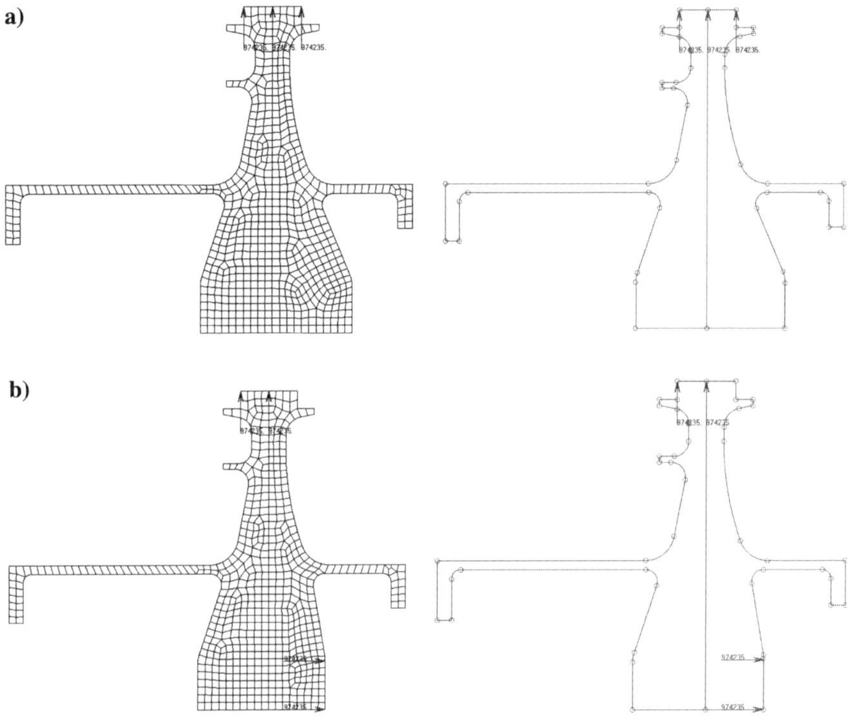

Fig. 20 a) Correct load application. b) Wrong load application.

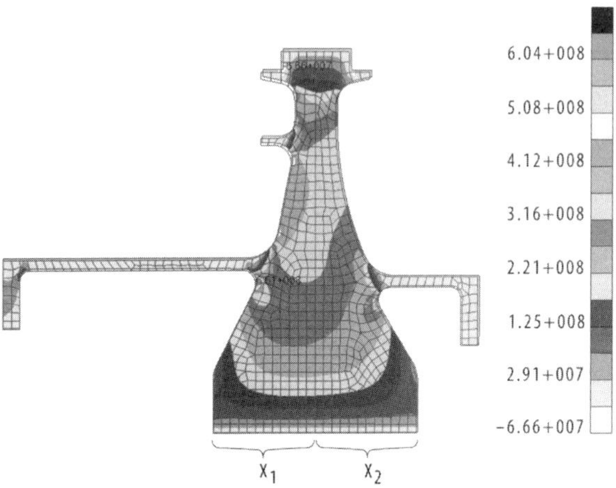

Fig. 21 Standard mesh of two-dimensional disc design with stress distribution. (See also color figure section at the back of the book.)

purposes. The geometry of the parameterized turbine disc model provides the mass of the disc at each design iteration step.

To prevent infeasible disc designs, a no-burst criterion was applied. A no-burst criterion prescribes a maximum allowable area weighted mean hoop stress, which prevents the disc from bursting.

The fatigue life is defined as the number of flight cycles before disc failure. The flight cycle needs to be defined for the calculation of the life. For obvious safety reasons, the fatigue life calculations need to assume realistic worst-case conditions. To obtain such worst-case life, each flight cycle is based on reaching the steady-state idle condition, the steady-state maximum takeoff (MTO) condition and reaching the steady-state cruise condition. The temperature and hoop stress distribution in the disc determine the resulting disc life. The disc is assumed to be of an isotropic nickel-based alloy. For the low cycle fatigue the Basquin equation (1) is used.

$$N_f(\sigma, T) = \left[\frac{1}{c(T)} \cdot \sigma(T)\right]^{1/m(T)} \tag{1}$$

The Palmgren–Miner rule (2) is used to combine the results from each flight condition. The Palmgren–Miner rule assumes that the total life of the disc can be estimated by adding the life consumed at each flight cycle:

$$\frac{1}{N_{f_{\mathrm{total}}}} = \sum \frac{1}{N_{f_i}} = \frac{1}{N_{f_{\mathrm{idle}}}} + \frac{1}{N_{f_{\mathrm{MTO}}}} + \frac{1}{N_{f_{\mathrm{cruise}}}} \tag{2}$$

where $N_{f_{\mathrm{total}}}$ is the predicted life in number of cycles.

E. Verification of Disc Analysis Tool Chain

It is important to verify the results of the disc analysis tool chain as the latter will be used in an automated environment. Such verification is even more important in the multipartner tool chain, as this partner's discipline experts might not assess the single disc analysis results, and so plausible, but incorrect data could remain undetected. As mentioned before, meshing is critical, and so it will be elaborated here. A large mesh (low number of elements) is efficient when many designs are to be evaluated, but a smaller mesh might produce more accurate results. Figure 21 depicts the computationally efficient standard (course) mesh of the two-dimensional disc design. Figure 21 also shows the calculated hoop stress for the actual conditions (such as temperatures, physical blade loading, etc.); the disc is subjected to in the collaborative design.

It will be shown that several regions at the disc exist where minimum life can occur (more details in the next subsection). A variable mesh is defined by the disc partner in Amsterdam, which is refined for each of those disc areas. Once these local refinements are defined manually, they can be applied automatically for all design variants concerned. The left part of Fig. 22 shows the results, the locally refined mesh, and the calculated stress distribution. The number of elements, the calculation time, and the life for both meshes are provided in Table 1.

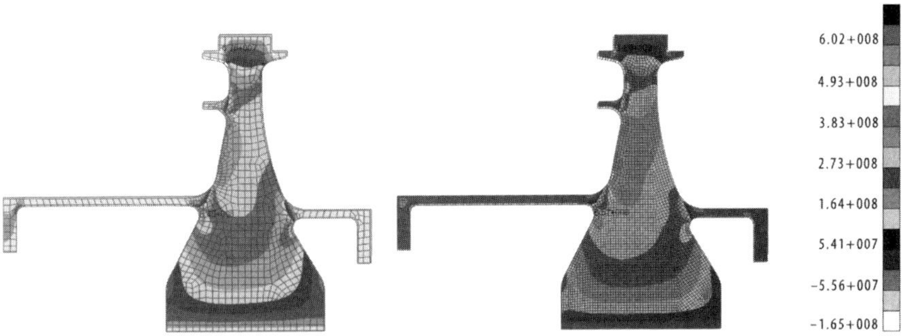

Fig. 22 Standard mesh of two-dimensional disc design with stress distribution. (See also color figure section at the back of the book.)

Table 1 seems to indicate that the coarse and variable meshes provide equally good results, that is, within the accuracy needed for the optimization, so that the computationally more efficient coarse mesh seems appropriate. When comparing these results with the result from the disc partner in Berlin, differences occurred. The results became comparable only after refining the mesh as shown in the right part of Fig. 22, with additional results included in Table 1. Because each partner used its own proprietary tool suite, the obtained similar results can be trusted for the collaborative design. In case the collaborative design progresses to subsequent design phases, more validation is needed, for example, comparing the calculated results with measured temperature distribution and deformation of the disc. Note that some of these verification data might be sensitive with respect to intellectual property rights. For this multipartner collaboration case study the fine mesh has been used. In an industrial setting the variable mesh could be refined in a tradeoff study to obtain the same accurate results as the fine mesh while consuming less computational resources. The experience gained from this project confirms the need for verification of tool results before applying them in automated tool chains. Incorporating the original computationally efficient but overestimated lifing results would have resulted in plausible but incorrect optimization results. It would be very hard for the engineers using the entire tool chain as a black box to detect such error [2].

Meshing standards are available for some domains. For example, the nine-part Eurocode [3] is available for the civil-engineering domain. The fact that such standards are utilized by many users provides some assurance that compliance

Table 1 Details of computational process modeling and solving of SCC for case 1

Mesh	No. of elements	Calculation time, s	Life
Coarse	665	2.8	59,823
Variable	827	3.4	59,975
Fine	4,058	18.2	52,801

with the standard will warrant reasonable results. Such public standards do not exist in the aeronautical domain and hence could not be referred to. Major aeronautic companies have their own proprietary procedures and processes.

A straightforward approach to obtain confidence in the calculated results would be to determine or estimate the error. Such additional work could be performed for a single design variant, prior to exploring the entire design space. Stein et al. [4] describe a method using adaptive modeling to determine a posterior estimate of the error for the physical properties of interest. Because of the additional work involved, such methods are only practical when supported by COTS tools. COTS tools will only incorporate such features when required by standards such as Eurocode [3], which is not the case yet.

The experience presented here on the importance of meshing and verification of the results also holds for the marine domain. Bergan and Skeie [5] found that meshing consumed up to 70 or 80% of the engineering effort, which further suggests that extending the COTS solution would be beneficial.

F. Optimization and Robust Design

The major steps for the optimization and robust design process are as follows:

1) sensitivity study—identification of those parameters that have a significant influence on the objective function;

2) optimization process—multidisciplinary and multi-objective optimization resulting in Pareto fronts;

3) surrogate modeling techniques—generating response surface methods in order to reduce analysis time instead of executing many time-consuming FE analyses; and

4) robustness and reliability assessment—the optimized designs identified at the Pareto front have to be assessed against their respective robustness and reliability.

1. Sensitivity Study—Disc Partner in Berlin

The objective is to identify the most relevant input parameters in order to reduce the numerical effort for the optimization algorithm. For this purpose, a design of experiments (DoE) using distance-optimized Latin hypercube was performed. Only geometry parameters were considered as input variables. All boundary conditions, for example, temperatures, heat-transfer coefficients, were kept constant. It was observed that the minimum life location varies with the actual values of the input variables. Hence, the cross section of the disc was partitioned in six regions. Depending on the values of the input variables, the location of the calculated minimum life occurs in one of those six regions. Figure 23 shows the arrangement of those regions. The percentage of how many samples, which have been used for the DoE, that will have their minimum life in a specific regions is also shown in Fig. 23. Some of the DoE samples have their minimum life outside these regions; therefore, the figures do not add up to 100%.

The relevance of the varied geometry parameters was evaluated using the coefficient of determination provided by the COTS tool [6]. Figure 23 shows that for the majority of the DoE cases the minimum life falls in the disc regions 3 and 1. The influence is illustrated for the four most relevant geometry parameters only, which are depicted in Fig. 24.

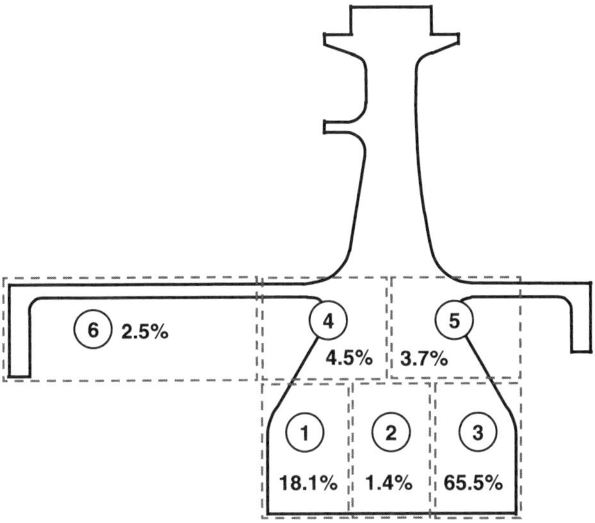

Fig. 23 Partitioning of disc cross section to identify location of minimum life resulting from DoE study.

Case 1 (column "global" in Fig. 25) takes all results into account. No care is taken where the minimum life is located. For that case "hor2" and "hor4" are indicated to be the most relevant geometry parameters.

The second case (column "Region1" in Fig. 25) considers only the DoE samples that give a minimum life located in region 1. That case gives the opposite result with parameter "hor1" and "hor3" to be the most influencing.

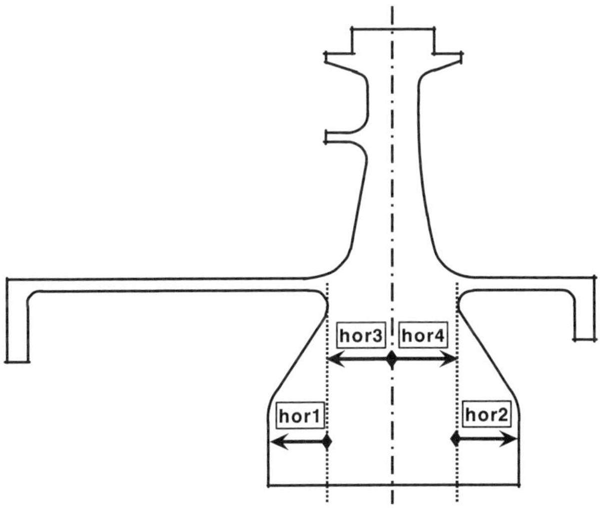

Fig. 24 Illustration of the four most relevant geometry parameters as a result of the sensitivity study.

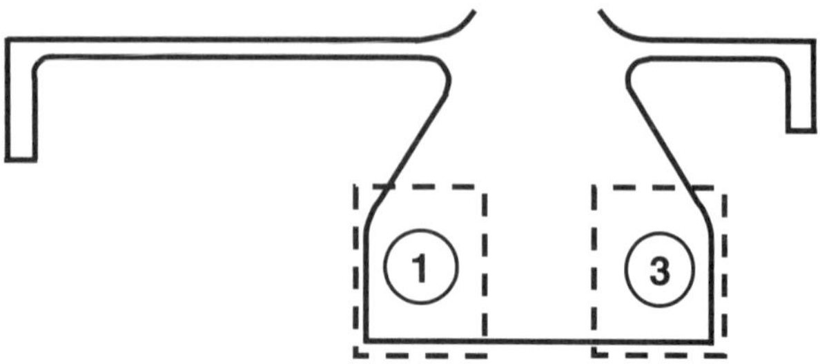

Geometry Parameter	global	Region1	Region3
hor1	3%	20%	0%
hor2	20%	2%	27%
hor3	7%	33%	8%
hor4	31%	13%	37%

Fig. 25 Results of DoE (linear coefficients of determination) considering different regions for predicted minimum life location life.

The last investigated case focuses on DoE samples that have their minimum life in region 3 (column "Region3" in Fig. 25). The outcome is similar compared to the global evaluation of the first case.

Those findings demonstrate how sensitive the results of a DoE might be regarding the range of the input parameters. It shows that care should be taken when concluding which parameters are the most relevant for an optimization task. Limiting the number of parameters too much can constrain the improvements obtained by the optimization. Therefore, in this investigation a higher number of parameters (seven) were considered for the optimization.

From a mathematical point of view, a simple DoE for a given range of the input (i.e., design) parameters cannot deliver the global sensitivities (gradients) in one run. The sensitiveness is a function of the input parameters and varies with the value of the input parameter because the objectives can have local extremes within the parameter range. A small range of design parameters, as for instance for a robustness analysis, can deliver the right sensitivity globally within this reduced range. The global extreme of the objective function then correlates with the local extreme. This can be easily overseen by just setting the range of the parameters (longer distance) and then performing the DoE. Nevertheless, a good knowledge of the physical behavior of a component in addition with the expertise of the engineer is needed for such a task.

2. *Optimization—Disc Partner Berlin*

Two objectives—predicted life and mass of the disc—are the subject of the optimization. Hence a multi-objective optimization had to be performed. A multi-objective evolutionary algorithm provided by the COTS tool [6] was employed. Figure 26 shows the calculated Pareto front.

A single design on the Pareto front had to be identified as the nominal design for the subsequent reliability analysis. Normally this design decision is made based on higher-level information, for example, customer requirements for a certain flight mission. A design with a nominal expected life of 35,280 cycles was chosen, which led to a mass of about 35 kg for the disc in this case. This nominal design was used for robustness and reliability analyses.

3. *Optimization—Disc Partner in Amsterdam*

As just mentioned, the parameterized two-dimensional disc has many independent parameters. To explain the method and the obtained results, only two parameters will be varied, X1 and X2; see Fig. 21. This choice will allow convenient visualization of the results. As design objectives, the disc weight and life are used, just as in the preceding section. The methods used will also hold for more parameters as well as for more design objectives. The optimization chapter provides more details of the methods used. The single engine-turbine-disc analysis tools used do not provide gradient information. In the case of complicated behavior of the design objective for the design space concerned, as suggested in the section on the sensitivity study, robust optimization methods, such as genetic algorithms, are deployed. Such algorithms require many design analyses, in the order of thousands or more, and so the response surface approach, described in the optimization chapter, is used.

Using an appropriate design of experiments, a sampling of the design space is performed. For each design a full analysis of the disc is conducted. Because all of these designs are independent, their respective analyses could be performed in parallel. Figure 27 shows the root-mean-square error (RMSE) for the various fit functions applied. The experience has shown that no single fit function is superior for all cases, and so the tool suite used provides several fit functions. From top to bottom in Fig. 27, these are smooth splines, second- and third-order polynomials, various kriging functions, and intersplines. The various kriging functions are classified by two letters. The first lowercase letter indicates a

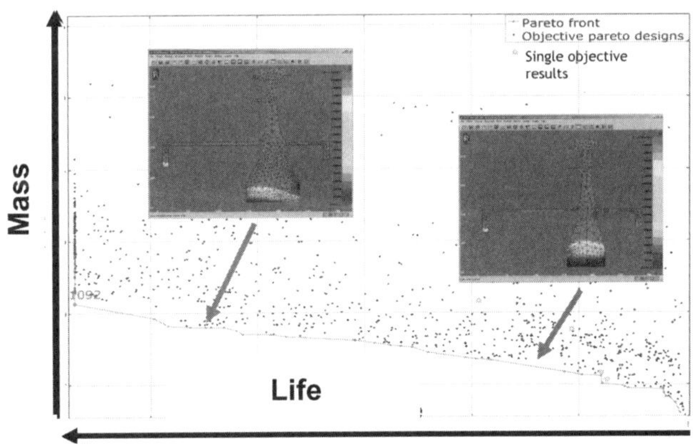

Fig. 26 Results of multi-objective optimization.

constant, linear, or quadratic regression function. The second capital letter indicates a constant, Gaussian, or exponential correlation function. A full explanation of the kriging approximation is provided in the optimization chapter, which also describes how the RMSE values have been determined. When choosing the best-fit functions, these approximation errors are sufficiently small for the disc optimization, so that using the response surface for the optimization algorithms is reasonable. Of course, the found optimum needs to be verified by using the full disc analysis tool chain.

Figure 28a shows the response surface for the life of the disc for the selected design space. The sea level in Fig. 28a indicates the minimum required number of cycles for the disc design. For the parameterized disc design, the mass will be minimal when the parameters X1 and X2 are minimal, that is, in the right area of Fig. 28a. As the original design was in the left compliant area, this example convincingly illustrates the need for robust optimizers. A gradient-based method would not have found an optimum in the disjoint right feasible area. However, because of the numerous design evaluations required, a genetic algorithm would not have been possible without the response surface approach. The response surface also allows finding the disc design with the maximum life, using a weight constraint. The result is shown in Fig. 28b. The resulting maximum life of 53,200 flight cycles is a significant 19% improvement of the baseline disc design of the second disc. This is achieved at a slightly lower disc weight. Even though the original disc was designed for maximum life, the MDO approach allows for a significant improvement. Such improvements when using MDO are typical and are confirmed by Chaput [7] and Rostrand [8]. The figure also shows how critical the definition of constraints is: a lower disc mass constraint would remove the left life peak.

As shown by the earlier optimizations, for real-world designs the optimal design will involve a tradeoff between various objectives and constraints.

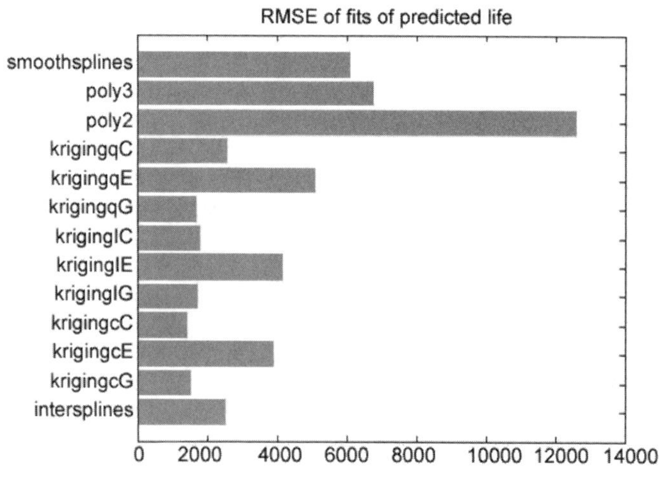

Fig. 27 Accuracy of various fit functions of life response surface.

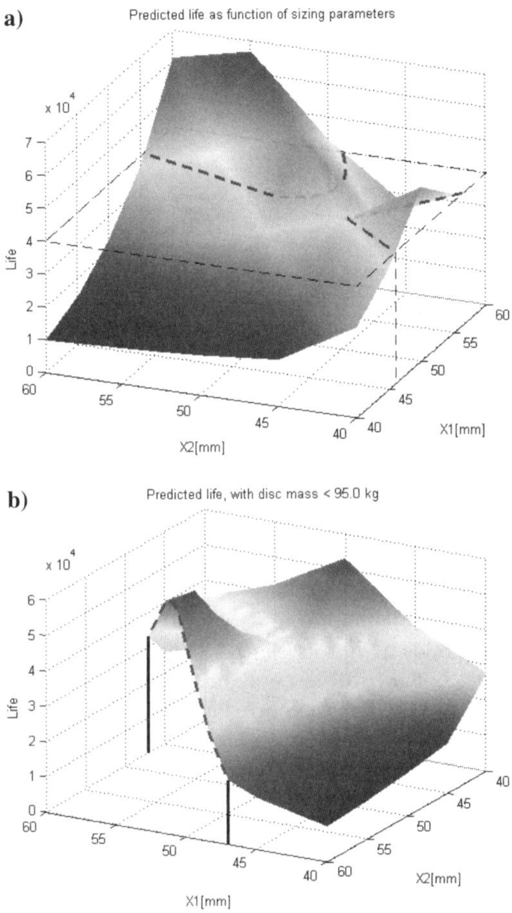

Fig. 28 a) Response surface of life for the defined design space of the two-dimensional disc. b) Response surface of life using a weight restriction. (See also color figure section at the back of the book.)

It is unlikely that the same design will be optimal for different objectives. As explained in the optimization chapter, Pareto optimization is a method for handling multi-objective optimizations. Figure 29 provides the resulting Pareto front for life (vertical axis) vs disc mass (horizontal axis). Figure 29b superimposes the Pareto front designs in the response surface of the mass. Only designs complying with the minimum life constraint of 40,000 flight cycles are shown. Starting with the lowest disc mass designs (which do not comply with the minimum life requirement), the first Pareto point starts at the lowest X1, X2 combination, the middle front side of Fig. 29. Adding mass by increasing X1 leads to complying designs, until the maximum in the right/forward feasible design area is obtained. From the Pareto front it can be seen that this occurs for a

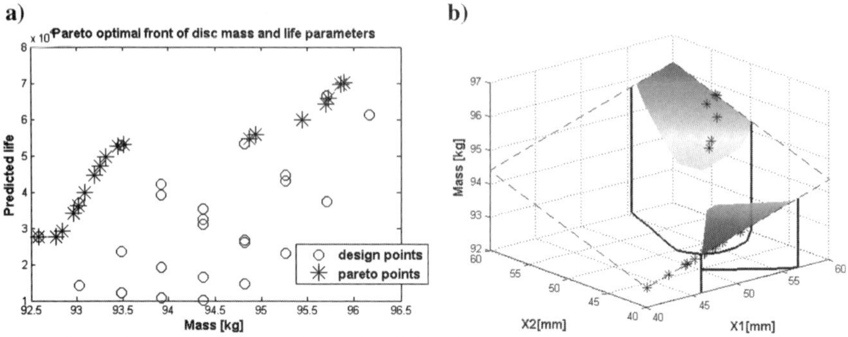

Fig. 29 a) Pareto front of life and weight objectives. b) Pareto designs plotted on the feasible design space.

mass of 93.5 kg. The disc designs with a mass between 93.5 and 95 kg are not Pareto optimal, as the mass increases but the life decreases. Only when the mass exceeds 95 kg, the life exceeds the previous value. The resulting designs are located in the aft feasible area of Fig. 29b. Both X1 and X2 increase for such designs. Such figures provide the disc designers with insight in the consequences of their design choices.

The described methods and tools can be extended with additional processes, for example, with a tool providing an estimate of the manufacturing or maintenance costs for the disc design to be used as additional objective or constraint. Such a tool with commercial or intellectual property rights restrictions cannot be supplied to other partners, but could be incorporated in the secured workflow of a multipartner collaboration.

4. *Robustness and Reliability*

One definition of robustness and reliability is the level of the probability of failure (POF) with regard to the standard deviation [6]. POF means that for a given number of samples, respectively, population (manufacturing parts, as example the discs), a specific percentage of these samples will not achieve the customer requirements as, for instance, the required life cycles of the disc. Therefore, a population of samples taking into account the variability and uncertainties of the design will be created and then assessed against the customer requirements. The less the POF will be the higher is the robustness respectively reliability of a design. Robustness investigations require lower sampling rates than for reliability analysis with a POF less than 10^{-6}. In other words, if only one sample out of 10^{-6} samples will not achieve the customer requirements, then the design is reliable. A reliability study was performed using the algorithm Adaptive Sampling on Adaptive Response Surfaces (ASonARS) implemented in the COTS tool used [6]. The most relevant geometry parameters and the boundary temperature values were varied (uncertainties and variability). Each of those parameters was sampled with an assumed statistical distribution of a truncated normal type.

The chosen design from the Pareto front (Fig. 26) with a nominal life of 35,280 cycles—without considering uncertainties and variability investigations—was checked for reliability regarding the expected life cycles. The predicted life was defined as a fictive requirement. Its POF was assumed to be less than 10^{-6}. For a real-world application the question needs to be answered: For which number of life cycles will that assumed POF be achieved?

The value for the limit-state function of the predicted life was set to 30,000 cycles in a first analysis. As such, an analysis is computationally expensive, and so only two cases are calculated. This limit leads to a POF of 3.0×10^{-5}, which does not fulfil the defined requirement.

A second analysis with a limit-state value of 28,000 cycles results in a POF of 1.3×10^{-8} (Table 2). This value exceeds the requirement clearly. Hence, the design decision made in the first step is not acceptable for the case that it is aimed to have 30,000 cycles with a POF less than 1.0×10^{-6}. A new design point with a higher nominal life has to be chosen from the Pareto front and its reliability checked again. The higher nominal life of the second design leads to a higher mass of the disc but can offer a higher level of reliability.

It is clear that the approach of checking the robustness or reliability after the multi-objective optimization might be very costly (Fig. 30a). For each design decision (point on the Pareto front), it is *a priori* unknown whether it fulfills a given robustness or reliability requirement.

Hence, it seems worth including the reliability information within the multi-objective optimization, that is, to perform a reliability optimization (Fig. 30b). Thus, each design point on the Pareto front fulfills the reliability requirements (given POF rate). The analysis effort however will significantly increase because for each design point a Monte Carlo simulation has to be performed. Surrogate modeling techniques and high-performance-computing strategies have to be developed to reduce this analysis effort.

VI. Design Framework Used at Disc Design Partner in Berlin

In the preceding the requirements are stated for an integrated design process in a collaborative multipartner environment. This subsection extends the principles of a generic framework for MDO and robust design. The development of the generic framework was based on three case studies demonstrating the increasingly complex achieved capabilities (data available online at http://www.vivaceproject.com):

1) The engine component level demonstrates the automated multidisciplinary optimization capability at a component level, for example, the high-pressure turbine engine disc.

Table 2 Results of reliability analysis

	Analysis 1	Analysis 2
Limit-state function (predicted life cycles)	30,000 cycles	28,000 cycles
POF	3.0×10^{-5}	1.3×10^{-8}

2) The engine assembly level extends the analysis process in order to be able to optimize assemblies. This level covers integration of individual engine components and their assembly in the context of integrator and supplier collaboration.

3) The last is collaborative MDO and RD: application of the developed tools, processes, and methods to a virtual enterprise engine case study within a multicompany environment

The principal design workflow within the generic framework is illustrated in Fig. 31. Parameterized design models (CAD) of the specific components or assemblies are analyzed (FE) and assessed against the design criteria and objectives (life cycles, weight) in an automated workflow controlled by an optimization or robust design instance. The process is distributed over different workstations with different operating systems and can be connected to a higher-level integrating process, as, for instance, the collaborative engineering main workflow of the integrator as described in Chapter 10.

An overview will be given with regard to the developed generic framework used for MDO and RD: the analysis, modeling, optimization, and robust design approaches, including also the collaborative engineering scenarios.

The basic generic requirements identified for a flexible integrated analysis process considering a collaborative engineering capability are the ability of

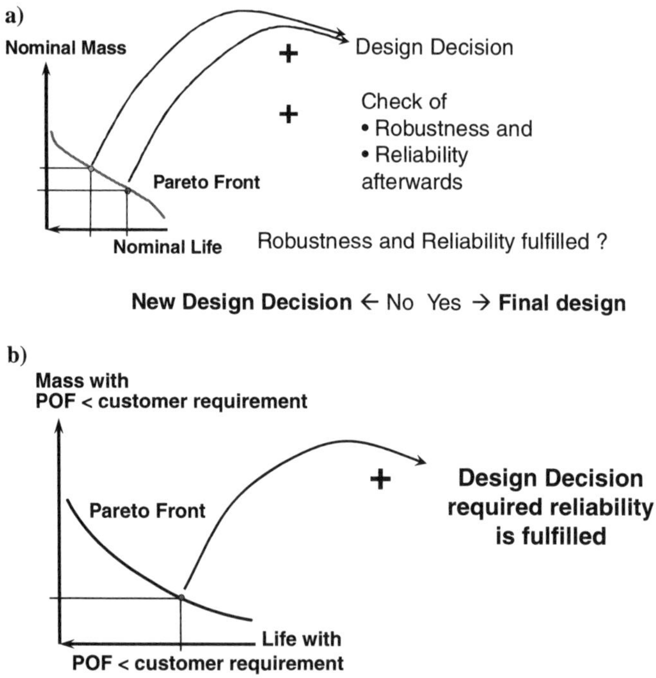

Fig. 30 a) Today's approach to find design with required reliability. b) Future approach to find a design with the required reliability.

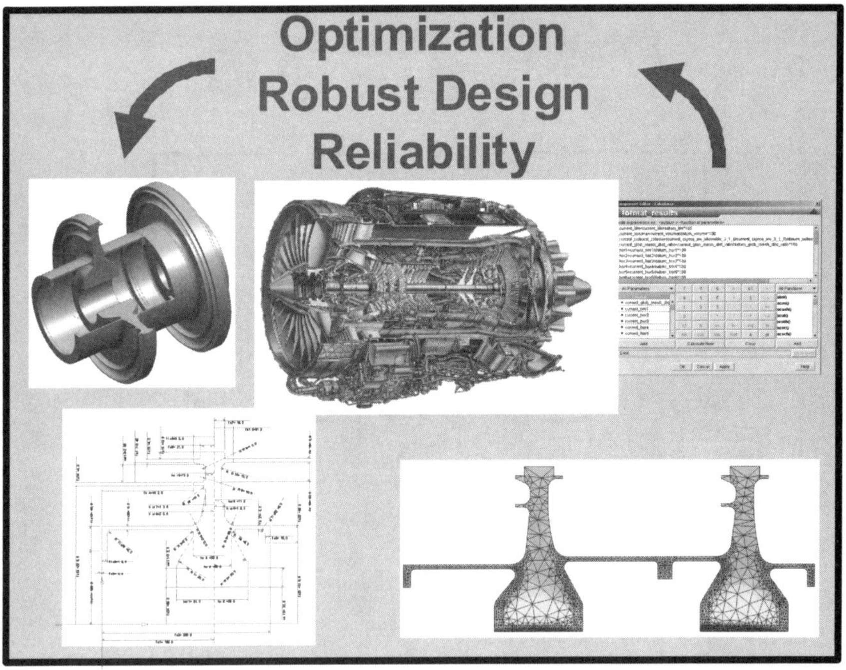

Fig. 31 Engine MDO and RD workflow.

flexible integration of various analysis tools, the easy change of the analysis and optimization process, and the possibility to distribute the design and analysis process across various company sites via secure connections. In addition, the ability to store the automated processes in a library (simulation data management) and the transferability of processes to other users considering platform independency and PLM integration possibility is also desirable.

Figure 32 shows a principal analysis process flow for MDO and RD implemented in the local framework. The process contains the access to a CAD system, the analysis tools (e.g., FE, CFD) and the results data evaluation tools for optimization and robust design purposes. The framework is able to run the tools on different workstations with different operating systems such as UNIX or Windows.

The process integration was achieved by using COTS software (data available online at http://www.simulia.com/products/sim_ept.htm) as described in Chapter 10. The necessary tools for the required automated analysis were implemented into the workflow. Figure 33 shows an example workflow for the integration of the CAD/CAE tools including CAD tool for parameterized geometry model (data available online at http://www.plm.automation.siemens.com/en_us/products/nx/index.shtml) coupled with an alternative workflow providing the ability to use the proprietary FE-tool SC03 (thermo mechanical analysis), or a COTS FE-tool.

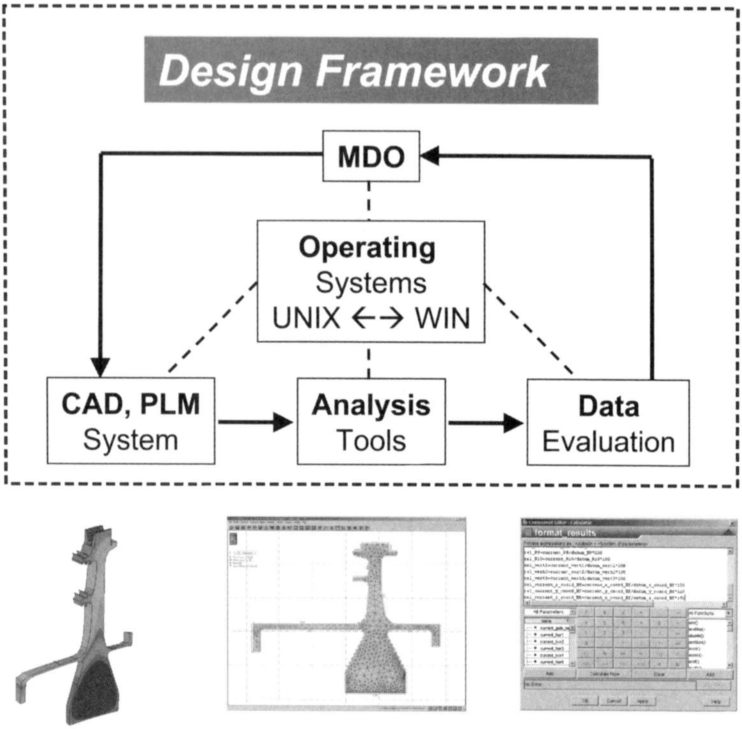

Fig. 32 Principles of the framework for MDO and RD.

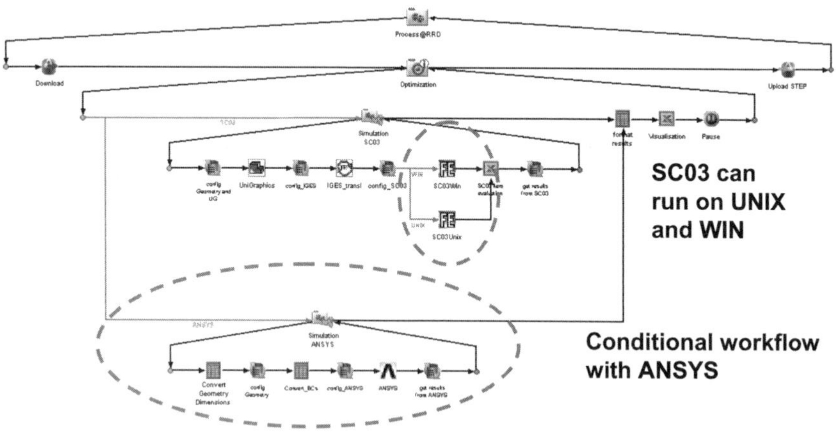

Fig. 33 Process integration for optimization and robust design on component level.

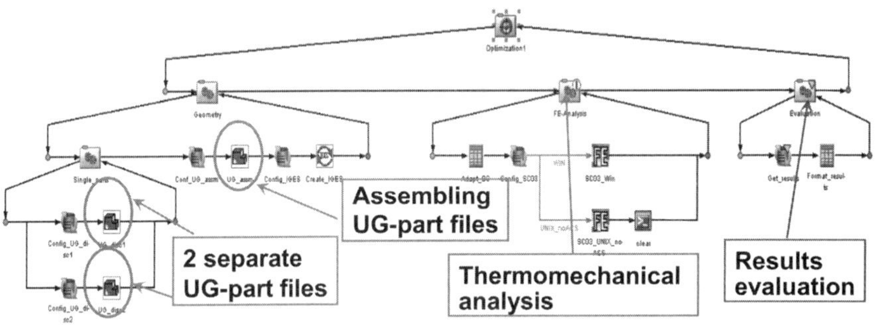

Fig. 34 Design process for assembly.

An additional workflow regarding the use of the proprietary FE tool running under the UNIX or Windows operating system is implemented.

In more detail, the workflow shown in Fig. 33 is controlled under the Windows system via a task manager, which can be used for DoE studies, optimization, robust design, statistical approaches, and so forth. Also the various parameters to be changed are controlled in the task manager. The parameters are mapped between the activities of the tool chain. Analysis loops can be grouped into modules, for example, automated analysis processes for components such as a disk or a blade. Modules and workflows can be stored securely in a library and are accessible to other users.

The development of the framework was based on the three case studies just mentioned. The thermomechanical analysis process was extended to allow the MDO and RD of an assembly in the context of an integrator and the supporting suppliers. This process is shown in Fig. 34. The geometry of the CAD part files can be changed individually by the process considering the access rights in a PLM system (thus considering the context of two different suppliers). The parts will be linked together via fixed or flexible interfaces. Subsequently, they will be analyzed, and the results will be assessed with regard to the design objectives (thus considering the role of the integrator).

Within the third case study, the framework was extended for collaborative engineering in a multicompany environment. As described in Chapter 10, access of the local design process to the main collaborative process flow controlled by the integrator requires a Web-based interface. The local design process used for the collaboration engineering is similar to the process shown in Fig. 33. The automated start of the local process is achieved by implementing special trigger components into the workflow (see Chapter 10).

VII. Design Framework Used at Disc Partner in Amsterdam

The disciplines and their supporting tools for the single engine-turbine-disc analysis, described in the preceding section, are combined in a tool chain [2] depicted in Fig. 35. The technical information on the tool chaining is provided separately in Chapter 10.

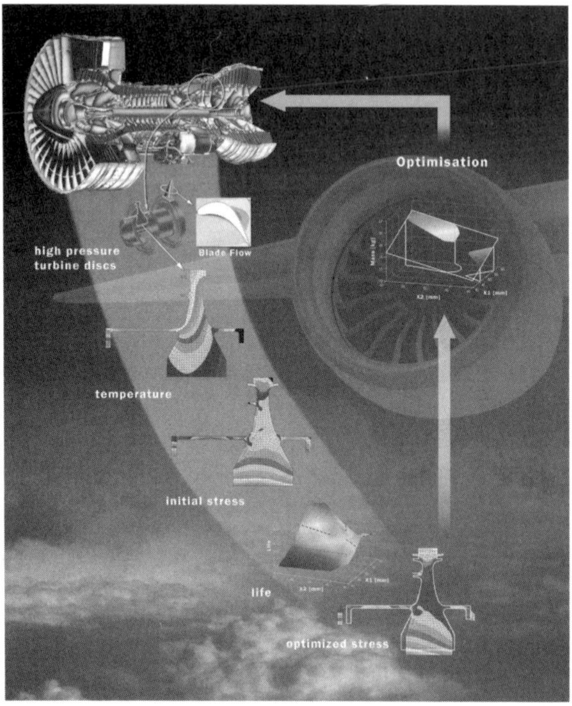

Fig. 35 Flexible tool chain to enable synergistic multidisciplinary design of a high-pressure turbine disc.

To allow an organization to dedicate its resources on engineering instead of on IT, most of the tools in the tool chain are COTS tools. The main exception is the lifing tool, for which the description has already been provided above. Such a mix of COTS and proprietary tools is typical for engine multidisciplinary design optimization. PLM solutions, as described in Chapter 12, can support the required automated conversion of information between the various tools involved. As proprietary tools might only be available on specific platforms, the chapter on the workflow engine explains how this requirement is taken into account. Because many engine disc design variants need to be analyzed, the set of tools was put into an automated top-level workflow, which invokes the automated workflow for this partner's single disc analysis. Even though the described tool chain excluded the computationally expensive computational-fluids-dynamics part for the blades and vanes performed by another partner (see preceding section), the analysis of a single disc design takes around 5 min. For some robust optimization algorithms numerous design evaluations are required, typically thousands or more. To increase the efficiency of the computations, meta-modeling or response surface approach is included in the tool chain.

VIII. Conclusions

The contributing partners in this engine case study have developed frameworks for MDO and RD considering a collaborative engineering capability. Each partner has set up its individual framework aligned with its company expertise, strategy, and a combination of COTS and proprietary tool suites. Different framework solutions were achieved by the different companies starting with simple design case studies demonstrating MDO and RD at component and assembly level and then progressing to extended design workflows. Thus online collaborative distributed design processes within an engineering consortium based on business case scenarios could be demonstrated. This in fact was one of the main project goals. In this context considered were design technologies and concepts, state-of-the-art optimization, robust design approaches, and coupled optimization processes between different companies.

Despite every partner having developed a different framework, it was possible to connect all companies together. However, there was no common solution; each partner had to find its own secure solution to get access to the main workflow. This was a major challenge for the contributing partners because the proprietary rights of each company had to be ensured. More information on the security is provided in Chapter 11.

A high maturity level of local design framework solutions could be achieved, ready for further exploitation and dissemination at each partner's company. In summary, a possible global integrating framework solution for collaborative engineering based on individual design framework solutions was demonstrated with regard to a technical solution and an architecture concept (VEC hub).

The innovative features of this concept include the demonstrated ability to perform an automated distributed multilevel and multidisciplinary MDO and RD design process within a consortium of different companies, using different design tool sets and analysis methods, but still considering all security aspects and proprietary rights of each company. These online collaborative workflows demonstrated at this time have been the most extensive ones within a European consortium.

Finally, described in this section has been yet another demonstration of excellent teamwork within the European community. The authors of this chapter wish to thank the whole team for their effort and good work.

Acknowledgments

The authors (disc design partner at Berlin) gratefully acknowledge the support from the companies DYNARDO and CADFEM for the optimization and reliability studies.

References

[1] Argüeles, P., et al., "Report of the Group of Personalities," *European Aeronautics: A Vision for 2020*," European Commission, Jan. 2001, http://ec.europa.eu/research/growth/aeronautics2020/en/personalities.html [retrieved July 2009].

[2] Kesseler, E., and van Houten, M. H., "Exploring Automatic Multi-Objective Turbine Disc Design for Virtual Engines," *ECCOMAS Multidisciplinary Jubilee Symposium*, Vol. 14 of *Computational Methods in Applied Sciences* edited by J. Eberhardsteiner, C. Hellmich, H. A. Mang, and Périaux, J., Springer, Netherlands, 2009, pp. 251–262.

[3] "Eurocode Part I Basis of Design and Actions on Structures," *Eurocode, Nine Parts (1991–1998)*, http://www.eurocode-resources.com/eurocodes/eurocodes.php [retrieved July 2009].

[4] Stein, E., Rüter, M., and Ohnimus, S., "Adaptive Finite Element Analysis and Modeling of Solids and Structures. Findings, Problems and Trends," *International Journal for Numerical Methods in Engineering*, Vol. 60, No. 1, 2004, pp. 103–138.

[5] Bergan, P. G., and Skeie, G., "Achievements and Challenges for Computational Mechanics Within the Offshore Industry," ECCOMAS Multidisciplinary Jubilee Symposium, 2008. The invited presentation was held on 20 Feb. 2008 from 9:20–9:45; see http://emjs08.tuwien.ac.at/fileadmin/mediapool-emjs08/Diverse/Programme_010208.pdf.

[6] "optiSlang—The Optimizing Structural Language for Sensitivity Analysis, Multidisciplinary Optimization, Robustness Evaluation and Reliability Analysis," ver. 2.1, Dynardo Software and Engineering GmbH, March 2007, www.dynardo.de [retrieved July 2009].

[7] Chaput, E., "Design Challenges in the Airbus A380 and A350 Projects," *ECCOMAS CFD 2006*, edited by P. Wesseling, E. Oñate, and J. Périaux, European Community on Computational Methods in Applied Sciences (ECCOMAS), Sept. 2006. E. Chaput was an invited speaker on 7 Sept. 2006 from 9:00–9:40; see http://eccomascfd2006.fyper.com/?menu=program.

[8] Rostand, P., "Status and Challenges of Aero Shape Design of Future Aircraft," *ECCOMAS CFD 2006*, edited by P. Wesseling, E. Oñate, and J. Périaux, European Community on Computational Methods in Applied Sciences (ECCOMAS), Sept. 2006, http://proceedings.fyper.com/eccomascfd2006/ [retrieved July 2009].

Chapter 10

Using the Virtual-Enterprise-Collaboration Hub for Distributed Engine Optimization

Holger Wenzel*

Engineous Software GmbH, Aachen, Germany

Fredrik Almyren†

Volvo Aero Corporation, Trollhättan, Sweden

Johannes Barner‡

MTU Aero Engines, Munich, Germany

Erik Baalbergen§

National Aerospace Laboratory (NLR), Amsterdam, The Netherlands

Bernd Meissner¶

Rolls-Royce Deutschland, Dahlewitz, Germany

and

Mats Lindeblad**

Volvo Aero Corporation, Trollhättan, Sweden

*Senior Engineer.

†Methods Development.

‡Engineering Systems.

§Senior Scientist.

¶Mechanical Methods.

**Collaboration and Product Information Strategy.

Abbreviations

ACL = access control list
ACS = application control system
ADP = airfoil design partner
API = application programmable interface
B2B = business-to-business connection
CAD = computer-aided design
CAE = computer-aided engineering
CORBA = common object requesting broker architecture
COTS = commercial-off-the-shelf
DCOM = distributed component object model
DDP1 = disk design partner 1 (first stage)
DDP2 = disk design partner 2 (second stage)
DMZ = demilitarized zone
FEM = finite element method
GUI = graphical user interface
HTTP = hypertext transfer protocol
HTTPS = secure HTTP
ID = identity
IPR = intellectual property rights
IT = information technology
JCR = Java Content Repository
JeCARS = Java Extendable Contents and Rights System
JEE = Java Enterprise Edition
LAN = local area network
LDAP = lightweight directory access protocol
MDO = multidisciplinary design optimization
MRO = maintenance, repair, and overhaul
O = organizational data
OASIS = Organization for the Advancement of Structured Information Standards
OEM = original equipment manufacturer
OWL = Web ontology language
P = product data
PDM = product data management
PKI = public key infrastructure
PLCS = product life-cycle support
PLM = product life-cycle management
R = reference data
RDL = reference data library
REST = representational state transfer
RPC = remote procedure calls
SaaS = software as a service
SME = small-to-medium enterprise
SOA = service-oriented architecture
SOAP = simple object access protocol
SSL = secured socket layer
TLAP = turbine life-assessment partner

URL = uniform resource locator
VE = virtual enterprise
VEC hub = virtual-enterprise-collaboration hub
W = work processes
WSDL = Web services definition language
W3C = World Wide Web consortium
XML = extensible mark-up language

I. Introduction

MOST complex products available today are made of parts and subsystems that are not designed by the enterprise that brings that product to the market, but by different, and often by many different, companies. If the constituent subsystems are not bought off the shelf, but are customized or specifically crafted for the given end product, it is often desirable for the original equipment manufacturer (OEM) to include the suppliers into the design process.

Traditionally the information exchange, which is necessary to enable each partner to perform the assigned design tasks, is done manually, by sending emails, data CDs, or printed documents. This is not only slowing down the design process considerably, but introduces many risks into the product development process. For example, it is not guaranteed that all partners work on a consistent data set at any given time.

Therefore, the need for design systems that can work within a multi-enterprise setup arises. Depending on the nature of the industry, one of two different scenarios can apply. In the extended enterprise scenario, there exists one company with superior strength and the power to force its own set of design tools and processes onto all other partners that want to take part in the distributed design process. These partners then need to connect to the collaborative infrastructure and adjust their own development tools and processes to the structure that the main player provides.

If, in contrast, a product is designed by a consortium of partners of similar strength, they can form a virtual enterprise (VE). A way to organize the information and work flow in such an entity is presented in this chapter.

Adopting this point of view, the multicompany, multinational, and multidisciplinary optimization of the two-stage high-pressure turbine of a jet engine, an example of which is described in Chapter 9, can be viewed as an optimization study in such a virtual enterprise. Here the virtual enterprise is a consortium of several partners of similar strength that is formed to pursue a common goal (see Fig. 1). Because of work share, risk share, and other business factors characterizing the virtual enterprise, it is very difficult even for a dominant partner to prescribe the computational and simulation environment, tools, and infrastructure. This presents a severe challenge to the implementation of any automated multidisciplinary optimization study, where all participating tools, owned by different partners in the consortium need to be linked together and be usable in an automated fashion. Tackling this challenge has been the objective behind the work described in this chapter, which is organized as follows.

The concept of the virtual-enterprise-collaboration hub (VEC hub) is described first. The VEC hub is an abstraction layer capable of integrating the different design systems located at the different virtual enterprise partners.

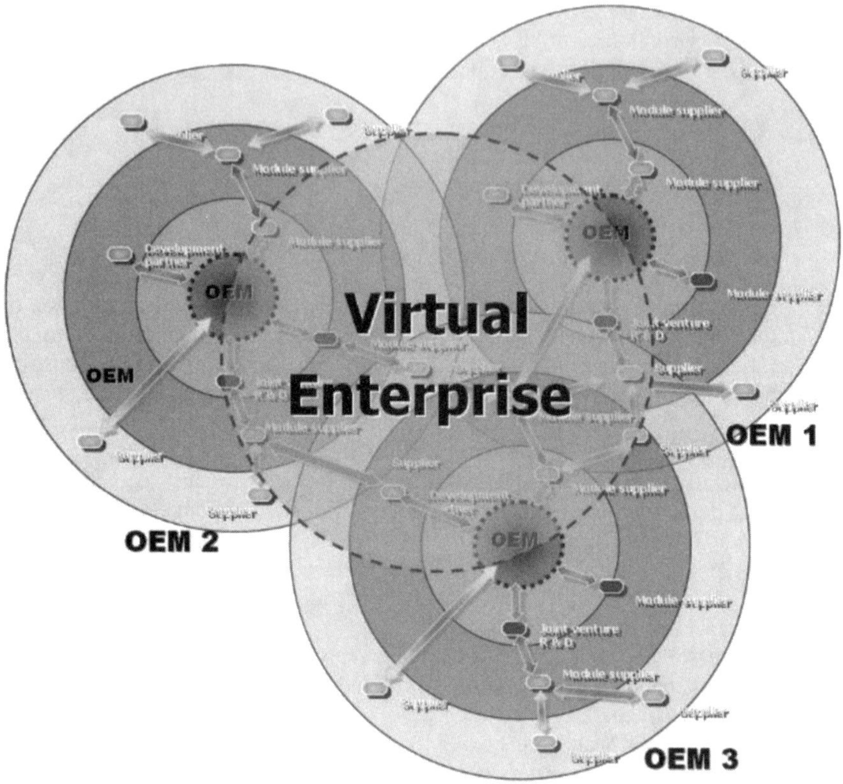

Fig. 1 Virtual enterprise.

The second part of this chapter details the actual realization of the VEC-hub concept that enables the distributed multidisciplinary engine optimization scenario (Chapter 9). Here the global workflow that drives the different aspects of the multidisciplinary design task is implemented directly at the VEC hub. During its execution, it calls the local optimization and simulation processes, which are described in Chapter 9. The different ways to achieve this coupling are also explained.

II. Description of the VEC Hub

Achieving an effective collaborative engineering design process in a VE implies a number of different technological, social, legal, and economical challenges.

Each partner possesses its own IT infrastructure and policies, with different operating systems, computer-aided design (CAD) systems, product data/lifecycle management (PDM/PLM) systems, and different suites of simulation tools. Also the way of working with these systems differs widely between companies.

The access to information located within the company is subjected to different security restrictions. Even the way to get hold of data from outside of the enterprises firewalls can be handled differently at each partner site.

Yet this information needs to be shared in an effective and secure manner, so that it is accessible in the common collaborative context and can be used in the design project.

The ownership of the data, methods, and knowledge must be managed in a precise fashion, so that at any time the ownership of the design tasks and parts are well defined. Any change of the information must be traceable, but the intellectual property of the partners that consists mainly of the expertise reflected in their design processes must be guaranteed.

A further challenge stems from the fact that during the lifetime of a product the partners of the VE can change. For instance, some partners would be taking part only in the design stage while others would be participating in the manufacturing phase or in the maintenance, repair, and overhaul (MRO) phase.

Mutual thrust between all participating companies plays a central role when setting up a collaborative engineering project in a VE. Trust must be established by nontechnical means, and it must be secured by legal contracts. (For more on security and trust, please refer to the dedicated security chapter.)

In this context, the objective of the VEC-hub concept is to define a service-oriented architecture (SOA) for the virtual enterprise, to support the collaboration at information-technology (IT) level, with the specific needs of the engineering related activities in all phases of the product life cycle in mind.

The VEC hub establishes a neutral collaborative platform, with respect to technical (vendor and system) internal and external aspects. Thus, each participating partner can connect to the VEC hub independently of their local IT infrastructure. The collaborative environment assets are available via commodity technology such as Web browsers or via an application programmable interface (API) implemented as Web services according to the Web services definition language (WSDL), see Sections II.B and III.B.4.

The VEC-hub concept defines a common, agile setup of partner roles, product information, work processes, and reference data for the collaborative project. It promotes partner-to-partner isolation through intermediate services that communicate using neutral data structures. Thus the intellectual property rights (IPR) of each partner are protected.

The solution base for the VEC hub is the integration and scale-up of emerging and commonly used software solutions with a focus on deployment and further development of open standards for context and content definitions such as ISO 10303, developed within ISO/TC184/SC4 (data available online at http://www.tc184=sc4.org), and its derivatives, as well as the OASIS product life-cycle support services (PLCS) [1]. This standards-based approach is chosen to ensure long-term review and reuse, independent of the software used for creation and updating of generated project and product information content.

By referring to and creating new standards when defining the VEC-hub concept, means for scalability and capability/service redundancy are secured. A common vocabulary defining the VEC-hub concept and its parts is developed.

A. Definition and Core Concepts of the VEC Hub

When a number of companies work together to produce a product, they can form a virtual enterprise that spans multiple enterprises; see Fig. 1. It is unlikely that the commercial-off-the-shelf (COTS) approach used by the multiple companies will be fully utilized, that is, the same tools would be used by each extended enterprise. Even if that were the case, these COTS tools would be configured differently, which complicates direct information exchange. The problem is further exacerbated because of a common practice where COTS tools are complemented by proprietary software, which constitutes some of the assets that each participant brings to the VE.

In such collaboration it is therefore essential to always know where to find approved data and how to notify other stakeholders of changes to it. If data are not managed properly, configuration management issues arising will consume a lot of time and resources during the product life cycle. By sharing data, configuration management can be accomplished with less effort during the full life cycle of the product, independent of the actual phase of the project.

The VEC-hub concept and the accompanying guidelines specify how a number of services collaborate and share data within a set of partner companies working together. The objective of the VEC hub is to be a partner-managed platform that is neutral. The use of services based on platform-independent application integration standards, namely, the WSDL, promotes this neutrality from a technical point of view. A core concept for realizing this functionality, without specifying details about which application/system/tool is providing it, is the use of World Wide Web consortium's (W3C) Web services [2].

1. Sharing of Common Data

The requirement of the collaboration project is to adopt a common agile setup of partner roles, product information, work processes, and reference data. However, in the context of the virtual enterprise, partners do not use the same infrastructure, neither do they have the same view on internal organization, product data, or processes. To enable collaboration, and to define WHO will do WHAT and HOW, the VEC-hub concept offers services to manage data common to the virtual enterprise, including product (P) information (WHAT), organizational (O) Roles (WHO), work (W) processes (HOW), and reference (R) data. All common project information for this specific collaboration is shared in a controlled manner among the partners of the virtual enterprise in the VEC hub. This arrangement is represented by the innermost area in Fig. 2.

2. Neutral Formats

Neutral data structures and formats are essential for enabling communication and usage of common data in a collaborative environment where partners' privacy and ownership of information must also be protected. Each of the organizations joining a virtual enterprise collaboration has its own internal infrastructure, for example, applications, operating systems, firewalls, and other security solutions. They also have means for managing their organization, their products, and their respective product development processes. In a work and risk-sharing

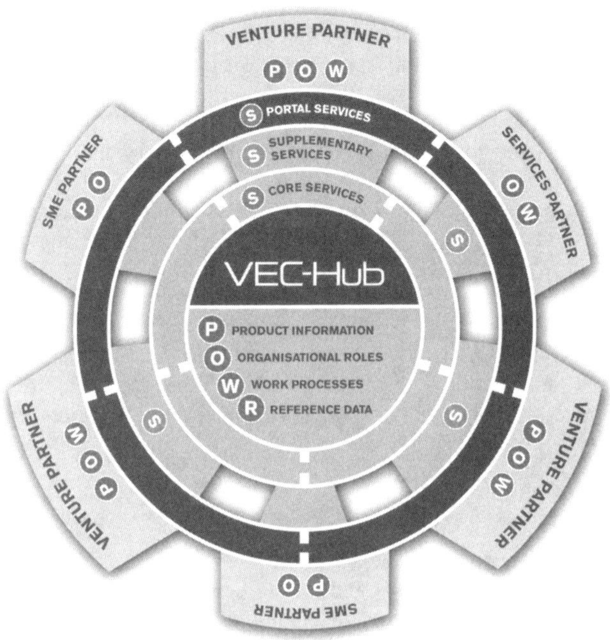

Fig. 2 VEC-hub concept.

environment where there is no dominant partner, the VEC-hub concept allows the partners to retain their infrastructure. In such an arrangement, the shared information is made usable for each partner by using neutral and, preferably, standardized data structures. Neutral formats support interoperability and also persistence. This persistence relates to long-term archiving, which is important to most businesses of today and tomorrow and especially to the aerospace sector, where the useful life of aircraft can span several decades.

3. *Service-Oriented Architecture*

The notion of scalability and agility needs to be addressed in order to define a neutral partnership collaboration platform. Wrapping the common shared data and processes as services helps to ensure that scalability and agility are in place. This service-oriented approach also promotes a partner-to-partner isolation through intermediate services.

The value of a service-oriented architecture (SOA) is that it provides a simple scalable paradigm for organizing large networks of systems that require interoperability to realize the value inherent in the individual components. In the VEC-hub specialization of the SOA reference models [3], the common data and processes are made visible through the VEC-hub core services represented by the inner segmented ring denoted as "Core Services" in Fig. 2.

Venture partners can share their internal product data, organization, and workflows by publishing them as services onto the VEC hub. This enables other partners of the virtual enterprise to see those services and interact with them while the providing partner still is in control of the usage and access to the service. The shared services are represented by the "S" between the outer ring called "portal services" and the inner area "core services" in Fig. 2.

A services partner is a partner specialized in providing software as a service (SaaS) on a pay-per-use basis allowing interaction with applications or processes through a Web service. The services partner does not have any product data internally as shown in the "Services Partner" area of Fig. 2.

Interaction with and visibility of the services provided by the VEC hub are enabled by the VEC-hub portal services represented by the outer segmented ring of "Portal Services" in Fig. 2. These services provide the authentication and authorization as well as the management of the VEC-hub services. They also enable access to the collaboration environment assets via commodity tools such as Web browsers.

B. Architecture

The VEC-hub concept is a specialization of a SOA just explained. It is a paradigm describing how to organize and utilize distributed capabilities that might be under the control of different ownership domains. Visibility, interaction, and effect are key concepts for describing the SOA paradigm.

The visibility concept refers to the capacity for those with needs and those with capabilities to be able to see each other. Awareness, willingness, and reachability are central to the visibility of a service. The VEC-hub services are defined using the WSDL notation. The service descriptions are publicly available. The services related to the data management are standardized as PLM/PLCS services within the Organization for the Advancement of Structured Information Standards (OASIS) [1]. For a service to be included in a VEC hub, the policies governing the service interactions should be documented in the service description of the service. Reachability is promoted through the VEC-hub service management services.

The second concept behind the SOA paradigm is interaction. It refers to the activity of using a capability. The structure and semantics of the information model along with the behavior model of a service are important issues to enable interaction. The interaction with the VEC-hub services is promoted by the service descriptions where methods, together with their arguments and their argument types, are defined.

The third concept, namely effect, refers to the return of information or the change in the state of entities (known or unknown) that are involved in the interaction. All interactions with the VEC-hub services are managed through the VEC-hub service management services. When a service is returning information, this information is routed through the VEC hub to enable traceability of service usage.

The SOA paradigm [3] does not itself address all of the concepts associated with ownership, ownership domains, and actions communicated between legal peers. To fully account for concepts such as trust, business transactions, authority, and delegation, etc., additional conceptual frameworks and architectural elements are required. The VEC hub specializes the SOA paradigm by defining the following services:

1) The VEC-hub portal services provide access management and security management end ensure control over services. They establish a service management to provide visibility of and interaction with services.

2) The VEC-hub core services provide services that are described and can be accessed in a standardized way in order to enable partners to share product data, organizational data, reference data and processes, including workflows.

3) In the VEC-hub concept supplementary services provide the means to enable access of services outside of the core set yet complying with the specified interface requirements. An example is a partner-supplied service that enables other partners to open an online interface into their own internally running processes and data storage while maintaining control of the access and usage of those services.

The standardized interfaces provide one application programmable interface (API) for applications to access the VEC-hub services and one graphical user interface (GUI) to enable accessing the VEC-hub services using commodity tools such as a Web browser.

The logical architecture, as described in Fig. 3, is divided into four layers.

Located at the top is the presentation layer. It provides a secure interface to the VEC hub where the VEC-hub services are exposed. Below the presentation layer resides the business logic layer. It organizes all of the services provided by the VEC hub, that is, it provides the visibility of the services along with the presentation layer. At the bottom is the data and infrastructure layer, where the networks and applications are hosted that are necessary for storing data and

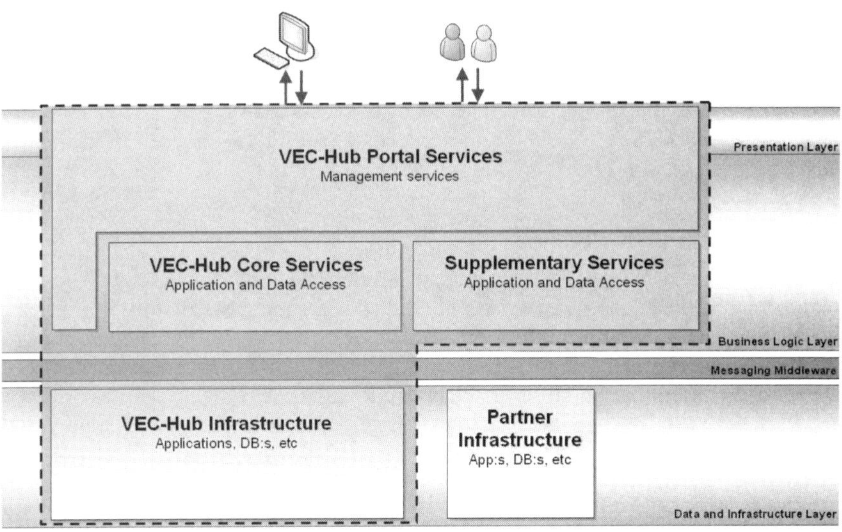

Fig. 3 VEC-hub logical architecture.

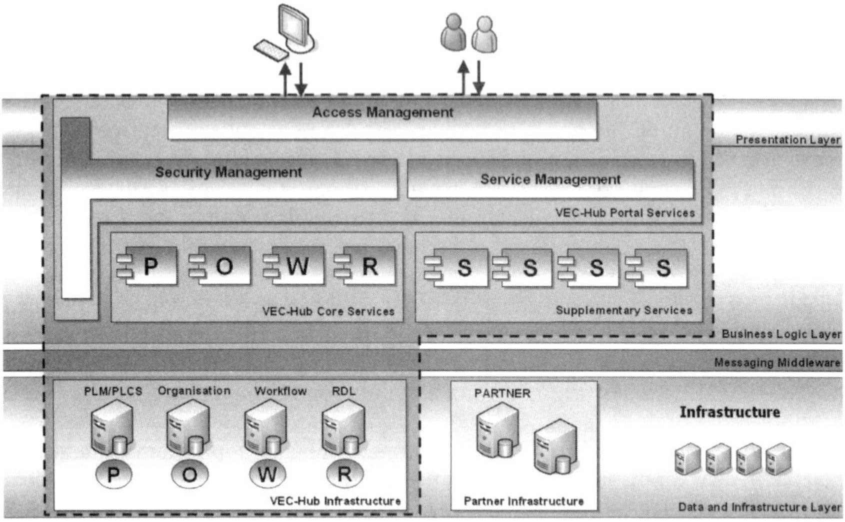

Fig. 4 Separation of service interfaces and infrastructure.

executing services. This layer is delivering the effect of a service. The messaging middleware layer between the data and infrastructure and the business layer mediates the communication between the services interfaces in the business layer and the corresponding infrastructure in the data and infrastructure layer. It enables the interaction with the service.

By separating the service interface from the actual infrastructure delivering the effect of a service, the VEC-hub logical architecture enables independent selection of service providers (see Fig. 4). This requires that the description of the service is represented in a standardized way. WSDL is used for describing the services in VEC hub.

Furthermore, the service interface needs to be standardized for the capability that it represents. PLCS Web services interface (see Fig. 4) is used in the VEC-hub core services for access to and handling of product data.

C. VEC-Hub Core Services

Core services represent the intrinsic characteristics of the VEC hub. They are mandatory in the sense that they define the product and its structure, the organizations/persons with their roles and access rights, the workflows such as change requests and dynamic design workflows, and finally, the classification of all of these previous concepts. These services are detailed next.

1. Product Information Services

The P part of the core services stands for product definition. Complying with the need for neutral formats as just discussed, the PLCS standard (ISO 10303-239) is the basis for the product and product-related representation in the VEC hub. The creation, updating, and deleting of product-relevant data are managed through PLCS/PLM Web services [4].

The key sets of PLCS/PLM services are the following:

1) *Part management services* cover management of parts and part structures.

2) *Document management services* cover services regarding management of documents.

3) *Change management services* cover services regarding controlled changes like managing engineering change proposals.

4) *Call-back services* manage subscriptions to and notifications of events in the VEC hub.

2. *Organizational Services*

Administrative services such as organization-related services, that is, the O part of the core services, are related to the organizations and persons participating in the collaboration in the VEC hub. The organizational services manage information about enterprises, users, and roles undertaken by other VEC-hub services, for example, the security services.

By defining organizations, users, and roles, the security services promote an authentication process ensuring that only dedicated users can login to the central hub. The same information is necessary to authorize the VEC-hub users to get access to services with access rights explicitly assigned to them. This is under the control of the user management and enterprise administration services that are implemented in the VEC hub.

For services provided by a venture partner or services partner in its partner-shared area, the providing partner still has the control of *who* is authorized to do *what* when interacting with the service.

For more information on security and authentication, please refer to the dedicated security chapter.

3. *Workflow Services*

The workflow-related services (the W part of the core services in Fig. 4) address the agreed-upon workflows of the collaboration.

These workflows can be at various levels of detail, from generic levels such as change request to detailed levels such as detailed design of a turbine disc, as discussed in this chapter.

The generic functionality provided here is management of workflows, that is, create, edit, and delete, start a remote workflow process, stop a remote workflow process, and monitor a remote workflow process (progress, status). Remote monitoring is important from usability point of view.

Exception handling is necessary, so that the user receives error messages from the workflow process, not only after it has finished, but also during the execution.

The file management is also part of the workflow services. The latter handle large input and output files, with the possibility that each individual workflow has multiple input and output files. Also, it is possible to feed parameters into the input files and extract output parameters from the result files. The security of the workflows needs to be at the same level as the data. This is established by treating the workflow models as data that are stored using the P services of the VEC hub.

4. *Reference Data Services*

The reference-data-related services, the R part of the core services in Fig. 4, are aiming at the local VEC-hub reference data, which can be referencing external sources or can be project specific.

Also called domain models the reference data itself, is defined in the reference data library (RDL). Products defined in the P part of the VEC hub will be checked against the RDL to ensure that the classification is valid.

This local reference data server is based on PLCS RDL server, a database that serves OWL classes for PLCS applications via a set of Web services.

5. *Portal Services*

The VEC-hub portal services enable interaction with and visibility of the services provided by the VEC hub. They are represented by the outer segmented ring of portal services in Fig. 2. These services provide authentication and authorization as well as management of the VEC-hub services. They also enable access to the collaboration environment assets via commodity tools such as Web browsers.

6. *Access Management*

The VEC-hub access services provide the functionality for the users to connect to the VEC hub and to interact with the service provided by the VEC hub. The access services are used in every access to the collaborative environment via the application programmable interface (API). This is the only interface by which access to services is provided, that is, the single interface to the VEC hub.

Depending on the organization and role of the user, as defined in the organizational services and validated by the security services, appropriate services are exposed as Web services in the API. The Web services offered by the VEC hub include a policy description regulating the access to and interaction with the services.

To provide easy access to the common collaboration assets, the VEC-hub access services provide a COTS-supported graphical user interface, built on top of the API.

7. *Service Management*

To provide a robust, maintainable, and performing VEC hub, the core services and the supplementary services just described need to be managed.

This is the task of the service management services. They are located directly at the VEC hub, but not visible to the end user. Only users with an administrative role can access these services.

Any service accessible at the VEC hub needs to be registered with the service registry, and subsequently the service repository can provide information about these services to the end users. All services need to be validated against the service validation services. Access to, and problems with using the services are recorded by the logging and execution managing services, so that the accounting manager can provide the information necessary for billing the customers that used the services.

Finally, the service access manager provides the means for accessing registered services. This part of the service management enforces the rule that no user can get access to services originally not accessible to him/her, even when attempting to call these through other services.

D. VEC-Hub Security

The virtual enterprise security requirements can be subdivided into requirements with respect to securing the VEC-hub areas, protecting the VEC-hub communications, VEC-hub services access control, and security requirements on specific VEC-hub Web services.

1. Securing VEC-Hub Areas

As shown in Fig. 5, the VEC-hub areas are the portal, core, and supplementary services areas while the partners' private areas reside outside of the VEC hub. These areas must be protected against access by non-virtual enterprise (VE) partners.

As any computer that is exposed to the Internet, the VEC-hub portal server must be installed on a *hardened* machine, where all feasible means to protect

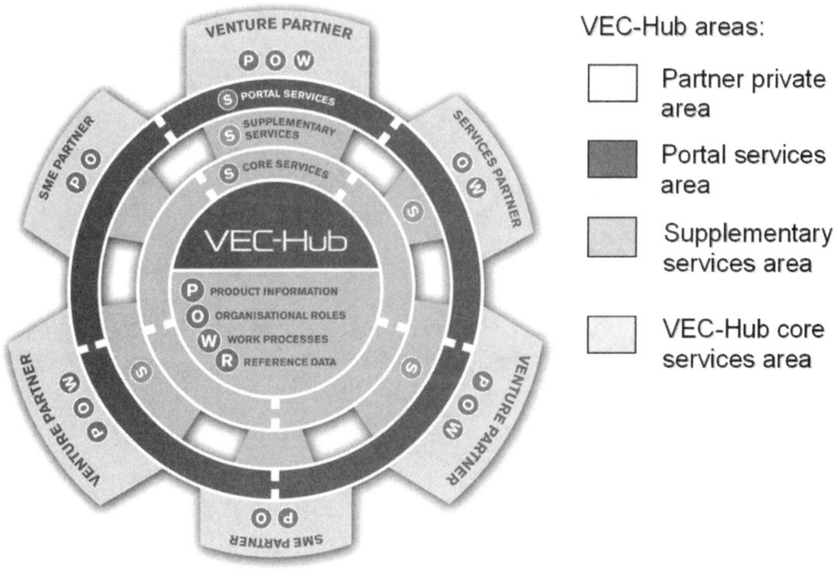

Fig. 5 VEC-hub areas.

the machine from malign attacks have been taken. The server should accept connections only from collaborating partners. Securing this area of the VEC hub is the responsibility of the VE or the service provider given the task to implement and maintain the VEC hub.

Securing the supplementary services is based on the premise that, in general, these reside in areas administered by the partners providing the supplementary services.

The interfaces to the supplementary services are normally exposed on hardened servers in the partner's demilitarized zone (DMZ). How this is achieved depends on the company's own IT policies and cannot be prescribed by the VE or the VEC-hub concept. Some examples on how this is implemented are given in Sections III.A–V of this chapter.

Securing each partner's private area (for usage of internal data and services, for example, due to licenses and available hardware) is that partner's own responsibility.

While implementing the VEC hub, it must be guaranteed that a successful attack at one partner's site cannot put the security of the VEC hub itself and the other partner sites to risk.

2. *Protecting VEC-Hub Connections*

Figure 6 shows the most often used ways to connect to the VEC hub. The security measures for each of these connection types are explained next.

In general, end users such as small-to-medium enterprise (SME) and venture partners will use their respective client applications in an insecure medium, the

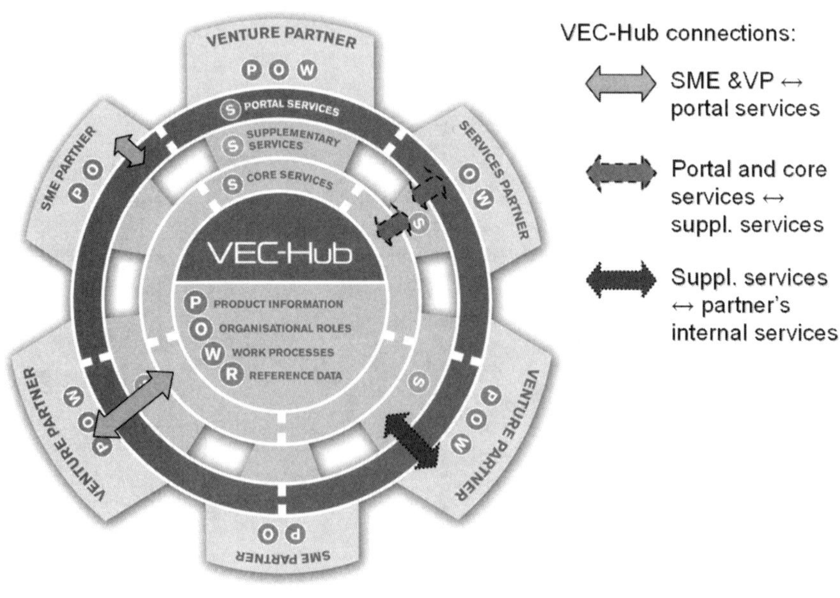

Fig. 6 VEC-hub connections.

Internet, to access the VEC hub. Hence, these connections must be secured. Also when using dedicated/leased lines as connection media, protecting these connections might still be appropriate.

Because, typically, the supplementary services are located in a zone administrated by a venture partner or services partner, such connection is commonly implemented using the Internet and must be secured. Securing the connection between the supplementary services administered by a partner and its own private area is this partner's own responsibility.

3. *VEC-Hub Access Control*

It is essential to set access rights to the VEC-hub services, only allowing certain trusted/mutually agreed users, certain groups, or companies to access a specific service, or at finer granularity, to access a part of a service. When access to the service has been granted, the service itself will also implement additional access control: for example, allowing only specified user or specified groups to make modifications, giving other users only read permissions, and so forth.

Identity management involves the management of user identities across all resources involved. It is established by an integrated system of business processes, policies, and technologies that enable organizations to facilitate and control their users' access to critical online applications and resources while protecting confidential personal and business information from unauthorized users. It represents a category of interrelated solutions that are employed to administer user authentication, access rights, access restrictions, account profiles, passwords, and other attributes supportive of users' roles/profiles on one or more applications or systems. (For more information please refer to the dedicated security chapter.)

4. *Authentication*

Users must authenticate themselves before being able to access the VEC hub. These credentials can consist of a simple user-name/password combination or a more advanced public key infrastructure (PKI) or a Web of trust [5].

When using user-name/password authentication, a strict password policy should be established. Many successful attacks on IT systems are caused by inadequate passwords. The VEC hub should define appropriate rules for the quality and renewal of passwords.

5. *Authorization and Role Management*

An appropriate authorization process ensures that only dedicated authenticated users can login to the central hub and that every VEC-hub user only gets access to information with access rights explicitly assigned to him/her.

For services provided by a VEC-hub partner in its partner-shared area, this VEC-hub partner must have the control of *who* is authorized to do *what* within this partner's service.

Specific requirements for access rights to the service that provides access to the data are listed next. The administration of access rights to the data supports the assignment of access rights to users and groups of users using access control lists (ACLs). Typically, this offers the possibility to set the permissions for accessing a single entity, or for accessing a collection of entities. The access rights include read, write, delete on data, plus execute on workflow items stored on the VEC hub. In different projects, different life cycles of a project, or in different collaborations, it should be possible to define different details of granularity of the permissions.

There should be at least the possibility to define roles such as user and system administrator. In addition, roles for project management might be required, for example, a role that allows the creation of a virtual project space in which a whole project is managed. This project management role can then edit the structure of the project while regular users are only allowed to work within this structure in accordance with the defined roles.

III. Global Optimization Workflow

The remainder of this chapter describes how the VEC-hub concept is utilized as the backbone of the multi-enterprise collaborative optimization of the high-pressure turbine as described in detail in Chapter 9. The workflow that drives the multisite, multicompany, multidisciplinary optimization is shown as a block diagram in Fig. 7. The interaction with the PLM/PLCS services at the VEC hub is indicated by the text blocks to the left of the diagram lines. Airfoil design partner (ADP) and disc design partner 1 (DDP1) have no direct connection to the VEC hub within their local design processes. For them the collaborative data and parameter management are handled by the main workflow. The technical details and the content of the local design processes are described in Chapter 9.

In this study the global optimization workflow is implemented directly at the VEC hub, using its core workflow services. The setup of this workflow was performed by the partner of the VE that was assigned the task of the global design workflow integration.

This partner starts the design process and, using the PLM/PLCS services of the VEC hub, updates the shared data repository with the mission parameters and the results from the preliminary design. This information is fed as a set of parameters into the multidisciplinary blade design process that is executed locally at the partner site of the ADP. The ADP offers this service as a partner-supplied service, which is registered with the VEC hub, as depicted in Fig. 1.

The result of the blade optimization is returned to the main workflow and published on the VEC hub, where it can be used in the subsequent step, that is, the design of the discs. This process is carried out by two partners, one for each stage of the high-pressure turbine. The interface between these two parts is the flange position where the rotating discs are in contact. Only this parameter needs to be shared between the two partners. At the same time, the flange position is an optimization parameter that is changed by an optimization loop across the two partner sites in order to achieve the required lifetime with the lowest possible mass. In Fig. 7 this loop is indicated as *flange position optimization loop*.

As with the ADP, no client is implemented that can connect to the PLM/PLCS-services of the VEC hub at the DDP1 site. Therefore, the necessary

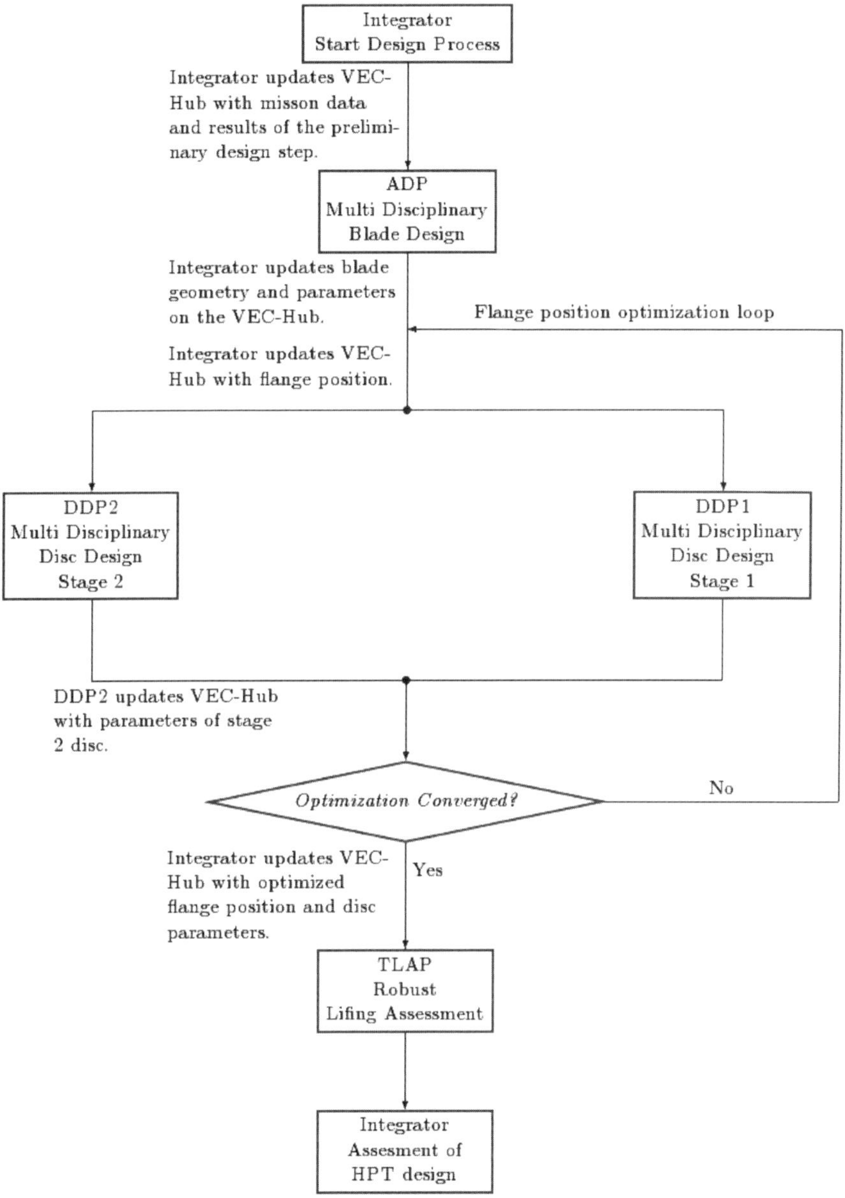

Fig. 7 Block diagram of the high-pressure turbine design process.

parameters are communicated from the global workflow to the local DDP process directly through the Web service, which the DDP1 exposes to the VEC hub.

By contrast, the local process at disc design partner 2 (DDP2) can get this information directly from the VEC hub, using the PLM/PLCS services in the local workflow. Here company-specific IT requirements prevented the

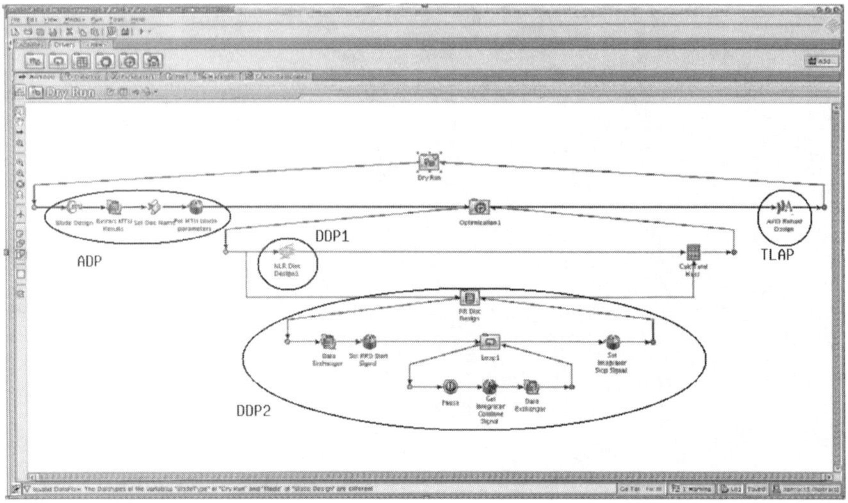

Fig. 8 Global control flow for the multisite MDO with the connections to the different partners of the VE.

direct exposure of services to the VEC hub. Therefore, a polling mechanism was developed that enabled DDP2 to participate in this multicompany optimization loop.

After finishing the cross-company disc optimization process, a robustness assessment of the life of the high-pressure turbine is executed by a process running at the site of turbine life assessment partner (TLAP).

Figure 8 shows how this execution sequence is implemented into the parametric workflow management tool [6] that provides the workflow services for this instantiation of the VEC hub.

Because the connections from the global optimization workflow to the local processes at the partner sites were the most demanding tasks in this aspect of the project, they are explained in the remainder of this chapter.

A. Connection to the Airfoil Design Partner

A prototype infrastructure was implemented to demonstrate the realization of the VE partners' minimal company security requirements based on an Internet connection as depicted by Fig. 9.

The aerodynamic optimization is offered as a Web service to the virtual enterprise by publishing the service definition by means of a WSDL file. To utilize the service, a SOAP (simple object access protocol) message according to the service definition is sent to the server at the ADP by the requesting partner (the integrator). The message contains the input data and is packed into an HTML request. After authorization by the providing partner's lightweight directory access protocol (LDAP) server, the request is forwarded to the internal SOAP server and verified. The server unpacks the input data and starts the optimization on a computer-aided engineering (CAE) compute server. The results are again packed into a SOAP data structure and sent back to the integrator.

The following security requirements have been implemented:

1) *Content security against unauthorized access:* Content security outside the providing partner is achieved by HTTPS encryption, while inside the partner it is implemented by access control on the reverse proxy.

2) *Authorization:* The authorization data are checked by the LDAP server containing the allowed service users. The prototype also shows the ease of implementation and maintenance of this authorization service, which is an additional important requirement beside security. (For more information on multicompany authorization, please refer to the security chapter.)

To achieve full security on the partners' side, the following requirements have also to be considered: content security against malicious content and authentication.

All of the just-mentioned requirements (except authentication) are taken into account for the security infrastructure concept described next. For example, a single sign-on infrastructure has to be agreed by all partners as it cannot be implemented independently.

Because the input/output data in SOAP messages are usually encoded, a conventional HTTP scanner will not detect malicious content hidden in extensible mark-up language (XML) data. As shown in Fig. 10, it is necessary to

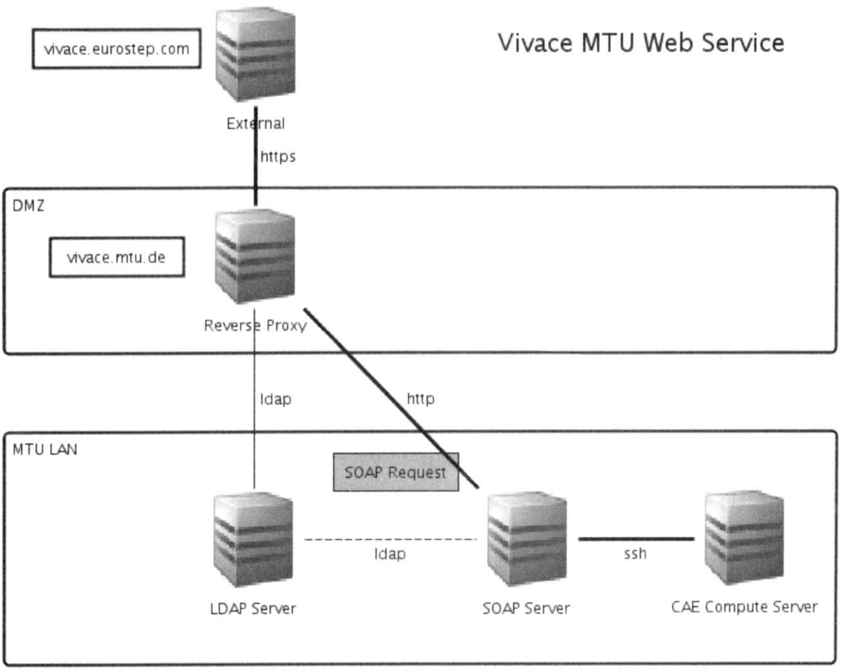

Fig. 9 ADP connection to the master workflow.

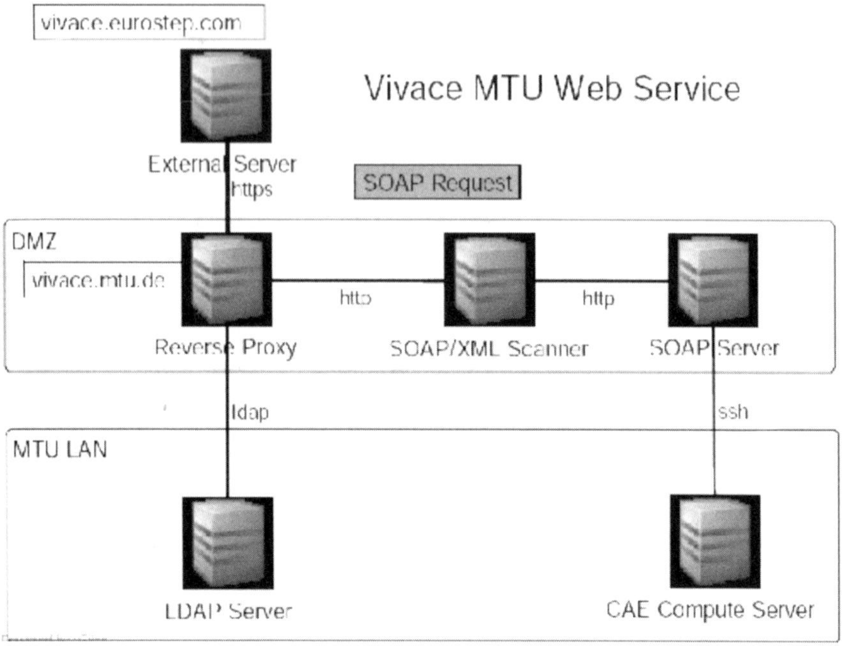

Fig. 10 Security infrastructure concept.

use a scanner that understands XML structures in a generic way and sends the raw element contents to a virus scanner. The known data formats (e.g., .gz, .zip, .exe, base64, etc.) can now be unpacked if necessary and scanned. In case of suspected virus, the SOAP request is rejected; otherwise, the message is forwarded unaltered to the SOAP server.

The SOAP server is now installed in the DMZ to reduce the vulnerability of the external accessible systems in the LAN. Special care has to be taken for the encapsulation of the computer server to prevent unauthorized access of the LAN.

B. Connection to the Disk Design Partner 1

1. Description and Access of the DDP1 Workflow

The DDP1 part of the whole engine use case workflow consists of a thermo-mechanical analysis of a high-pressure turbine disc. The DDP1 workflow defines and controls the execution of the different tools needed within this analysis. For a turbine disc with given geometry and loaded by a (simplified) cyclic loading, structural responses such as predicted life and no-burst criteria are calculated. The turbine disc is subjected to high thermal loading caused by hot gas flow and to centrifugal forces caused by the attached turbine blades. The disc analysis is reduced to an axially symmetric model where a two-dimensional geometry is sufficient for the analysis and optimization [7, 8]. (The full description of the MDO process is provided in Chapter 9.)

The flow incorporates structural design and analysis COTS tools, including a CAD tool for geometry manipulation, a postprocessor for FEM modeling, a finite element analysis tool for thermal and stress analysis, and a proprietary lifing tool based upon the Basquin equation and the Palmgren–Miner rule. The concept of the flow and the tools and data files involved are shown in Fig. 11. It starts at the design table and ends with the lifing values. The flow can be used, for example, in an external optimization loop by including a redesign tool that controls iterative application of the flow, thereby setting or modifying the design table.

The CAD tool runs on a Windows PC, and the analysis and postanalysis tools run on a Linux PC. A shared file system among the Windows PCs and Linux PCs is used so that all input and output files involved can be located in the same directory. The input variables, including design variables and other parameters defining the design case, are specified in a text file. This file might contain the geometry parameters, mesh densities, loading conditions, as well as bounds on the design variables and constraints as needed to set up the analysis model. The parameters are specified in a particular common data format (Loadproc) that is commonly used by DDP1 for storing and exchanging engineering data between different analysis tools. The design variables steering the geometry are transferred to the CAD tool by means of a spreadsheet file (design table), under the control of a Python code. Model data that do not change during the design and analysis process are available from a specific set of data files.

The flow is available as a stand-alone tool, that is, as a dedicated application. The application displays an engineering-oriented and easy-to-use graphical user interface (see Fig. 12). It allows the specification of design values for the disc such as width and displays the resulting stress distribution and the predicted life after execution of the chain. The engineer might repetitively specify a different value and examine the results on his own screen, without having to explicitly

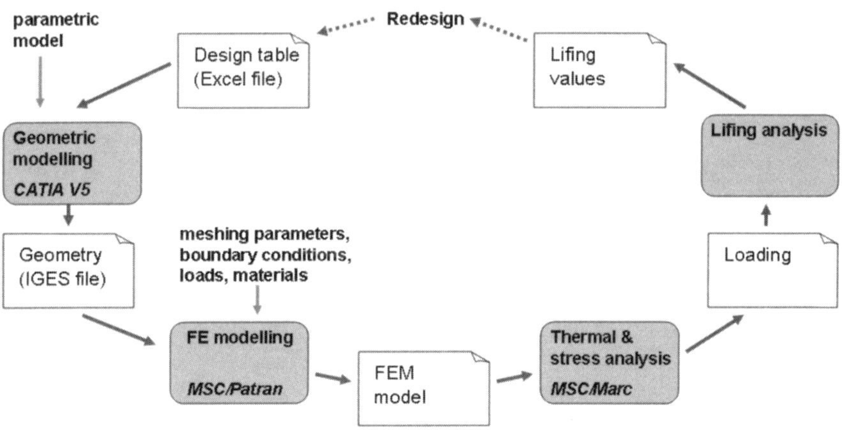

Fig. 11 Concept flow for thermomechanical analysis on a high-pressure turbine disk, based on a parametric model of a simplified disc.

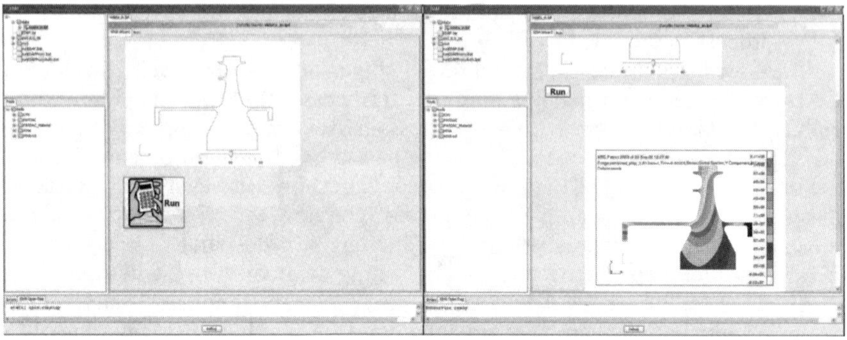

Fig. 12 Stand-alone application of the DDP1 thermomechanical analysis tool chain via a dedicated application in which most parameters have been fixed. Left is the initial state, and right is the result of the completed calculation cycle.

take care of starting the applicable tools on the various platforms in the system. The same flow is also made available as a component in the global workflow.

The flow is implemented using a combination of DDP1 middleware for tool-chain automation and Java and Web technology. The middleware [9] enables the definition of a workflow, which comprises a description of the use of a series of tools and the manipulation of data sets involved in a tool chain. The latter is a graph representation of a process (flow) including tools, data containers for storing and passing data files among tools, and connections among tools and data containers. The tool chain facilitates clear and unambiguous definition of a flow and controlled and integrated execution of the tools and management of data involved in the flow. Figure 13 shows a tool chain corresponding to the concept flow from Fig. 11. Once a tool chain is defined, the middleware controls the execution of the tool chain and manages the data files involved.

The middleware enables stand-alone as well as embedded use of a tool chain. In the stand-alone case, an engineer might run the tool chain directly. This is possible via either the middleware's GUI or a dedicated application. Essentially, the GUI gives access to the tool chain and its individual constituents in *white-box*

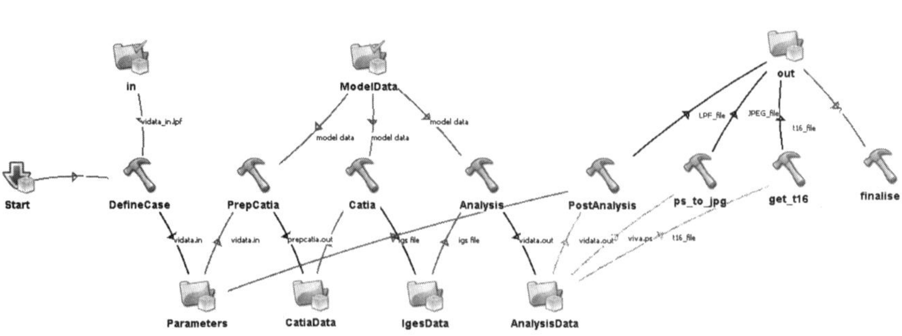

Fig. 13 Flow for thermomechanical analysis on a high-pressure turbine disk implemented as a tool chain.

mode, as shown in Fig. 13. The GUI allows the user to explore the tool chain interactively, to zoom into any subchains and tools, to control its execution up to the level of individual tools, and to monitor the process during and after the execution of the chain. The GUI also enables the user to use the tool chain as a *black box* operated via a single icon. The dedicated application provides use of the same tool chain also in a black-box mode only, as shown in Fig. 12. In this case, the end user first specifies all of the inputs, next runs the tool chain without further handling of the tool chain or any individual steps and intermediate results, and finally is provided with results.

In the case of embedded use, the tool chain is provided for use through an API from within applications, for example, within a higher-level (global) tool chain or workflow, such as the VEC-hub global workflow. The API is based on Web services technology.

The middleware facilitates use of a tool chain inside as well as outside the scope of a company's network. In the latter case, the tool chain can be used remotely, outside the local area network in which the tool chain is defined and executed. Remote execution is accomplished using state-of-the-art Java and Web-interfacing technology. Remote interactive application of tool chains with individual tools still running in batch mode is possible through the middleware's Web version of the GUI. Remote embedded as well as interactive application of the tool chain is possible through Web service operations. For example, in the global workflow, the turbine disc predicted life calculation tool chain runs on a set of dedicated Linux and Windows systems. The tool chain is available to the participating and collaborating partners for interactive as well as embedded use over the Internet as a Web service, with appropriate security measures. The global workflow running outside DDP1 executes the tool chain, thereby providing input data and receiving output data, through a Web service operation. The same Web service operation can be used as stand-alone, in an interactive session through a dedicated client program as shown in Fig. 12. The Web version of the middleware's GUI makes it possible to view and operate a possibly remote tool chain in white box mode, displaying the tool chain as shown in Fig. 13.

2. *Local and Remote Access to the DDP1 Workflow*

In summary, the middleware supports various levels of controlled and integrated use of tools and tool chains, varying from stand-alone use of loosely coupled tools and manual manipulation of data files involved, to highly integrated and automated use from within dedicated engineering applications and workflows, as depicted in Fig. 14.

The tools and the tool-chaining middleware are available as a Web application served by the open source application server controlled by an open source Web server located in the so-called project zone in the DDP1 IT network. This setup is shown in Fig. 15. The project zone is part of the demilitarized zone of the DDP1 network, with its own security policy implemented via firewall settings. In this project zone, secure access to the tool chain over the Internet is accomplished via secure HTTP (HTTPS) and protected using a valid user name/password combination.

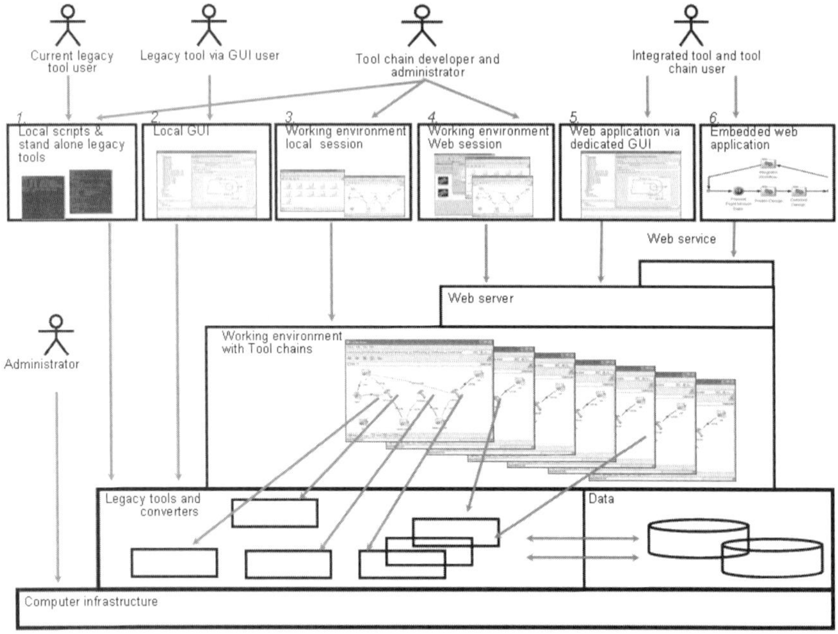

Fig. 14 Different levels of using tools and flows, varying from uncontrolled, stand-alone use of individual tools (1) to highly automated, embedded use from within dedicated applications (5), and umbrella workflows (6).

3. DDP1 Local Middleware Setup

The basis for the implementation of tool chains is an DDP1 middleware utility for storing and applying knowledge about application of engineering tools and tool chains [9]. The DDP1 middleware utility supports wrapping of the heterogeneous legacy tools. That is the legacy tools are left as such and are encapsulated with a thin layer of code, called *tool wrapper*. The legacy tools are usually COTS, developed in house, or provided by partners or the customer, into easy-to-operate uniform tool objects. Thus tool wrapping makes it possible to integrate a legacy tool without having the tools source code or specific tool libraries available. It also facilitates the inclusion of tools that have been certified for specified usage in aircraft design, and that would require revalidation when being modified or recompiled. The tool objects facilitate usage of the legacy tools in a uniform way. The tool objects also provide uniform building blocks to facilitate the definition and usage of tool chains. Tool chains allow series of tools to be executed repetitively and iteratively, according to formalized scenarios, for example, in parameterized modeling, optimization, and parameter variation studies. The DDP1 middleware support automated use of tools and tool chains, either interactively (i.e., directly by the engineers) or embedded in a larger chain as implemented in the global optimization workflow. In both interactive and embedded use, the tools and tool chains can be operated remotely in secure way, achieved through state-of-the-art Web technology.

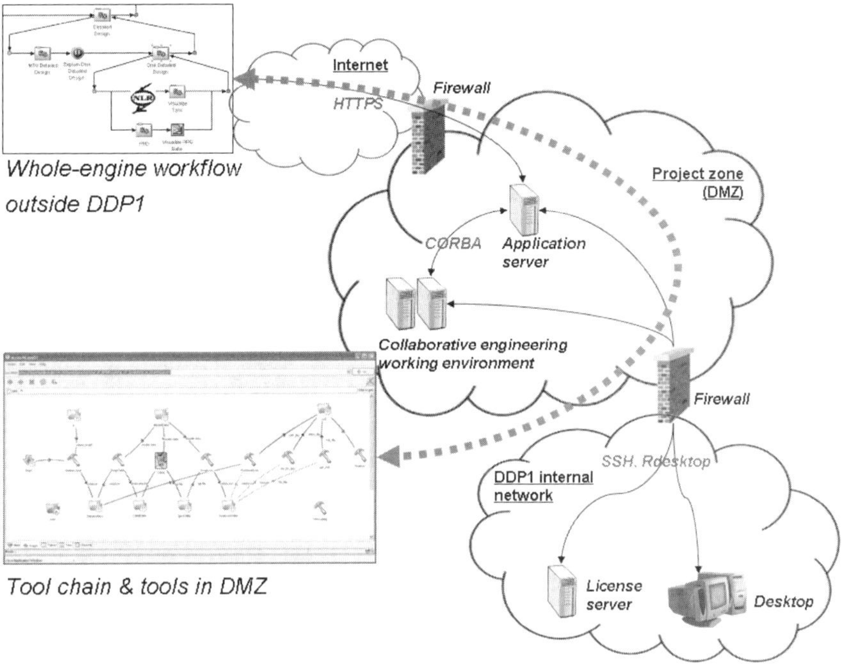

Fig. 15 Remote invocation of the DDP1 thermomechanical analysis tool chain from the global whole-engine workflow.

4. *Description of Web Technology Used to Connect DDP1*

As described before, flexible local, remote, and embedded access to tools and tool chains is achieved via a combination of state-of-the-art Java and Web technology. This technology provides programmatic interfaces enabling possibly dispersed software components to interact over the Internet in a secure and controlled way. Key technologies applied in the realization of the DDP1 flow, and in particular in making the DDP1 flow accessible, are Java Content Repository (JCR) [10] and RESTful Web services [11]. As indicated at the top left in Fig. 14 (labeled 1), a tool chain's client being a human operating the tool chain interactively via the Web-based GUI or a dedicated client program, or a workflow that coordinates the processes at the various partners, applies the tool chain via a Web service operation, which interacts with the tool chain middleware via a JCR interface. The JCR, Web services, and REST technologies are described in the subsequent subsections.

JCR. JCR is a standard programmatic interface (API) for accessing content repositories in a uniform way [10]. A content repository is a hierarchical content (or data) store with support for, among other things, structured and unstructured data, binary as well as textual contents, full-text search, filtering, versioning, transactional read/write access, fine-grained access control, content categorization, and contents event monitoring. Today, many applications, including Web applications and document management systems, use content repositories to manage data.

From an end user's perspective, a collaborative engineering working environment such as the VEC hub (or DDP1 workflow) can be considered a content repository, with content being the objects contained in the working environment, such as files, tools, and tool chains as well as any objects stored in tool chains and data containers. The JCR interface enables Java applications to communicate with a working environment in a simple and well-defined way in embedded form, without making use of the GUI component of the working environment's own user interface.

The implementation of the JCR interface of the DDP1 tool chaining middleware replaces the GUI package in the user interface. From the working environment's perspective, the JCR implementation mimics the GUI, which receives commands and inputs from, and returns outputs to the JCR operations instead of the GUI end user. The JCR interface is part of the Java Extendable Contents and Rights System (JeCARS) [12] framework. JeCARS uses an open-source implementation of JCR [5].

Web services. In a multipartner collaborative engineering working environment, the computational resources, including the legacy tools, are commonly located in the local-area networks of the partners involved. An important goal for flexible collaboration in a virtual enterprise is to make the computational resources available as independent services that are accessible without knowing the details of the actual implementation and underlying platforms of the services. To establish the required remote access to tools and tool chains across company boundaries, while achieving a proper level of security, and to use standard and platform independent technology, the choice was made to use Web services. A comparison between available technologies for accomplishing distributed applications, leading to the choice of Web services instead of solutions based on common object requesting broker architecture (CORBA) and distributed component object model (DCOM), is given in [13].

The notion of Web services is defined by the W3C. Today, World Wide Web technology is used more often for communication between possibly distributed software components and hence for accomplishing distributed applications. Based on lessons from the past, and developments and demands with respect to the Web, the W3C established Web services. This technology provides software applications with programmatic interfaces, enabling interoperability between different applications running on different platforms and in different frameworks. Web services facilitate the combination of software and services from different companies to form integrated services. Web services use open standards and protocols, such as HTTP, SOAP, and XML. HTTP is the hypertext transfer protocol, commonly used to transfer information on Internet and the World Wide Web. SOAP is a protocol for transferring XML-based messages over networks, commonly using HTTP or its secure variant: secure HTTP or simply HTTPS. XML, the extensible mark-up language, is a general-purpose language for encoding and serializing information. XML is commonly used for sharing structured information among systems, possibly over network connections and the Internet.

Web Services comprise a set of tools that can be applied in several ways. The three most common ways are RPC (remote procedure calls), SOA (service-oriented architecture), and REST (representational state transfer).

In the RPC approach, the interface (API) of a service is well defined in terms of (distributed) function or method calls. Details of these calls, including the service bindings to specific protocols, are published to the clients as a WSDL description, in XML format. The WSDL description enables clients to interact with the service. Web services also define service broker technology. Service providers can register services with a service broker. Service requesters can use the service broker to find a particular service and retrieve the services WSDL description in order to subsequently invoke the operations implemented by the service.

In the SOA approach, Web services are used to implement a distributed system in the style of a SOA [3]. This software architecture defines the use of individual, loosely coupled services on a computer network to support the implementation of an integrated software system comprising the services. The services communicate through messages rather than function or method calls. The message-oriented interface is also published through WSDL.

In the REST approach, clients interact with a service through a well-defined protocol, usually HTTP or a similar protocol. The interface to the services is defined in terms of small set of standard operations (e.g., HTTPS get, post, put, delete). These operations can be defined using WSDL, describing SOAP messaging over HTTP or HTTPS. Instead of function and method calls and messages, this approach is based on interaction with stateful resources.

Support for the application of Web services technology in the development of distributed services and applications is emerging. For example, the Java development environment NetBeans enables software developers to have a first Web service implementation for their application operational quickly.

End use of the collaborative engineering working environment, including the execution of tool chains, is provided as a Web service, implemented on top of the JCR package. The key Web service operation is that of running a tool chain. This tool chain has a unique starting point and is defined to run its constituent tools in a specified order, automatically, and in batch mode. The input files are passed to the tool chain via the data container named in, and the output files are produced in the data container out. The tool chain from Fig. 13 complies with these requirements.

Measures are taken to establish a level of security required for collaborative environments across company borders. Data exchange over the Internet is secured through the use of HTTPS when accessing a working environment. Authentication is achieved by the need for a client to authenticate itself with a server using a valid user name/password combination before being able to access the contents of a working environment. In addition, authorization is accomplished by applying a user and group-based access control to the objects in a working environment.

RESTful Web services. The tool-chain Web service is implemented as a RESTful Web service, using HTTPS and user authentication to implement the demanded level of security. REST is an architectural approach based on a stateless client-server architecture [11]. REST exploits existing Web technology and protocols, such as HTTP and URLs. It is based on the notion of resources, that is, sources of specific information, which are identified using URLs. Clients and servers communicate via a standard interface (HTTP) to perform operations on resources. The operations are specified through stand-alone requests, implying stateless interaction between requests.

An early implementation of the tool-chain Web service was based on the SOA Web services approach. Further development however took place as a RESTful Web service. The most important reason for adopting REST is its simplicity for the developer of the service. REST supports the developer in making the services available through existing, out-of-the-box, and proven Internet technology, protocols, and standards. This makes it easy to achieve and maintain, for example, the required security and to deal with the security measures, such as firewalls and proxy servers, of each of the collaborating companies. Moreover, the popularity of providing Web services based on REST is growing (data available at http://www.programmableweb.com). Several large-scale Web applications such as Google, Amazon, Yahoo, and eBay have adopted REST to make their resources available.

IV. Connection to the DDP2

At this partner, the IT environment complies with the company's security policy and does not allow an active publishing of the services to the VEC hub. Instead a polling mechanism needs to be used, so that the master workflow can trigger this DDP2 workflow execution.

The component "Trigger RRD" in the master workflow updates a file in the data repository at the VEC hub, with the confirmation that all necessary information is available for the disc 2 design partner to start its design process. The internal workflow at this partner queries the file at specified intervals and executes the internal design task only if the start signal is given in that file. After completion of the local analyses, the results are uploaded. In addition, a separate file is uploaded to the VEC hub, which contains the information that the DDP2 internal workflow has finished. The internal workflow uses the API interface at the VEC hub to retrieve and upload the necessary data automatically.

While the internal workflow is running at the partner's site, the main workflow at the VEC hub executes a similar loop in parallel. It is called "Wait for RRD Finish" as shown in the lower part of Fig. 16. It makes a query to the file containing the updated information uploaded by the partner's workflow once the disc 2 optimization process has completed. The master workflow continues only if it finds the required information in that file. When the master workflow calls the partner's subflow in the subsequent iteration, it automatically resets the updated information to "notupdated." The master workflow always requires the updated information from the partner's optimization process before it continues with the next iteration.

Using this construction, a complete separation between the local processes at DDP2 and the VEC hub is guaranteed. The polling loop of the local process can be stopped at any time locally, and the internal processes can no longer be accessed from the outside.

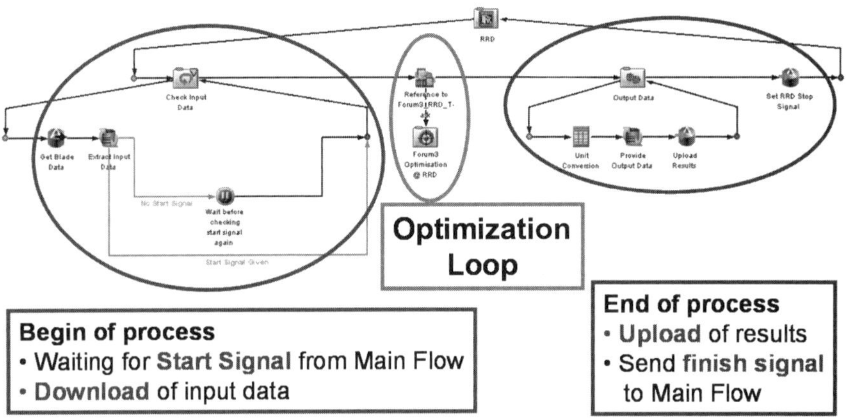

Fig. 16 DDP2's framework for multidisciplinary optimization and robust design of disc 2. Trigger components are used for starting and finishing the local design process.

V. Connection to the Turbine Life Assessment Partner

At this partner's site an application control system (ACS) is implemented similar to the one used at the VEC hub. This system builds upon Java Enterprise Edition (JEE) certified commercially available application servers [14, 15]. To couple the turbine life-assessment partner (TLAP) into the design process, a direct business-to-business (B2B) connection between the VEC hub and the TLAP is used. This B2B connection uses the security features of these application servers.

The ACS takes an XML component model as input and dispatches work to the local computers assigned to run this model. The solution provides Internet-based connectivity and security features to support secure ACS-to-ACS communications across a public or private network. It gives the enterprise partners the ability to create and execute a federated model where different components of the model reside in physically remote locations. The dependencies between partners' internal security process are minimized by using third-party security certificates and real-time application level authorization. Figure 17 illustrates an overview of this architecture. This way of coupling partner processes provides the capability to indicate that a particular model can be used by a remote site/user; include a remote model into a federated model; execute the federated model, making the remote execution transparent to the local workflow; ensure that all resources (both local and remote) are available for a federated model; link to remote sites, and any other job execution resources; include network infrastructure commonly used when interfacing across public networks such as firewalls and network gateways; use HTTPS for encrypted communications; use digital certificates for site-independent user authorization; and perform application-level authorization for remote requests.

When building the federated model, a remote service is represented in the model by an instance of the Proxy component. This component acts as a

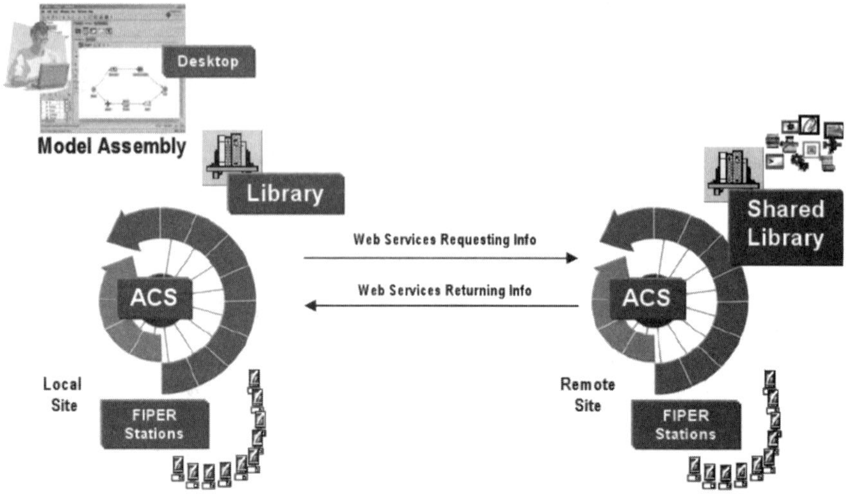

Fig. 17 Business-to-business model execution.

placeholder for a remote service. The inputs and outputs of the Proxy exactly match those required by the remote model. No other information about the remote model is known to the local model (e.g., the implementation of the remote service is hidden; only the inputs and outputs are known).

A Proxy begins as a generic placeholder for a remote service. It needs to be configured specifically for the service to be used in the model. The Proxy editor allows the performing of such a configuration. The B2B execution provides the capabilities for the local ACS to recognize remote services and use them to configure a Proxy when necessary.

The Proxy configuration depends on Web services that retrieve service information from the remote ACS. These Web services retrieve necessary information such as model names, inputs, and outputs for which the model creator is authorized.

After setting the inputs to the desired values, the model is run on the local ACS just like any other model. The Proxy is executed by the local ACS invoking a Web service to call the remote ACS in order to run the remote service. The results of that execution are returned from the remote ACS and become the outputs of the Proxy in the local model.

It is assumed that prior to the creation of shareable models a business arrangement would have been made between the enterprises that will define the level of security required for communications between any two partners' systems. Given that different business agreements will result in different security requirements, this solution allows flexibility in the implementation of user authentication and authorization.

All Web service communications are transmitted using secured socket layer (SSL) protocol. This prevents third-party interception and interpretation of the data. Contained within the encrypted Web service messages are the local user and local system names of the requestor. To further ensure a no-unauthorized use of its resources occurs, the remote system might also require the inclusion of a third-party digital certificate for identity verification.

As Web service requests are received by the remote system, the local user ID is mapped to a remote user ID. This mapping can be accomplished in two ways:

1) At system level, all user IDs from the local system are mapped to a single remote user ID, making all valid users from the local system authorized to use models on the remote system.

2) At user level, each user ID from the local system is mapped to a unique remote user ID, making each local user having unique credentials for executing models on the remote system.

The user authentication and authorization steps are performed for each request, whether model information is being requested for inclusion in a Proxy or for a model execution request.

VI. Summary and Conclusions

Our research into multisite, multicompany MDO, described in this chapter, shows that in the virtual enterprise there is no single IT solution that fits the specific needs of every participating partner.

Because of differences in policies, security, and IT infrastructure, requirements differ widely between the participating companies. This heterogeneity extends to the design systems and design tools used by each partner. Any integration attempt that requires substantial changes in the partners' respective local work processes is likely to fail. The reason is that these work processes have been developed over a long period of time and reflect very closely the design methods and philosophy of each partner company. Moreover, these processes often constitute the added value that each partner brings to the virtual enterprise.

For these reasons, this work proposes a loose coupling of the local design systems, where the existing processes are wrapped into Web services and exposed to the VEC hub. It is demonstrated that this approach provides an integrated solution that is acceptable to all. In addition, this approach enables the flexibility necessary in dealing with the different requirements of each partner. In the presented study four partners were integrated, and each of them used a different coupling mechanism: ADP password-protected Web service using a reverse proxy; DDP1 password-protected Restful Web service; DDP2 polling mechanism using secured documents on the VEC-hub system; and TLAP proprietary B2B connection with authentication by encrypred digital certificates.

Another challenge facing the virtual enterprise is the need for common data to be shared rather than be sent from point to point upon request. In our work this functionality is provided by the PLM/PLCS services of the VEC hub. In the presented MDO study this way of data sharing provided the foundation for multicompany, multidisciplinary workflow execution in a reliable and repeatable manner. The multipartner workflow is resilient. In case a partner workflow fails, the top level multipartner workflow can recover from such events. Using the proposed common shared data structure guarantees that for each execution of either shared or local workflows all used data are in a consistent state.

Last, but not least, the VEC-hub concept, described in this chapter, provides a framework where both the workflow and the data-centric aspects of these types of collaborative engineering projects are integrated and the foundation for the security infrastructure is laid out. With this security architecture, only authorized users have access to the VEC hub, and only they could access exclusively that part of the information that the individual partners have given them access to.

References

[1] Smith, J., Mason, H., Kreiler, C., and McRae, M., "OASIS Product Life Cycle Support (PLCS) Technical Committee," Organization for the Advancement of Structured Information Standards (OASIS), http://www.oasis-open.org/committees/tc_home.php?wg_abbrev=plcs [retrieved July 2009].

[2] Christensen, E., Curbera, F., Meredith, G., and Weerawarana, S., "Web Services Description Language (WSDL) 1.1," World Wide Web Consortium (W3C), March 2001, http://www.w3.org/TR/wsdl [retrieved Jan. 2009].

[3] MacKenzie, C., Laske, K., McKabe, F., Brown, P., and Metz, R., eds.,"Reference Model for Service Oriented Architecture 1.0," Organization for the Advancement of Structured Information Standards (OASIS), Aug. 2006, http://www.oasis-open.org/committees/download.php/19679/soa-rm-cs.pdf [retrieved Jan. 2009].

[4] "PLCS Webservices," Product Life Cycle Support (PLCS), Eurostep Group, http://www.plcs-resources.org/plcs_ws/index.html [retrieved Jan. 2009].

[5] "Public Key Infrastructure," Wikimedia Foundation, http://en.wikipedia.org/wiki/Public_key_infrastructure [retrieved Jan. 2009].

[6] "FIPER," Dassault Systèmes, 2009, http://www.simulia.com/products/fiper.html [retrieved July 2009].

[7] Kesseler, E., et al., "Empowering Engine Engineers: Advancing the State-of-the-Art in Collaborative Multinational Multidisciplinary Engine Design," *First CEAS European Air and Space Conference*, Berlin, 10–13 Sept. 2007, CD-ROM published by Council of the European Aerospace Societies (CEAS), Paper number 2007-118.

[8] Kesseler, E., and van Houten, M. H., "Exploring Automatic Multi-Objective Turbine Disc Design for Virtual Engines," *ECCOMAS Multidisciplinary Jubilee Symposium*, Vol. 14 of *Computational Methods in Applied Sciences*, edited by J. Eberhardsteiner, C. Hellmich, H. A. Mang, and J. Périaux, Springer Netherlands, 2009, pp. 251–262.

[9] Baalbergen, E. H., and van der Ven, H., "SPINEware—A Framework for User-Oriented and Tailorable Metacomputers," *Future Generation Computer Systems*, Vol. 15, No. 5–6, Oct. 1999, pp. 549–558.

[10] "JSR 170: Content Repository for Java^{TM} Technology API," Java Community Process (JCP), Sun Microsystems, Inc., http://jcp.org/en/jsr/detail?id=170 [retrieved Jan. 2009].

[11] Fielding, R. T., "Architectural Styles and the Design of Network-Based Software Architectures," Ph.D. Dissertation, Dept. of Computer Science, Univ. of California, Irvine, CA, 2000, http://www.ics.uci.edu/~fielding/pubs/dissertation/top.htm [retrieved July 2009].

[12] "JeCARS—Java Extendable Contents and Rights System," SourceForge, Inc., Mountain View, CA, http://jecars.sourceforge.net [retrieved Jan. 2009].

[13] Alexiou, A., Baalbergen, E. H., Kogenhop, O., Mathioudakis, K., and Arendsen, P., "Advanced Capabilities for Gas Turbine Engine Performance Simulations," *ASME Turbo Expo 2007: Power for Land, Sea and Air*, Montreal (Canada), 14–17 May 2007.
[14] Wujek, B, Koch, P., McMillan, M., and Chiang, W., "A Distributed, Component-Based Integration Environment for Multidisciplinary Optimal and Quality Design," AIAA Paper 2002-5476, Sept. 2002.
[15] "An On Demand PLM Infrastructure Framework for the Future," IBM Product Lifecycle Management, IBM Corp., Armonk, NY, 2007, ftp://ftp.software.ibm.com/solutions/plm/doc/content/bin/plm_on_demand_plm_infrastructure_framework_for_the_future.pdf [retrieved Jan. 2009].

Chapter 11

Securing Virtual-Enterprise Collaboration

Michel Kamel*

Paul Sabatier University, IRIT, Toulouse, France

Ernst Kesseler†

Katern 09, Amsterdam, The Netherlands

Jean Paul Buu-Sao‡

Transglobal Secure Collaboration Program, Toulouse, France

Abdelmalek Benzekri§

Paul Sabatier University, IRIT, Toulouse, France

Rene Wiegers¶

National Aerospace Laboratory (NLR), Amsterdam, The Netherlands

François Barrère**

Paul Sabatier University, IRIT, Toulouse, France

Bert Schultheiss††

National Aerospace Laboratory (NLR), Amsterdam, The Netherlands

Romain Laborde**

Paul Sabatier University, IRIT, Toulouse, France

and

Erik H. Baalbergen††

National Aerospace Laboratory (NLR), Amsterdam, The Netherlands

*Doctor.

†Director.

Abbreviations

A = action
AA = attribute authority
ABAC = attribute-based access control
AC = attribute certificate
CIEEMG = Commission Interministérielle pour l'Etude des Exportations de Matériels de Guerre (Interministerial Commission for the Study of Export of Military Materials)
DAC = discretionary access control
DMZ = demilitarized zones
DRM = digital rights management
DTD = document type definition
IBAC = identity-based access control
IdP = identity provider
IRIT = Institut de Recherche en Informatique de Toulouse
IT = information technology
LDAP = lightweight directory access protocol
NLR = National Aerospace Laboratory
O = organizational
OASIS = Organization for the Advancement of Structured Information Standards
OrBAC = organization-based access control
P = process
PDP = policy decision point
PEP = policy enforcement point
PKI = public key infrastructure
PMI = privilege management infrastructures
R = resource
RADIUS = remote authentication dial in user service
RBAC = role-based access control
S = subject
SAML = security assertion mark-up language
SCE = secure collaborative environment
SME = small-to-medium enterprise
SOA = service-oriented architecture
SOAP = Simple Object Access Protocol
SP = service provider
SSO = single sign on

[‡]Information Systems Architect.
[§]Professor.
[¶]R&D Engineer.
[**]Associate Professor.
[††]Senior Scientist.

SWIFT = Society for Worldwide Interbank Financial Telecommunication
TSCP = Transglobal Secure Collaboration Program
VE = virtual enterprise
VEC hub = virtual-enterprise-collaboration hub
VIVACE = value improvement through a virtual aeronautical collaborative enterprise
XACML = extensible access control mark-up language
XML = extensible mark-up language
W = work
WAYF = where are you from
WFE = workflow engine
WSDL = Web services definition language

I. Introduction

THE concept of virtual enterprise (VE) is a direct consequence of the evolution of networks and tools that promote collaborative work as technical enablers. From the demand site, aircraft design and manufacturing involve significant financial investment of up to 10 billion Euros for new large commercial aircraft like the Airbus A380 or the smaller Boeing 787 [1, 2]. To share these investments and the accompanying risk, the first tier suppliers become risk-sharing partners. Consequently, these partners need to be involved from the start and need to share their critical assets, that is, their designs and their design tool suites. The result is a virtual enterprise, providing the demand pull to complement the technology push. More information on collaborative design is provided in Chapters 9 and 10, with Chapter 9 concentrating on the engine design used also as an example in this chapter and Chapter 10 concentrating on the enabling technology. The main example discussed in this chapter is based on this virtual enterprise. To demonstrate the generic applicability of the approach, in the last section another enabling technology for the virtual enterprise is discussed.

Currently, organizations need to exchange resources and competencies in different areas in order to complete the common objectives. Virtual enterprises offer the different partners a way to access resources available at different sites separated geographically; this form of collaboration will allow the participating organizations to reach new business opportunities by becoming partners and realize mutual benefits as a result of collaboration with other organizations. Building the infrastructures needed for the communication between the partners brings many security issues when administrators have to unlock some of the security doors of their information systems. When deciding to build a VE, the participating organizations must deploy a shared information technology infrastructure, ensuring the secure exchange of resources and competencies between their respective sites. Many strategies can be defined regarding this situation (including the centralization, the decentralization, or the regrouping of resources within a defined protected zone). One of the tendencies today is to utilize advanced network technologies, letting each participating organization define and manage the users having the rights to exploit the VE's resources.

Thus, each participating organization remains the manager of its resources and keeps control of them. Each organization keeps its own security policy and extends it when the collaboration with partners necessitates it. Each organization will manage its local users' credentials and attributes for authentication and authorization purposes.

Bultje et al. have enumerated characteristics of VEs [3] and state that a VE is characterized by a unique identity and a cooperation based on trust and information technologies shared between the members. As a consequence, a VE is the result of a trust relationship explicitly specified within a closed context, which is the common project/activity. Trust allows the addressing of scalability in managing security tasks such as authentication, authorization, and access control security services; it is becoming a necessity within VEs and provides the means to establish distributed and secure collaborative environments where organizations put their critical resources at the disposal of their partners.

As a consequence, VE collaborative environments should provide enough flexibility for people and security services to further allow each organization to control its assets at any step of the collaborative process.

The rest of the paper is organized as follows. Section II discusses the concept of trust within VEs in the context of the Transglobal Secure Collaboration Program (TSCP). Section III identifies the security constraints within a VE environment. Section IV describes the proposal access control architecture that responds to the identified security constraints. Section V details the "workflow engine" scenario adopted for the proof of concept. Section VI describes the prototype that we have deployed for building the collaborative environment based on the concepts of identity federation and privilege management infrastructures (PMIs). Section VII proposes an enhancement for this solution. Section VIII discusses the concept of the virtual-enterprise-collaboration hub (VEC hub), which also needs to rely on our access control solution. Finally, Section IX concludes and states our future work.

II. Trust

A. Trust in Virtual Enterprises

A VE can only work when all participating organizations and all of their cooperating personnel trust each other. When the collaboration of the virtual enterprise is extended to include the earlier parts of the product life cycle, like the various phases of product design, this implies that sensitive information needs to be shared as well. Also, access to design tools needs to be granted to the partners of the virtual enterprise. Such design information and the design tool suite are the core assets of each participating organization resulting in the need for trust.

One characteristic of the business environment in which each virtual enterprise operates is that a partner in a current collaboration can be a competitor in the next program. Consequently, a prerequisite for successful deployment of the virtual enterprise is creating and maintaining sufficient trust between all of the organizations involved. This section will elaborate on trust by providing some definitions of trust and discuss some implications for the virtual enterprise including positioning the work described.

On the one hand, the general opinion in organization science relates trust between organizations with highly positive effects on performance [4]. On the other hand, the temporary character of the collaboration within any virtual enterprise could also impede the trust needed to function.

Trust can be approached from different perspectives, such as economics, social sciences, and organization science. Three common elements in trust are as follows:

1) The first is interdependence between trustor and trustee, that is, the completion of the trustor's task depends on prior actions or coordination with the other person or the trustee [5].

2) Trust provides a way for coping with the risk or uncertainty in the interactions between trustor and trustee. The trustor needs to take a (potentially) risky commitment with the trustee [6].

3) The expectation that the vulnerability of the trustor resulting from the acceptance of the risk will not be taken advantage of by the other person, that is, the trustee [4].

The various perspectives just mentioned divide trust into the following three classes:

1) Calculative trust is based on the notion that trust involves a calculation by the trustor between the costs and the benefits of trust [7]. Calculative trust is used in economics and game theory.

2) Value-based trust assumes that all individuals involved share common (institutionalized) values from which the trust is derived [8]. As the virtual enterprise can have partners with different backgrounds (roles, nationalities, cultures, and jurisdictions), this notion implies the need to instill such common values in the virtual enterprise. Value-based trust is used in economics, organizational, and management fields.

3) Expectation-based trust assumes that an expectation held in common between all partners concerned can structure their behavior in a predictable way, which then forms the basis of trust. This notion comes from sociology but is used in the organizational field [9]. Expectation-based trust captures the notion of the incremental increase of trust, that is, the trust of the trustor in the trustee increases (or is eroded) based on previous actions of the trustee. Note that such expectation can come from corporate culture, so that it can be applicable in a virtual enterprise.

The last perspective on trust relates to the various levels it operates on:

1) Trust at the micro-level is between people using their individual assessments.

2) Institutional-based trust relies on formal, legitimized structures to guarantee trust. It is needed for trust across group boundaries, across geographical boundaries, and when many interdependent actions are involved [9]. All of these conditions hold for the virtual enterprise. Institutional trust in business relationships requires an extended period of experience. During this time, knowledge about the prospective partner (trustee) is accumulated directly by the trustor or through a trusted third partner, for example, accrediting authority or certifying authority.

3) System trust derives trust from the reliable functions of systems outside the realm of the persons involved. This is summarized in the statement "each person trusts on the assumption that others trust" [6]. This form of trust implies a form of auditing of the trust system allowing all to keep trusting it.

We are interested in demonstrating the feasibility of providing system-based trust in an aeronautical virtual enterprise environment. It is expectation based, as the used framework forms the basis of the expectations. A scalable framework is needed to shorten the required built-up period for the institutional trust for it to be deployable in the agile virtual enterprise. To provide assurance that all partners maintain conformance to their expected behavior needed to maintain trust, the framework needs to be audited regularly. The framework also needs to provide technical means to ensure collaborating partners keep complying with the agreed trust processes. The latter is also referred to as information technology (IT) security.

For the provision of IT security, there are multiple IT-security options. For our security demonstration, TSCP (data available online at http://www.tscp.org/) has been selected among others as it specifically addresses the aeronautical domain. The next section will elaborate on TSCP and its approach to trust.

B. TSCP Approach to Trust

TSCP is a consortium grouping large aerospace and defense organizations around the globe, which objective is to set up an industry architecture aimed at supporting large-scale collaboration programs, while ensuring compliance with regulations and mitigating risks. This context of broad collaboration spanning across multinational regulation environments is further stressed out by the following considerations (we follow the terminology on "components of trust", which has been introduced in the preceding section):

1) *Challenges on risk commitment:* A characteristic of the supply chain in the aerospace and defense industry is the number of tiers involved (depth) and the large regional repartition of these tiers (breadth). With respect to risk commitment, what guarantees that trustor (e.g., information owner) will be ready to take risk commitments with trustees who he barely (or any) knows, and who are furthermore located in potentially different jurisdictions from his owns? The risk that the trustor needs to take increases with the sensitivity of the information exchanged, and therefore a lack of trust jeopardizes the operation of the supply chain.

2) *Challenges on value-based trust:* Some core TSCP participants alone, who are culturally from the civilian sector, are exposed to issues regarding value-based trust, as they see their own information acquire the dual-use status (dual use refers to products and technologies that are normally used for civilian purposes but that might have military applications): handling of such dual-use information requires adapting the culture, training, processes, and toolset, with associated cost impact. It is not surprising that the expansion of the collaboration beyond organization and regional boundaries raises even further issues to the applicability of value-based trust.

3) *Challenges on expectation-based trust:* Expectation-based trust is commonly used in the aerospace and defense industry in context of the sharing of information, whereby the owner of the information emits the expectation that the information is handled by all appropriate parties in a way that corresponds to his intent. Issues arise when the collaboration spans across organizations and

regions: how can the information owner indicate the intent of use in unambiguous way. If data-tagging techniques are chosen, how to ensure that the data-tag appears to all parties, and how to ensure that all parties understand the tags in a similar way? The risk of misinterpretation, even in a nonmalicious way, of the intent is increasingly larger when crossing organizational and regional boundaries.

The additional exposure to risks, as a result of the expansion of collaboration across organizations and regional boundaries, is the common theme that we have covered so far. TSCP has been formed with the objective to help address these challenges, by taking the following approach:

1) There is a need to impose institution-based trust across the industry.

2) There is a need, within each organization, to translate the institutionalized trust onto system trust, whereby the processes and mechanisms of the organization contribute to implement the institutionalized trust.

1. Institution-Based Trust: A Hierarchy of Policy Authorities

The formal, legitimized structures for guaranteeing trust are the many policy authorities that each individual organization, participant to the aerospace and defense collaboration, need to recognize. Policy authorities are organizational structures that own policies, that is, which are in the position to define, promote, and force the enforcement of policies. Policy authorities impose themselves at different levels, whether pan-national, national, regional, or corporate.

As an example, a French aerospace and defense organization having to comply with export-control policies (data available online at http://www.kiet.re.kr/kiet/export/images/Session1-2Michel_e.pdf) would need to recognize the following policy authorities: the organization's own corporate security division, the national authority for export control (the "Commission Interministérielle pour l'Etude des Exportation de Matériels de Guerre," CIEEMG, in our example), the European Council (European authority), and the Wassenaar Arrangement (international authority regulating export control for conventional arms and dual-use goods and technologies).

The preceding example shows some additional complexities: among all of the relevant policy authorities, the difficulty is to identify which ones apply to the context of the collaboration, given the various jurisdictions upon which each collaboration partner depends. The result of this determination can produce some form of aggregate of the relevant policy authorities. All of this needed to be done before any actual work could be performed, which is to figure out how to apply the policy, both procedurally (via processes, run by human beings) and systemically (run by systems infrastructure).

This problem is sufficiently complex to (already) deserve multidisciplinary team work to figure it out: legal, process, technology, product, and organizational knowledge is needed in order to produce the aggregate of all of the policies that need compliance for a given collaboration context. Once the applicable policies are identified, how to enforce those and how to verify and prove compliance?

Given the complexity just illustrated, and in order to ensure trust among all of the collaboration participants, the option to set up an institution solely aimed to establishing trust has been put forward. The structure, rules of engagement, mission, domains of competence, and operation of this institution are beyond the scope of this discussion.

It suffices to mention that such an institution-based trust (or systems trust) cannot function without an appropriate audit regime. The audit regime defines the methodology to demonstrate compliance and thus ensures that all participants comply with the appropriate policies, without having to disclose the exact details of how they implement compliance, by requiring all participants to be audited by accredited auditors.

Another role of this institution is to ensure the establishment of trust in a scalable manner across all of the collaboration participants. This is important in aerospace where a system integrator is complemented by many suppliers, ranging from a select set of first-tier suppliers to many third-tier suppliers, potentially not even related to the aerospace domain. As an example, for the A380 the number of suppliers totals 452 (data available online at http://www.airframer.com/aircraft_detail.html?model=A380), with many third-tier suppliers. This scalability can be achieved with multilateral trust, whereby the institution is a trusted tier with whom each participant needs to establish trust in a bilaterally way: establishment of trust becomes scalable as each individual participant needs to perform the heavy-duty qualification only once, with the trusted tier. Then trust is established transitively or indirectly between all participants in the virtual enterprise. This requires the institution to become a policy authority by itself, with authority to dictate policies that all participants of the industry, regardless of their competitive characteristics, need to comply with. This model of institution-based trust by the mean of a trusted tier has been successfully proven, within the financial industry, where competitive organizations (e.g., the financial institutions) rely on a trusted tier (SWIFT, the Society for World-wide Interbank Financial Telecommunication) for the definition of common standards and the service to convey financial data across financial institutions. This model has been in operation for the last 25 years.

2. *System-Based Trust*

Trusting systems (run by software) for the enforcement of designated policies is a problem that has been tackled with, ever since the first operating systems and still is alive nowadays. The challenges behind this vast objective are addressed in an incremental way, following the technology evolution. Capable of firstly enforcing very simple policies, systems have evolved with the capability to cope with incrementally more complex policies, and yet, we currently observe that not all policies can be fully enforced by systems solely: a large part of procedural means is still currently required. Nevertheless, one can observe that the evolution of the capability to enforce incrementally more complex policies follows the evolution of the various access control mechanisms, which are largely covered by the literature. This evolution is summarized next:

1) Discretionary access control (DAC) [10] based-systems could enforce simple policies that can be modeled as matrices of permissions for designated subjects.

2) Role-based access control (RBAC) [11] systems still enforce policies that can be modeled as matrices of permissions, with the difference from DAC that the matrices do not contains subjects directly, but an abstraction (role) that can

be used to group subjects bearing the same characteristic together; in some cases, the concept of inheritance helps to simplify the role of management by establishing hierarchies of roles, which present the caveat to not always accurately model the reality of business roles.

3) Attribute-based access control (ABAC) [12] systems represent a quantum leap forward by introducing the concepts of rule set, run-time selection, instantiation and evaluation of the rule-set; this allows a better enforcement of real-world policies, as long as these can be formalized as rule-set, which is not always possible. In addition to a simple rules-engine, some systems add artificial-intelligence mechanisms, such as inference mechanisms and/or the capability to handle fuzzy logic (in addition to predicate logic).

4) Organization-based access control (OrBAC) [13] systems start new grounds in leveraging outcome of social and organizational sciences (and not solely computer sciences as for the systems just seen) by introducing constructs, such as context or temporal constraints, that are better suited to enforcing real-world policies.

TSCP is evaluating several complementary mechanisms to bring systematic enforcement of policies. Theses mechanisms include rules-based policy engines capable of supporting at least ABAC, complemented by enterprise digital rights management (DRM) mechanisms, which allow information assets to be protected regardless of their location.

At the time of writing (mid-2009), combination of ABAC and enterprise DRM has not yet been concretized by industrial-strength products (COTS), as existing products still pose limits in terms of expressiveness of the rules-set and still lack interoperability. An ABAC prototype that has been used to test a simplified intellectual property protection policy will be detailed in the next sections.

Even if systems capable of fully enforcing the type of policies that need to be enforced (e.g., intellectual property or export control policies) were available, which again is not the case as of this writing, additional challenges will need to be addressed in order to ensure a full system-based trust, like these that follow:

1) *Issue of precision:* How does one prove that a given system enforces a given policy with a sufficient precision?

2) *Issue of interoperability:* How does one ensure that executable policy expressions are exchanged between different enforcement systems in an interoperable way?

3) *Issue of policy-expression equivalence:* How does one prove that two different interpretations (or implementations) of a same policy (possibly using a different policy language) are equivalent?

4) *Issue of policy selection:* In situations of multiple policy candidates, how does one ensure that the correct set of policies have been selected for evaluation?

5) *Issue of policy composition:* In situations where multiple policies need to be combined, how does one ensure the correctness of the combination?

III. Collaborative Environments and Security Constraints

The secure collaborative environment (SCE) that we have to deploy in order to let the potential partners collaborate and share their critical resources and competencies securely has to integrate the VE characteristics [3]. We need to deploy an interconnection solution providing the security services needed and the level of trust requested by the potential partners. Thus, it should allow each organization, member of the VE, to keep control of its assets, that is, control of the people participating to the VE and also control of the resources put at their disposal. However, if each organization keeps control of its assets, it has also legal responsibilities for these assets. The importance of a contract is that it fixes the legal VE framework and brings elements for the establishment of the trust relationship between the participants.

This means that each company has to open its information system to one of its partners and let them access its critical resources without losing them or losing control on them. Many security constraints arise from such interconnection and need to be treated. The major security constraints are as follows: consideration of the heterogeneity problem regarding authentication systems that can be used on the different partner sites; definition and adoption of a unique accreditation mechanism to be deployed across different organizations, which could be the institute referred to in the preceding section; offering a single-sign-on (SSO) service to avoid users having to remember multiple and different accreditations given by different authorities; simplification of security management by providing efficient solutions and tools; and ensuring trust and compliance among all participating partners.

Consequently, a SCE must provide the tools allowing the following:

1) *The specification of the access control policy:* To implement the policy, the specification language must be flexible enough so that it can take into consideration any concept or characteristic related to security issues.

2) *Each organization to keep control of its assets:* Organizations providing human competencies (the users' home organizations) must be able to decide who can participate in the VE. Consequently, those organizations that have both authority and responsibility over their human resources within the VE environment should manage the processes of authenticating and accrediting their people.

3) *To treat the problem of interoperability between the different authentication and accreditation solutions adopted within organizations* (different technologies and perhaps not compatible with one another).

To respond to these security constraints and provide the organizations with the means to build the SCE, we propose an access control solution combining the concepts of ABAC, privilege management infrastructures (PMIs), and identity federation. First, we specify access control policies using the ABAC model based on the users' attributes. Then, we guarantee the authority and responsibility of each organization towards its assets using the concept of PMIs offering a language for expressing the access control policies and an authorization engine for enforcing the authorization decisions. Finally, the interoperability problems are treated by using the concept of identity federation.

IV. Access Control Architecture Proposal

A. ABAC Model and Access Control Policies

Classical models such as the identity-based access control (IBAC) model [10] and the RBAC model do not offer the means to express the resources' attributes in addition to the users' attributes. Thus, we are adopting the ABAC model, which allows the expressing of access control policies based on "attributes." An attribute is any characteristic, relevant in terms of security, associated with a subject (i.e., a user), an action (i.e., read or write), a resource (i.e., a shared document), or the environment (i.e., when, where, in which conditions). This approach brings to the access control solutions more flexibility and extensibility. It is more adequate to the needs of open (heterogeneous and interoperable) and distributed systems where the users' identities are no more the unique characteristic taken into consideration for access control. This means, we are not giving "Bob" access to the shared document anymore because he is "Bob," but because he has specific privileges authorizing him to access the document (his role within his home organization for example).

B. Access Control Implementation

ABAC access control policies offer a high flexibility of specification. However, the associated access control mechanisms must be able to control who assigns the attributes values.

The X.509-v4 [14] standard has developed the concept of the PMIs. The major component of this specification is the attribute certificate. An attribute certificate, signed by an attribute authority (AA), generalizes the identity certificates managed within the public key infrastructures (PKIs) (data available online at http://www.ietf.org/rfc/rfc3280.txt) by linking an attribute (a user's identity or its role) to a public key. The concepts of PMIs and AAs are very interesting in the context of VEs because they offer the means to control the organization's authority and responsibilities:

1) The organizations providing human resources by acting as AAs can assign attribute values to their employees and give a guarantee to other organizations providing resources about the fact that they have assigned these attributes to these employees.

2) The mechanisms of access control to resources can filter the attributes depending on the AAs that have assigned them to users.

The choice of "who is the attribute authority" depends on a strategic level, based on the concepts of both authority and responsibility of partners. It must be defined within the contract.

C. Harmonizing the Information of Authentication and Accreditation

Because a VE is composed of independent partners, each partner can have its own technology for managing/storing authentication and accreditation information that could be different from the one used by its partners. Authentication is the ability to prove that a user or application is genuine, and accreditation is the act of giving credentials and rights to users depending on their roles within the enterprise. For example, authentication can be carried out by means of username/password couples, digital certificates, biometric techniques, etc. using different protocols such as remote authentication dial-in user service (RADIUS) (data available online at http://www.ietf.org/rfc/rfc2865.txt), Kerberos (data available online at http://www.ietf.org/rfc/rfc1510.txt), etc.

Traditionally, a site providing a service (SP, service provider) implements the authentication and accreditation mechanisms. Each time a user wishes to access a

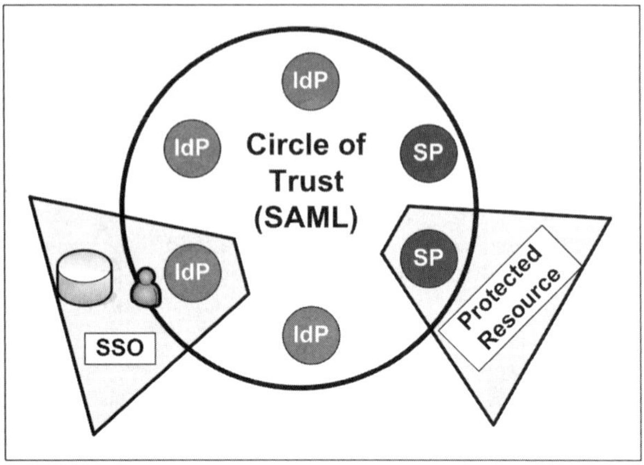

Fig. 1 Identity federation seen as a circle of trust.

new resource provided by a new SP, a new user account should be created and managed at the SP. As a user might belong to an organization that is different from the SP organization, some information should be exchanged between the organizations' administrators. (The SP asks for information about the user, creates the user account, and sends connection information to the user hoping that the user has the necessary technology on its workstation.) In addition, the user has to remember a number of passwords equal to the number of SPs managing the resources that the user wishes to access, which makes the tasks of the users and the administrators more complex.

A new concept, called *identity federation* (Fig. 1), can be adopted to harmonize the information of authentication and accreditation between partners.

Identity federation is the sharing of digital identities to enable applications in different security domains to work together securely. The federation enables users and applications to work seamlessly as if they were part of the same security domain, while in fact the security domains remain largely independent. Identity federation can be seen as a circle of trust between security domains where each domain accepts accreditations from the other domains.

V. "Workflow Engine" Descriptive Scenario

We will explain our approach and detail our solution for access control within virtual collaborative environments based on a proof-of-concept demonstrator. This demonstrator was initially defined by industrial partners, which realized the virtual enterprise described in Chapters 9 and 10 (data available online at http://www.vivaceproject.com) complemented by the TSCP (data available online at http://www.tscp.org). The implementation was successfully demonstrated for the first time in the VIVACE Forum 3 [15].

As a representative aeronautical use case, the whole engine use case was chosen to validate our approach and solution for access control within virtual collaborative environments. The whole engine use case is set up as described in

Chapters 9 and 10 to demonstrate how new technological aids could assist such multipartner collaborations.

A. Proof of Concept

The proof of concept described in this section is based on part of the whole engine use case. It demonstrates our access-control solution based on identity federation and attribute-based access control. Examples of collaboration in the whole engine use case were identified and captured in a simplified, but still representative collaboration scenario, which was used as a basis for the proof of concept.

In the resulting proof of concept, multiple partners are collaborating on a shared design of an engine turbine disk. The collaboration involves the following partners: an integrator, who provides an initial design of the engine turbine disk and receives the result from the subcontracted work performed by the first tier suppliers; a designer, who, based on the initial design, provides a parameterized detailed design of the engine turbine disk; an analyzer, who analyzes the detailed design and provides an analysis report; and a validator, who validates the analysis and provides a validation report.

In a real-life scenario, engineers at the different partner sites would have specialized engineering tools at their disposal (MSC Patran, data available online at http://www.mscsoftware.com/assets/MSC_DS_Patran.pdf) and (MSC Nastran, data available online at http://www.mscsoftware.com/assets/MSC_DS_MSCNastran.pdf). The results of the work at each partner site are depicted in Fig. 2. Chapter 9 provides an in-depth explanation of the activities involved.

The next section describes how the collaboration proceeds in the proof of concept.

B. Collaboration Setup

We assume that a VE has been set up by the partners for the collaboration. To support the collaboration, the whole collaborative job (i.e., the use case) is mapped onto a workflow. The workflow is a recipe of the successive tasks (or activities) to be performed collectively by the partners on behalf of the collaborative job. A shared data repository is used for secure data sharing.

As an interested party, the integrator is responsible for defining the applicable workflow (also called the "orchestrator" of the workflow).

The workflow defines a schedule, in which each task is assigned a start and end date, a performer (i.e., partner), preconditions (i.e., conditions that need to be fulfilled before the task can be started, such as presence of required inputs), and results (e.g., outputs).

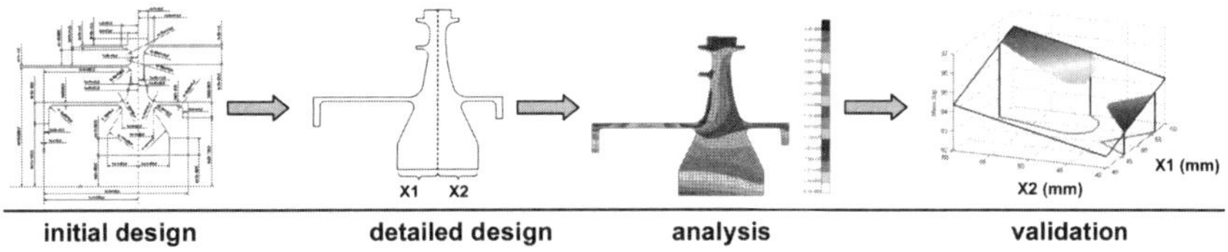

Fig. 2 Engine turbine disk design.

A "conductor" takes care of the actual implementation and coordination of the execution of the workflow. The conductor triggers the responsible "performers" (engineers/designers performing their respective tasks) based on the schedule and the job's present status. In its simplest form, the conductor is a human being playing the role of process operator. He or she either regularly inspects the schedule and its status and is prompted automatically by e-mail on status updates, or manually maintains the schedule and contacts the performers by phone and e-mail to trigger tasks and to stay informed on progress and status. In our case, however, the conductor is an automated system, more precisely a workflow management system at a central location. This system triggers the performers by e-mail, monitors the shared repository for particular data being available as expected results of triggered tasks, and maintains the schedule accordingly.

When triggered by e-mail to perform a task, a performer retrieves the applicable inputs from the shared repository, performs the task, and finally stores the applicable outputs to the shared repository. Any relevant status updates (e.g., the task is finished) can be communicated to the conductor, preferably with no explicit administrative actions by the performer. In the proof of concept, the conductor monitors the shared data repository to determine if a performer has completed his part of the job. As a result, a performer's share of the job is considered finalized when the results have been uploaded to the shared data repository. The performer is responsible for uploading the results. The conductor concept is depicted in Fig. 3.

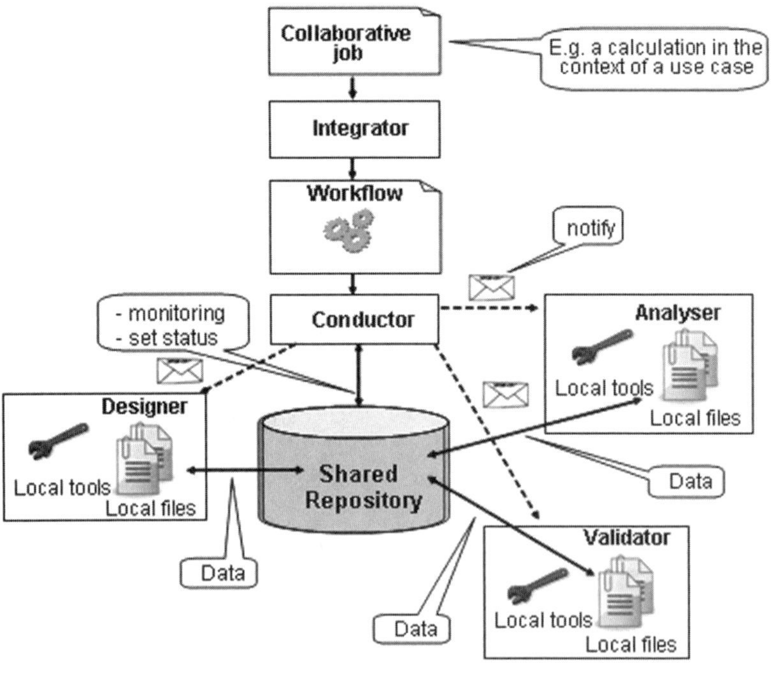

Fig. 3 Conductor-coordinated workflow.

C. Workflow Implementation

In our proof-of-concept demonstrator, the workflow was implemented as an automated system, called WFE (workflow engine). Each of the conductor tasks (monitoring the progress of the workflow, triggering of the proper performer at the right time, and setting the right status to the shared document at the right time) is implemented as a tool using Java technology. These tools were connected in a tool chain according to the workflow design, using proprietary tool-chaining middleware [16].

D. Proof-of-Concept Demonstration Setup

For ease of demonstration, the partners' computers are implemented on VMware (data available online at http://www.vmware.com/) virtual machines. A virtual machine can be described as a simulated computer that behaves as a separate computer but runs in parallel with the host computer. This allows the running of multiple simulated computers on just one single host computer. All of the virtual machines are installed on one single laptop running Windows XP in combination with VMware Server. Each virtual machine runs Linux as its operating system. A static Internet protocol address (IP address) is assigned to each virtual machine, and a local network is set up on the laptop connecting all virtual machines. Sendmail is used as the mail server, and KMail (data available online at http://userbase.kde.org/Kmail) is used as the mail client, allowing reception of local mail. The result is a simulation of the entire virtual enterprise and its computers on a single machine.

To download documents from and upload documents to the shared repository, a Mozilla Firefox web browser is used. The demonstration setup is depicted in Fig. 4.

The setup of the collaborative job, including the conductor workflow is shown in Fig. 5. It is assumed that the orchestrator has finished, so that the virtual enterprise is in place. Data are exchanged through and collected in a central shared data repository, which is accessible by the partners. The chain of activities is defined and controlled with the help of the conductor workflow, which coordinates the activities at the various partners by sending e-mail and monitoring the contents of the shared data repository. The conductor workflow runs only locally on the conductor's system. The demonstration scenario is as follows (note that the numbers correspond to the arrow labels in Fig. 5, and that numbers relating to a document status change are not depicted in the figure):

1) The conductor sets the initial status of the shared design document to "start." As a consequence, only the integrator has permission to access and change the document.

2) The integrator places the initial design of the engine turbine disk on the shared repository.

3) The conductor notices that the document has changed and updates the status of the document to "await-design." As a consequence, only the designer has permission to access and change the document.

4) The conductor notifies the designer by e-mail of the availability of the initial design.

5) The designer downloads the initial design from the shared repository.

6) Based on the initial design, the designer creates a detailed design. When finished, the designer uploads the detailed design to the shared repository.

7) The conductor notices that the document has changed and updates the status of the document to "await-analysis." As a consequence, only the analyzer has permission to access and change the document.

8) The conductor notifies the analyzer by e-mail of the availability of the detailed design.

9) The analyzer downloads the detailed design from the shared repository.

10) The analyzer analyzes the detailed design. When finished, the analyzer uploads the resulting analysis report to the shared repository.

11) The conductor notices that the document has changed and updates the status of the document to "await-validation." As a consequence, only the validator has permission to access and change the document.

12) The conductor notifies the validator by e-mail of the availability of the analysis report.

13) The validator downloads the analysis report from the shared repository.

14) The validator validates the analysis. When finished, the validator uploads the resulting validation report to the shared repository.

15) The conductor notices that the document has changed and updates the status of the document to "validated." As a consequence, all parties involved are allowed to download the shared document. As the design is validated, changes to the shared documents are no longer allowed. (It is frozen.)

16) Finally, the conductor notifies all parties involved by e-mail of the availability of the final document.

The shared repository is secured by our access control solution. Access to the shared repository is only granted to authenticated users providing the right attribute values as defined by the policy protecting the shared document.

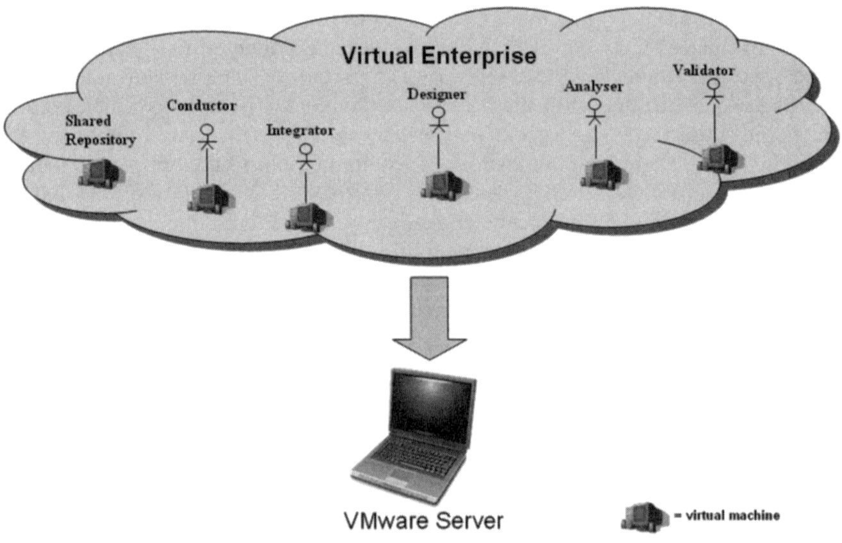

Fig. 4 Demonstration setup.

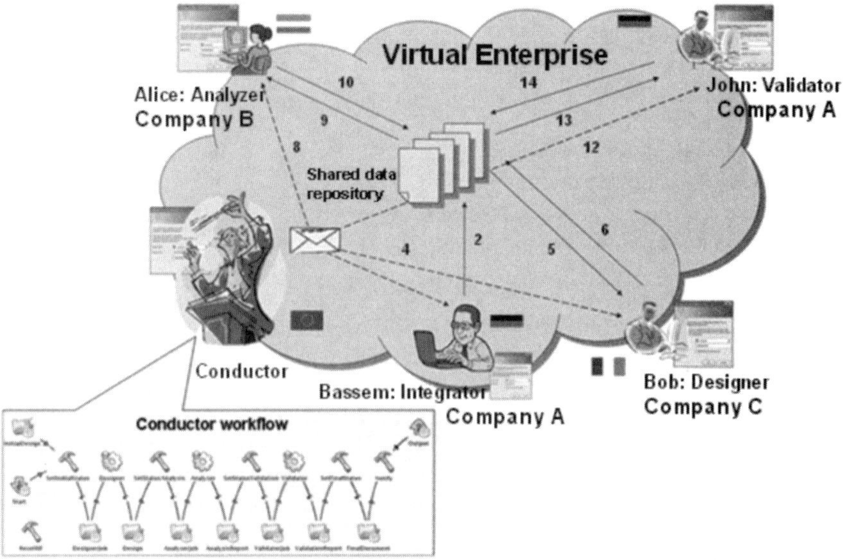

Fig. 5 Setup of the collaborative engineering job.

In reality, a scenario such as that just described can involve multiple iterations before the design is completed. For example, if anomalies are detected during analysis stage, the document will be returned to the designer in order to adjust the design accordingly. Support of these types of iterations would only entail a more detailed specification of the supporting workflow, requiring no adjustments to the access-control solution.

E. Collaboration Formalization

The collaboration between these different companies (Fig. 5) is formalized by a *contract* stating the following:

1) Company A provides *integrators* and *validators*. This means that a user, coming from company B and having the role *validator*, will not be authorized to act as a *validator* within this SCE, and thus his request will be rejected. Thus, company A is an AA for the *integrator* and *validator* values of the *role* attribute.

2) Company B provides analyzers,

3) And company C provides designers.

The contract does not define who should perform the job within each company; thus, each company is responsible for managing its people (managing their credentials and assigning specific roles to them). We consider that the agreement defined and signed between the participating companies is also an attribute given by the different companies to their employees. In our scenario, the value of the *veagreement* attribute is *vecontract*.

Thus, the access control policy related to our scenario consists of five rules corresponding respectively to *integration policy*, *design policy*, *analysis policy*, *validation policy*, and *end policy*. S, A, and R represent the subject, the action and the resource, respectively. We are providing policies in plain English and using logical connectives (see Appendices for more details on logical connectives).

1- role(S) = integrator ∧ veagreement(S) = vecontract ∧ (name(A) = read ∨ name(A) = write) ∧ name(R) = specification-document ∧ status (R) = start

This means that a user can access the document *specification-document* with read or write permission if the user has the role *integrator* defined within the contract *vecontract* and the document status is *start*.

2- role(S) = designer ∧ veagreement(S) = vecontract ∧ (name(A) = read ∨ name(A) = write) ∧ name(R) = specification-document ∧ status (R) = await-design

This means that a user can access the document *specification-document* with read or write permission if the user has the role *designer* defined within the contract *vecontract* and the document status is *await-design*.

3- role(S) = analyser ∧ veagreement(S) = vecontract ∧ (name(A) = read ∨ name(A) = write) ∧ name(R) = specification-document ∧ status (R) = await-analysis

This means that a user can access the document *specification-document* with read or write permission if the user has the role *analyzer* defined within the contract *vecontract* and the document status is *await-analysis*.

4- role(S) = validator ∧ veagreement(S) = vecontract ∧ (name(A) = read ∨ name(A) = write) ∧ name(R) = specification-document ∧ status (R) = await-validation

This means that a user can access the document *specification-document* with read or write permission if the user has the role *validator* defined within the contract *vecontract* and the document status is *await-validation*.

5- role(S) = any ∧ veagreement(S) = vecontract ∧ name(A) = read ∧ name(R) = specification-document ∧ status (R) = validated

This means that a user can access the document *specification-document* with read-only permission if the user has the role *any* (here, we are using the concept of role hierarchy where the role *any* is a subordinate role of *integrator*, *designer*, *analyzer*, and *validator*, that is, any permission affected to *any* is also affected to *integrator*, *designer*, *analyzer*, and *validator*) defined within the contract *vecontract* and the document status is *validated*.

Thus, we are using an attribute called *status*, which will be modified by the WFE. This mechanism allows the WFE to activate or deactivate the access control rules. It is fruitful to use the same mechanisms of access control to protect the document/service allowing only the WFE to modify the *status* of the document. So, we add the following rule:

6- Role(S) = WFE ∧ veagreement(S) = vecontract ∧ (name(A) = read ∨ name(A) = write) ∧ name(R) = status- specification-document.

By letting the conductor manage the status attribute of the shared document, it is guaranteed that the shared document is updated in the proper order and according to the agreed workflow. For example, this prevents the undesired scenario of an analyzer analyzing an incomplete design. The analyzer is only granted access to the shared document when the conductor notices that the designer has completed the detailed design and changes the status attribute of the document to "await-analysis." It also prevents the undesired scenario of a designer updating a design that is already being validated by a validator. Because the conductor has set the status attribute of the shared document to "await-validation," users with role attribute values other than "validator" are not allowed access to the shared document.

VI. Prototype

The secure environment we have deployed exploits two technologies: one providing a harmonization of authentication and accreditation information (data available online at http://shibboleth.internet2.edu/) and the other one giving us the requested PMI (data available online at http://sec.cs.kent.ac.uk/permis/).

A. Adopted Identity Federation Architecture

The Internet2 consortium has defined a standard-based open-source middleware software, which provides Web single sign on across or within organizational boundaries (data available online at http://shibboleth.internet2.edu/). It provides mechanisms for users to access remote resources via authentication at their home site and authorization via a set of user attributes provided by the home site. It distinguishes between two entities: the identity provider (IdP) and the service provider (SP). When a user is authenticated against its IdP, the IdP informs the SP by sending him a proof-of-the-user authentication and its accreditations. This software uses the security assertion mark-up language (SAML) (data available online at http://www.oasis-open.org/committees/tc_home.php?wg_abbrev=security), defined by OASIS (data available online at http://www.oasis-open.org/home/index.php), as a protocol for exchanging the authentication and accreditation information between IdPs and SPs.

This software includes a framework that allows the members of a VE to authenticate once to the VE (via their home site), and then to move between the different computer systems and services of the VE that might be physically located at different members sites, without needing to authenticate again.

B. Adopted PMI Architecture

The PMI [17] is developed by the Information Systems Security Research Group of the University of Kent (data available online at http://sec.cs.kent.ac.uk/permis/). It is mainly composed of two entities: the PDP (policy decision point) making access control decisions according to a given policy and the PEP (policy enforcement point) implementing the decisions taken by the PDP.

It provides a higher-level policy and authorization framework for the creation and management of a distributed authorization infrastructure using X.509 standard attribute certificates (ACs) to hold users roles. It implements hierarchical RBAC as defined in the NIST RBAC model, whereby superior roles inherit the privileges of subordinate roles.

Administrators at various sites in a VE can allocate roles (or other attributes) to their users, in the form of X.509 attribute certificates, and store these in a local (or remote) lightweight directory access protocol (LDAP) directory. The manager of a VE resource determines the authorization policy for the target resources under his control. This authorization policy is written in XML, according to a defined document type definition (DTD), and stored in the target resource.

When a user attempts to access a target resource, the application gateway (typically a Web server) traps the user's request; authenticates the user in an application dependent way (e.g., X.509 client certificates, username/password, or SSO, etc.); and then calls the decision engine to find out if the user is authorized to perform the requested action on the target or not.

Access control decisions are made according to the authorization policy currently in force, the roles currently valid for the user, the target being accessed, the action being requested, and the environmental variables such as the time of day.

System administrators write policies, specifying which roles have which privileges and what kind of credentials will be recognized by the adopted PMI. The policy will be used for all reasoning regarding authorization. Attribute administrators will issue credentials containing users' attributes, telling what roles the users have.

However, the adopted PMI being initially conceived to implement RBAC policies only recognizes attributes that are the users' roles. Actually, as roles are implemented, it is possible to specify any attributes to the users. But, it does not recognize resources attributes, which are required in our use case.

Thus, we extended the adopted PMI architecture so that we can specify and implement ABAC policies. This new version takes into consideration any possible attribute of the managed resources (in our example, the attribute *status*).

For example, the *integration-policy* rule is as follows:

$\text{role}(S) = \text{integrator} \wedge \text{veagreement}(S) = \text{vecontract} \wedge (\text{name}(A) = \text{read} \vee \text{name}(A) = \text{write}) \wedge \text{name}(R) = \text{specification-document} \wedge \text{status}\ (R) = \text{start}$

In the adopted PMI architecture, it will be expressed as follows:

```
<TargetAccess>
  <RoleList>
    <Role Type="PMIRole" Value="integrator"/>
    <Role Type="veagreement" Value="vecontract"/>
    <Role Type="status" Value="start"/>
  </RoleList>
  <TargetList>
    <Target Actions="GET, POST">
      <TargetDomain ID="TargetDomain1"/>
    </Target>
  </TargetList>
</TargetAccess>
```

where **TargetDomain1** is the domain including the specification document (our protected resource) and which has been already defined within the policy.

C. Identity Federation and the PMI Integration

In our scenario, the authentication on the IdP site is realized using the classical authentication module of Apache (data available online at http://www.apache.org/). The users' accreditations (here, *role* and *veagreement* attributes) are stored within an LDAP repository; these attributes are protected within attribute certificates. By signing these certificates, the administrator of the organization, acting as the IdP, is considered as the AA. The resources' attributes (here, *status* attribute) are specified in an XML file. Each element provides the name of the resource and the set of attributes that the resource has. During the phase of access control on the SP side, only the attributes signed by trusted AAs are considered.

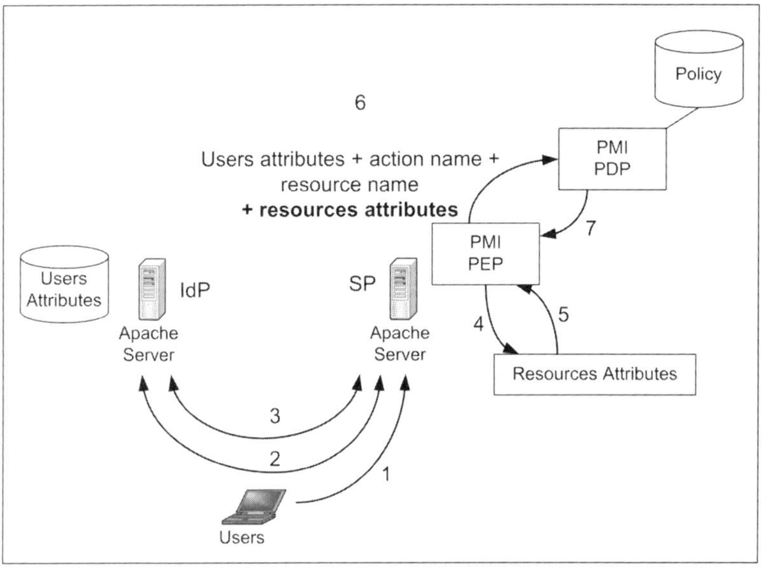

Fig. 6 Architecture of the secured collaborative environment.

Figure 6 illustrates the different messages exchanged when a user (real person or software) requests access to a protected resource:

1) The user requests access to resource R hosted by service provider SP.

2) The SP module redirects the user to its identity provider IdP to authenticate itself. When SP knows only one IdP, this redirection is done automatically. In general, when SP knows many IdPs, the redirection is done thanks to a WAYF (where are you from) module.

3) When the user is authenticated, IdP generates a SAML assertion that contains a statement indicating the user has been authenticated and the user's credentials are within an attribute certificate. This assertion is sent, through the user, to the SP module that decodes the SAML assertion and forwards the credentials to the adopted PMI architecture.

4) The PEP gets the attributes of resource R.

5) The PEP sends the request to the PDP. The request includes the attributes of the user, the attributes of the resource R, and the name of the action.

6) The PDP takes its decision according to the access control policy. For each attribute, it checks if the AA is known and if the attributes are valid. Then, the decision is returned to the PEP.

7) Based on the decision of the PDP, the PEP informs Apache so it accepts or refuses the user's request.

8) The user is granted access to the resource R.

VII. Enhancing the Access Control Solution

Although the access control solution combining authentication, attributes propagation (data available online at http://shibboleth.internet2.edu/) and

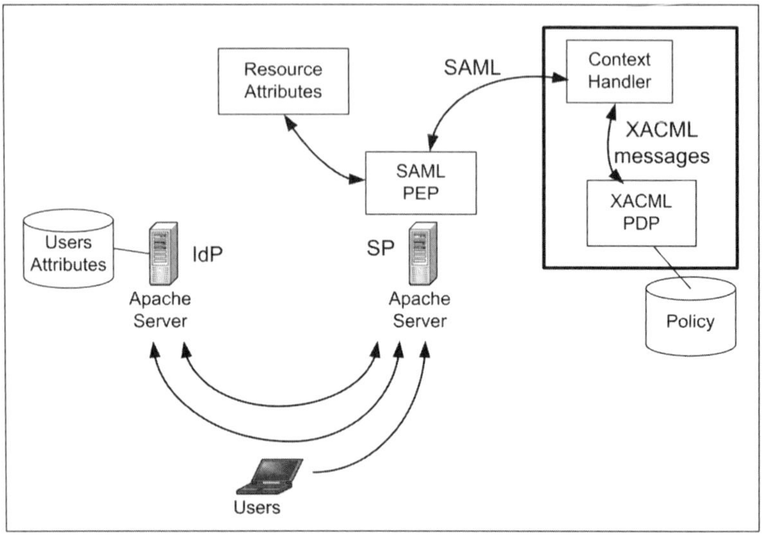

Fig. 7 SAML/XACML access control solution.

authorization engine (data available online at http://sec.cs.kent.ac.uk/permis/) is efficient and has been demonstrated and validated based on a real-life scenario, it has some weaknesses that are related to the adopted PMI architecture. In fact, the PMI architecture uses a proprietary policy language, the PMI exchanges between PEP and PDP are realized through a proprietary language, and PEP and PDP are actually implemented on the same workstation.

So, in order to provide solutions to these weaknesses, the access control architecture must provide a standard-based policy language and a standard protocol for the exchanges between PEPs and PDPs. For such reasons, we have turned towards the OASIS standards SAML and XACML (extensible access control mark-up language) in order to define a PMI based only on standards (data available online at http://www.oasis-open.org/committees/xacml/).

Figure 7 illustrates the different messages exchanged when a user (real person or software) requests access to a resource, when this enhancement is adopted.

This solution is being developed (data available online at http://www.irit.fr/Eguipe-SIERA-), and a first prototype has been finalized.

In addition to the described scenario, our access control solution can be used to ensure security services within other collaborative environments. In the next section we introduce the virtual-enterprise collaboration hub concept and discuss how it might benefit from our solution to resource access control.

VIII. Securing the Virtual-Enterprise-Collaboration Hub

Considered in this section is an alternative solution for implementing our secure collaboration approach: the virtual-enterprise-collaboration hub (VEC hub) described in Chapter 10. Compared to the solution discussed in preceding sections, the solution discussed in here does not follow the traditional Web-based approach to secure collaboration. Instead, the solution is based on a service-oriented architecture (SOA), which makes partner proprietary tools available as services. These services allow authorized partners to access the functionality of these tools remotely without exposing the tools themselves. In addition to the Web-browser-based solution described in the preceding section, the SOA architecture also faces the challenge of securing inter-Web-services communication. This section discusses the specific security challenges of this intercompany collaboration architecture.

The VEC-hub is a concept of a framework that supports intercompany collaboration in virtual enterprises (data available online at http://www.VEC-Hub.org).

The concepts as introduced in preceding sections have also been applied to the VEC hub, where the following holds:

1) Each organization member of the virtual enterprise should manage the process of authenticating and accrediting their people. As a result, federated identity management was introduced.

2) Each organization member of the virtual enterprise should be in control of the services of their own enterprise provided to the virtual enterprise. A single-sign-on service will be required to avoid that users have to remember multiple accreditations given by different authorities.

A. Virtual-Enterprise-Collaboration Hub Concept

The objective of the VEC-hub concept is to define a SOA for the virtual aeronautical enterprise providing a basic level of service functionality promoting the requirements on the collaborative solution: a common agile setup of partner roles, product information, work processes, and reference data for the collaboration project; a neutral partnership collaboration platform with respect to technical (vendor and system), internal and external aspects; partner-to-partner isolation via intermediate services communicating using neutral data structures; and accessibility via commodity technology such as Web browsers to collaboration environment assets.

The VEC hub forms a foundation for the operational environment of the virtual enterprise. This gives everyone, within the heterogeneous IT environment of the partners, the ability to securely and traceably find, consolidate, and publish information and collaborate in a common context.

The concept is shown in Fig. 8. The outer dashed circle of the conceptual layout represents the portal front for the VEC-hub area of interest providing (S)ervices for access, management, and security. Outside of this area each collaboration peer maintains its private view on internal (P)roduct information, (O)rganizational roles, and (W)ork processes. Three main kinds of collaboration peers have been identified: venture partners that are capable of providing internal (P), (O), and (W) assets for collaboration; small-to-medium enterprise (SME) partners, with mainly commodity tools available for collaboration; and services partners that provide supplementary (S)ervices supporting the collaborative workflows.

The inner dashed circle labelled core services represents the core (S)ervices of the VEC hub that orchestrate the processes and information storage. The interior

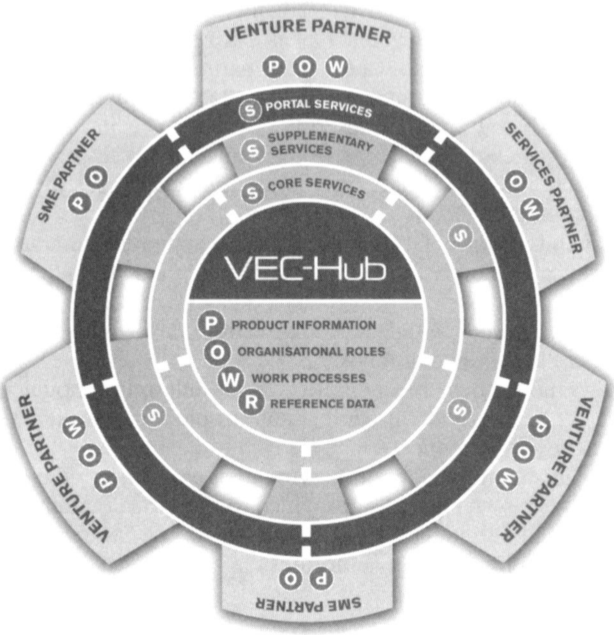

Fig. 8 VEC-hub concept.

green area represents the common view on (P), (O), (W), and (R)eference data for the collaborative work.

The area between the dashed circle and circle labelled core services is called the shared area of the VEC hub. Here directly integrated partners can communicate internal (P), (W), (O) information via supplementary (S)ervices under control by the VEC-hub portal services.

The VEC hub concept specification also covers issues related to trust, security, and the legal aspects in setting up a virtual enterprise and collaborating through a VEC hub.

B. Security Requirements for the VEC Hub

The security requirements can be subdivided into requirements with respect to securing the VEC-hub areas, protecting the VEC-hub communications, and VEC-hub services access control. As shown in Fig. 8, the areas to be secured are the portal services and core services areas and the supplementary services area, while the partners' private areas reside outside of the VEC hub. These areas must be protected against access by non-virtual enterprise partners.

Measures to secure VEC-hub portal services (hosted by the VEC-hub core services provider) and core services (provided by core service partners) areas are protecting the area using firewalls, for example, only allow connections from collaboration partners, and using servers with up-to-date security fixes installed.

Securing the VEC-hub portal and core services is the responsibility of the virtual enterprise or the service partner providing the service.

In general, the supplementary services reside in areas administered by the partners providing the supplementary services. Securing the VEC-hub supplementary services is the partner's responsibility, but important for VEC-hub's security. Measures to secure supplementary services areas are applying demilitarized zones (DMZ) protected by firewalls and using servers with up-to-date security fixes installed.

Securing each partner's private area (for usage of internal data and services, e.g., due to licenses and available hardware) is that partner's sole responsibility. The VEC-hub connections to be secured are the following:

1) First is the end-users' client applications (e.g., a Web browser) of SME partners and venture partners accessing the VEC hub. In general, an insecure medium such as the Internet will be used. Hence, these connections must be secured. Even when using dedicated or leased lines as the connection mechanism, protecting these connections might still be appropriate.

2) Next comes the connection between the portal and core services and the supplementary services. Because typically, the supplementary services are located in a zone administrated by a venture partner or services partner, again this connection is typically implemented using the Internet and must be secured.

3) Securing the connection between the supplementary services administered by a partner and this partner's private area is this partner's own responsibility. This connection might not be an Internet one, but rather an internal partner connection.

The third VEC-hub security subject is access control. It must be possible to set access rights to the VEC-hub services to only allow certain users, certain groups or companies to access a specific service. In some cases it might be appropriate to control the access to a part of a service at a finer granularity. Furthermore, when access to the service has been gained, the service itself will also implement access control: for example, allow certain user or certain groups to make modifications, give other users only read permissions, and so forth.

Role-based access control means that permissions to perform certain operations are assigned to specific roles. Role-based access control reduces the complexity and cost of security administration in large networked applications and is an important security concept for the VEC hub. When more fine-grained access control is desirable, the already introduced attribute-based access control could be applied.

Managing access includes managing the identities. Identity management involves the management of user identities across all resources involved. It is established by an integrated system of business processes, policies, and technologies that enable organizations to facilitate and control their user's access to critical online applications and resources—while protecting confidential personal and business information from unauthorized users. It represents a category of interrelated solutions that are employed to administer user authentication, access rights, access restrictions, account profiles, passwords, and other attributes supportive of users' roles/profiles on one or more applications or systems.

Single sign on is a desirable feature for the VEC hub. In the context of the VEC hub, single sign on means, for example, that the different stand-alone or Web applications all turn to the same authentication server for user credentials.

Compared to an intra-organizational implementation of single sign on, the VEC hub faces a greater challenge because there are two or more organizations with different policies and solutions for authentication and authorization. Implementation of single sign on for a virtual enterprise requires trust among the partners (see Section II), and includes federated identity management (see Sections II and IV), and application of security frameworks (data available online at http://shibboleth.internet2.edu/) (see Section VI). For the VEC hub, the Web services are defined using the Web services definition language (WSDL, data available online at http://www.w3.org/). End-users can use Web browsers to connect to the VEC-hub portal. However, also Web clients can use that to communicate with the VEC-hub services using XML messages that follow the Simple Object Access Protocol (SOAP) standard (data available online at http://www.w3.org/). The security framework must support this SOAP client-server communication. A single-sign-on solution based on browser redirects will not suffice.

Storing data in encrypted form is an additional measure to make sure that only specific authorized users can gain access to the data and even that system administrators would not have access to the information contained in these stored files.

Another security requirement is related to Web service authentication. The VEC hub must offer certain Web services the possibility to ensure that the offered Web service is really the Web service it claims to be, including its version. This requirement is justified as using the wrong Web service inadvertently, or the wrong version of one, might harm both the VEC-hub content and the expected use in the remote application.

Finally, there should be a session log containing audit information of "who did what and when." The log files should be suitable for "after the fact" damage assessment. Failed log-in attempts shall be logged with the appropriate details. Intrusion detection software could be used to detect and log break-in or break-in attempts. The details of the log files and of access to these log files (who has access, rules for usage of log data, etc.) must be defined in agreement between the VEC-hub partners. The agreement shall take into account the legal details (e.g., what is allowed to be logged).

IX. Conclusions

Presented in this chapter is the work we have accomplished within the context of the collaboration between the value improvement through a virtual aeronautical collaborative enterprise (VIVACE) project and the TSCP. Our objective was to provide an access control solution to be deployed within a virtual collaborative environment. This solution must implement the recommendations given by the TSCP concerning trust and identity management and consequently allowing a secure collaboration between the participating organizations.

We have shown that a secure collaborative environment can be defined and deployed so that organizations can join this network, share their resources and competencies, and respond to new business opportunities. The access control architecture that we have proposed manages the critical resources in the distributed environment, allows each organization to stay independent, and provides the organization with technologies to manage its own people and assets.

To respond to the security constraints related to virtual enterprises, we have proposed an access control solution based on three concepts. The first concept is access control models based on attributes offering some flexibility during the phase of the specification of the access control policies. The second one is the privilege management infrastructures controlling the management of these attributes, and the third concept is identity federation in order to harmonize the authentication and accreditation mechanisms. Our access control proposal solution has been used to secure the workflow generated by the actual real-world requirements, as derived from collaboration in a virtual enterprise and described in Chapters 9 and 10. This demonstrates the proof of concept of the described approach, but, in order to respond to the scalability criterion, the solution should be based on standards. We propose an enhancement for our solution based on the SAML and the XACML standards defined by OASIS.

The VEC hub followed the same approach as our access control solution (see Chapter 10). However, our access control solution has not been tested with a Web-services-based SOA, only on an architecture based on Web browser redirects.

Current TSCP results focus on secure e-mail and secure data exchange. Future TSCP work will focus on secure collaboration. Therefore, our work on securing the virtual-enterprise collaboration fits well within the future work plan of the TSCP.

Appendix: Description of a Logical Connective

In logic, a *logical connective*, also called a truth-functional connective, logical operator, or propositional operator, is a logical constant that represents a syntactic operation on a sentence, or the symbol for such an operation that corresponds to an operation on the logical values of those sentences. A logical connective serves to produce a compound sentence from one or two other sentences. The truth value of the resultant compound sentence is determined by the truth values(s) of the one or two other sentences. Consequently, a logical connective can be seen as a function, which, when applied to sentences as arguments whose values are true or false, returns in turn the value true or false; consequently, logical connectives are called truth-functional connectives.

The basic logical operators are as follows: negation (not) ($\neg$ or $\sim$), conjunction (and) ($\wedge$ or &), disjunction (or) ($\vee$), material implication (if…then) ($\rightarrow$, $\Rightarrow$ or $\supset$), and biconditional (xnor) ($\leftrightarrow$ or $\leftrightarrow$).

References

[1] Kesseler, E., and Vankan, W. J., "Taking Collaborative Engineering to the Sky, Formation Flying with Knowledge Management," *European Conference for Aerospace Sciences (EUCASS)* (Moscow, Russia), 2005, http://www.onera.fr/eucass/2005/system-integration.html [accessed July 2009].

[2] Pritchard, D., and MacPherson, A., "Industrial Subsidies and the Politics of World Trade: The Case of the Boeing 7e7," *The Industrial Geographer*, Vol. 1, No. 2, 2004, pp. 57–93, http://igeographer.lib.indstate.edu/pritchard.pdf [accessed July 2009].

[3] Bultje, R., and van Wick, J., "Taxonomy of Virtual Organisations, Based on Definitions, Characteristics and Typology," *VOnet Newsletter*, Vol. 2, No. 3, 1998, pp. 7–21.

[4] Lane, C., and Bachmann, R. (eds.), *Trust Within and Between Organisations*, Oxford Univ. Press, Oxford, U.K., 2002, pp. 3–30.

[5] Dasgupta, P., "Trust as Commodity," *Trust: Making and Breaking Cooperative Relations*, edited by Gambetta, D., Basil Blackwell, Oxford, U.K., 1988, pp. 49–72.

[6] Luhmann, N., *Trust and Power*, Wiley, Chichester, U.K., 1979, pp. 27–37.

[7] Chiles, T. H., and McMackin, J., "Integrating Variable Risk Preferences, Trust and Transactions Cost Economics," *Academy of Management Review*, Vol. 21, No. 1, 1997, pp. 73–99.

[8] Fukuyama, F., *Trust: The Social Virtues and the Creation of Prosperity*, Faber and Faber, London, 1995.

[9] Zucker, L. G., "Production of Trust: Institutional Sources of Economic Structure 1840–1920," *Research in Organisational Behaviour*, Vol. 8, 1986, pp. 53–111.

[10] Harrison, M., Ruzzo, W., and Ullman, J., "Protection in Operating Systems," *Communication of the ACM*, 1976, pp. 461–471.

[11] Ferraiolo, D. F., Sandhu, R., Gavrila, S., Kuhn, D. R., and Chandramouli, R., "Proposed NIST Standard for Role-Based Access Control," *ACM TISSEC*, Vol. 4, No. 3, 2001, pp. 222–274.

[12] Wang, L., Wijesekera, D., and Jajodia, S., "A Logic-Based Framework for Attribute Based Access Control," *2nd ACM Workshop on FMSE*, 2004, http://csis.gmu.edu/faculty/jajodia-fmse-2004.pdf [retrieved July 2009].

[13] Abou El Kalam, A., El Baida, R., Balbiani, P., and Benferhat, S., "Organisation Based Access Control," *Policies for Distributed Systems and Networks*, Como, 01/01/03-31/12/03, edited by H. Lutfiyya, J. Moffett, and F. Garcia, Inst. of Electrical and Electronics Engineers, New York, 2003, pp. 120–131.

[14] "Information Technology-Open Systems Interconnection-The Directory: Public-Key and Attribute Certificate Frameworks," *ITU-T Recommendation X.509(V4)*, 2000, ISO/IEC FDIS 9594-8, http://www.iso.org/iso/catalogue_detail.htm?csnumber=53372 [accessed July 2009].

[15] Parchem, R., Nagel, M., Arendsen, P., Wenzel, H., De ploli, G.P., Söderberg, S., et al., "Engine MDO and Robust Design in the Virtual Enterprise," VIVACE Forum-3, http://www.vivaceproject.com/content/forum3/forum3.php [retrieved July 2009].

[16] Baalbergen, E. H., and van der Ven, H., "SPINEware—A Framework for User-Oriented and Tailorable Metacomputers," *Future Generation Computer Systems*, Vol. 15, Elsevier Publishers, Amsterdam, The Netherlands, 1999, pp. 549–558.

[17] Chadwick, D., Zhao, G., Otenko, S., Laborde, R., Su, L., and Nguyen, T-A., "Building a Modular Authorisation Infrastructure," *UK e-Science All Hands Meeting 2006*, National e-Science Centre, Oxford, U.K., 2006, http://www.allhands.org.uk/2006/proceedings/papers/677.pdf [retrieved July 2009].

Chapter 12

EDMRF to Ensure Partners' Collaboration for Multidisciplinary Optimization

Frederic Féru*

EADS, France

Thomas Nguyen Van† and Pierre-Laurent Frossard†

Snecma, France

Ernst Kesseler‡

Katern 09, Amsterdam, The Netherlands

and

Pascal Guellec§

AIRBUS Cimpa, France

Abbreviations

CAD = computer-aided design
CAE = computer-aided engineering
CAX = computer-aided activities (X)
COTS = commercial-off-the-shelf
CR = consolidated repository
DEX = data exchange set
DM = domain model
EDM = engineering data management
EDMRF = engineering data management reference framework
EMI = enterprise modeling and integration
GUI = graphical user interface

---

*Senior Expert.

†Project Manager.

‡Director.

§PLM Expert.

IM = information model
NIST = National Institute of Standards and Technology
PCM = product context management
PDM = product data management
PLCS = product life-cycle support (STEP – ISO10303-239)
PLM = product life-cycle management
SDM = simulation data management
SE = simulation engineer
SLM = simulation life-cycle management
STD = standard
STEP = standard for the exchange of product data
VIVACE = value improvement through a virtual aeronautical collaborative enterprise
WF = workflow
XML = extended mark-up language
XPDL = XML process definition language

I. Introduction

CURRENT industrial collaboration in various domains suffers from the heterogeneity and the diversity of software, exchanged data, and of the heterogeneous software and data management. This holds both between partners as well as along all activities of the product life cycle. Collaborative product development is hampered by the lack of interoperability and integration of partners systems [1, 2]. This lack of interoperability incurs significant costs. The next paragraphs provide some illustration of interoperability costs for several domains.

For the construction industry in the United States, a conservative estimate puts the costs at 15.8 billion US$ per year [3] for capital facilities (i.e., nonresidential buildings). Of these costs 60% are incurred during facility operation, 16% during facility planning and engineering, and the remainder during the construction phase. Despite this huge cost estimate, the authors in [3] report that significant interoperability costs remain beyond the analysis, and so this estimate has to be considered as conservative. The current estimated costs constitute already a significant part of the 374 billion US$ total value of the facilities put in place. The major interoperability cost categories are mitigation costs (49%) and avoidance costs (42%) with the costs of delay accounting for 9%. The planning and design phase (typically one to two years) and construction phase (typically one to three years) are short compared with the aeronautic domain. The 30–50 years of operation and renewal are comparable. To illustrate the tool diversity in this domain, this industry has 35 different types of computer-aided design (CAD) systems in use [3]. The domain is characterized by many small firms, from the 700,000 individual firms 80% have less then 10 employees [3].

For the U.S. car industry, the author in [4] estimates that the lack of interoperability accounts for additional costs of 200 to 400 million US$ per vehicle program. Much of these costs are borne by the 300,000 small- and medium-size enterprises. In another estimate [5], the annual interoperability costs for the U.S. automotive industry are estimated at 1 billion US$, with mitigation costs at 90% comprising the vast majority of the costs. The 9% delay costs exclude the cost borne by the customer, for example, caused by the delayed product availability. Interoperability delays are estimated at four months out of an average 36-month development time. To illustrate the tool diversity, the system integrators use four different CAD systems, and the first-tier suppliers use an additional two different CAD systems for a total of six, with the lower-tier suppliers adding yet another eight additional different CAD systems.

For the aeronautical industry less published data are available. The National Institute of Standards and Technology (NIST) estimates that for a major aircraft manufacturer a yearly 1.5 billion US$ cost was in large measure caused by interoperability costs [6]. By any measure interoperability costs are a substantial burden, which creates opportunities to address the current practice.

This new organizational setup for collaborative product development (also called virtual enterprise) [7, 8] has significant implications for the product development. These new collaborative ways of developing technologies and aeronautic systems produce changes in the ways aeronautic systems are designed, produced, operated, maintained, and disposed of. Collaborative engineering is defined as the application of team-collaboration practices to an organization's total product development efforts [9]. It builds, upon the systems engineering, project/program management foundations of primarily in-house cross-functional product development teams introduced by concurrent engineering [10].

Cooperation between companies in a competitive domain implies that it is impossible for a system integrator at the top of the value chain to impose a single solution to all partners involved. Consequently, collaboration has to be based on federating capabilities instead of enforcing a single solution. It turns out that improving the communication efficiency and the interpretability between systems becomes necessary. Consequently, there is a need for an integrated reference framework for the product development that improves control of the aforementioned properties. The definition of an intermediary platform ("middleware") for improved support of these capabilities appears necessary to ensure a more efficient (or productive) partnership.

There are issues on the support of collaboration related to the capability to define collaborative contexts between various activities and multiple partners. Indeed, the definition of common methods and rules are also an important aspect to identify the characteristics of the "integrated product development" [11, 12]. From the practical experience in the aeronautical domain and from literature study, the following research streams are derived:

1) The definition of architectural laws for collaborative structure, defining the transactions and interactions that exist between collaborating systems and components of the collaborative architecture [13].

2) The definition of a collaborative platform for data dissemination to elaborate structural models for the collaboration and migration between heterogeneous systems, which allows defining the characteristics attached to the product development process so as to define major components and the framework layers to support them [10, 14].

3) The definition of a data reference framework for the creation of associative environments to define how a collaborative environment is implemented and what it is composed of, which defines the structure of the integrated view based on the definition of collaborative objects and also considers the association between heterogeneous environments and sustains the definition of the migration between them [15].

In this context of a collection of heterogeneous tools and methods to manage multipartners data and environment, the engineering data management (EDM) is presented as a collaborative framework based on the major requirements of the following:

1) *Noninvasive framework:* Enterprises have already an "in-house" environment running with their own processes. The final target is not to replace existing environments but to provide an interoperable middleware that integrates existing tools and methods.

2) *A standards-based communication:* The EDM framework targets domain (activities) interoperability using a standards-based semantic reference.

3) *Services to provide information in context:* The EDM framework defines an information model providing the usage context and a domain model characterizing data involved in product development activity.

In this chapter, we identify and define the necessary components to be linked for an efficient communication of product design data in the virtual enterprise. The definition of a multilayer architecture is the basis for the progressive differentiation of data. This aspect of progressive differentiation of data is built upon the notion of context. We then present the context notion defined for the aeronautic industry. Moreover, this kind of architecture is also a requirement for the application of data standardization. We finally propose the illustration of this architecture with a classical case study of a design/simulation loop. This case study forms the basis to evaluate the efficient application of the product data standardization for the collaboration in industry and especially the use of STEP (standard for the exchange of product data) [16, 17].

II. Virtual Enterprise and Collaborative Environments

A. Definition of a "Collaborative Context"

Collaboration between two distinct engineering systems and/or subsystems is the practice of combining functions and characteristics of a set of both systems and subsystems so as to produce a single unified system that satisfies some overall need of an organization [18]. Collaboration is the main way to allow team members to manage processes, data exchange, and consistency from the early stages of a project. The collaboration then consists of analyzing and coordinating business entities to ensure the consistency of the system.

Then, a collaborative framework can provide an integrated view of the product, namely, a product reference frame for the project partners. Using such a strategy of a common framework ensures that the product integration is no longer performed by one partner but by all of the partners. As soon as each partner is updating the collaborative view, the latter must be straightforwardly made available to the other partners. The constant refinement of the common product environment is a condition to perform better collaboration. Teams are no longer working alone on the design of a module but perform it within the whole product environment and in interaction with the work of the other partners [1, 19].

In such an integrated environment, the integration of processes and activities is also a factor to improve collaboration and to decrease time step between engineering studies. To ensure to concurrently work with a clear project breakdown, it is necessary to organize groups of designers who share a set of common design items. Their work should be coordinated automatically to avoid redundancies (duplicated design environments and data) and inconsistencies (loss of attributes, three-dimensional models shaded off).

B. Collaborative Typology for Systems Collaboration

Li et al. [7] analyzes the viewpoints associated with CAD activities and especially the collaboration using three-dimensional representations. His work is based on the statement of systems heterogeneity, of the use of tools and major functions that need to exist in such systems. To better characterize the environment, he has adopted a collaborative description defining two approaches: the horizontal and the hierarchical collaboration. Multidisciplinary optimization can then be interpreted as the result of the combination of this collaborative typology. This characterization can be extended in our case.

1. Description of the "Horizontal" Collaboration

The horizontal collaboration corresponds to the exchange and integration of data between equivalent tools. Usually used at the same process level and for the same activity, this collaboration corresponds to the exchange of data packages that are integrated by partners and teams to rebuild the common design environment of the product. To date, this collaboration is performed in two ways: the first is supported by the use of translators and the second by the use of standards for data description.

The use of data translators allows the transformation of data output of a system into data input of another. Data are pre- and postprocessed in order to convert them from one data structure to another. Using such a method requires using $(N^2 - N)/2$ translators (with the hypothesis of N partners collaborating with N different engineering tools) [20] (see Fig. 1). This method is the most costly and constraining one because it obliges each partner to develop an interface to communicate with the others partners in the two ways (import and export of data).

The use of a standardized file description between partners is the most flexible and adaptable solution. It ensures that all partners are exchanging data using the same data structure and then communicate exactly the same kind of data. Data

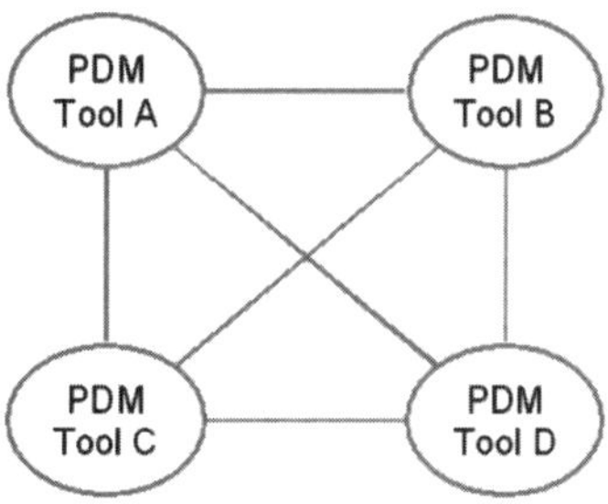

Fig. 1 Tool interoperability without standard.

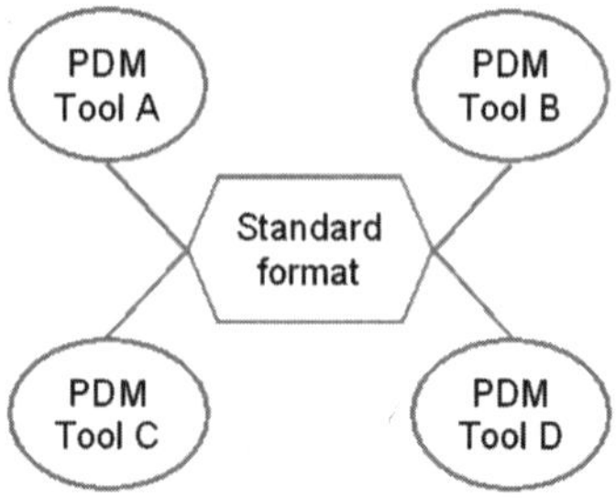

Fig. 2 Tool interoperability with standard use.

incoming from a partner can then be exploited by all of the others without the need to convert them into different format [21]. Using such a method ensures less effort to set up a collaboration and ensures the communication between engineering tools. Using the preceding example, if we consider N partners, only $2N$ translators are required for the communication of engineering data (see Fig. 2).

2. Description of the "Hierarchical" Collaboration

The hierarchical collaboration designates collaboration between teams from different domains and disciplines. It concerns the definition of links and dependencies between upstream and downstream activities applying methods to integrate engineering data from a domain to another. Currently, data management principles are mainly used for design activities [with product data/life-cyle management systems (PDM/PLM)], and collaboration using different domain tools and engineering data are not well developed. The development of new systems such as simulation data management systems tends to change this fact. As numerical simulation is strongly linked to design, the need to integrate the various data structures occurs. They are currently performed using translation codes to gather necessary information. Particular descriptions and attributes such as simulation parameters are then still not integrated. Considering the migration between heterogeneous categories of engineering tools, integration principles of horizontal collaboration remain applicable [22].

C. Industrial Gap to Reach a Better Collaboration

By analyzing collaboration with data management systems in literature and regarding their implementation and fitting with industrial system, we can notice that some studies are still not well developed or do not exist. Major authors focus on the development of the exchanges between tools (that is quite essential) [23, 24] or have more architecture considerations [25]. Indeed, in industry, efforts are made on these topics, but, as we have identified, they have to evolve on two major topics:

1) The first is the data "semantic broker": A recent evolution towards new technologies of data management has highlighted that data semantic is not enough decomposed and characterized in terms of entities to be applicable in

multiple cases and/or migrated in different tools [26]. Some efforts have to be made in terms of the following:

a) *Semantics for exchange:* Indeed, some methods already exist. However, industrial practices make that each time a new partnership is considered; a new definition of the formal semantics has to be specified. The main solution to shrink the time taken by this proposal and its corresponding developments is to define a common semantic reference frame compliance with the product definition for any available environments.

b) *Semantics for redesigning environments:* The environments proposed by tools (such as the comparison of PDM and SDM, simulation data management systems) are not composed in the same manner, and their redesign is not submitted to the same semantic rules. Their implementation should consider the definition of mapping between the different existing semantic objects.

c) *Data referential and integration:* A data referential provides the essential structure of data and the template for redesigning its structure. Then, integration is in charge of the interpretation of the data regarding the template to rebuild the data structure in a given system.

2) The second topic is to consider rules and methods within data management systems and their application for collaboration. Data management practices and methods can be assimilated as processes to rebuild data from different environments [27]. Two topics are emerging:

a) system supports for integration and rules implementing the collaboration upon such structures, where the need is to define more clearly the supporting systems that ensure the management of relationships and dependencies at different levels of granularity across domains, the synchronization process of tasks and data flows with process modeling across subprocesses, and the inclusion of configuration management and the dynamic definition of structures to accommodate multiple views;

b) the architectural considerations to build the collaborative exploitation and interoperability upon systems with the support for integration with other applications, the definition of a standardized exchange infrastructure between the components and with external modules, and the definition of a modular architecture so that components can be easily plugged in and out of the system.

D. Definition of the High-Level Objectives for a Collaborative Platform

A collaborative platform appears now as a necessary element to allow data flows between applications used in companies. Their role would be ideally to integrate and associate data created and managed during design and engineering activities. Considering data management systems, integration and association means that all partners' data can be merged to define the whole product. Indeed, considering the data management systems, all engineering definitions and attributes of the product can be integrated and used in different systems. As the number of attributes linked to the product is also increasing with the evolution of data management requirements and data exchanges, data management between partners needs more complete interfaces. As a consequence, there exists also an increasing need to provide management on data processed (e.g., managed, integrated) by these systems.

This contributes to the definition of integrated infrastructures for the design of collaborative environments and multiview application (design, simulation) [1, 7]. The major objective of an integrated infrastructure is to provide efficient data management systems such as those proposed by classical PLM systems but in extending them to collaborative environments. Integrated infrastructure must also provide shared spaces for the different partners, using dedicated work spaces for the different activities (e.g., work space for design, simulation). That way, collaborative platform must allow tools interoperability and data integration in the most meaningful way. The platform must provide services to create requests on activities and define interenterprises workflows.

Following the thought of [12] on collaborative infrastructures, they must create networked activities using product data representation (reference framework) provided by the collaborative workspace. For the aeronautic industry, an integrated infrastructure must comply with six major requirements:

1) Common data reference framework is used to store and retrieve product data in its entire life cycle. (For example, simulation activities currently lack the concept of referential and the association of the product design within all simulation activities does not exist.) This should also allow the consistency of data while considering product's characteristics. (Presently, the consistency of product configuration between design and simulation is implicit and defined by engineers who need it.)

2) Management of information between partners is ensured by standardization of collaborative data; this should define the meaningful way to manage data. Models should describe the attributes attached to the product and must manage the attributes links between partners' systems.

3) PLM functions must define the major actions on data and guarantee the management all along the life cycle.

4) Data context has to allow retrieving the context for data use. This defines necessary elements (e.g., process, activity, data, parameters, and models describing the product) for used data.

5) Providing data associativity has to define links between data created for activities.

6) Definition of flows and processes connections (engineering requests, engineering validation) must determine the different ways to access partners' processes.

1. Collaboration Within a Product Development Process

To achieve its definition, a product is refined throughout different activities (depending of the development stage). Indeed, at a functional definition phase, the product is not shaped yet, and project team members try to clarify requirements and specifications regarding the customer needs. Later in the development process, the product will be designed by an engineer through a design environment. Then the product will be submitted for validation in a calculation loop performed by a simulation engineer. Regarding this typical case, collaboration is present along the whole product development process and thus critical for its definition. As aeronautical products are complex and developed within large partnerships, the number of people involved in the product definition increases at every level of the life cycle. Generally, using processes specific to a single enterprise, the collaboration lacks of synchronization between established and rigid individual processes. Harmonization of processes is a priority step for defining a new way to develop a product between many partners. Main issues of collaborative processes can be then highlighted:

1) *Activities scheduling and synchronization:* A wrong scheduling of activities delays the product definition and costs to the company.

2) *Workflow controlling:* Without a good view of the process progress, identifying the blocking points is difficult.

3) *Concurrent engineering and collaborative design enabling:* Working difficulties occur with an unclear workflow definition.

2. *Collaboration and Synchronization Between Tools*

Regarding the virtual-enterprise context where partnerships are set up everyday, it is problematic to work on the same final product definition using many different commercial-off-the-shelf (COTS) environments. Even within a company, a product definition is processed through different environments such as design or simulation applications. Both are used to define the same product, but they do not deal with the same data or methods. Hence, it is important to synchronize everyone's work to optimize the product definition. Indeed, the number of tools and methods tends to infinite. So many heterogeneous tools and environments make the exchange between applications more difficult. The resulting issues of this new way of project organization and management are as follows:

1) *Standardization of product structure:* Data might be lost or recreated if they are not stored where they are supposed to be.

2) *Definition of information to be shared with partners:* Without a clear mapping of the data to be shared, time is lost when trying to retrieve good information.

3) *Synchronization between partners' view of product structure:* If engineers from different companies work on different versions of the product, the product might have a wrong applicable definition.

III. Technological Guidelines and Drivers

A. From Previous Studies to an Original Concept

In the early 1980s, there was little interest in the idea of enterprise reengineering or enterprise modeling, and the use of formalisms and models was generally limited to some aspects of application development within the information systems community [28]. Many of these legacy systems were and are still mission critical: as businesses change to address the competitive pressures of today and tomorrow, these systems must also change. The problem is not just the complexity of technologies of yesterday or of today: open architecture environments, client/server systems, CASE tools, object-oriented development, etc. The problem is broader; organizations are undergoing rapid changes as they reengineer to compete. Organizations and systems must be designed for change [29, 30]. The subject of "architecture" was acknowledged at that time; however, there was little definition to support the concept. This lack of definition has primarily been developed in the concept of "framework for information systems architecture." Although from the beginning, it was clear that it should have been referred as a "framework for enterprise architecture." This enlarged perspective now begins to be generally understood as a result of the relatively recent and increased worldwide focus on enterprise "engineering." "The framework" as it applies to enterprises is simply a logical structure for classifying and organizing the descriptive representations of an enterprise. These descriptions are significant to the management of the enterprise as well as to the development of the enterprise's systems [18, 31].

One of the main challenges to solve is the use and the integration of the framework in the collaborative context of aerospace projects. It means to identify the business scope, the type of data managed, the associated processes, the type of contractual relationships, and the physical architecture. Many implementations have been done such as association of middleware using Web services, association of models in integrated environment, creation of distributed business processes, or platforms using enterprise-modeling approaches [12]. Middleware is a solution based on a thin server and a strong client solution. It ensures, through Web services, the accessing of the data. Sudarsan et al. [32] propose another solution based on models association in order to create a PLM structure that is a support framework for product information. Expected benefits rely on the capability to access, store, serve, and reuse product information all along the life cycle.

At this level, enterprise modeling and integration (EMI) is identified as a necessary approach to result in an efficient collaboration. This consists of defining business processes and the explicit/implicit connectors that exist between them. After Vernadat [15], the enterprise integration consists of "breaking down organizational barriers" and the enterprise modeling consists of "making models of the structure, behavior, and organization of the enterprise." In one sense, enterprise modeling and integration consist of defining models regarding different levels of the enterprise to characterize the processes exploited. A next step consists of organizing these models so as to exploit the available connectors between processes and integrate them (modeling and state-of-the-art definitions of Shunk [33] and examples of use by Vernadat [15]).

B. Approaches to "Design" Collaborative Environments

Although collaborative environments were limited to design environment a few years ago, such an environment use is now extended to processes identified in the enterprise (from design to manufacturing) and to the different stages of a project (from preliminary to service).

This is moreover true if we consider the approaches of Fuh and Li [1] and Li et al. [7] who propose mechanisms for distributed and integrated product development and engineering. Such mechanisms rely on the association of two major approaches for product development:

1) A *static approach* is used to associate the static elements of the product definition (e.g., ID, name, maturity, version, exchange date). This is designed with classical modeling tools such as UML (unified modeling language) [14, 34].

2) A *dynamic approach* is relevant for a process/workflow definition. It defines the activities sequences that describe the actions on data. This is mainly designed with classical modeling tools such as IDEF0 (ICAM definition language) [33].

The definition of a collaborative architecture must use both approaches [35]. Each environment is submitted to its own functioning laws. The objective is to create networked activities using product data representation and more especially a reference frame for the product in the collaborative workspace [12].

C. Approach for Architectural Design of the Collaborative Environment

While analyzing the need of the framework architecture, we noticed there were systems interacting with each other and with the collaborative framework. This has caused us to define a model of architecture layers. Figure 3 represents the results and shows the integration of the layers in the industrial system.

1) The *operational layer* corresponds to "end-user" environment in which activities are performed. Composed of operational tools, this layer is grouped and included in a rigid infrastructure and possesses its own processes and methods.

2) The *process layer* is the guarantor of activities' evolutions. Composed mainly of workflow tools, this layer is the backward environment that provides to the operational layer the instructions for activities' sequence.

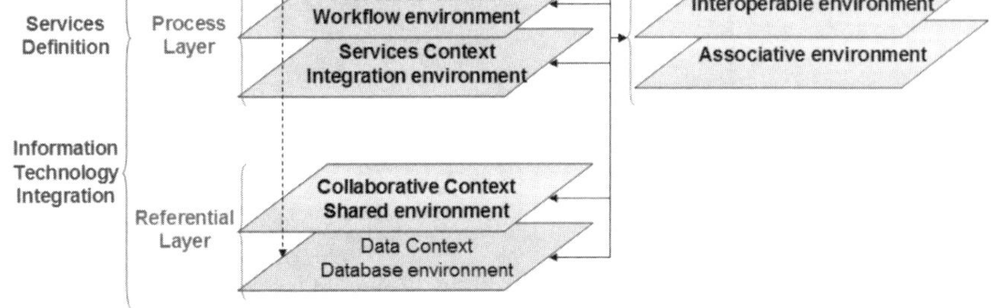

Fig. 3 Representation of system orchestration.

3) The *referential layer* is the common representation and interpretation of the phenomenon that happens during project. Providing a unified view on product for activities and partners, it is acknowledged as the shared working environment.

4) The *means layer* interprets the environment in which individual activities and actions are performed. Composed of information systems and translators, it ensures the design and redesign of environments and application domains.

IV. EDM Framework

A. Introduction to the Engineering "Context Definition"

"Context of something consists of the ideas, situations, events, or information that relate to it and make it possible to understand fully" [36]. In literature, context is seen as a snapshot of an object and any related information that is relevant to its management and manipulation along its life cycle [37]. More generally, context is not an object on its own. Context is rather a snapshot of relations between objects that are relevant to each other and their interactions [38]. Context is then an unstable concept that relies on objects and interactions in between.

Figure 4 shows an example of an implementation in a well-known PDM application. Figure 4 shows the relations between objects to create the context. The user is directly linked to a context that is the union of an organization, a role, and a project. Then authorizations of the user are defined under the role, the organization, and the user.

B. Business Context and Collaborative Context

Engineering context is a system that we can decompose [39].

Figure 5 introduces the engineering context typology on which we have based our approach. The main objects of the typology are as follows:

1) Business context is related to the execution of a business activity in a specific environment. This context takes into account the tool used to perform a specific activity on a specific product and with a specific role.

2) Collaborative context is related to the partnership and orchestration of work within the collaboration. This context takes into account the project leading a specific activity on a specific product with a specific role.

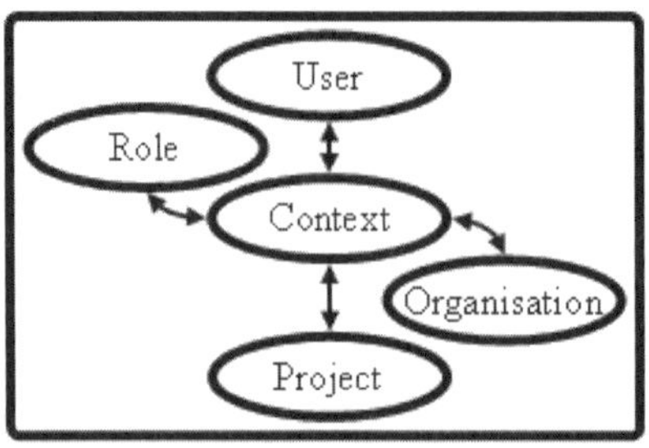

Fig. 4 PDM context implantation example.

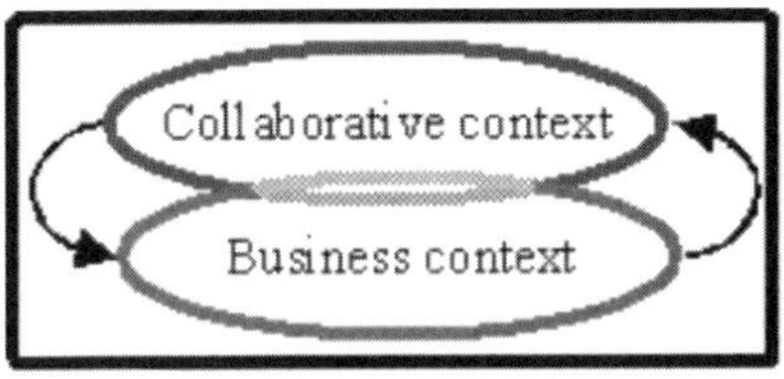

Fig. 5 Engineering context typology.

Both contexts are strongly linked and complementary to define an engineering context.

C. Engineering Context Building

We can identify the important and relevant objects to set the engineering context in a collaborative process. As collaboration begins with a project, it appears essential to consider it. Then, we identified mandatory objects to define the engineering context. The product and the process applied to it (what is performed on what part!) are essential to define the engineering activity. The role is considered as essential for the context definition. Indeed, a specific activity is performed by the person who possesses correct skills and knowledge. (A design activity is performed by a design engineer and a simulation activity by a simulation engineer.) Directly linked to the role, the user is the real performer of the activity. A mapping is needed between users and roles. Then, two final objects are identified to clearly define an engineering activity. We consider the tool the activity has been performed with (e.g., CAD for design, CAE for simulation). Finally, we consider the resources the performer can use during his work. Resource can be of any type (specific data translator, physical resource, and so on).

Figure 6 shows the seven identified objects that compose our contextual approach of engineering in collaboration. The "L" line represents the collaborative

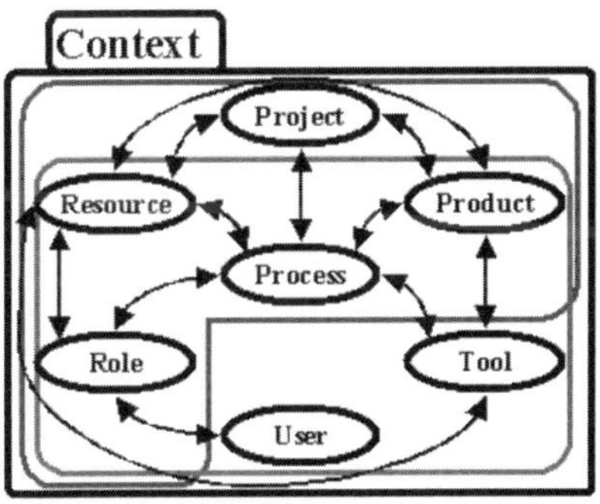

Fig. 6 Context systemic breakdown.

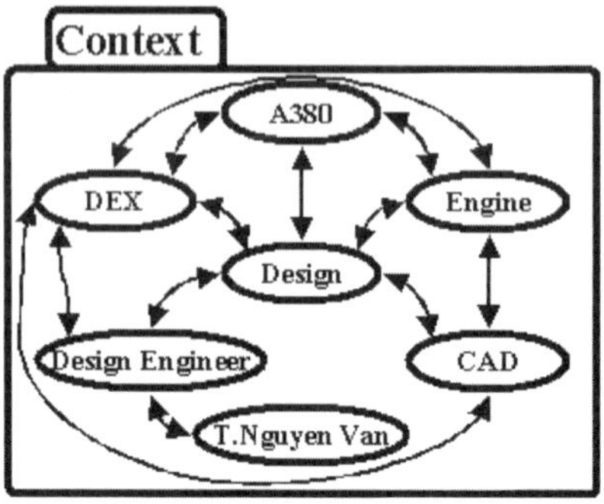

Fig. 7 Example of a context instantiation.

context and the rectangle line the business context from the context typology. Links between the objects are explained as follows:

1) *Project-context:* This is the top level of the structure. The product is created within a project and a process is run on a product within this project.

2) *Product-context:* It is the final product that is to be sold. It is managed by a project, and modification and manipulation are possible only through processes.

3) *Process-context:* This defines the possible applicable activities on a product regarding a project and its progression. The process is also linked to a tool, a role, and resource.

4) *Tool-context:* This element represents the gate that permits the access to a specific role of a product for a specific process in a project.

5) *Resource-context:* This represents any other systems, information, data type, data exchange set (DEX) (data available online at http://www.oasis-open.org/committees/plcs/fag.php), or product that is crucial for the performance of a process. It can be related to any other object of the context. In the context, it is a list of objects related to the context. It can be seen as a package that contains all needed information and objects relevant for the corresponding context.

6) *Role-context:* This element of the context is also the key of the activities applicable on product within a project and opened to a specific user. It represents the link between the organization of the virtual enterprise and the organization of a project.

7) *User-context:* This final element is also the one that establishes the navigation through the context. Indeed, by connecting onto a platform, the user chooses his role to access the projects, the processes, the products, and so on. Users are classified regarding the organizational hierarchy.

Figure 7 shows an example of an instantiation of context. The user T. Nguyen Van is mapped to the role of design engineer. Within the project A380, he performs the process "design" on the product "engine." He uses a CAD tool to perform the modification and a special DEX (data available online at http://www.oasis-open.org/committees/plcs/fag.php) is used to retrieve correctly the relevant data.

These attributes are mandatory for a good context definition. For example, if we take off one of the guidelines, the product, for instance, here the engine, it appears that without this information multiple cases are possible. The problem is that there is no specification on what product this modification takes place. Hence, the designer might modify a wrong part or, worse, have access to data he should not.

D. Engineering Data Management Framework "Blocks"

Regarding the European Project value improvement through a virtual aeronautical collaborative enterprise (VIVACE) context, the EDM framework has been developed. This product data management (EDM) relies on six main entities (see Fig. 8):

1) *Applications (Apps):* These are operational tools that are used to perform specific engineering activities such as product data management (PDM), simulation data management (SDM), or any computer-aided activity (CAX).

2) *Information model (IM):* This is an object broker on which is based the EDM server and which contextually defines the product, process, resources, and links in between.

3) *Domain model (DM):* This model represents the needs for attributes or data for each domain of expertise such as design or simulation and characterizes a view of the product.

4) *Consolidated repository (CR):* This provides the persistency for consolidated data of the virtual aircraft. It does not replace the repository of legacy application but provides a consolidated repository to share the published data based on the Standard for the Exchange of Product model data (STEP) (Standards ISO TC184-SC4) [16, 17].

5) *Product context management (PCM):* The collaborative interface allows users to browse collaborative data and processes.

6) *Workflow (Wf):* This engine ensures the right achievement of activities regarding a product in a process and including team members and resources. This allows a relevant monitoring of the product all along its definition phase.

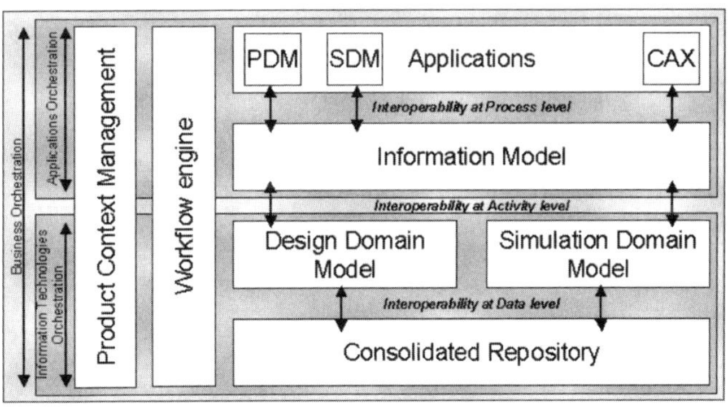

Fig. 8 EDM framework.

E. Functional Overview of the Framework

Let us take the example of export data in order to identify the data flow (see Fig. 9). The design engineer receives a change request in the PCM (1) and launches the workflow (2) to carry out a design change. Once done, he saves the data both in his environment and in the collaborative environment (3). The information model breaks the native data attributes in order to map them to the standardized common definition (4). The mapping consists of associating native attributes to the standard (STD) attributes used in product life-cycle support (PLCS) (STEP ISO10303-239 [40]). Thanks to this mapping that ensures the associativity between data through domain models, data are changed in both simulation (Sim.) and design (Des.) domain models (5 and 5′). For each domain model, the product definition is rebuilt to fit to the common definition (6 and 6′). The last activity is then the storage of data within the collaborative consolidated repository (7 and 7′).

F. Definition of the Data Structure Mapping—Domain Models Mapping

Based on a common database, commercial applications can rebuild their own environment or update the EDM one. To do so, Web services (symbolized with arrows) are used to send or retrieve information based on the XML standard. This technology permits communication between applications and ensures coherency of data structures by defining the mapping between EDM and commercial applications. Hence the tools' data models are different and specific to the COTS they are for. Those specific data models can be mapped and then instantiate a master data model (see Fig. 10).

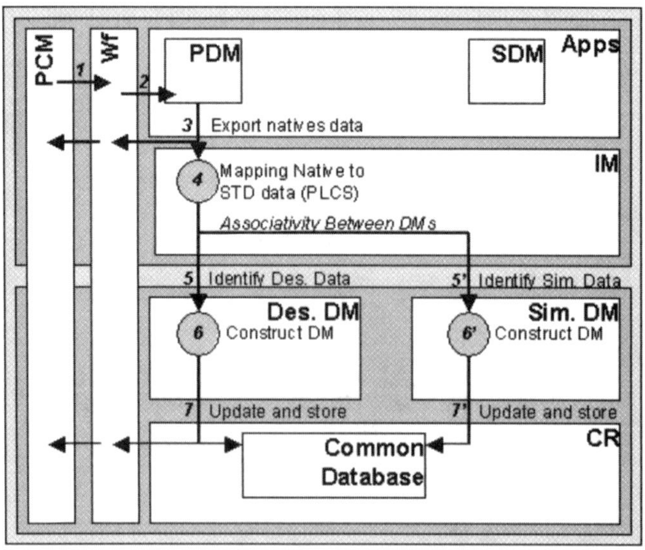

Fig. 9 Data export use case.

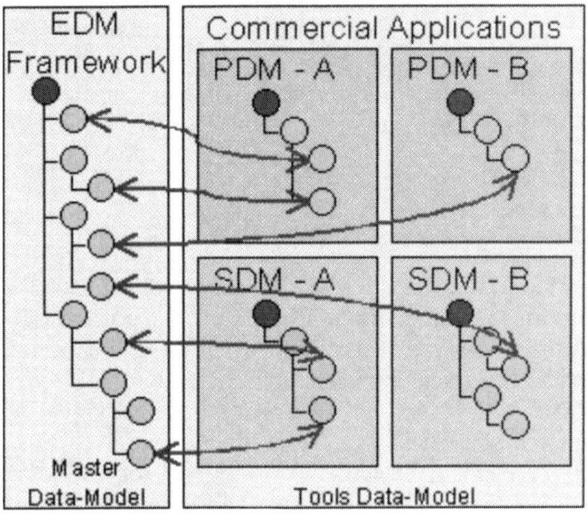

Fig. 10 Data mapping.

G. EDM Graphical User Interface (GUI) for Control—"Implemented" Information Model

The collaborative application is called PCM and possesses seven interfaces:

1) User workspace—lists user's tasks

2) Execution view—displays tasks details

3) Project overview—displays workflow overview

4) Workflow management—manages the workflow and is composed to three interfaces

5) Workflow administration—manages processes and activities

6) User and groups management—manages users and groups; mapping management—manages mapping between users, groups

7) Processes roles; and context setting—sets the navigation context

This interface represents the implementation of the information model we have created to manage the overall engineering contexts and the collaborative environment. The top layer of the interface is common to each interface. It is composed of tabs corresponding to the application interfaces. It displays details of the user navigation: the navigation context. The navigation context contains the user name, the activity he is performing, the product he works on, the workflow this activity takes part in, and the role of the user. If this navigation context is not set, execution view and project overview are not available. The user can set this navigation from any of the interfaces but the users and groups management and the mapping management ones.

1. User Workspace

The first interface the user sees after login to the PCM is the workspace (Fig. 11). It contains information about activities that he has been assigned. In

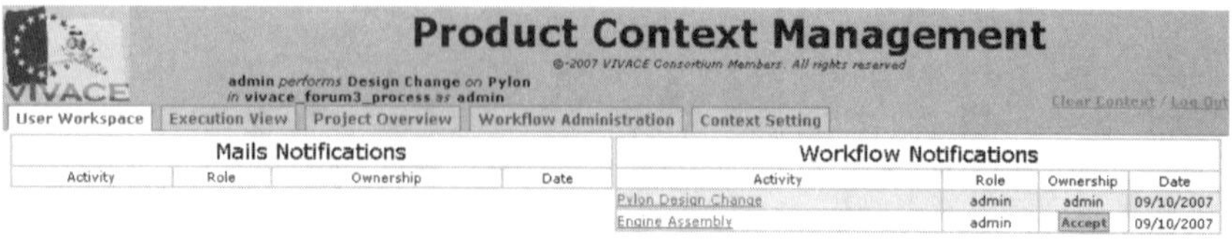

Fig. 11 PCM—user workspace interface.

this workspace, the user can switch the interface by using the available tabs. Name of the activities are also a link. This link sets the user navigation context and directs the user to the execution view to see the activity details.

For each activity, the notion of ownership informs the user if he has accepted the activity or not. If the button "Accept" is displayed, the activity is not yet accepted and cannot be performed until it is accepted. If an activity is assigned to a role and not yet accepted, it appears in the workspace of every user that is linked to the role. Once the activity is accepted, it only appears in the workspace of the user that accepted it. Actions that are completed or not yet activated are not displayed.

2. *Execution View*

This second interface (Fig. 12) is the contextual view of the engineering task the user has to perform. With navigation context, the execution view displays all of the activity details. The first line highlights the process, the product, the role, the ownership, the status, and the resources.

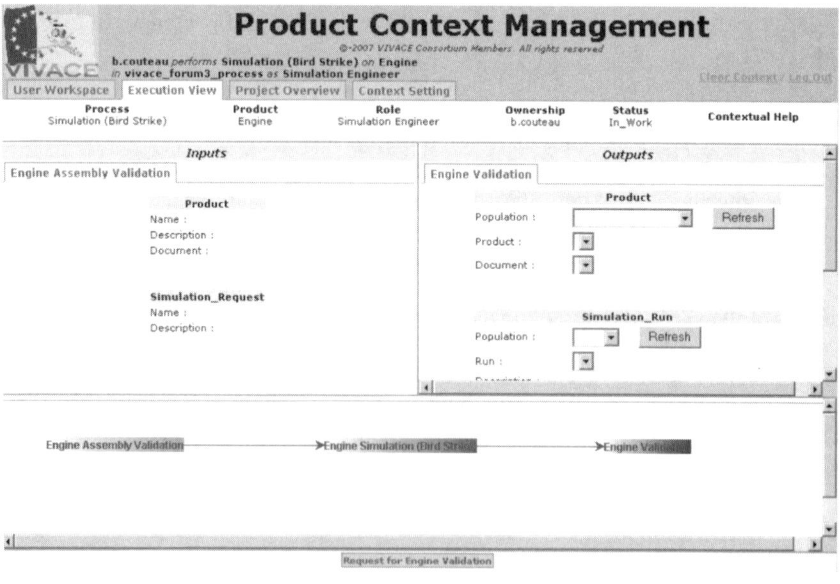

Fig. 12 PCM—execution view interface.

Then two frames display inputs and outputs of the activity. In each frame tabs correspond to the number of inputs or outputs. Inputs and outputs depend on the workflow design. Each input contains details, requirements, information, and data that are relevant for the current task. The outputs tabs have to be filled in by the user when he performs the activity. He creates outputs with every data, information, and requirements that have been created by performing the task and that are relevant for the collaboration and for the corresponding following activities.

Under outputs/inputs frames is displayed a short view of the workflow the activity is part of. Indeed, the current activity is displayed with the previous activities on its left and the following activities on its right. Numbers of outputs/inputs tabs and numbers of previous/following activities are equals.

Finally, the lowest line contains the activities validation buttons. It corresponds to the outputs paths that are available from this activity. If there are two following activities that are necessary (i.e., "and split"), there is one button. If there is a choice, each output corresponds to a button. Clicking on a button automatically closes this activity and starts the corresponding one.

3. *Project Overview*

This third interface (Fig. 13) enables the user to visualize the overall workflow he is part of by performing the activity. The displayed workflow corresponds to the workflow that is set in the navigation context. It is composed of two main frames: the workflow view and the activity details table.

Based on JAVA technologies, the workflow viewer enables a dynamic navigation on the workflow. The user can zoom, hide/show nodes (activities), hide/show relations, and manipulate the viewer to fit his vision and what he

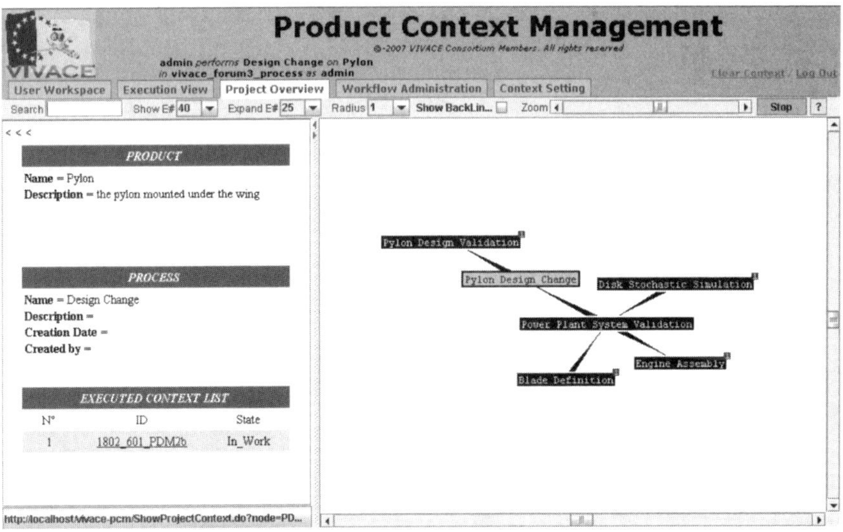

Fig. 13 PCM—project overview interface.

wants to see. Clicking on an activity displays automatically its details in the left frame: the activity details table.

The activity details table contains relevant information on the activity. The product and the process the activity is based on are displayed. A list of instantiated activities shows if the activity is currently running, accepted, not yet started, or already completed. Clicking on the ID of an instantiated activity directs the user to the execution view to see the details. If the activity is completed, the execution view display inputs and outputs that have been filled in. If the user is not the owner of the activity (i.e., user has not accepted the activity), buttons in execution view are disabled.

This interface is also available if the navigation context only contains the name of a workflow. Execution view will not be displayed, but the project overview tab is displayed. This is possible only from the context setting interface.

4. *Process Management*

This interface (Fig. 14) enables loading of workflow definitions from XML process definition language (XPDL) files, instantiating one or more of these workflow definitions, and managing the running activities.

This interface is directly linked from the JAVA Web application (PCM) and the workflow engine. It displays the list of all activities that are running, waiting to be accepted, or completed. The user also has access to the closed workflows to clear the workflow engine database.

The activities table displays the name of the activity, the workflow it is part of, and the state of the workflow and of the activities and the assigned user of the activity. Clicking on the activity name sets the navigation context and directs the user to the execution view to visualize the activity details.

If a problem happened at run time, the administrator can here resolve the problem and manage the workflow engine.

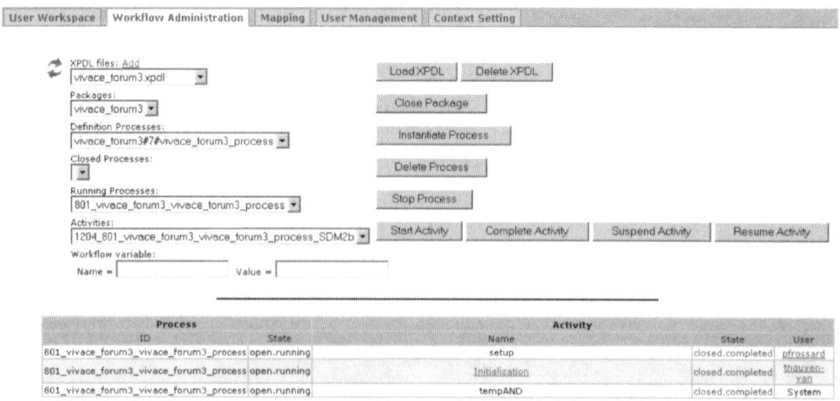

Fig. 14 PCM—process management interface.

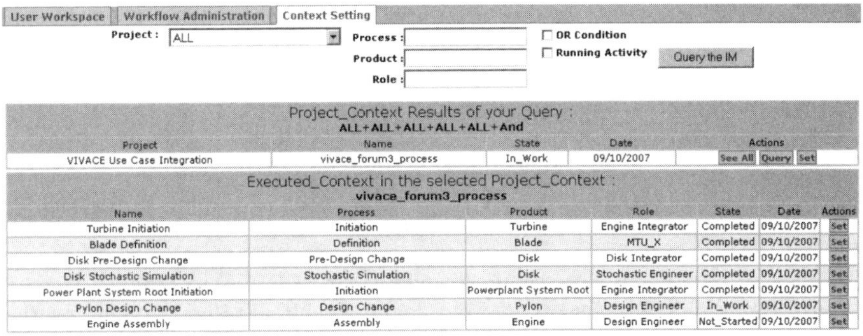

Fig. 15 PCM—context setting interface.

5. Context Setting

This last interface (Fig. 15) enables users to browse the existing engineering tasks that have been performed or that are currently being run. This consists of a query to the collaborative database by filling in some fields: project, process, product, role, and state of the activity. The first result is a list of workflows that corresponds to the query. From this list of workflows, all activities (completed or currently being run) are available. The user has the possibility to apply the query on this activities list, or to see them all.

On the workflows list the user can also set the navigation context to browse the workflow within the project overview interface (Fig. 13). The user can also set the navigation context by selecting one of the activities from the activities list. The user is redirected to the execution view interface (Fig. 12).

H. Application Interactions to Ease the Context Implementation

Figure 16 represents the work we have produced to specify the interactions between the different interfaces. This sequence diagram shows how to set the context by selecting the characteristics of the engineering context. It also introduces the concept of navigation context. The navigation context stores the relevant information of the engineering context the user is currently browsing. It helps to construct the different interfaces by providing the path to retrieve the information to display on every interface. This sequence diagram shows also where to find the information. On this diagram are represented two components of the collaborative framework: the information model and the domain model. But we also define the interactions with the workflow engine.

V. Scenario Description and Test Case Run

A. Scenario Overview

To implement the whole framework in an industrial context, a scenario has been set up. This scenario has to show all of the points discussed in this chapter. It then takes into account the collaborative design and concurrent engineering as well as the interoperability between applications and the different layers of activities. Regarding the VIVACE context, this scenario is a close-up onto the operational part of the global workflow. Before building it, the first assumption has been made on the industrial purpose it can have. Let us take two elementary technical components P and E that are assembled together into an assembly named A ($P + E = A$). Each component is a product from two different companies B and C. The company A in charge of the component P is also in charge of the integration between both components, which is then the final assembly A. Each company uses its own applications regarding the design and the calculation of each component. As presented in Fig. 17, the workflow must be simple enough to be seen at a program level and detailed enough for the operational manager. The shaded boxes at the bottom of Fig. 17 represent subworkflows. In those subworkflows, the first activity is carried out through another application. This is where the workflow has to control from an external application, while the other activities are driven by each company manager from the main interface.

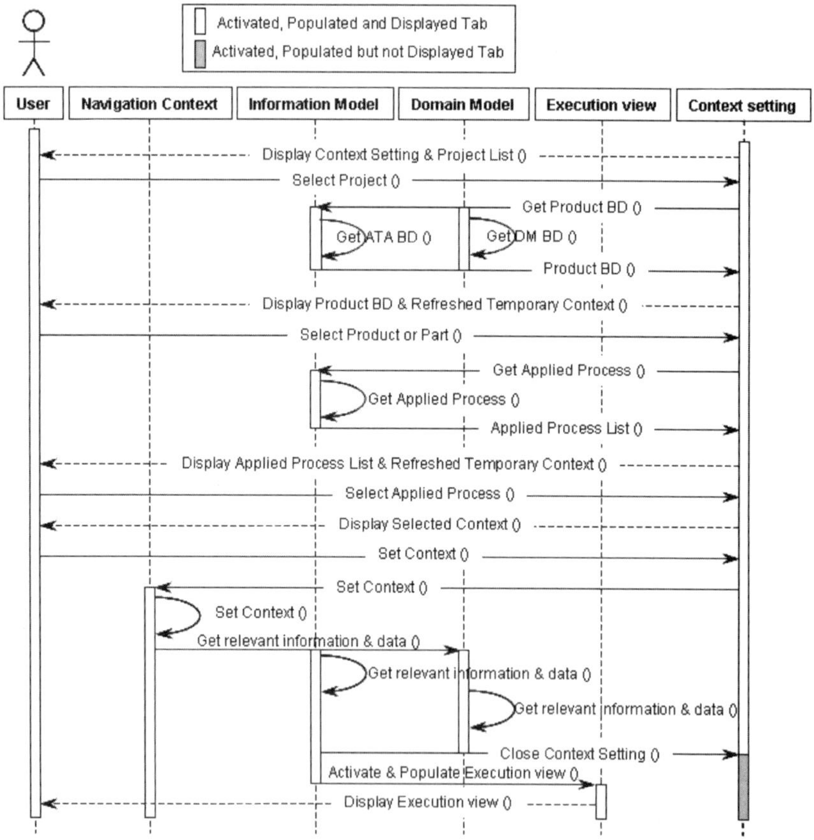

Fig. 16 UML sequence of setting a context.

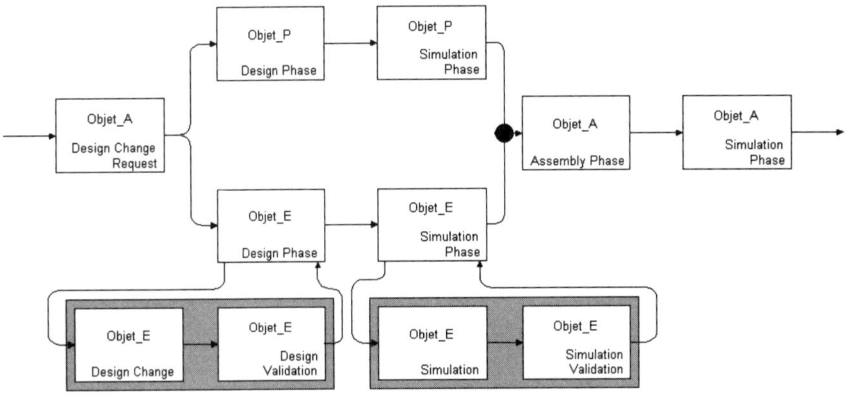

Fig. 17 Proposed scenario.

Furthermore, as it is specified in the VIVACE project, this scenario takes into account two different fields that are the design part and the calculation part for each component. The simulation activities need different kind of data other than those of the design activities. Indeed, for the simulation of a component, the simulation engineer (SE) needs meshes of the component and load cases for the simulation run. Last, this activity produces results that are the formal element for the validation of the simulation phase. To come back to the collaborative aspect of this scenario, it can be seen that both P and E components' changes are run in parallel and independently form each other. Saying that, the assembly phase of the final assembly A has the role to retrieve both components before launching the assembly operation. This means enables the synchronization of the system for the components.

B. Scenario Implementation

To implement this scenario in the EDM framework, the information model helps to map the basic workflow concepts "As Planned" and "As Executed." This implementation relies on the three following objects (see Fig. 18):

1) *Context definition* is composed of a process applied to product; this is the basis element for the workflow implementation. They also can be assembled together to create module, representing subworkflow.

2) *Project context definition* is the instantiation of context definition and is linked between each other; they represent the "As Planned" workflow. From this schema, the built XPDL file will be used for the instantiation of the workflow.

3) *Executed context definition* is the sequence of instantiated project context definition; they represent the "As Executed" workflow.

To manage all of the applications together, the workflow engine has different behaviors regarding the current activity. As shown in Fig. 19, the monitoring and controlling of workflow can be represented as this activities sequence. A workflow (described in an XPDL file) is first instantiated within the workflow

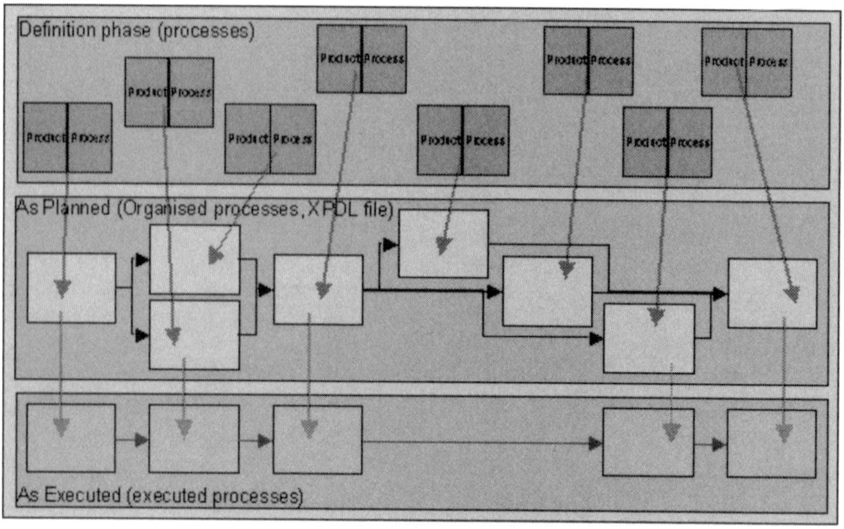

Fig. 18 Scenario implementation.

engine and is launched automatically (1). Corresponding to the first activity, a new task is sent to the manager. By accepting it, the manager launches the activity (2). Through the PCM interface, the manager creates a simulation request by gathering product and process information from the information model. When the simulation request is filled, the manager completes the activity (3). Automatically, a new activity is instantiated (4), and the simulation request is sent to the SDM environment (5). The SE starts by retrieving product structure.

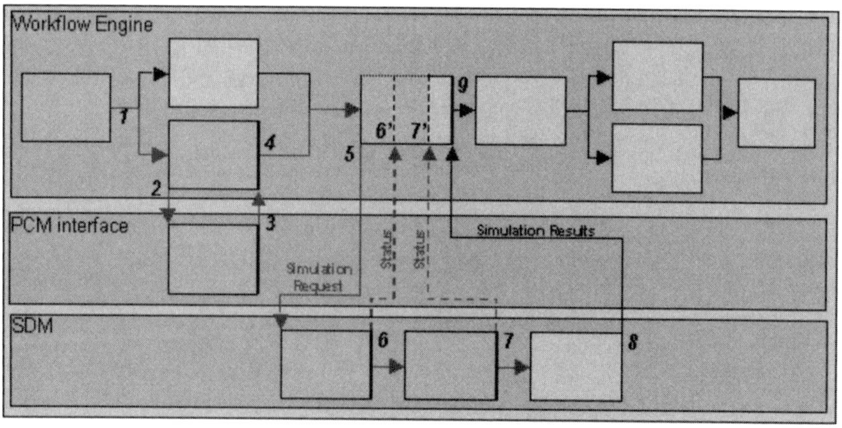

Fig. 19 Scenario use case.

At the end of this activity, a status is sent to the high-level workflow (6). This status updates automatically the activity progress (6′). This allows the manager to follow a lower-level activities progress. The status transfer between both applications is based on a JAVA client. Then the SE begins to simulate the behavior of component. When the SE estimates that the simulation loop is over, a new status is sent to update the workflow engine (7 and 7′). In the next activity, the SE elaborates the simulation results report that will complete the simulation process (8). The validation of those results is then activated (9).

VI. Conclusion

The communication between partners and activities in industry suffers from issues concerning the communication between tools and systems and concerning the contextual definition for processes and engineering environments. In the preliminary stages of projects, interoperability and integration of engineering systems are essential to set up an efficient partnership and to ensure the time decrease of conceptual stages. Here, we have addressed the following issues for the aeronautic industry:

1) The increase of communication and semantic problems is related to the increase of the heterogeneity of systems and tools; it generates misunderstandings and wrong interpretations of data and information conveyed in information technology systems.

2) The rationalization of collaborative processes using multiple environments and product references. The dissociation and nonassociativity of these systems cause duplication and redundancies problems. Processes are no more guided upon the implementation of environments but by the need to perform the activity. Problems then appear while attempting to create a complete environment with available data.

Regarding these aspects, the contribution of the VIVACE Project on the development of a methodological referential and on the implementation of a collaborative platform for multipartners and multi-engineering answers the major constraints presented in industry by the following:

1) The development of the engineering interoperability supports the definition and interpretation of information conveyed in the engineering environments with the development of interoperability between engineering systems, the development of interoperability product/process, and the development of interoperability between engineering data.

2) The definition of integrated environments supports the collaborative and referenced view on the product using the development of shared workspaces, where domains of engineering are dependent but separated, and the definition of reference models using the digital mock-up. Our work was mainly oriented on this solution because it is the convergence point of multiple activities.

3) The definition of engineering contexts: The assimilation of conceptual domains is related to a defined typology of environments determined by the objectives of the activity and the needs of the engineering environment.

References

[1] Fuh, J. Y. H., and Li, W. D., "Advances in Collaborative CAD: the State-of-the Art," *Computer-Aided Design*, Vol. 37, No. 5, 2005, pp. 571–581.

[2] Valckenaers, P., Van Brussel, H., et al., "On the Design of Emergent Systems: an Investigation of Integration and Interoperability Issues," *Engineering Applications of Artificial Intelligence*, Vol. 16, No. 4, 2003, pp. 377–393.

[3] Gallaher, M. P., O'Connor, A. C., Dettbarn, J. L., and Gilday, L. T., "Cost Analysis of Inadequate Interoperability in the US Capital Facilities Industry," Information Technology and Electronics Office, U.S. National Institute of Standards and Technology, NIST GCR 04-867, Gaithersburg, MD, Aug. 2004, http://www.bfrl.nist.gov/oae/publications/gcrs/04867.pdf [retrieved July 2009].

[4] National Coalition for Advanced Manufacturing (NACFAM), "Exploiting E-manufacturing: Interoperability of Software Systems Used by US Manufactures," Feb. 2001, Washington, DC, http://www.mel.nist.gov/div826/msid/sima/Final-ReportSummary.PDF [retrieved July 2009].

[5] Brunnermeier, S. B., and Martin, S. A., "Interoperability Costs in the US Automotive Supply Chain," *Supply Chain Management Journal*, Vol. 7, No. 2, 2002, pp. 71–82.

[6] "Metrology Interoperability Project," NIST Manufacturing Engineering Lab., 2008, http://www.mel.nist.gov/programs/raisp.pdf [retrieved July 2009].

[7] Li, W. D., Lu, W. F., et al., "Collaborative Computer-Aided Design–Research and Development Status," *Computer-Aided Design*, Vol. 37, No. 9, 2005, pp. 931–940.

[8] Pardessus, T., "The Multi-Site Extended Enterprise Concept in the Aeronautical Industry," *Air & Space Europe*, Vol. 3, Nos. 3–4, 2001, pp. 46–48.

[9] Beckett, R. C., "Determining the Anatomy of Business Systems for a Virtual Enterprise," *Computers in Industry*, Vol. 51, No. 2, 2003, pp. 127–138.

[10] Huifen, W., Youliang, Z., et al., "Feature-Based Collaborative Design," *Journal of Materials Processing Technology*, Vol. 139, Nos. 1–3, 2003, pp. 613–618.

[11] Mervyn, F., Senthil Kumar, A., et al., "Developing Distributed Applications for Integrated Product and Process Design," *Computer-Aided Design*, Vol 36, No. 8, July 2004, pp. 679–689.

[12] Perrin, O., and Godart, C., "A Model to Support Collaborative Work in Virtual Enterprises," *Data & Knowledge Engineering*, Vol. 50, No. 1, 2004, pp. 63–86.

[13] Lehmann, H., and Gallupe, B., "Information Systems for Multinational Enterprises–Some Factors at Work in Their Design and Implementation," *Journal of International Management*, Vol. 11, No. 2, 2005, pp. 163–186.

[14] Aziz, H., Gao, J., et al., "Open Standard, Open Source and Peer-to-Peer Tools and Methods for Collaborative Product Development," *Computers in Industry*, Vol. 56, No. 3, 2005, pp. 260–271.

[15] Vernadat, F. B., "Enterprise Modeling and Integration (EMI): Current Status and Research Perspectives," *Annual Reviews in Control*, Vol. 26, No. 1, 2002, pp. 15–25.

[16] "Overview and Fundamental Principles," ISO 10303-1, International Organization for Standardization, Geneva, Switzerland, 1994, http://www.iso.org/iso/catalogue_detail?csnumber=20579.

[17] Technical Committee 184 for Industrial Automation Systems and Integration, Subcommittee 4 for Industrial Data, *ISO TC184/SC4/WG10 N326-2000-11-08–PROPOSED ISO TC184/SC4 Standing Document*, International Organization for Standardization, Geneva, Switzerland, 2008.

[18] Dustdar, S., and Gall, H. "Architectural Concerns in Distributed and Mobile Collaborative Systems," *Journal of Systems Architecture*, Vol. 49, Nos.10–11, 2003, pp. 457–473.
[19] Rosenman, M. A., and Gero, J. S., "Purpose and Function in a Collaborative CAD Environment," *Reliability Engineering & System Safety*, Vol. 64, No. 2, 1999, pp. 167–179.
[20] Ruland, D., and Spindler, T., "Integration of Product and Design Data Using a Metadata- and a Rule-Based Approach," *Computer Integrated Manufacturing Systems*, Vol. 8, No. 3, 1995, pp. 211–221.
[21] Zha, X. F., and Du, H., "A PDES/STEP-Based Model and System for Concurrent Integrated Design and Assembly Planning," *Computer-Aided Design*, Vol. 34, No. 14, 2002, pp. 1087–1110.
[22] Oh, Y., Han, S.-h., et al., "Mapping Product Structures Between CAD and PDM Systems Using UML," *Computer-Aided Design*, Vol. 33, No. 7, 2001, pp. 521–529.
[23] Lee, C. K. M., Lau, H. C. W., et al., "Development of a Dynamic Data Interchange Scheme to Support Product Design in Agile Manufacturing," *International Journal of Production Economics*, Vol. 87, No. 3, 2004, pp. 295–308.
[24] Leong, K. K., Yu, K. M., et al., "Product Data Allocation for Distributed Product Data Management System," *Computers in Industry*, Vol. 47, No. 3, 2002, pp. 289–298.
[25] Lim, E.-P., and Chiang, R. H. L., "The Integration of Relationship Instances from Heterogeneous Databases," *Decision Support Systems*, Vol. 29, No. 2, 2000, pp. 153–167.
[26] Rousset, M.-C., and Reynaud, C., "Knowledge Representation for Information Integration," *Information Systems*, Vol. 29, No.1, 2004, pp. 3–22.
[27] Goranson, H. T., "Architectural Support for the Advanced Virtual Enterprise," *Computers in Industry*, Vol. 51, No. 2, 2003, pp. 113–125.
[28] Bradley, P., Browne, J., et al., "Business Process Re-Engineering (BPR) – A Study of the Software Tools Currently Available," *Computers in Industry*, Vol. 25, No. 3, 1995, pp. 309–330.
[29] Nurcan, S., and Rolland, C., "A Multi-Method for Defining the Organizational Change," *Information and Software Technology*, Vol. 45, No. 2, 2003, pp. 61–82.
[30] Quattrone, P., and Hopper, T., "What Does Organizational Change Mean? Speculations on a Taken for Granted Category," *Management Accounting Research*, Vol. 12, No. 4, 2001, pp. 403–435.
[31] Montreuil, B., Frayret, J.-M., et al., "A Strategic Framework for Networked Manufacturing," *Computers in Industry*, Vol. 42, Nos. 2–3, 2000, pp. 299–317.
[32] Sudarsan, R., Fenves, S. J., et al., "A Product Information Modeling Framework for Product Lifecycle Management," *Computer-Aided Design*, Vol. 37, No. 13, 2005, pp. 1399–1411.
[33] Shunk, D. L., Kim, J.-I., et al., "The Application of an Integrated Enterprise Modeling Methodology—FIDO—to Supply Chain Integration Modeling," *Computers and Industrial Engineering*, Vol. 45, No. 1, 2003, pp. 167–193.
[34] Merlo, C., and Girard, P., "Information System Modeling for Engineering Design Co-ordination," *Computers in Industry*, Vol. 55, No. 3, 2004, pp. 317–334.
[35] Bergman, R., and Baker, J. D., "Enabling Collaborative Engineering and Science at JPL," *Advances in Engineering Software*, Vol. 31, Nos. 8–9, 2000, pp. 661–668.

[36] Akman, V., and Surav, M., "Steps Toward Formalizing a Context," American Association for Artificial Intelligence, Fall 1996, http://www.cs.bilkent.edu.tr/~akman/jourpapers/aimag/aimag1996.pdf [retrieved July 2009].
[37] Dey, A. K., Abowd, G. D., and Salber, D., "A Conceptual Framework and a Toolkit for Supporting the Rapid Prototyping of Context-Aware Applications," 2001, http://www.cc.gatech.edu/fce/ctk/pubs/HCIJ16.pdf [retriieved July 2009].
[38] Dourish, P., "What We Talk About When We Talk About Context," School of Information and Computer Science, University of California, Irvine, CA, 2004, http://www.ics.uci.edu/~jpd/publications/2004/PUC2004-context.pdf [retrieved July 2009].
[39] Simon, H. A., *The Sciences of the Artificial*, MIT Press, 1969.
[40] "Industrial Automation Systems and Integration—Product Data Representation and Exchange—Part 239: Application Protocol: Product Life Cycle Support," ISO 10303-39:2005, Cambridge, MA, 1969, http://www.iso.org/iso/catalogue_detail.htm?csnumber=38310 [retrieved July 2009].

VIVACE Partners

This appendix contains the names of all of the partners that have been involved in the VIVACE project. The country distribution of the partners is provided graphically in Fig. A.1 as well as in Table A.1. The list is split in four parts, one for each type of organization. Tables A.2–A.5 list the 39 companies, 7 research institutes, 12 universities, and 5 third-tier suppliers, respectively.

Table A.1 Distribution of VIVACE partners geographically and per type of organization

Country	Companies	Research institutes	Universities	Third-tier suppliers	Total
Belgium	3	1	—	—	4
France	14	2	1	4	21
Germany	7	2	1	—	10
United Kingdom	7	—	6	—	13
Greece	—	—	1	—	1
Italy	4	—	2	—	6
Netherlands	—	1	—	—	1
Norway	1	—	—	—	1
Portugal	—	1	—	—	1
Spain	2	—	—	1	3
Sweden	2	—	1	—	3

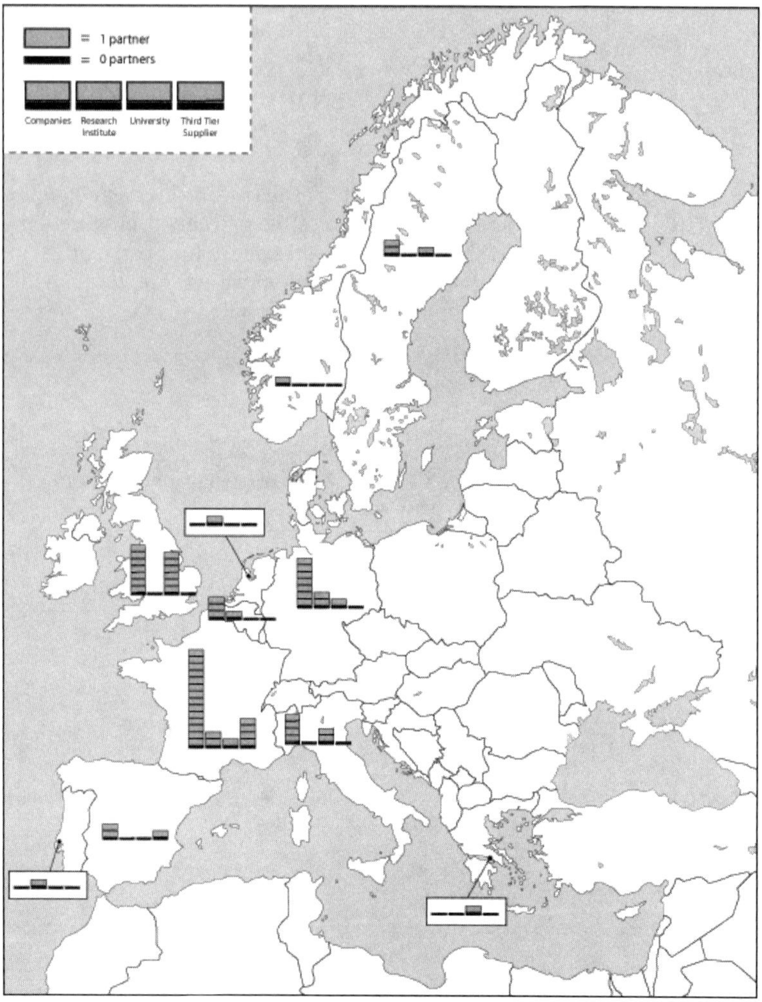

Fig. A.1 Geographic overview collaboration partners per organization type.

Table A.2 VIVACE company names and country of origin

Companies' participant name	Country
Airbus France	France
Airbus Deutschland GmbH	Germany
Airbus SAS	France
Airbus UK, Ltd.	United Kingdom
Ajilon Engineering	France
Alenia Aeronautica S.p.A.	Italy
ARTTIC	France
Assystem Ltd., UK	United Kingdom
Avio S.p.A.	Italy
BAE SYSTEMS (Operations) Ltd.	United Kingdom
CIMPA GmbH	Germany
Dassault Aviation	France
Dassault Systèmes	France
EADS CCR	France
EADS Deutschland GmbH	Germany
Empresarios Agrupados Internacional, S.A.	Spain
EPM Technology AS	Norway
ESOCE NET	Italy
Eurocopter SAS	France
Eurostep Group AB	Sweden
Hewlett-Packard Limited	United Kingdom
Hydro-Control-Steuerungstechnik GmbH	Germany
Industria de Turbopropulsores, S.A.	Spain
ISIGHT Software SARL	France
Leuven Measurements and Systems International N.V.	Belgium
Messier-Dowty Limited	United Kingdom
MSC Software GmbH	Germany
MTU Aero Engines GmbH	Germany
Rolls-Royce Deutschland, Ltd. and Co KG	Germany
Rolls-Royce plc	United Kingdom
Samtech S.A.	Belgium
Snecma	France
Techspace Aero S.A.	Belgium
Thales Avionics	France
Thales Avionics Electrical Systems S.A.	France
Transcendata Europe, Ltd.	United Kingdom
Turbomeca S.A.	France
Volvo Aero Corporation	Sweden
Xerox Italia S.p.A.	Italy

Table A.3 VIVACE research institutes' names and country of origin

Research institutes' participant name	Country
CENAERO	Belgium
Centre de Recherche et de Formation avancée en Calcul Scientifique	France
Deutsches Zentrum für Luft - und Raumfahrt e.V.	Germany
Office national d'Etudes et de Recherches Aérospatiales	France
Stichting Nationaal Lucht - en Ruimtevaartlaboratorium	Netherlands
Technische Universität Hamburg-Harburg represented by TUHH-Technologie GmbH	Germany
UNINOVA - Instituto de Desenvolvimento de Novas Tecnologias	Portugal

Table A.4 VIVACE universities' names and country of origin

Universities' participant name	Country
Cranfield University	United Kingdom
Imperial College of Science, Technology and Medicine	United Kingdom
UPS/IRIT, Université de Toulouse 3 Paul Sabatier Institut de Recherche en Informatique de Toulouse	France
Luleaa University of Technology	Sweden
National Technical University of Athens	Greece
Politecnico di Milano	Italy
Politecnico di Torino	Italy
Queen's University Belfast	United Kingdom
The University of Nottingham	United Kingdom
The University of Warwick	United Kingdom
Universität Stuttgart	Germany
University of Manchester	United Kingdom

Table A.5 VIVACE third-tier suppliers' names and country of origin

Third-tier suppliers' participant name	Country
Dassault Data Services	France
Ibérica del Espacio, S.A.	Spain
Intespace	France
Oktal	France
Teuchos	France

Index

Note: Page numbers in *italic* represents figures. Page numbers in **bold** represents tables.

Supporting Materials

Many of the topics introduced in this book are discussed in more detail in other AIAA publications. For a complete listing of titles in the Progress in Astronautics and Aeronautics Series, as well as other AIAA publications, please visit www.aiaa.org.

AIAA is committed to devoting resources to the education of both practicing and future aerospace professionals. In 1996, the AIAA Foundation was founded. Its programs enhance scientific literacy and advance the arts and sciences of aerospace. For more information, please visit www.aiaafoundation.org.

Chapter 2

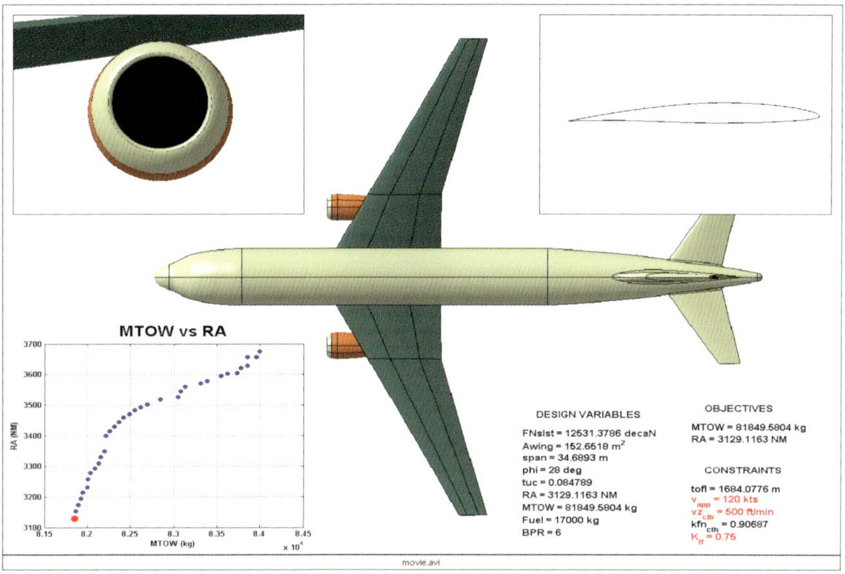

Fig. 23 Visualization of the Pareto frontier, including parametric geometry and constraints activation.

Chapter 4

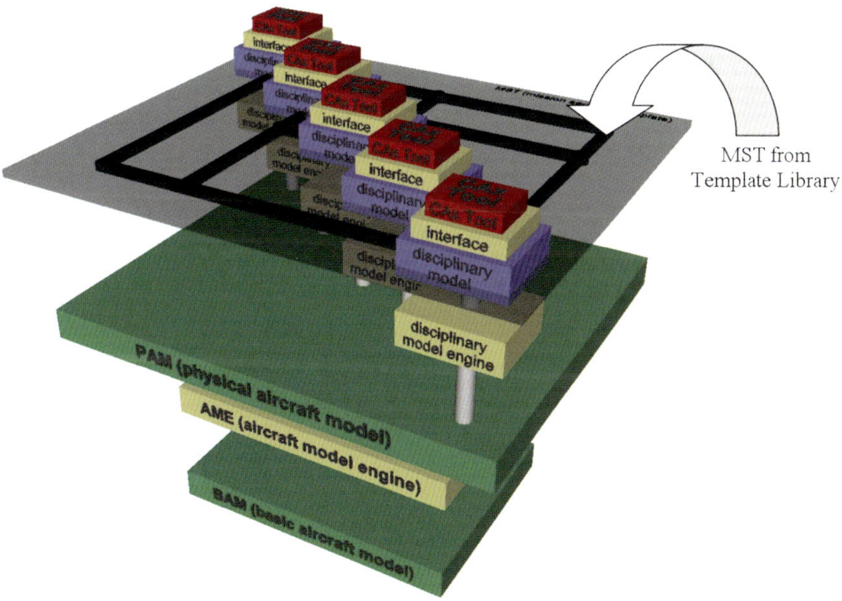

Fig. 5 Interplay of framework and mission module.

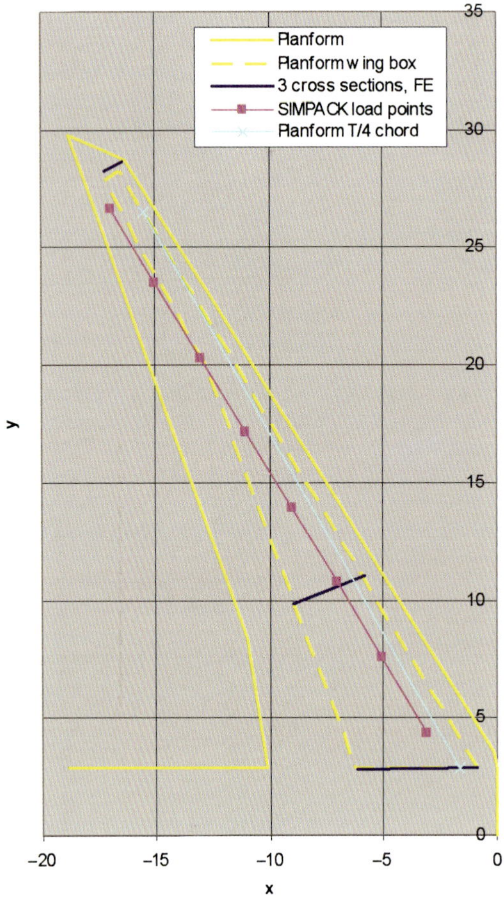

Fig. 14 Geometric data of the wing, x–y plane, MBS and BEA layouts.

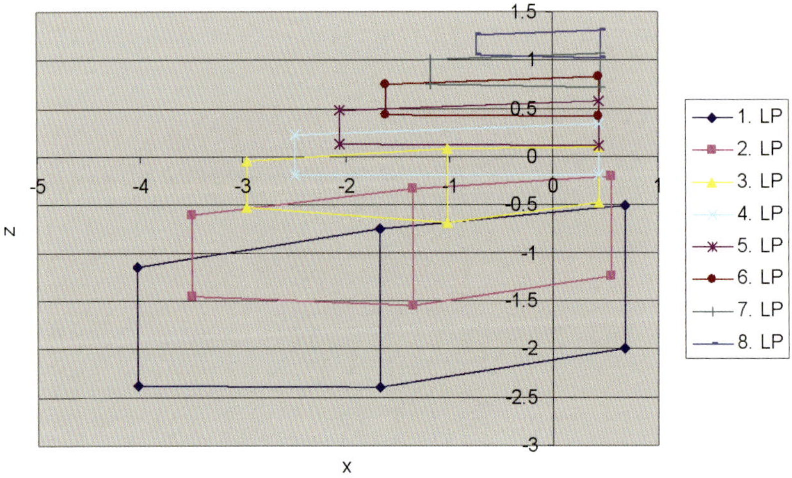

Fig. 15 Cross section for the eight load points (LP) in the wing coordinate system (y axis along quarter-chord).

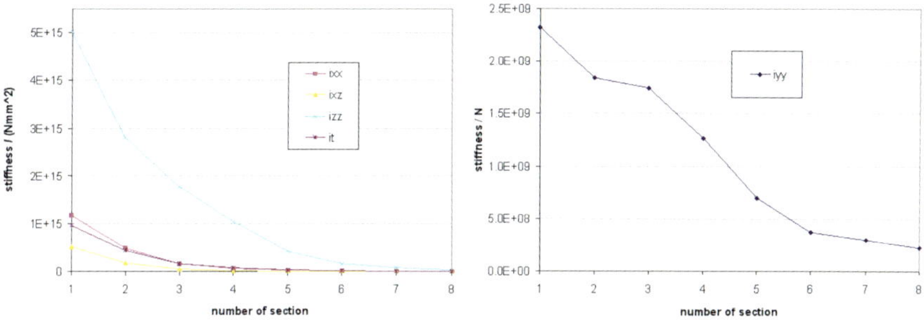

Fig. 19 Stiffness vs number of sections.

Chapter 5

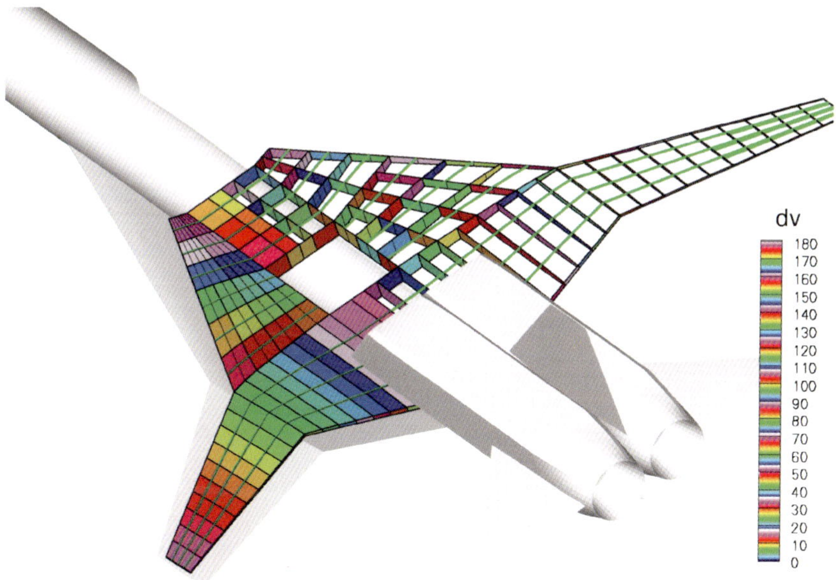

Fig. 5 Finite element mesh and design area distribution for the wing box of a supersonic business jet [4, reprinted with permission of NLR].

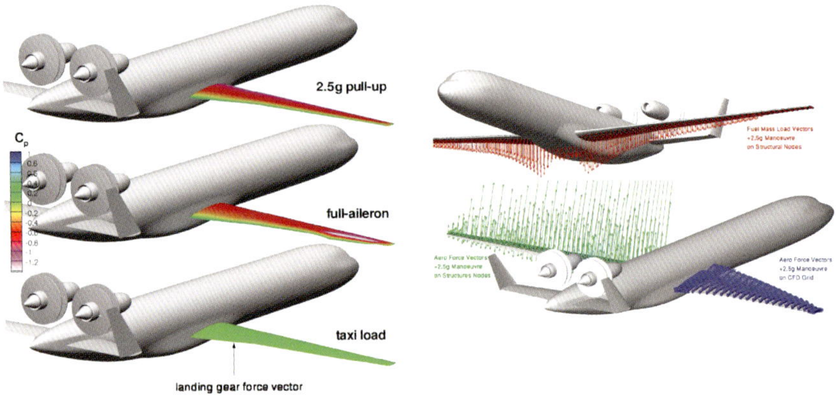

Fig. 6 Aerodynamic surface-pressure loads computed with CFD (left-hand side) and aero and inertial fuel load vectors mapping to the structural nodes (right-hand side) for an innovative low-fuel-burn aircraft concept.

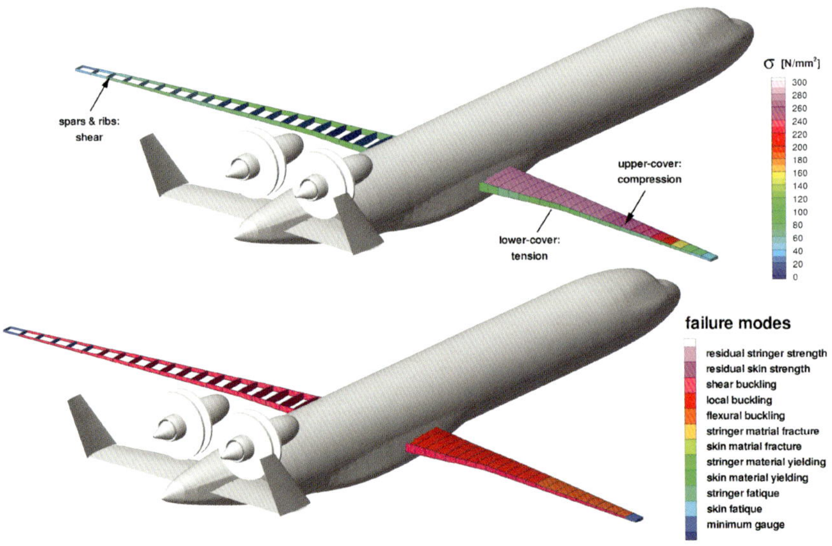

Fig. 8 Allowable stresses and critical panel failure modes for an innovative low-fuel-burn aircraft concept.

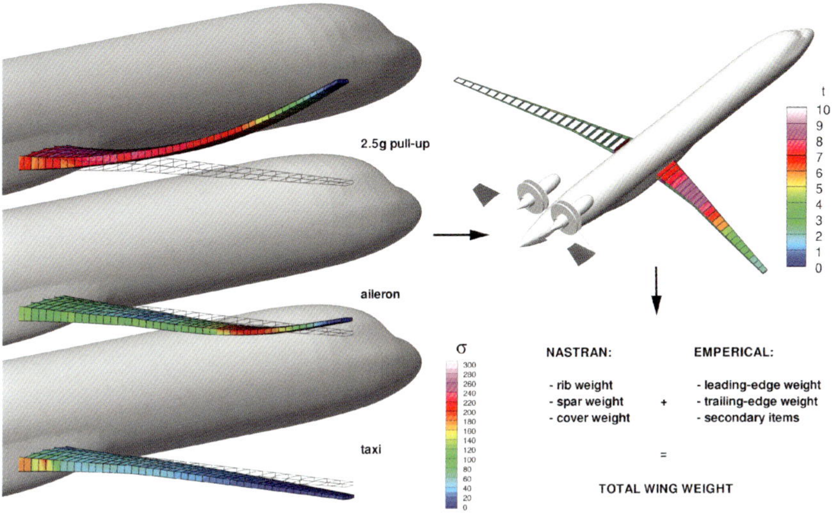

Fig. 9 Von Mises stresses and optimum material thickness distribution for an innovative low-fuel-burn aircraft concept.

Fig. 10 Technique based on solving the Euler equations for cruise drag assessment of a supersonic business jet wing/body configuration [4, reprinted with permission of NLR].

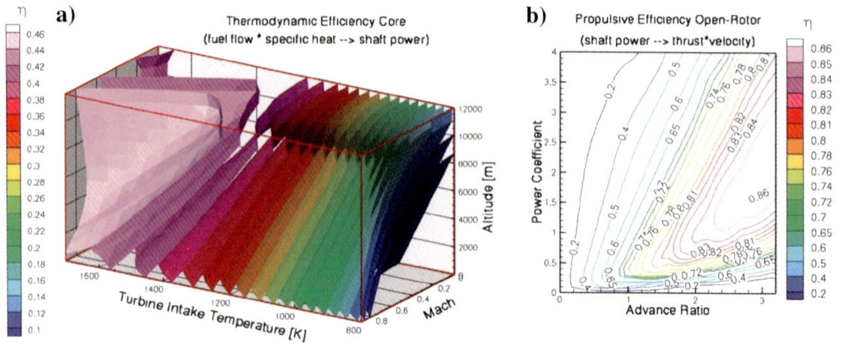

Fig. 11 Example of a data deck of a turboshaft engine driving a counter-rotating open rotor: a) the core engine thermodynamic efficiency and b) the rotor propulsive efficiency chart.

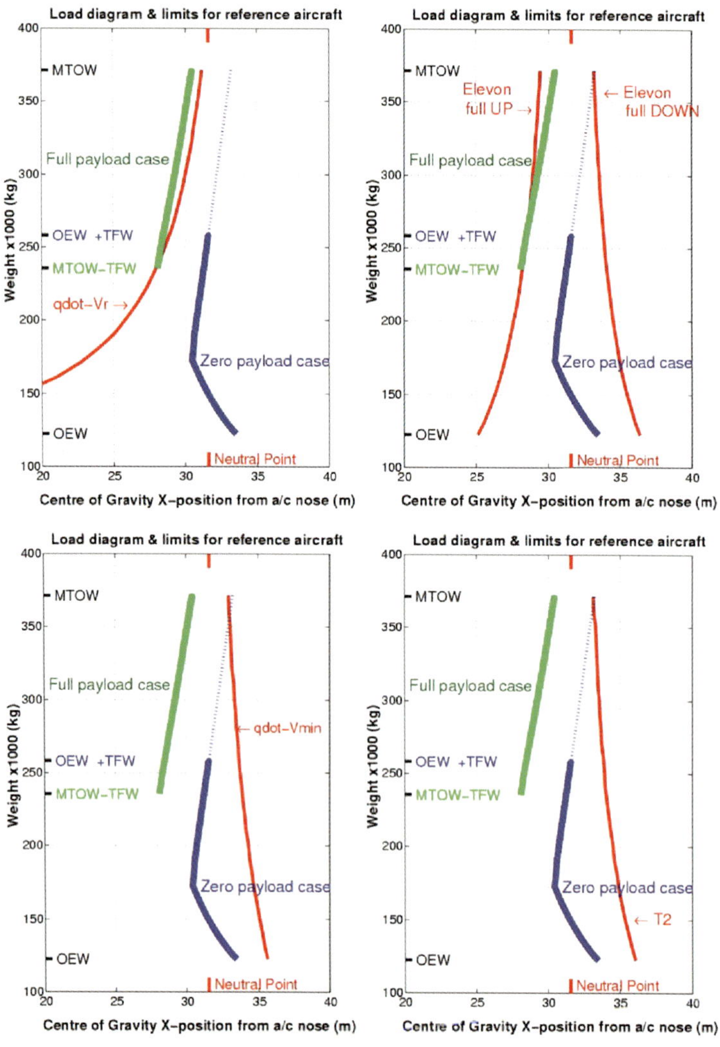

Fig. 12 Weight and balance check for the blended-wing/body configuration [1, reprinted with permission of NLR].

Chapter 6

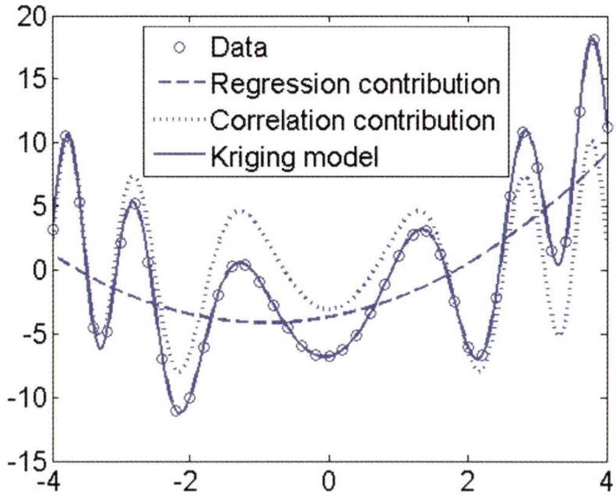

Fig. 1 Plot of a simple kriging approximation and the different contributions to the model.

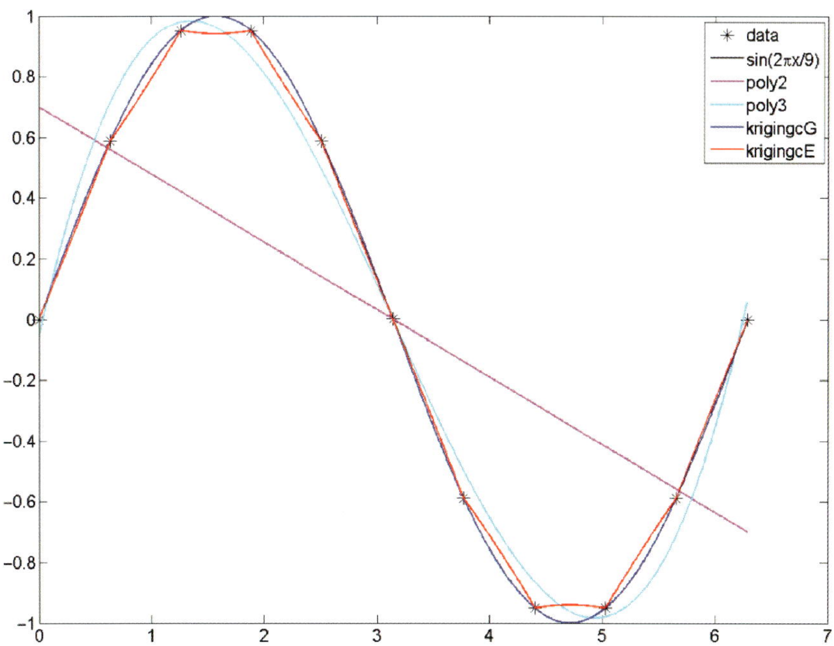

Fig. 4 Several models fitted to a period of the sine function.

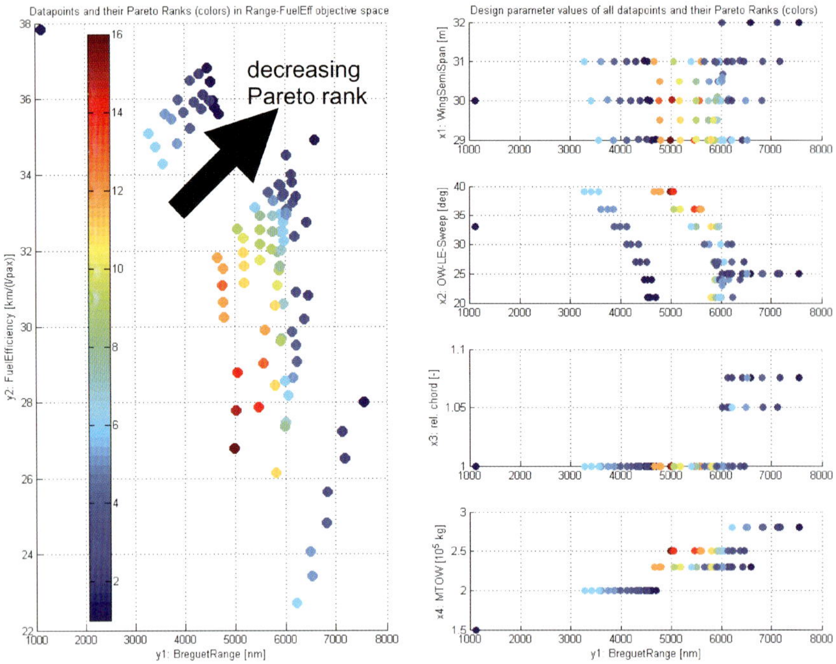

Fig. 13 Range and fuel results in the 99 design points in objective space (left) and in parameter space (right), colored by their Pareto rank. (The rank 1 points have the highest range and highest fuel efficiency values.)

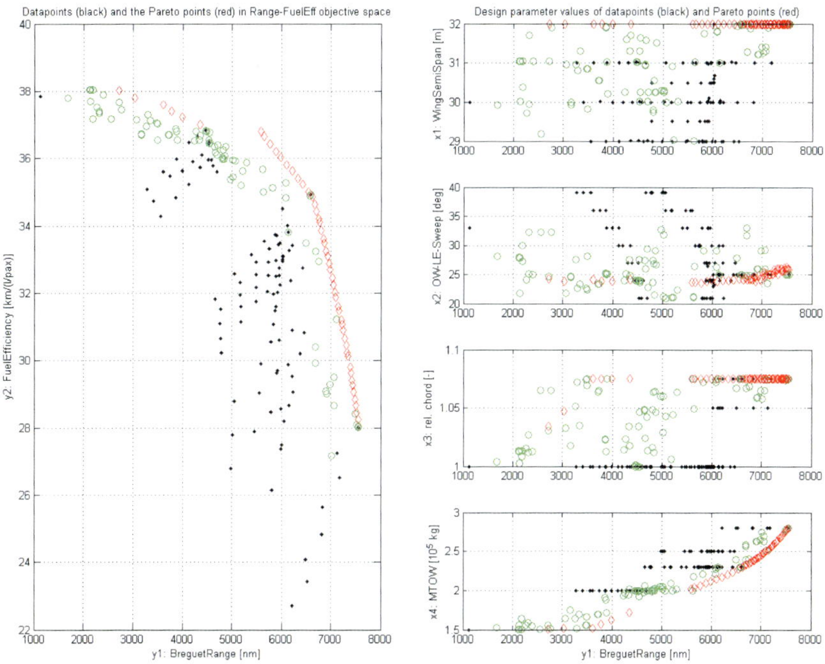

Fig. 14 Design points of data set (·), population after 3 generations (○, ●), and Pareto front after 100 additional generations (◇, ◆) for maximum range vs maximum fuel efficiency found with the kriging-linear-Gauss and kriging-constant-exponential meta models, respectively, for range and fuel efficiency. Results presented in objective space (left) and in the range-parameter subspaces (right) for each of the four design parameters.

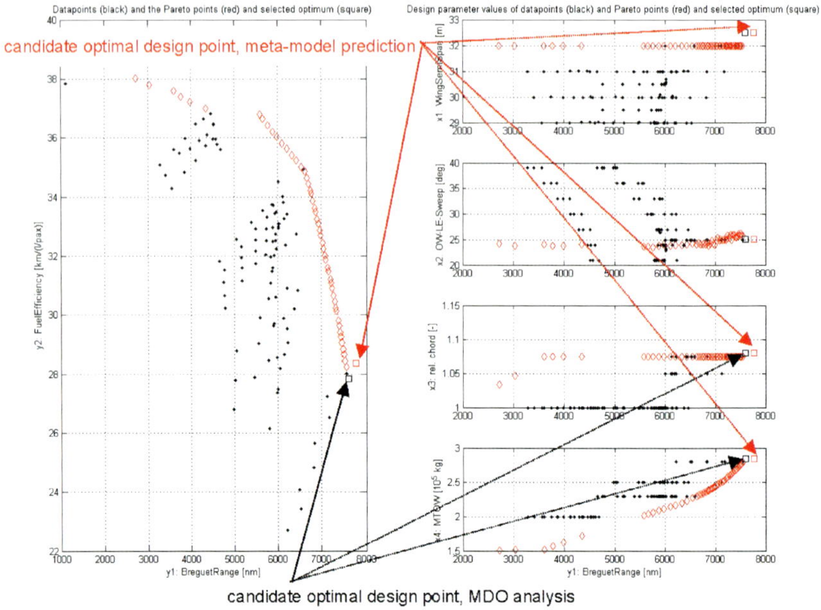

Fig. 15 Pareto front found with initial meta-model (◆, ◇), data set (·), and MDA analysis and meta-model predictions for candidate optimal design point (□).

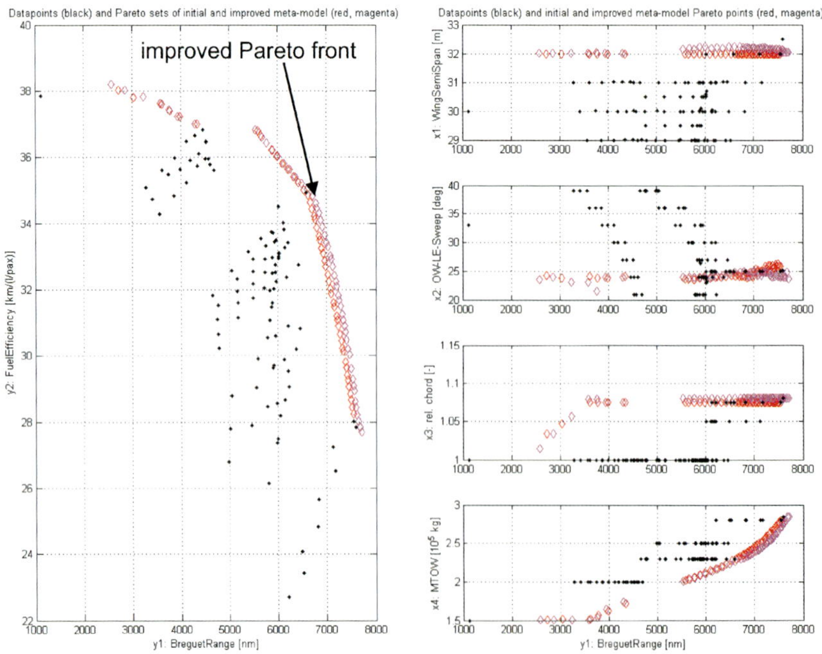

Fig. 16 Pareto fronts found with the initial meta-models (◇, ◆) and the improved meta-models (shifted diamonds).

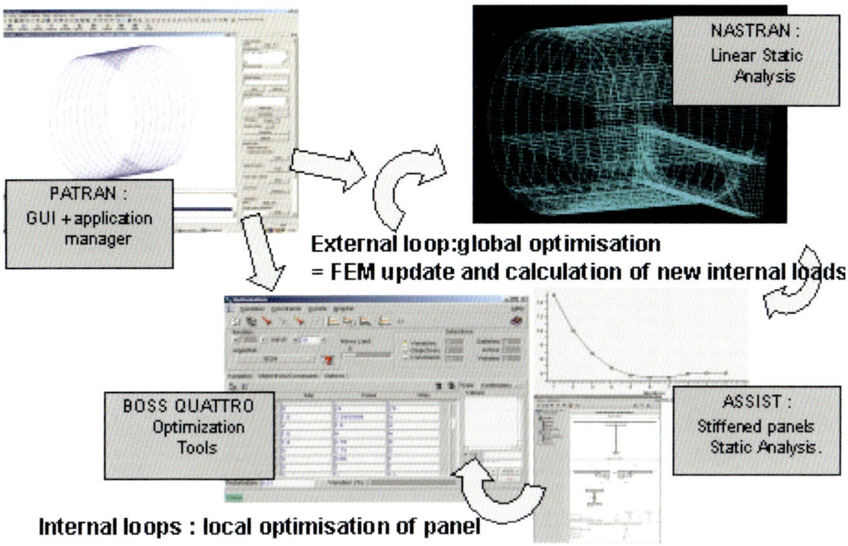

Fig. 13 STIFFOPT principle and COTS used.

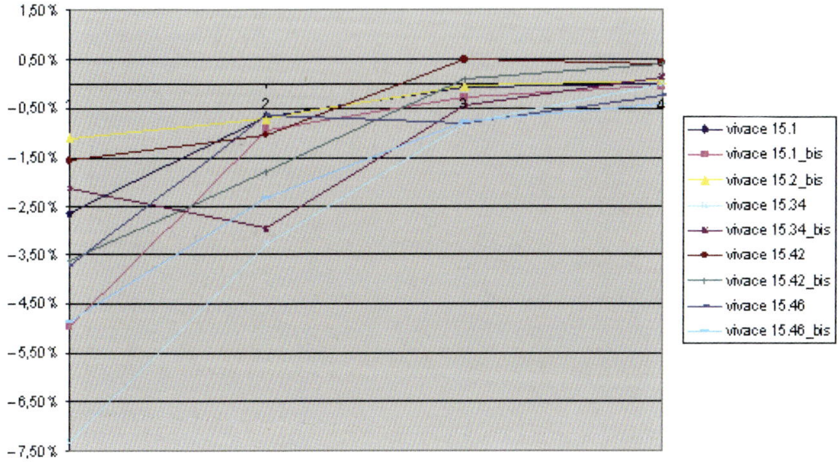

Fig. 29 Convergence of case 1.

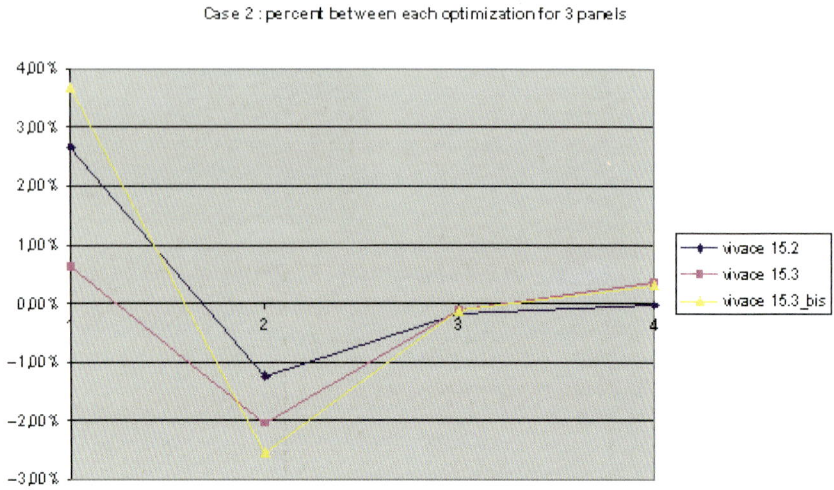

Fig. 30 Convergence of case 2.

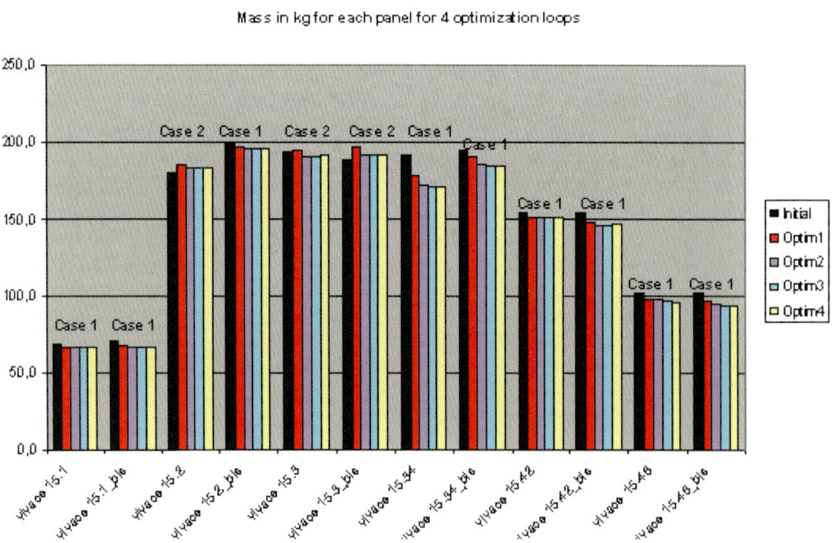

Fig. 31 Convergence history panel per panel.

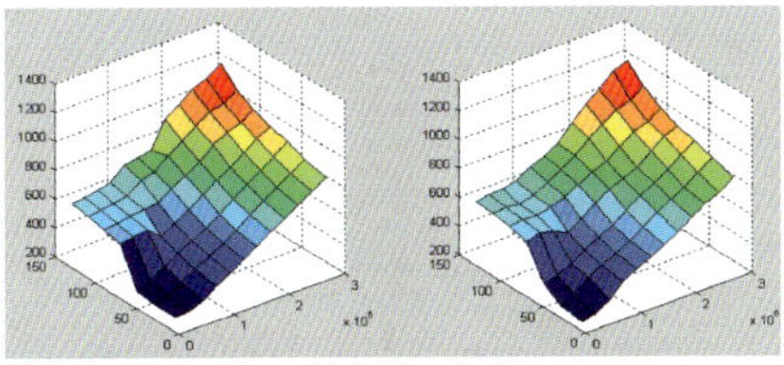

Superstiffener section area 132/5520

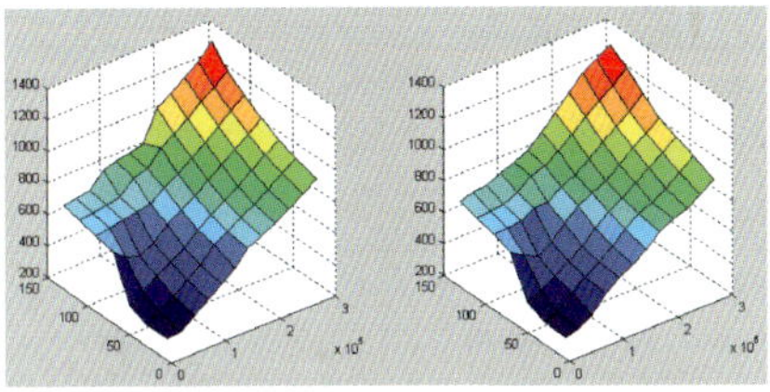

Superstiffener section area 158/5220

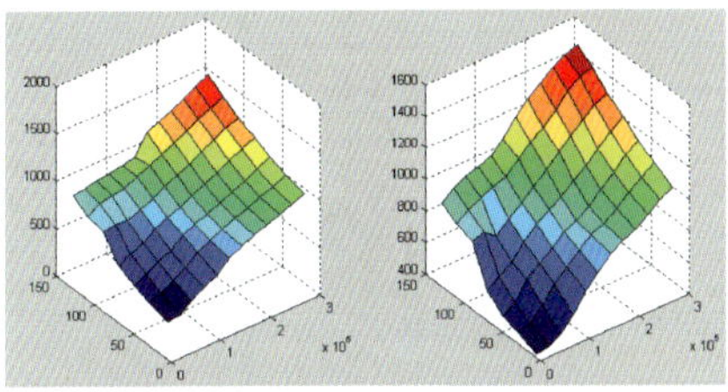

Superstiffener section area 191/3300

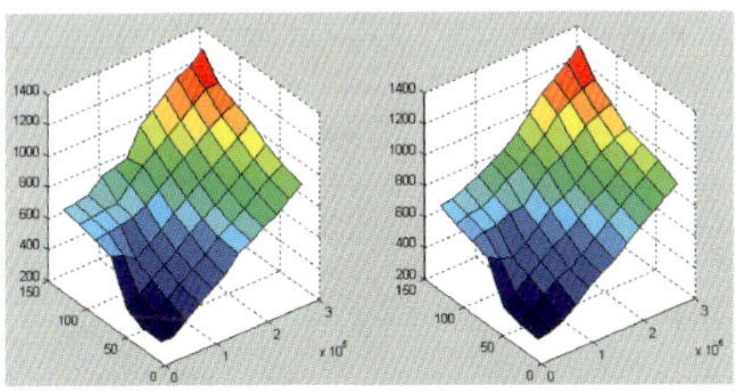

Superstiffener section area 154/3430

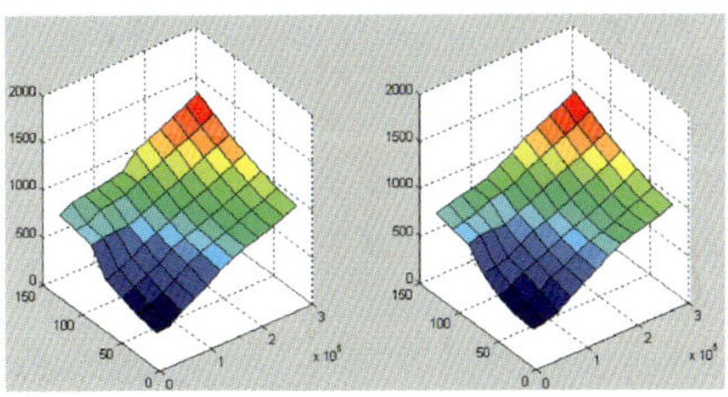

Superstiffener section area 180/5220

Fig. 39 Response surface for section area found for various (pitch, curvature radius) couples.

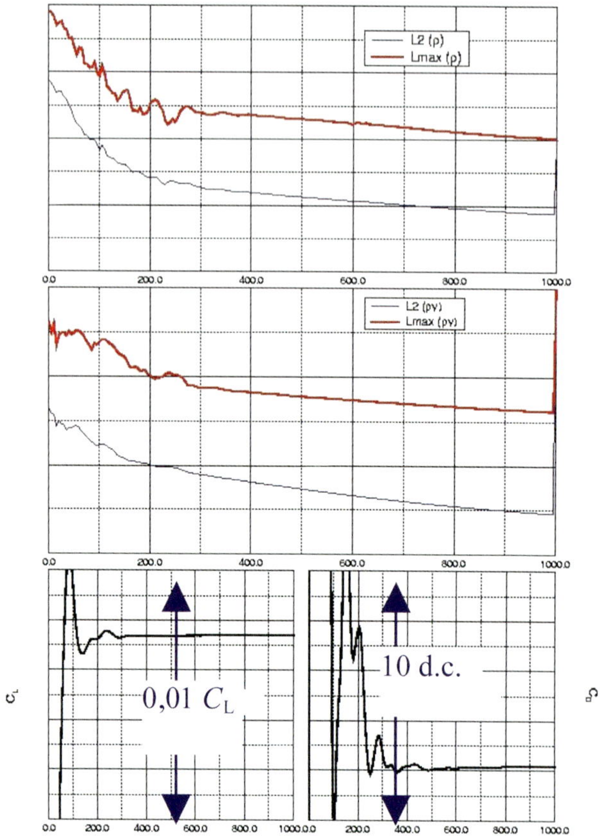

Fig. 8 Convergence history of computation on baseline configuration.

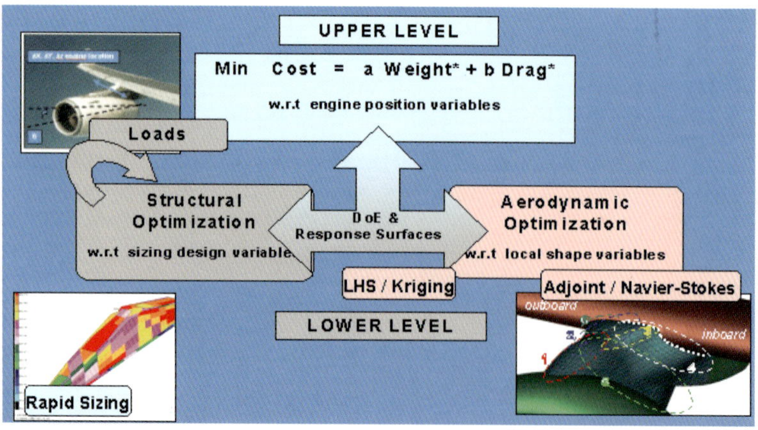

Fig. 12 Illustration of the multilevel and multidisciplinary optimization process.

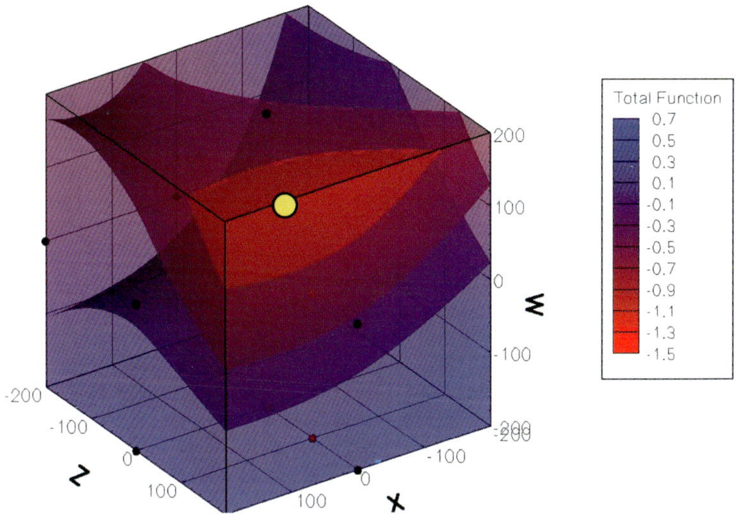

Fig. 18 MDO demonstration.

Chapter 9

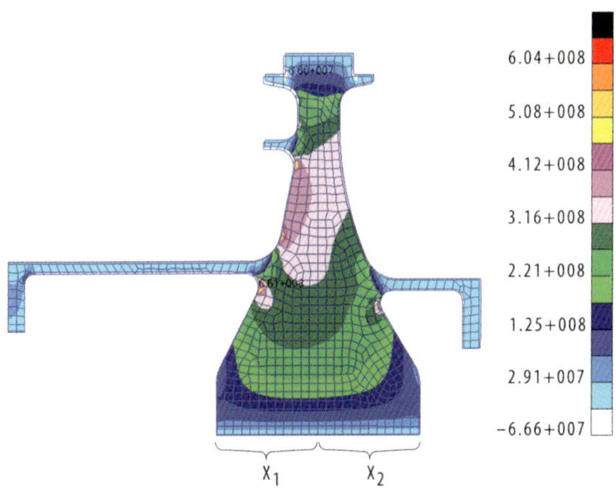

Fig. 21 Standard mesh of two-dimensional disc design with stress distribution.

Fig. 22 Standard mesh of two-dimensional disc design with stress distribution.

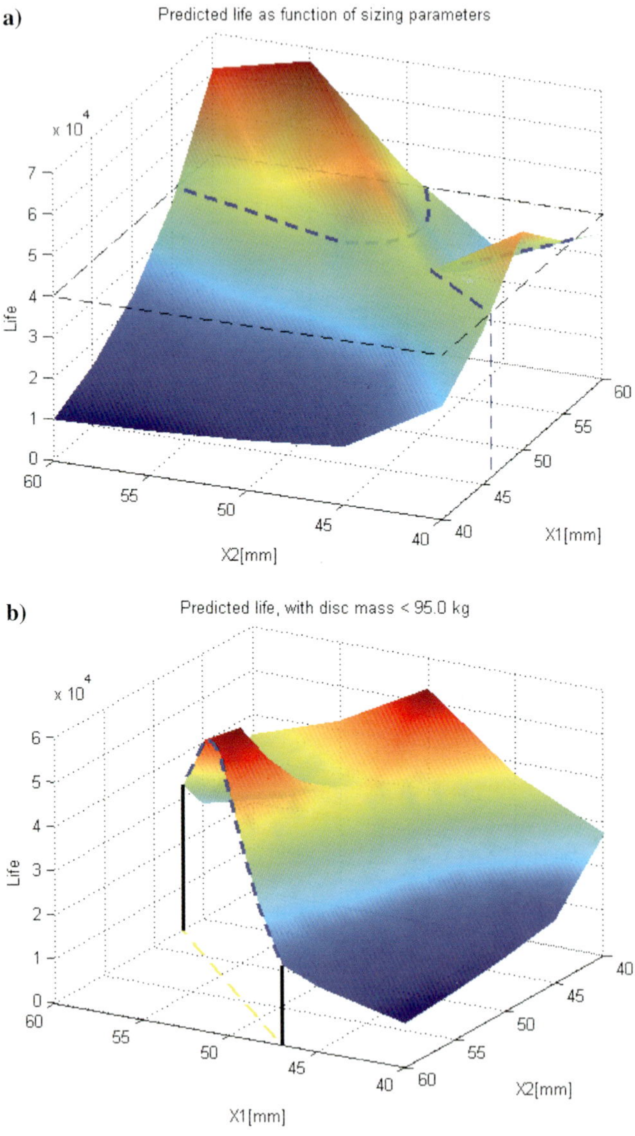

Fig. 28 a) Response surface of life for the defined design space of the two-dimensional disc. b) Response surface of life using a weight restriction.